AF863311

Protecting Critical Infrastructure

Series Editors

Simon Hakim
Erwin A. Blackstone

For further volumes:
http://www.springer.com/series/8764

Robert M. Clark · Simon Hakim · Avi Ostfeld
Editors

Handbook of Water and Wastewater Systems Protection

Springer

Editors
Robert M. Clark
9627 Lansford Drive
Cincinnati, OH 45242, USA
rmclark@fuse.net

Simon Hakim
Department of Economics
Temple University
Philadelphia, PA 19122, USA
simon.hakim@temple.edu

Avi Ostfeld
Department of Civil and Environmental Engineering
Technion – Israel Institute of Technology
32000 Haifa, Israel
ostfeld@tx.technion.ac.il

ISBN 978-1-4614-0188-9 e-ISBN 978-1-4614-0189-6
DOI 10.1007/978-1-4614-0189-6
Springer New York Dordrecht Heidelberg London

Library of Congress Control Number: 2011935004

© Springer Science+Business Media, LLC 2011
All rights reserved. This work may not be translated or copied in whole or in part without the written permission of the publisher (Springer Science+Business Media, LLC, 233 Spring Street, New York, NY 10013, USA), except for brief excerpts in connection with reviews or scholarly analysis. Use in connection with any form of information storage and retrieval, electronic adaptation, computer software, or by similar or dissimilar methodology now known or hereafter developed is forbidden.
The use in this publication of trade names, trademarks, service marks, and similar terms, even if they are not identified as such, is not to be taken as an expression of opinion as to whether or not they are subject to proprietary rights.

Printed on acid-free paper

Springer is part of Springer Science+Business Media (www.springer.com)

We would like to dedicate this book to our wives Susan Clark, Galia Hakim, and Yael Ostfeld and to our children and grandchildren.

Acknowledgement

We would like to acknowledge, in memoriam, Dr. Paul Seidenstat who was a pioneer in the field of urban economics, an advocate of protecting societies' critical infrastructure, and who materially contributed to this effort. We would also like to acknowledge the individuals and institutions who contributed to this book and the men and women who are diligently working to protect critical infrastructure throughout the world.

Contents

1 **Securing Water and Wastewater Systems: An Overview** 1
Robert M. Clark, Simon Hakim, and Avi Ostfeld

2 **Water/Wastewater Infrastructure Security: Threats and Vulnerabilities** 27
Laurie J. Van Leuven

3 **EPA Drinking Water Security Research Program** 47
Hiba S. Ernst, K. Scott Minamyer, and Kim R. Fox

4 **Drinking Water Critical Infrastructure and Its Protection** 65
Rakesh Bahadur and William B. Samuels

5 **Wastewater Critical Infrastructure Security and Protection** 87
Rakesh Bahadur and William B. Samuels

6 **Protecting Water and Wastewater Systems** 103
Randy G. Fischer

7 **Spatial Distributed Risk Assessment for Urban Water Infrastructure** 119
Michael Möderl and W. Rauch

8 **US Water and Wastewater Critical Infrastructure** 135
Robert M. Clark

9 **Microbial Issues in Drinking Water Security** 151
Eugene W. Rice

10 **Rapid Detection of Bacteria in Drinking Water and Wastewater Treatment Plants** 163
Rolf A. Deininger, Jiyoung Lee, and Robert M. Clark

11 **Chlorine Residual Management for Water Distribution System Security** 185
Jeanne M. VanBriesen, Shannon L. Isovitsch Parks, Damian E. Helbling, and Stacia T. McCoy

12 **Biosensors for the Detection of *E. coli* O157:H7 in Source and Finished Drinking Water** 205
Mark D. Burr, Andreas Nocker, and Anne K. Camper

13 **Guidelines, Caveats, and Techniques for the Evaluation of Water Quality Early Warning Systems** 229
Dan Kroll

14 **Protecting Water and Wastewater Systems: Water Distribution Systems Security Modeling** 247
Avi Ostfeld

15 **Protecting Consumers from Contaminated Drinking Water During Natural Disasters** 265
Craig L. Patterson and Jeffrey Q. Adams

16 **Cyber Security: Protecting Water and Wastewater Infrastructure** 285
Srinivas Panguluri, William Phillips, and Patrick Ellis

17 **Real-World Case Studies for Sensor Network Design of Drinking Water Contamination Warning Systems** 319
Regan Murray, Terra Haxton, William E. Hart, and Cynthia A. Phillips

18 **Enhanced Monitoring to Protect Distribution System Water Quality** 349
Zia Bukhari and Mark LeChevallier

19 **Testing and Evaluation of Water Quality Event Detection Algorithms** 369
Sean A. McKenna, David B. Hart, Regan Murray, and Terra Haxton

20 **Water Infrastructure Protection Against Intentional Attacks: The Experience of Two European Research Projects** . . . 397
Cristiana Di Cristo, Angelo Leopardi, and Giovanni de Marinis

21 **Utility of Supercomputers in Trace-Back Algorithms for City-Sized Distribution Systems** 419
Hailiang Shen and Edward McBean

22 **Water/Wastewater Infrastructure Security: A Multilayered Security Approach** 435
Laurie J. Van Leuven

23 **Vulnerability of Water and Wastewater Infrastructure and Its Protection from Acts of Terrorism: A Business Perspective** . . . 457
Dave Birkett, Jim Truscott, Helena Mala-Jetmarova, and Andrew Barton

About the Editors 485

About the Principle Contributors 487

Name Index 497

Subject Index 501

Contributors

Jeffrey Q. Adams National Risk Management Research Laboratory, Water Supply and Water Resources Division, USEPA, Cincinnati, OH, USA, adams.jeff@epamail.epa.gov

Rakesh Bahadur Science Applications International Corporation Center for Water Science and Engineering, McLean, VA, USA, bahadurr@saic.com

Andrew Barton GWMWater, Horsham, VIC, Australia; University of Ballarat, Ballarat, VIC, Australia, andrew.barton@gwmwater.org.au

Dave Birkett Truscott Crisis Leaders, Wembley Downs, WA, Australia, dbirkett@crisisleaders.com

Zia Bukhari American Water, Voorhees, NJ, USA, zia.bukhari@amwater.com

Mark D. Burr Center for Biofilm Engineering, Montana State University, Bozeman, MT, USA, mark_b@erc.montana.edu

Anne K. Camper Center for Biofilm Engineering, Montana State University, Bozeman, MT, USA, anne_c@erc.montana.edu

Robert M. Clark 9627 Lansford Drive, Cincinnati, OH, USA, rmclark@fuse.net

Rolf A. Deininger School of Public Health, The University of Michigan, Ann Arbor, MI, USA, rad@umich.edu

Giovanni de Marinis Water Engineering Lab (L.I.A.), Department of Mechanics, Structures and Environmental Engineering (Di.M.S.A.T.), University of Cassino, Cassino, Italy, demarinis@unicas.it

Cristiana Di Cristo Water Engineering Lab (L.I.A.), Department of Mechanics, Structures and Environmental Engineering (Di.M.S.A.T.), University of Cassino, Cassino, Italy, dicristo@unicas.it

Patrick Ellis Broward County Water and Wastewater Services, 2555 West Copans Road, Pompano Beach, FL, USA, pellis@broward.org

Hiba S. Ernst US Environmental Protection Agency, National Homeland Security Research Center, Cincinnati, OH, USA, ernst.hiba@epa.gov

Randy G. Fischer Division of Public Health, Nebraska Department of Health and Human Services (NE DHHS), Lincoln, NE, USA, randy.fischer@nebraska.gov

Kim R. Fox US Environmental Protection Agency, National Homeland Security Research Center, Cincinnati, OH, USA, fox.kim@epa.gov

Simon Hakim Center for Competitive Government, Fox School of Business & Management, Temple University, Philadelphia, PA, USA; Department of Economics, Temple University, Philadelphia, PA, USA, simon.hakim@temple.edu

David B. Hart National Security Applications Department, Sandia National Laboratories, Albuquerque, NM, USA, dbhart@sandia.gov

William E. Hart Sandia National Laboratories, Albuquerque, NM, USA, wehart@sandia.gov

Terra Haxton National Homeland Security Research Center, U.S. Environmental Protection Agency, Cincinnati, OH, USA, haxton.terra@epa.gov; Haxton.Terra@epamail.epa.gov

Damian E. Helbling Department of Environmental Chemistry, Swiss Federal Institute of Aquatic Science and Technology (Eawag), Duebendorf, Switzerland, Damian.Helblig@eawag.ch

Dan Kroll Hach Homeland Security Technologies, Loveland, CO, USA, dkroll@hach.com

Mark LeChevallier American Water, Voorhees, NJ, USA, mark.lechevallier@amwater.com

Jiyoung Lee Division of Environmental Health Sciences, College of Public Health, Ohio State University, Columbus, OH, USA, jlee@cph.osu.edu

Angelo Leopardi Water Engineering Lab (L.I.A.), Department of Mechanics, Structures and Environmental Engineering (Di.M.S.A.T.), University of Cassino, Cassino, Italy, a.leopardi@unicas.it

Helena Mala-Jetmarova GWMWater, Horsham, VIC, Australia; University of Ballarat, Ballarat, VIC, Australia, helena.jetmarova@gwmwater.org.au

Edward McBean School of Engineering, University of Guelph, Guelph, ON, Canada, emcbean@uoguelph.ca

Stacia T. McCoy Department of Civil and Environmental Engineering, Carnegie Mellon University, Pittsburgh, PA, USA, staciatmccoy@gmail.com

Sean A. McKenna National Security Applications Department, Sandia National Laboratories, Albuquerque, NM, USA, samcken@sandia.gov

K. Scott Minamyer US Environmental Protection Agency, National Homeland Security Research Center, Cincinnati, OH, USA, minamyer.scott@epa.gov

Michael Möderl Institute of Infrastructure, University of Innsbruck, Innsbruck, Austria, Michael.Moederl@uibk.ac.at

Regan Murray National Homeland Security Research Center, U.S. Environmental Protection Agency, Cincinnati, OH, USA, murray.regan@epa.gov; Murray.Regan@epamail.epa.gov

Andreas Nocker Centre for Water Science, Cranfield University, Cranfield, Bedfordshire, UK, andreas.nocker@gmail.com

Avi Ostfeld Department of Civil and Environmental Engineering, Technion – Israel Institute of Technology, Haifa, Israel, ostfeld@tx.technion.ac.il

Srinivas Panguluri Shaw Environmental & Infrastructure, Inc., 5050 Section Avenue, Cincinnati, OH, USA, srinivas.panguluri@shawgrp.com

Shannon L. Isovitsch Parks Environmental Science and Sustainable Technology Division, Alcoa, Inc., Pittsburgh, PA, USA, Shannon.Parks@alcoa.com

Craig L. Patterson National Risk Management Research Laboratory, Water Supply and Water Resources Division, USEPA, Cincinnati, OH, USA, patterson.craig@epamail.epa.gov

Cynthia A. Phillips Sandia National Laboratories, Albuquerque, NM, USA, Phillips.Cynthia@epamail.epa.gov

William Phillips CH2MHILL, 3011 SW Williston Road, Gainesville, FL, USA, bill.phillips@ch2m.com

W. Rauch Institute of Infrastructure, University of Innsbruck, Innsbruck, Austria, Wolfgang.Rauch@uibk.ac.at

Eugene W. Rice National Homeland Security Research Center, U.S. Environmental Protection Agency, Cincinnati, OH, USA, rice.gene@epa.gov

William B. Samuels Science Applications International Corporation Center for Water Science and Engineering, McLean, VA, USA, william.b.samuels@saic.com

Hailiang Shen School of Engineering, University of Guelph, Guelph, ON, Canada, shenh@uoguelph.ca

Jim Truscott Truscott Crisis Leaders, Wembley Downs, WA, Australia, jtruscott@crisisleaders.com

Jeanne M. VanBriesen Department of Civil and Environmental Engineering, Carnegie Mellon University, Pittsburgh, PA, USA, jeanne@andrew.cmu.edu

Laurie J. Van Leuven Seattle Public Utilities/U.S. Department of Homeland Security (DHS), FEMA, Washington, DC, USA, laurie.vanleuven@gmail.com; Laurie.VanLeuven@fema.gov

Chapter 1
Securing Water and Wastewater Systems: An Overview

Robert M. Clark, Simon Hakim, and Avi Ostfeld

1.1 Introduction

There is a general, and growing, awareness that urban water systems are vulnerable to both manmade and natural, but unpredictable, threats and disasters such as droughts, earthquakes, and terrorist attacks. Other natural disasters that can effect water supply security and integrity include major storms such as hurricanes and flooding. Earthquakes and terrorist attacks have many characteristics in common. They are almost impossible to predict and can cause major devastation and confusion. Several recent earthquakes centered in urban areas such as the earthquake that struck Kobe City, Japan, in 1995 have demonstrated the disastrous effect that earthquakes can have on urban water systems. Terrorism is also a major threat to water security, and recent attention has turned to the potential that these attacks have for disrupting urban water supplies. In the United States, government planners have been forced to consider the possibility that the nation's critical infrastructure, including water systems, may in fact be vulnerable to terrorism. The President's Commission on Critical Infrastructure Protection concluded that the nation's water supply system might be vulnerable to certain biological agents (Clark and Deininger, 2001). The *Public Health Security and Bioterrorism Preparedness and Response Act of 2002* (US Congress, 2002) has intensified the focus on water security research in the United States. After the attacks of September 11, 2001, the US Environmental Protection Agency (EPA) developed a *Homeland Security Strategy* (USEPA, 2004). Its intent was to enhance national security and protect human health and the environment. Much of the research conducted as a result of these directives is presented in this book (Ernst et al., Chapter 3, this volume).

In addition to urban water supply natural and manmade threats are important issues for urban wastewater systems. There are approximately 16,255 publicly

R.M. Clark (✉)
9627 Lansford Drive, Cincinnati, OH 45242, USA
e-mail: rmclark@fuse.net

R.M. Clark et al. (eds.), *Handbook of Water and Wastewater Systems Protection*, Protecting Critical Infrastructure, DOI 10.1007/978-1-4614-0189-6_1,
© Springer Science+Business Media, LLC 2011

owned treatment works (POTWs), and 100,000 major pumping stations in the United States. According to Bahadur and Samuels (Chapters 4 and 5, this volume) damage to the nation's wastewater facilities or collection systems could result in loss of life; catastrophic environmental damage to rivers, lakes, and wetlands; and contamination of drinking water supplies. In addition damage to the nation's wastewater systems could result in long-term public health impacts, destruction of fish and shellfish production, and disruption to commerce and the economy.

This book contains insights and recommendations from a group of internationally recognized experts who review the state of the art in protecting water and wastewater systems from natural and manmade threats. These experts address the following issues:

- Problems in protecting water and wastewater systems.
- The consequences of not protecting these systems.
- The state of the art in protecting water and wastewater systems.
- Alternative solutions that might be employed to address water and wastewater security problems.

Contributed chapters from US and international experts will cover the following areas:

- Overview of the current state of water supply and wastewater system security and the ability to respond to threats and disasters.
- Characteristics of the water supply and wastewater systems in the United States.
- Chemical and microbiological threats for water system contamination.
- Monitoring for natural and manmade threats in drinking water systems.
- Modeling contaminant propagation and contaminant threats in drinking water distribution systems.
- Case study applications.
- Distribution system modeling, SCADA systems, security hardware, and surveillance systems.
- Institutional and management issues in responding to natural and manmade threats.
- Progress in developing techniques and approaches for natural and manmade threat response in water and wastewater systems since September 11.

1.2 History of Water Supply Vulnerability

According to Gleick (2006) the recorded history of attacks on water systems dates from 4,500 years ago. Urlama, King of Lagash, and his son Illater cut off the water supply to Girsu, a city in Umma, during the period 2450–2400 BC. In New York in

1748 an angry mob burned down a ferry house on the Brooklyn shore of the East River. It is reported that this act was revenge for unfair allocation of East River water rights. Small groups attacked small dams and reservoirs in the 1840s and 1850s in the eastern and central United States due to concerns about threats to health and to local water supplies. In the Owens Valley of California between 1907 and 1913 farmers repeatedly dynamited the aqueduct system being built to divert their water to the growing city of Los Angeles.

In New York City (New York Times, 1986), low levels of plutonium were found in the drinking water (on the order of 20 fCi). The usual background is below 1 fCi. However, a person would have to drink several million liters of water to acquire a lethal dose estimated at about 100 μCi. A femtocurie is nine orders of magnitude smaller than a microcurie (Clark and Deininger, 2000). Another case was the contamination of salad bars in Dalles, Oregon, by the Rajneeshee religious cult, using vials of *Salmonella typhimurium. S. typhimurium* is a highly toxic bacteria frequently carried by birds. The cult also contaminated a city water supply tank using *Salmonella*. A community outbreak of salmonellosis resulted in which at least 751 cases were documented in a county that typically reports fewer than 5 cases per year. The cult apparently cultured the organisms in their own laboratories (Clark and Deininger, 2000; Gleick, 2006).

In terms of natural threats, water shortages and droughts have led to crises and disasters throughout history and in many parts of the world. Drought may affect both developing and developed countries and according to the UN's Office of Foreign Disaster Assistance no other natural disaster has caused as many displaced persons in the 20th century. For example, a drought in the Great Plains in the United States in the 1930s caused serve economic hardship in Missouri, Kansas, Nebraska, Oklahoma, South Dakota, and Arkansas. The Great Plains also experienced droughts in the 1950s, 1970s, and 1990s. Drought affects more people than any other natural hazard; earthquakes and terrorism can affect water security in modern urban communities. According to Bruins (2000), Israel included Arab villages to receive water from the National Carrier System in order to limit the potential posed by terrorists. Water played an important role in the Peace Treaty that Israel and Jordan signed on October 26, 1994, and to this point the worst case scenarios have not materialized over water disputes in the Middle East. With the advent of global climate change and the anticipated increase in droughts in some locations, there is concern that water scarcity might become the basis for future wars.

Unlike droughts which are described as a creeping phenomenon the damage associated with earthquakes is concentrated in time and space. In 1906 an earthquake in San Francisco caused numerous pipes to rupture and caused drowning of dozens of residents when broken water pipes flooded the Valencia hotel. It was impossible to control the firestorms that spread through the area, and entire buildings exploded in a huge firestorm during which the temperature was reported to reach 2000°F (1093.2°C). In 1995, a major earthquake directly hit the city of Kobe, Japan. The quake lasted 20 s and 4,069 people died, 14,679 were injured, and 222,127 people were moved into evacuation shelters. There were 67,421 fully collapsed structures of which 6,985 were burned to the ground and there was a city-wide power failure and a nearly city-wide water supply failure (Clark and Deininger, 2001). Floods and major storms can pose a threat to water system security. Patterson and Adams (Chapter 15, this volume) describe the problems associated with recovery from Hurricane Katrina.

Until September 11, 2001, terrorism in the United States was not generally regarded as a serious threat because of the nation's military strength, relative geographic isolation, and secure borders. However, recent attacks against targets within the United States by domestic and foreign terrorists forced many government planners to consider the possibility that the nation's critical infrastructure may, in fact, be vulnerable to terrorist attacks. In response to this concern, the President's Commission on Critical Infrastructure was formed to evaluate the vulnerability of the water and wastewater infrastructure to internal and external terrorism. The rapid proliferation of telecommunication and computer systems, which connect infrastructures to one another in a complex network, compounds this vulnerability (Clark and Deininger, 2000).

Vital Human Services include community water supply systems on local and state levels. In terms of public administration, water supply systems are generally governmental in nature. However, each supply system tends to be highly localized. Failures in one community may have little direct impact on other communities, although the problems and vulnerabilities may be similar. Water supply systems are vulnerable to the full range of terrorist threats including physical attack and cyber and biological terrorism.

The potential of bioterrorism as a threat to public safety is becoming increasingly apparent. For example, two epidemics of smallpox occurred in Europe in the 1970s. Each outbreak resulted from one infected individual. An aerosolized anthrax discharge from a Russian bioweapons facility in 1979 resulted in 77 cases of anthrax and 66 deaths. It is estimated that the release probably lasted no more than a few minutes and the weight of the aerosols released may have been as little as a few milligrams (Clark and Deininger, 2000; Gleick, 2006).

1.3 Threats from Earthquakes

It is the authors' opinion that many of the approaches adopted for earthquake response would be useful in responding to a terrorist attack. Specific examples are discussed below. During the San Francisco earthquake of 1906, which had a magnitude of 8.3 on the Richter scale, approximately 3,000 people lost their lives. A devastating fire swept through the city which caused more destruction than the immediate effects of the earthquake itself. As a consequence of that experience engineers today strive to build water systems characterized by strength, flexibility, and redundancy. Water systems survived much better during the Loma Prieta and Northridge earthquakes, averting the kinds of catastrophic losses experienced in the San Francisco earthquake (Clark and Deininger, 2001).

1.3.1 The Loma Prieta Earthquake

The Loma Prieta earthquake that struck on October 17, 1989, had a reading of 7.1 on the Richter scale. It caused 62 deaths and damaged over 18,000 homes. The earthquake caused water pipes to break in some areas, particularly in places with older cast iron pipes and in areas known as liquefaction zones, where loose saturated sandy soil is prone to intensified ground shaking. A reservoir with an earthen dam and a treatment plant were damaged primarily by earthquake-generated wave action. However, water distribution facilities were largely left intact.

1.3.2 The Northridge Earthquake

The Northridge earthquake of January 17, 1994, had a reading of 6.7 on the Richter scale and although smaller in strength than the Loma Prieta earthquake struck a heavily populated sector in urban Los Angeles causing 57 deaths as well as the loss of 14,600 homes. Overall, the Northridge earthquake impacted more households and businesses than any other disaster in recent US history. Two major wastewater treatment facilities suffered significant damage due to liquefaction. Aboveground water storage tanks suffered damage due to failures at their bases (buckling and tearing), and roof structures and pipe joints failed. The earthquake jolts uncoupled the fittings causing hundreds of breaks in the water distribution system. Some areas were without water or power and advisories to boil water went out to areas impacted by pipe failures. Water agencies made full use of mutual aid agreements and brought in repair crews from around the state. Within 10 days, all water main breaks were repaired and the treatment plants were back in service.

1.3.3 Kobe City Earthquake

At 5:46 am on January, 17, 1995, the Southern Hyogo Prefectural Earthquake (the Great Hanshin-Awaji Earthquake), the first major quake to directly hit a Japanese urban area, inflicted heavy damage on cities and their surrounding areas in the Hanshin-Awaji region. The jolt, which lasted barely 20 s, took 4,569 lives in Kobe City alone and virtually reduced the harbor to a pile of rubble.

Some of the existing facilities that proved to be effective during the earthquake included emergency shut-off valves, a remote telemetry/telecontrol system, and earthquake-resistant pipes. Some of the unexpected incidents that resulted from the earthquake were severe traffic jams, dire shortage of water, a lack of water wagons, frequent pipe breaks, and very slow progress in restoring water from the city's various sources.

Based on this experience the city made drastic revisions to its community disaster prevention plan that prescribes how each organization should act when disaster strikes. The new plan stipulates the role to be played by volunteers, those vulnerable to disasters, community residents, and businesses.

1.3.4 Technological and Institutional Adaptation

Water management in California is unique because of the complexity of its water delivery system and provides an enlightening as to how states might deal with security threats. Three main aqueducts supply water to the more than 16 million inhabitants in the southern part of the state where most of the population lives. However, most of the rain and snow falls in the northern half of the state. For example, the average annual precipitation in the north is over 760 mm (30 in.), while the south receives only 50–360 mm (2–14 in.). Recurring disasters, including earthquakes, and their effect on water systems have spurred emergency planning in California. These experiences are leading to new approaches to emergency response that include inter-organizational coordination among various agencies that will help the water industry cope even more effectively with future emergencies. The success of these developments is illustrated by comparing the events that took place during the San Francisco earthquake to the events during the Loma Preita and Northridge earthquakes (Clark and Deininger, 2001).

1.3.4.1 Technological Adaptations

As a consequence of these experiences the water utilities in earthquake zones in California have developed innovative technologies to mitigate the impact of future earthquakes. For example, engineers at the East Bay Municipal Utilities District (EBMUD) in Oakland, California, devised a unique alternative for transporting large amounts of water across a known earthquake fault. They developed a specially constructed flexible polyurethane hose with a large diameter (up to 12 in.) which can be stored for long periods of time. In an emergency, a small crew using light transport vehicles can deploy the hose in a matter of minutes. The hose can be used to bridge breaks in water mains or to bring large volumes of water from one part of the water system into another part. Different types of fittings allow fire trucks to connect to the hose and to add branch pipelines with a smaller diameter. EBMUD has identified key water distribution pipes that cross faults and are expected to fail during certain earthquake scenarios. Following an earthquake, prepositioned valves will allow crews to close off and isolate a broken section of pipe. Crews can then attach the polyurethane hose to prepositioned connections in undamaged sections of the original pipe, thereby restoring flow in the water distribution system.

1.3.4.2 Institutional Adaptations

The California state government has adopted a system of standardization that encourages cooperating agencies to use common terminology, a common functional management template, a standard for liaison relationships between cooperating agencies, a mutual aid system, and clearly defined governmental roles. California water utility agencies have learned to partner with government and private agencies to devise mutual aid and mutual assistance plans, to produce collaborative emergency planning guidance documents, and to arrange for reliable communications during emergency response.

Other collaborative efforts for emergency response include the work of the California Utilities Emergency Response Association, in which water utilities may coordinate with electricity, gas, telecommunications, and pipeline utilities. The purpose of the Water Agency Response Network is to identify the need to help each other in an emergency. The Water Agency Response Network links the Emergency Operations Centers of the member agencies with one another. Many public agencies incorporate amateur radio backup communication. In Los Angeles the distribution of potable water has been delegated to the fire departments in an emergency. These partnerships have developed through time and experience and have demonstrated an attempt to work together in an emergency or disaster and could provide a template for emergency response to a terrorist attack.

1.4 Vulnerable Characteristics of US Water Supply Systems

The President's Commission on Critical Infrastructure Protection identified several features of US drinking water systems that are particularly vulnerable to terrorist attack. For example, community water supplies in the United States are designed to deliver water under pressure and generally supply most of the water for fire-fighting purposes. Loss of water or a substantial loss of pressure could disable fire-fighting capability, interrupt service, and disrupt public confidence (Clark and Deininger, 2000).

This loss might result from a number of different causes. Many of the major pumps and power sources in water systems have custom-designed equipment and in case of a physical attack it could take months or longer to replace them. Sabotaging pumps that maintain flow and pressure or disabling electric power sources could cause long-term disruption (Clark and Deininger, 2001).

Many urban water systems rely on an aging infrastructure. Temperature variations, large swings in water pressure, vibration from traffic or industrial processes, and accidents often result in broken water mains. Planning for main breaks is usually based on historical experience. However, breaks could be induced by a system-wide hammer effect, which could be caused by opening or closing major control valves too rapidly. This could result in simultaneous main breaks that might exceed the community's capability to respond in a timely manner, causing widespread outages. Recognizing this vulnerability, water systems have been incorporating valves that cannot be opened or closed rapidly. However, many urban systems still have valves that could cause severe water hammer effects.

Interrupting the water flow to agricultural and industrial users could have large economic consequences. For example, the California aqueduct, which carries water from northern parts of the state to the Los Angeles/San Diego area, also serves to irrigate the agricultural areas in mid-state. Pumping stations are used to maintain the flow of water. Loss of irrigation water for a growing season, even in years of normal rainfall, would likely result in billions of dollars of loss to California and significant losses to US agricultural exports.

Another problem associated with many community water systems is the potential for release of chlorine to the air. Most water systems use gaseous chlorine as a disinfectant, which is normally delivered and stored in railway tank cars. Generally, there is only minimal protection against access to these cars. Accidental release of chlorine gas could cause injury to nearby populations.

1.5 The Threat of Terrorism to Urban Water Systems

Unlike the earthquake experience there has never been a successful terrorist attack on an urban water system, and until recently terrorism in the United States was not generally considered to be a serious threat. The President's Commission on Critical Infrastructure was formed to evaluate the vulnerability of the nation's infrastructure to internal and external terrorism. The Commission identified water supply systems as vulnerable to the full range of terrorist threats including physical attack and cyber and biological terrorism.

1.5.1 Bioterrorism and Chemical Contamination

A major concern with regard to water supplies is the potential of bioterrorism as a threat to public safety. The US Army Combined Arms Support Command evaluated 27 agents for the potential for "weaponization." Seven of the 27 agents are listed as having the potential for being "weaponized" and 14 others are listed as either possible or probable weapons. A number of these organisms are listed as definite or probable threats in water (Clark and Deininger, 2000).

The President's Commission concluded that there is a credible threat to the nation's water supply system from certain known biological agents. In addition, newly discovered or emerging pathogens may pose a threat to water supply systems. One such pathogen was isolated during a US Environmental Protection Agency (USEPA) study in Peru.

Several chemical agents have also been identified that might constitute a credible threat against water supply systems. Although much is known about chemical and biological agents dispersed in air, almost nothing is known about these agents in potable water.

The amount of material needed to deliberately contaminate a water source (such as a reservoir or aquifer) is large and generally exceeds what an individual or small group of terrorists could easily acquire, produce, or transport. However, contaminants introduced into a distribution system would be less susceptible to dilution and would reside in the system for shorter times, thus diminishing the effects of disinfectants and chemical decomposition and oxidation.

1.6 Countermeasures Against Terrorism

As illustrated by the California and Japan experiences, there are several steps that a water utility can and should take to protect itself against the sudden catastrophic effect of an earthquake. These approaches include both technological changes and establishment of institutional mechanisms that will assist in mitigating against the potential damage that might occur from such an event. Based on this experience the authors believe that there are also several steps that a water utility can take to protect against terrorist threats. These steps will be discussed in terms of physical countermeasures, chemical countermeasures, and institutional countermeasures.

1.6.1 Physical Countermeasures

Access to a free water surface such as existing in a water reservoir should be eliminated. For example, the ventilation devices in a reservoir must be constructed in such a way so as to prevent contamination of the reservoir. The intakes, pumping stations, treatment plants, and reservoirs should be fenced to secure them against casual vandalism. Beyond that, intrusion alarms should be installed to notify the operator that an individual has entered a restricted area. An immediate response might be to shut down a part of the pumping system until the appropriate authorities determine that there is no threat to the system.

An important extension of the security concept against terrorist attack would be the planning and construction of separate water lines that are fed from a protected water supply source, which would only be activated during an emergency. Many of the older cities in the United States have separate water lines that have been installed for fire protection in heavily developed downtown areas. These water lines might be upgraded for possible use to supply the population with safe water during emergency conditions. Such proactive planning for water security, including the continuous maintenance and monitoring of chlorine residual in the water, would help to ensure the safety of most water supply systems. Nevertheless, it is of vital importance that system planners and managers be constantly on the alert to prohibit deliberate sabotage of municipal water supply systems.

1.6.2 Sensor Networks

Among the different threats to a water distribution system a deliberate chemical or biological contaminant injection is the most difficult to address, both because of the uncertainty of the type of the injected contaminant and its consequences and because of the uncertainty of the location and injection time. In principle, a pollutant can be injected at any water distribution system connection (node) using a pump or a mobile pressurized tank. Although backflow preventers provide an obstacle to such actions, they do not exist at all connections and at some might not be functional.

An online contaminant monitoring system (OCMS) is considered (ASCE, 2004; AWWA, 2004) as the major tool to reduce the likelihood of a deliberate contaminant chemical or biological intrusion. An OCMS should be designed to detect random contamination events and to provide information on the location of the contaminants within the system, including an estimation of the injection characteristics (i.e., contaminant type, injection time and duration, concentration, and injected mass flow rate). Once the type of the contaminant and its characteristics are discovered, a containment strategy can be implemented to minimize the pollutant spread throughout the system and to suggest for the system's portions that need to be flushed.

However, although an OCMS is recognized as the appropriate solution to cope with a deliberate contaminant intrusion, much of the basic scientific and engineering knowledge needed to construct an effective OCMS is only partially available: (1) the monitoring/sensors instrumentation tools required to accomplish the detection task, (2) knowledge of the injected contaminants' impacts on public health, and (3) modeling capabilities of sensors locations.

1.7 Cyber Security

Growth in the use of the Internet throughout the world, since the 1990s, has dramatically changed the way that both private sector and public sector organizations communicate and conduct business. Although it was originally developed by the US Department of Defense, the vast majority of the Internet is owned and operated by various entitles in the public and private sector. It is becoming increasingly recognized that all countries need to prepare for the potential of debilitating Internet disruptions. Therefore in the United States the Department of Homeland Security (DHS), at the Federal level, has been assigned to develop an integrated public/private plan for Internet recovery, should it be impaired. The US Government Accountability Office (GAO) was asked to (1) identify examples of major disruptions to the Internet, (2) identify the primary laws and regulations governing recovery of the Internet in the event of a major disruption, (3) evaluate DHS plans for facilitating recovery from Internet disruptions, and (4) assess challenges to such efforts (USGAO, 2006).

The GAO found that a major disruption to the Internet could be caused by

- A cyber incident (such as a software malfunction or a malicious virus)
- A physical incident (such as a natural disaster or an attack that affects key facilities)
- A combination of both cyber and physical incidents

Recent cyber and physical incidents have, in fact, caused localized or regional disruptions but have not caused a catastrophic Internet failure. The GAO report presents several examples of major interruptions of the Internet which will be summarized briefly in this chapter.

1.7.1 Laws and Regulations Governing the Internet

Current Federal laws and regulations addressing critical infrastructure protection, disaster recovery, and telecommunications infrastructure provide broad guidance that applies to the Internet. It is not clear, however, how useful these regulations and authorities would be in helping to recover from a major Internet disruption. For example, key legislation on critical infrastructure protection does not address roles and responsibilities in the event of an Internet disruption. Other laws and regulations governing disaster response and emergency communications have never been used for Internet recovery.

1.7.2 Internet Recovery

The DHS has begun efforts to develop an integrated public/private plan for Internet recovery, but, according to GAO, these efforts are not complete or comprehensive. Specifically, DHS has developed high-level plans for infrastructure protection and incident response, but the components of these plans addressing the Internet are not complete. The department has started a variety of initiatives to improve the nation's ability to recover from Internet disruptions, including working groups to facilitate coordination and exercises in which government and private industries practice responding to cyber events. However, progress to date on these initiatives has been limited, and other initiatives lack time frames for completion and the relationships among these initiatives are not evident. Therefore, the government is not yet adequately prepared to effectively coordinate public/private plans for recovering from a major Internet disruption.

Key challenges to establishing an Internet recovery plan are as follows:

- The diffuse control of the many networks making up the Internet and the private sector ownership of core components
- A lack of consensus on DHS' role and a clear understanding as to when the department should get involved in responding to a disruption
- Legal issues affecting DHS' ability to provide assistance to restore Internet service
- Reluctance on the part of the private sector to share information on Internet disruptions with DHS
- Leadership and organizational uncertainties within DHS

Until these challenges are addressed, it is anticipated that DHS will have difficulty in being a focal point for helping the Internet recover from a major disruption.

1.7.3 Examples of Internet Interruption

The following five examples were cited in the GAO report to illustrate the breadth and depth of both natural and manmade disasters that could have a major effect on electronic communications (USGAO, 2006).

1.7.3.1 Case Study – The Slammer Worm

On Saturday, January 25, 2003, the Slammer worm infected more than 90% of vulnerable computers worldwide within 10 min of its release on the Internet. It exploited a known vulnerability for which a patch had been available since July 2002. Slammer caused network outages, canceled airline flights, and caused automated teller machine failures. The Nuclear Regulatory Commission confirmed that the Slammer worm had infected a private computer network at a nuclear power plant, disabling a safety monitoring system for nearly 5 h and causing the plant's process computer to fail. The worm reportedly also affected communications on the control networks of at least five utilities by propagating so quickly that control system traffic was blocked. On Monday, January 27, the worm infected more networks when US and European business hours started. Cost estimates on the impact of the worm range from $1.05 billion to $1.25 billion.

However, responses to Slammer worm were rapid. Within 1 h, Web site operators were able to filter the worm and block the main communication channel that the worm was using. This helped control the spread of the worm.

1.7.3.2 Case Study – A Root Server Attack

On Monday, October 21, 2002, a coordinated denial-of-service attack was launched against all of the root servers in the Domain Name System around the world. Two root server operators reported that traffic was three times the normal level, while another reported that traffic was 10 times the normal level. The attacks lasted for approximately 1 h and 15 min. While reports of the attack differ, they all agreed that at least nine of the servers experienced degradation in service and seven failed to respond to legitimate network traffic and two others failed intermittently during the attack.

The response to these attacks was handled by the server operators and their service providers. According to experts the government did not have a role in recovering from the attack.

1.7.3.3 Case Study – The Baltimore Train Tunnel Fire

On July 18, 2001, a 60-car freight train derailed in a Baltimore tunnel, housing fiber-optic cables for seven of the largest US Internet service providers. The resulting fire burned and severed fiber-optic cables, causing backbone slowdowns for at least three major Internet service providers. Interruptions to service were sporadic. For example, users in Baltimore did not suffer disrupted service, while users in Washington

D.C. did. In addition, there were selected impacts far outside the disaster zone. The US embassy in Lusaka, Zambia, experienced problems with e-mail. Two of the service providers restored service within 2 days and despite the outages caused by the fire, the Internet continued to operate.

The affected Internet service providers handled the recovery and city officials also worked with telecommunications and networking companies to reroute cables. Federal and local government efforts to resolve the disruption consisted extinguishing the fire, maintaining safety in the surrounding area, and rerouting traffic.

1.7.3.4 Case Study – The September 11, 2001, Terrorist Attack on the World Trade Center

On September 11, 2001, terrorists crashed two commercial airplanes into the World Trade Center, which led to the deaths of nearly 3,000 people and the destruction of 12 buildings and physically damaged one of the Internet's most important hubs. The local communications infrastructure (including facilities, critical computer systems, and fiber-optic cables that ran under the ruined buildings) was disrupted. The attack also disrupted electrical power in Lower Manhattan. Back-up power systems were used by local telecommunications facilities until they ran out of fuel or batteries and had to shut down their operations. Repairs to key infrastructure centers were delayed because of structural concerns for buildings and government-ordered evacuations.

The attack disrupted local financial and communications systems, which led to the closing of financial markets for up to 1 week, and interrupted Internet connectivity to several universities, medical colleges, and hospitals and to the city government's official Web site. Internet service providers in parts of Europe lost connectivity and there were Domain Name System disruptions in South Africa due to interconnections in New York City. However, in general Internet functions were largely back to normal within 15 min, and there were no widespread connectivity issues, thereby demonstrating the flexibility and adaptability of the network. Internet operators rerouted traffic to bypass the physical damage in lower Manhattan.

The federal government's efforts in restoring Internet service included facilitating communications and providing logistical support. The government also secured the area and provided military transport to the New York area for key telecommunications personnel while commercial air traffic was shut down.

1.7.3.5 Case Study – Hurricane Katrina

On August 29, 2005, Hurricane Katrina significantly damaged and in some cases destroyed the communications infrastructure in Louisiana, Mississippi, and Alabama. According to the Federal Communications Commission, the storm resulted in outages for over 3 million telephone customers, 38 emergency 9-1-1 call centers, hundreds of thousands of cable customers, and over 1,000 cellular sites. The Coast Guard's computer hub in New Orleans dropped off-line, resulting in no computer or Internet connectivity to all coastal ports within the area. This lack of Internet service caused Coast Guard units to resort to communicating with telephones and fax machines.

A substantial number of the networks that experienced service disruptions recovered relatively quickly. According to the Federal Communications Commission, commercial carriers restored service to over 80% of the 3 million affected telephone customers within 10 days of Hurricane Katrina. Despite the overall devastation caused by Katrina, the hurricane had minimal affect on the Internet.

Private sector representatives stated that with the exception of the Federal Communications Commission (which coordinated provision of some governmental resources and information), coordination with the government was limited. Virtually no assistance was received from the Federal government and it was reported that requests for assistance, such as food, water, fuel, and secure access to facilities, were denied. The Stafford Act (which authorizes such assistance) does not include for-profit companies.

1.7.4 Cyber Attacks in the Public Sector

Clark and Knake (2010) have explored the potential for cyber attacks from unnamed adversaries on institutions in the United States. Consistent with the GAO report they have concluded that the civilian sector is highly vulnerable to such an attack. Results from the conflict between the Republic of Georgia and Russia in 2008 provide an example of the damage that can result from cyber attacks. There was physical fighting between Russia and Georgia; however, before fighting broke out cyber attacks were launched against Georgian government sites in an attempt to cut off Georgia from connecting to the Internet. As a result of these attacks the Georgian banking sector shut down its servers. Consequently, Georgia's banking operations were paralyzed and credit card systems crashed followed by the mobile phone system. Clark and Knake (2010) contend that the United States is not effective in defending cyber attacks especially in the banking and electrical utility sectors. Clearly, drinking water and wastewater utilities are heavily dependent upon electrical power. They cite the Slammer Worm case study, described earlier, which slowed controls on a power grid. The "worm" attack in combination with a programming "glitch" in a widely used Supervisory Control and Data Acquisition (SCADA) System slowed utilities response to a falling tree that created a power surge in Ohio. The surge resulted in a power outage that encompassed eight states, two Canadian provinces, and 50 million people. The Cleveland water system was left without electricity causing their pumps to fail and placing the utility in a near crisis. The authors cite a deliberate hacker attack launched against an electrical system in Brazil with similar results.

The American Water Works Association (AWWA Streamlines, 2010a) reported that a Belarus computer security has identified a virus, called Stuxnet, which attacks SCDA systems through a vulnerability in Microsoft Windows. It has been reported that most of the affected are in India, Indonesia, and Iran. It has been characterized as a virus, a worm, and a Trojan. The American Water Works Association has also reported that several SCDA software programs used by utilities are raising cyber security concerns (AWWA Streamlines, 2010b). The US Department of Homeland Security has issued an alert concerning the vulnerability of VT Scada software and server system to unauthorized access. To date no one has been attacked according to the developer.

1.7.4.1 The "Stuxnet" Virus

According to news reports (http://www.foxnews.com/scitech/2010/11/26/secret-agent-crippled-irans-nuclear-ambitions/), in June (2010) a Belarus-based company doing business in Iran discovered a highly sophisticated computer worm called "Stuxnet." Stuxnet is an incredibly advanced, undetectable computer worm that probably took years to construct and was designed to jump from computer to computer until it found its specific target which, in this case, was Iran's nuclear enrichment program. Iran's nuclear enrichment program is seemingly impenetrable. For security reasons, it is constructed several stories underground and is not connected to the World Wide Web. Therefore the virus had to make its way through a set of unconnected computers. It had to adapt to security measures until it reached a computer that could bring it into the nuclear facility. It was designed in such a manner that when it found its target, it would secretly manipulate it until it was so compromised that it ceased normal functions. After achieving its goals it would have to destroy itself without leaving a trace.

The virus was apparently successful in finding its target which was both of Iran's nuclear enrichment facilities. It entered the operating systems at both facilities and then modified itself when it was discovered. What is especially interesting is that the nuclear facilities in Iran run an "air gap" security system, meaning they have no connections to the Web, making them secure from outside penetration. Stuxnet was apparently designed on the assumption that someone working in the plant would take work home on a flash drive, acquire the worm, and then bring it back to the plant.

It is instructive to examine what the virus was able to do after it entered the operating systems for both facilities. After defeating the security systems the worm ordered centrifuges to rotate extremely fast and then to slow down precipitously damaging the converter, the centrifuges, and the bearings, and corrupting the uranium in the tubes. At the same time it confused Iran's nuclear engineers and left them wondering what was wrong, because computer checks showed no malfunctions in the operating system. It is estimated that this penetration went on for more than a year, leaving the Iranian program in chaos and that the worm grew and adapted throughout the system. When a new worm entered the system, it would adapt and become increasingly sophisticated. The source of the virus has not been identified but the evidence points to institutions with highly sophisticated cyber war capability. This example is very instructive for the water and wastewater industry because the type of equipment and processes utilized in these industries is very similar to the type of equipment used in the chemical processing industry.

1.8 Material to Be Included in This Book

Larry Mays (2004) of Arizona State University edited a book intended to summarize the state of the art of the knowledge in securing water and wastewater systems. It was an excellent overview. However, much progress has been made since that time and this book attempts to summarize the current state of the art in water and wastewater systems security.

States (2010) published a book under the auspices of the American Water Works Association which intended to be a compilation of developments in water and wastewater systems since September 11, 2001. His goal was to provide a practical reference document for use by drinking water and wastewater managers and operators for dealing with homeland security and general emergency response.

1.8.1 Current State of Water Supply and Wastewater Systems Security: An Overview

Van Leuven (Chapter 2, this volume) examines why water infrastructure is so critical to our society and identifies the hazards that could threaten and disable an entire system. She provides illustrations of the vulnerabilities and the potential consequences of an intentional attack on a water system and provides an approach to making vulnerability assessments.

Ernst et al. (Chapter 3, this volume) describe research being conducted by the US Environmental Protection Agency (EPA) which is the lead US sector-specific agency responsible for water security. Research conducted by EPA's National Homeland Security Research Center (NHSRC) supports the agency's Goal 2 "Clean and Safe Water – Ensuring drinking water is safe" and its mission of providing drinking water treatment plants with tools and methodologies to improve water security and recover as quickly as possible should a chemical, biological, or radiological event occur. The research also has multiple benefits in optimizing treatment operations and improving water quality.

Bahadur and Samuels (Chapters 4 and 5, this volume) describe the nature of water and wastewater systems in the United States. They discuss the general nature of water and wastewater systems and why an intentional attack against this critical infrastructure would be problematic for the citizens they serve. In addition, they suggest several approaches to minimizing the vulnerability of these systems.

Fischer (Chapter 6, this volume) describes the state of Nebraska's Drinking Water Security Program. The program has set the following goals: (1) to encouraging public water systems to secure their facilities to the greatest extent possible, (2) training public water system personnel to develop an effective emergency response plan, (3) developing a sense of cooperation and teamwork among all emergency responders that ensures effective action in the wake of a disaster, and (4) meeting and producing a video for law enforcement personnel to educate them in the particulars of crime scene evidence related to public water systems.

Möderl and Rauch (Chapter 7, this volume) from the Institute of Infrastructure, University of Innsbruck, Technikerstr in Innsbruck, Austria, suggest a new approach for managing the risk for critical infrastructure vulnerability. They suggest that a methodology is developed where the effect of functional changes of a component are computed by means of a hydraulic simulation and expressed in terms of indicator values. When this is done for each individual component of the entire system, spatial information on the intrinsic vulnerability of the system is generated. VulNet is a software tool that performs these computations and also the subsequent assessment of the vulnerability. The methodology has been tested for five water supply systems (WSSs) and one urban drainage system (UDS). It was demonstrated that the spatial information of the intrinsic vulnerability of WSSs offers significant information on critical sections of the supply system and indicates also how the situation can be improved, e.g., vulnerabilities occur if different demand areas (e.g., separated by a river) are not properly connected. By strengthening these connections, vulnerabilities are reduced. The application of the method using VulNet is suggested as a valuable tool for managers and operators of water utilities to improve the performance of their system and to consider system vulnerability in rehabilitation planning. Additionally, an alpine region including five municipalities were chosen to evaluate the public drinking water supply security. A methodology was developed to identify, on a regional basis, zones with high risk by merging information on vulnerability and four potential natural hazards. The methodology aids water management to make decisions on which sites of the WSS should be chosen for preventive measures.

1.8.2 Characteristics of Water and Wastewater Systems in the United States

According to Clark (Chapter 8, this volume) substantial water supply and wastewater infrastructure has been constructed in the United States including extensive storage and distribution facilities especially in the West and Southwest. Drinking water in the United States is regulated under the Safe Drinking Water Act of 1974, and the Federal Water Pollution Control Act or Clean Water Act of 1948 is the principal law that regulates the pollution discharged into the nation's streams, lakes, and estuaries. There are over 162,000 water systems in the United States that meet the federal definition of a public water system. It is estimated that there are 980,000 miles (1.6×10^6 km) of distribution system pipes. There are 16,024 publicly owned treatment plants in the United States and all but 200 provide secondary treatment. In many older cities sanitary sewage and storm water runoff are collected in a single sewage system and are vulnerable to sanitary sewer overflow during peak rainfall events. The USEPA's 2003 Needs Assessment found that the nation's water systems will need to invest $276.8 billion over the next 20 years in order to continue to provide safe drinking water to their consumers. The Clean Watershed Needs Survey (CWNS) 2004 Report to Congress by the USEPA estimated a need of $202.5 billion (2004) for wastewater treatment and collection facilities.

1.8.3 Chemical and Microbiological Threats for Water System Contamination

According to Rice (Chapter 9, this volume) the presence of microbial pathogens in a water supply following a disaster poses an urgent threat to public health. There is an extensive amount of literature available on the classical waterborne pathogens, but by contrast there is a limited amount of information on the overt bio-threat or bio-warfare agents which could be introduced into a water system. A bio-terrorism incident in a municipal drinking water system would have the potential for causing widespread disease and disruptions of vital public services which could affect large segments of the population. He reviews recent developments for assessing the role of microbial pathogens which have the potential for being used as bio-threat agents when intentionally introduced into a water system.

According to Deininger et al. (Chapter 10, this volume) current microbiological standards are focused on a single group of indicator organisms for the bacteriological safety of drinking water. Although the current standards of water quality have eliminated massive outbreaks of waterborne disease, a question has been raised about the adequacy of the standard drinking water quality to prevent waterborne illnesses. The present HPC method using R2A agar is known to be the most sensitive test for enumerating the bacteria from treated water; the test takes 7 days to complete. The authors propose an ATP bioluminescence assay allows an estimation of bacterial populations within minutes and can be applied on a local platform. Their research indicates that the test they have developed could estimate bacterial populations might occur in a practical and timely manner during a contamination event.

VanBriesen et al. (Chapter 11, this volume) discuss the importance of maintaining chlorine residuals in treated water to protect drinking water consumers and to provide protection against small-scale intrusions. Monitoring these residuals is important for operational control and has the potential for providing early warning of contaminant intrusions. In order to use online real-time chlorine detection as part of a security system, a utility must have an accurate map of their distribution system along with corresponding operational parameters in order to assess vulnerabilities. Further, they must have a predictive model of chlorine concentrations throughout the system under many different dynamic scenarios. This enables prediction of expected chlorine at sensor locations and thus determination of "alarm" conditions. Finally, to counter the possibility of low chlorine residual concentrations, some distribution systems have installed chlorine booster stations. The authors evaluate the steps a utility can take from initial vulnerability assessment through installation and operation of chlorine sensors and boosters. For security reasons, simulated distribution systems are used in examples rather than an actual case study.

1.8.4 Monitoring for Natural and Manmade Threats in Water and Wastewater Systems

Burr et al. (Chapter 12, this volume) discuss the potential for the development of biosensors for warning about potential contamination to streams and watersheds focusing on biosensors to detect *Escherichia coli* O157:H7. They conclude that biosensors will not be attractive to the water industry until it has been demonstrated in pilot studies that they can be operated over long periods of time with minimal operator expertise, can be integrated into systems that process water volumes on the liter to cubic meter scale, achieve two to three orders of magnitude improvement in detection limits, and produce responses that are unambiguous.

According to Kroll (Chapter 13, this volume) a number of studies have shown that the utilization of multi-parameter monitoring has the potential to indicate the presence of a wide variety of harmful agents in water at levels that would be protective of human health. He discusses the key elements that should be considered when choosing and deploying such systems and presents a number of criteria and considerations for the selection and deployment of these systems. These criteria can form the basis for successful selection and deployment of early warning systems for water. As the analytical science behind these systems progresses, they will increase their ability to satisfy all of these factors. As the state of the industry stands today there are systems available that do a good job of addressing all of the criteria, but progress will continue.

1.8.5 Modeling Contaminant Propagation and Contaminant Threats

Ostfeld (Chapter 14, this volume) describes a water distribution system as an interconnected collection of sources, pipes, and hydraulic control elements (e.g., pumps, valves, regulators, and tanks) delivering consumers prescribed water quantities at desired pressures and qualities. The behavior of a water distribution system is governed by (1) the physical laws that describe the flow relationships in the pipes and the hydraulic control elements, (2) the consumer demands, and (3) the system's layout. Interest in modeling flow and water quality in water distribution systems stems from three types of circumstances: use of waters from sources with different qualities in a single distribution system serving as a "treatment facility" to mix and convey them, with a blend supplied to its consumers. Simulation and optimization algorithms for modeling water quality in distribution systems are needed by designers, utilities, and regulating agencies for a number of purposes: (1) planning and design of networks and facilities, (2) real-time operation, (3) monitoring design and operation, (4) simulation of contamination events, and (5) guidelines establishments for planning, design, operation, and monitoring. Water quality simulation modeling is aimed at studying the changes of water quality substances in time and in space within the distribution system. The need for optimization exists whenever the solution to a problem is not unique. Common examples for optimization needs in modeling water quality in water distribution systems are design, operation, chlorine control, monitoring, calibration, and since the September 11 events in the United States – water security. The author describes issues related to water security within the context of water distribution systems modeling and highlights future needs and challenges in this area.

1.8.6 Case Study Applications

Patterson and Adams (Chapter 15, this volume) describe EPA's Disaster Recovery Plan and the steps that the water industry (water utilities, government agencies, non-governmental organizations, academia, and consultants) is taking to tackle potential threats to safe drinking water and drinking water infrastructure. A case study of EPA emergency response efforts after Hurricane Katrina is provided to bring the impact of major natural disasters on public water systems into focus. Government agencies including the EPA are supporting the development of small drinking water treatment technologies to bring timely relief to devastated communities. EPA research is focusing on household devices, mobile treatment systems, and disinfection processes as described to protect consumers from contamination in drinking water wells, tanks, and distribution systems. US government agencies including the EPA are planning ahead to provide temporary supplies of potable water to communities during emergencies. EPA is supporting the development of small drinking water treatment technologies to bring timely relief to devastated communities.

1.8.7 Distribution System Modeling, SCADA Systems, Security and Surveillance Systems

Panguluri et al. (Chapter 16, this volume) discuss an area that may represent major vulnerability in the nation's critical infrastructure. Early assessments of water and wastewater systems found no evidence of an impending "cyber attack" which could have a debilitating effect on the nation's critical infrastructures. However, more recent studies have demonstrated that publicly available computer equipment and hacking software could be used to infiltrate and take control of the computer centers at Defense Department, as well as power grids and 911 systems in nine major US cities. There are many other well-known hacking incidents that have targeted the military and other critical infrastructure. Since these studies have been publicized, many research organizations operating under various mandates have undertaken efforts to understand the complex infrastructure interdependencies especially between water/wastewater infrastructure and the energy infrastructure (electric, oil, and gas). Four major categories of infrastructure interdependencies (physical, cyber, geographic, and logical) have been identified as they apply to the water/wastewater infrastructure. In addition, the proliferation of information technology (IT) for organizational efficiency and the increased use of automated monitoring and control systems (e.g., Supervisory Control and Data Acquisition (SCADA) systems) for operational efficiency by the water and wastewater utilities have created additional cyber vulnerabilities that need to be appropriately addressed. The authors cite an incident that occurred at the Maroochy Shire Sewage Treatment Plant in Queensland, Australia, in which a disgruntled employee hacked into the SCADA system for the plant causing approximately 212,000 gallons of raw sewage to spill out into local parks, rivers, and the grounds of a nearby hotel. The authors discuss current approaches (relevant standards and vendor initiatives), their key elements, and provide a summary of the recently developed sector-specific cyber security roadmap. Examples are presented that document various successes and challenges faced by the water and wastewater sector to meet the requirements of these standards and achieve the goals identified in the sector-specific roadmap.

Murray et al. (Chapter 17, this volume) discuss the strategic placement of sensors throughout the distribution network which is a key aspect of designing a Contamination Warning System. There has been a large volume of research on this topic in the last several years, including a study that compared 15 different approaches to solving this problem. The authors focus on the sensor placement methodologies that have been developed by EPA's Threat Ensemble Vulnerability Assessment (TEVA) Research Team, which is composed of researchers from EPA, Sandia National Laboratories, the University of Cincinnati, and Argonne National Laboratory. This team has developed TEVA-SPOT – the Threat Ensemble Vulnerability Assessment Sensor Placement Optimization Tool – a collection of software tools that can help utilities design sensor networks. Case studies are presented using TEVA-SPOT and open challenges for application of sensor network design to large-scale real-world drinking water systems are discussed.

1.8.8 Institutional and Management Issues in Responding to Natural and Manmade Threats

Bukhari and LeChevallier (Chapter 18, this volume) believe that physical hardening of a drinking water plant does not eliminate vulnerabilities at the plant or in the distribution systems, which can extend over hundreds of miles. A comprehensive approach is required to protect distribution system water quality by employing technologies that facilitate "real-time" feedback and provide tools to indicate an early warning of unanticipated changes in water quality. The approach being evaluated by the US Environmental Protection Agency in their Water Security Initiative consists of integrating multi-streams of information (i.e., water quality, syndromics, eye witness, law enforcement, etc). The authors using the Water Security Initiative as the platform discuss a conceptual model that is capable of integrating information from various technologies (i.e., Automatic Meter Readers capable of backflow and leak/tamper detection) in distribution system pipes to convey multi-streams of information to software-assisted alarms, which can then integrate information from hydraulic models to trigger automated sampling at strategically selected sites for laboratory-based verification of intentional or accidental contamination events. Following water quality aberrations, the utility needs to initiate a Consequence Management Plan (CMP). The CMP needs to be a "living document" designed in a manner that is intuitive, self-explanatory, and is capable of guiding utilities through the most appropriate data collection, analysis, communication, and mitigation steps. The authors provide an overview of the wide and varied processes that a utility needs to navigate to return an impacted system back to normal operation as quickly and safely as possible.

1.8.9 Developing Techniques and Approaches for Natural and Manmade Threat Response

McKenna et al. (Chapter 19, this volume) discuss event detection systems that provide online analysis of water quality data for identification of significant water quality events. Two different online algorithms are discussed that utilize multivariate data from two monitoring locations in an operating water distribution network. The data are split into training and testing sets and parameter identification is completed on the training data prior to application on the testing data. Water quality events are added to the testing data sets as perturbations from the measured water quality using 11 different event strengths. The resulting receiver operating characteristic curve areas quantify the relationship between probability of detection and false detections at the time step scale. Additionally, the proportion of events containing at least one detection is measured. Results show that both algorithms are capable of reliably detecting events that change the background water quality by 1.5 times the standard deviation of the water quality signal while limiting the false-positive results to 3–4% of the time steps. Trade-offs in the delay to detection versus the number of false-positive results are examined in the context of the event length.

According to Di Cristo et al. (Chapter 20, this volume) in the last few years many interesting studies have been devoted to the development of technologies and methodologies for the protection of water supply systems against intentional attacks. However, the application to real systems is still limited for many different economic and technical reasons. She and her colleagues from The Water Engineering Lab (L.I.A.) of the University of Cassino (Italy) were involved in two research projects financed by the European Commission in the framework of the European Programme for Critical Infrastructure Protection (E.P.C.I.P.). Both projects had as a common objective to provide guidelines for enhancing security in water supply systems with respect to intentional contamination risk and they were developed in partnership with large Italian Water Companies. They present a general procedure for protection systems design of water networks. In particular, the procedure is described through the application to real water systems, characterized by different size and behavior.

Shen and McBean (Chapter 21, this volume) have developed a contaminant source identification procedure intended to protect water distribution systems that have to be both rapid and able to incorporate uncertainties, when identifying possible intrusion nodes (PINs). PINs identification has two major issues, the false-negative rate (failure to identify the true ingress location) and the false-positive issue (falsely identifying a location which is not the true ingress location). A data mining procedure is described and applied, which involves mining an off-line-built database, to select PINs that possess first detection times within $\pm m$ from the online sensor first detection time. The "m" value is a statistical characterization of the array of events of the offset values between online sensor first detection time under uncertainty and the one corresponding to the same intrusion event stored in the off-line database; with "m" selected, issues of controlling false negatives and positives are addressed. The approach described herein is made possible through the power of parallel computing in supercomputers, which demonstrates huge potential by simulating scenarios simultaneously. The online data mining procedure, i.e., the PINs identification, is integrated into a geographic information system toolkit for rapid emergency response. In the case studies, simulation of scenarios is reduced linearly to the number of processors applied. Results show that increasing the number of scenarios in the database can provide input to compute the "m" value, always reduce the false-negative rate of each sensor, and usually reduce the number of false-positive PINs.

Van Leuven (Chapter 22, this volume) discusses the need for a multilayered security approach for protecting critical water and wastewater infrastructure that includes policies, procedures, plans, protective countermeasures, training, exercises, relationships with intelligence agencies, and response capabilities. In this chapter Van Leuven (Chapter 22, this volume) identifies common elements that drive security investments encompassing everything from a calculated risk-based approach to the gut reactions of operators who understand the consequences of a significant asset failure. She describes available countermeasures and physical security investments designed to deter, delay, detect, assess, and respond to security incidents. Van Leuven (Chapter 22, this volume) concludes with a synopsis of recommended programmatic components to ensure a comprehensive, multilayered security approach to protecting drinking water and wastewater systems.

According to Birkett et al. (Chapter 23, this volume) water and wastewater infrastructure has been subject to attacks and threats since ancient times. Following the terrorist attack on the Twin Towers in New York in 2001, there has been increased interest in examining new approaches for ensuring adequate protection to water and wastewater infrastructure. The investigators propose a unique approach to mitigating threat levels by introducing the concept of crisis leadership and crisis control. This methodology is illustrated by regularly practicing plans and procedures in the form of scripted crisis exercises. There are four major types of exercises which display processes, roles, and responsibilities with an accent on planning and documentation. Water and wastewater agencies which adopt these strategies will survive and produce a resilient organization. This chapter provides an overview of a preparedness and recovery framework suitable for water industries worldwide.

1.9 Summary and Conclusions

Urban water systems are vulnerable to both manmade and natural, but unpredictable, threats and disasters such as droughts, earthquakes, and terrorist attacks. Although there have been no recorded attempts in the modern era of attacks against urban water supplies there have been many natural disasters such as earthquakes which have had catastrophic effects on water systems. Much can be learned from these events and the subsequent preparedness for these events. Terrorism is also a major threat to water security and recent attention has turned to the potential that these attacks have for disrupting urban water supplies. The *Public Health Security and Bioterrorism Preparedness and Response Act of 2002* (US Congress, 2002) intensified the focus on water security research in the United States. After the attacks of September 11, 2001, the US Environmental Protection Agency (EPA) developed a *Homeland Security Strategy* (USEPA, 2004). Its intent was to enhance national security and protect human health and the environment. A particular area of concern for water systems is cyber security even though this area of vulnerability was not addressed in these documents. A report by the US Government Accountability Office has highlighted some of the vulnerabilities associated with possible attacks against the Internet and Panguluri et al. (Chapter 16, this volume) present some possible solutions. Much of the research conducted as a result of these directives is reviewed in this book through contributed chapters by US and international experts.

References

American Society of Civil Engineers (ASCE) (2004) "Guidelines for designing an online contaminant monitoring system."

American Water Works Association (AWWA) (2004) "Security guidance for water utilities."

American Water Works Association-Streamlines (2010a) "Virus exploits USB vulnerability to reach SCDA systems." July 27, 2010, Vol. 2, No. 19, November 2, 2010a.

American Water Works Association-Streamlines (2010b) "Cyber alert for SCADA program." Vol. 2, No. 26, November 2, 2010b.

Bruins, H.J. (2000) "Proactive Contingency Planning vis-à-vis Declining Water Security in the 21st Century." *Journal of Contingencies and Crisis Management*, Vol. 8, No. 2, pp. 63–72.

Clark, Robert M. and Deininger, Rolf A. (2000) "Protecting the Nation's Critical Infrastructure: The Vulnerability of US Water Supply Systems," *Journal of Contingencies and Crisis Management*, Vol. 8, No. 2, pp. 73–80.

Clark, R.M. and Deininger, R.A. (2001) "Minimizing the Vulnerability of Water Supplies to Natural and Terrorist Threats," in the *Proceedings of the American Water Works Association's IMTech Conference* held in Atlanta, GA, April 8–11, pp. 1–20.

Clarke, Richard A. and Knake, Robert K. (2010) Cyber War: The Next Threat to National Security and What to Do About It. Harper-Collins, New York, NY, pp. 18–21.

Gleick, P.H. (2006) "Water and Terrorism," *Water Policy*, Vol. 8, pp. 481–503.

Mays, L.W. (2004) "Water Supply Security: An Introduction," in Water Supply Systems Security, edited by Larry W. Mays. McGraw-Hill: Two-Penn Plaza, New York, NY. pp. 1.1–1.12.

New York Times (1986) White House water cut off temporarily. July 10, p. 16.

States, S. (2010) "Security and Emergency Planning for Water and Wastewater Utilities". American Water Works Association. 6666 West Quincy Avenue, Denver, CO 80235-3098.

US Congress (2002) Public Health Security and Bioterrorism Preparedness and Response Act of 2002: Public Law 107–188. http://thomas.loc.gov/cgi-bin/query/z?c107:H.R.3448.ENR:. Accessed 03 March 2010.

United States Government Accountability Office (USGAO) (2006) Internet Infrastructure: DHS Faces Challenges in Developing a Joint Public/Private Recovery Plan. GAO-06-672.

US Environmental Protection Agency (2004) Water Security Research and Technical Support Action Plan. Publication number EPA/600/R-04/063. http://www.epa.gov/nhsrc/pubs/600r04063.pdf. Accessed 02 March 2010.

Chapter 2
Water/Wastewater Infrastructure Security: Threats and Vulnerabilities

Laurie J. Van Leuven

2.1 Introduction

The nation's critical infrastructure is made up of thousands of networks, pipelines, roads, conduits, and facilities; some are connected and some are isolated structures. Most of these critical systems are reliant on the full functionality of one or more other critical systems to ensure ultimate delivery of essential services to the public. Protecting these services requires a multilayered security program tailored for each system. Protective measures in the form of policies, procedures, and security investments can help reduce risks to critical infrastructure. The first step in developing a comprehensive security program is to recognize threats and each asset's vulnerabilities. This chapter will describe why drinking water and wastewater systems need to be protected, what threats to consider, and identify the vulnerabilities that increase risks and leave assets susceptible to an attack or large-scale system failure.

Utilities provide essential services to people 24 h a day, 7 days a week, and their services are essential to keeping communities healthy and economically viable. People rely on the constant delivery of drinking water and the collection, conveyance, and treatment of wastewater. The public uses water for the most basic human needs. Vital networks and businesses, industries, hospitals, other utilities, agriculture, and manufacturing industries are dependent on water systems. Water systems are also essential to recovery efforts following any natural disaster and for maintaining the standard of living for our everyday lives.

The systems responsible for delivering such fundamental commodities in the United States have long been identified as critical infrastructure. Drinking water and wastewater systems are both grouped into the Water Sector, one of 18 critical infrastructure sectors recognized by homeland security experts and officials as vital systems and networks that need to be protected (HSC, 2007). The Department of Homeland Security (DHS) designates the Environmental Protection Agency (EPA)

L.J. Van Leuven (✉)
Seattle Public Utilities/U.S. Department of Homeland Security (DHS), FEMA, Washington, DC, USA
e-mail: laurie.vanleuven@gmail.com; Laurie.VanLeuven@fema.gov

R.M. Clark et al. (eds.), *Handbook of Water and Wastewater Systems Protection*, Protecting Critical Infrastructure, DOI 10.1007/978-1-4614-0189-6_2,
© Springer Science+Business Media, LLC 2011

as the lead agency overseeing the Water Sector, which includes both drinking water and wastewater utilities. The Water Sector's goal is to recognize and reduce risks to infrastructure and support practices that build and maintain system resiliency (USDHS, 2010).

As of 2006, there were approximately 160,000 public drinking water utilities and more than 16,000 wastewater utilities in the United States. A high percentage of the population receives potable water and sanitary sewer service from these utilities, approximately 85 and 75%, respectively (USDHS and USEPA, 2007). The disparate ownership of the nation's water infrastructure, consisting of private, municipal, and special purpose districts, spreads across thousands of jurisdictions from coast to coast. The level of preparedness to which these independent systems could prevent or recover from a catastrophic incident varies greatly.

The wide range of dependencies on water systems increases the consequence of system outages through cascading impacts such as the effects on public health, the ability of first responders to provide emergency services, economic losses, and damage to the confidence of the American people (USDHS, 2007b). The assets necessary to keep water systems functioning are so vital that destruction or incapacity of these systems could debilitate national security, economic security, and public health or safety (USDHS, 2007a).

This chapter will examine why water infrastructure is so critical, identify the hazards that could threaten and disable an entire system, and illustrate the vulnerabilities and the potential consequences of an intentional attack on a water system. Other issues to be discussed are the drivers for security improvements and physical security countermeasures available to prevent security incidents and to protect against, prepare for, and respond to large-scale water system failures.

2.2 Why Secure Water Infrastructure?

Water systems are vulnerable to a variety of natural and human-caused threats. In the past decade, growing concerns about critical infrastructure becoming potential targets by terrorist attacks in the United States have contributed to a new dimension of security threats to utilities.

Utilities make up a considerable portion of the nation's critical infrastructure. Three sectors identified by DHS as critical infrastructure can be distinguished as utilities: water, telecommunications, and energy (USDHS, 2006). These utility sectors are all highly reliant on one another for their operations and in some instances they are co-located at the same geographic location (i.e., hydroelectric dams, pipes secured to bridges, and telecommunications antennas on water tanks and standpipes). In addition, an outage in any one of these sectors could have a significant impact on the other 17 critical infrastructures.

People may question the need to secure pump stations, water storage facilities, treatment plants, or pipelines. The simple answer is that the negative consequences of an intentional attack are too great to ignore. A significant attack on a water system could result in widespread illness or casualties. A denial of service scenario could affect critical services such as firefighting and health care and could disrupt other dependent sectors such as energy, transportation, and food and agriculture. Most people recognize how devastating these consequences could be. However, they might question the likelihood of an attack, postulating that "the system has never been attacked before, why would it be attacked now?" The problem with this perspective is that the threats and risks to water systems are on the rise, due to an evolving threat environment.

The severe consequences of an attack on critical infrastructure and the significant interdependencies among so many sectors are enough to provide a motive to terrorists. An intentional attack on a water system would certainly spread fear and anxiety throughout society. One intentional successful attack anywhere in the country could lead to panic. People could easily become afraid to drink the water flowing out of their taps. Citizens living in other areas of the country would begin asking, "How safe is my drinking water?"

In addition to motive, the opportunity exists, since there are so many potential targets. There are literally thousands of water or wastewater assets that could be exploited by a determined terrorist. It is simply impossible to secure everything. There are also known interests by terrorist organizations to experiment with weapons of mass destruction. History has proven that attempts have been made by terrorists to contaminate drinking water systems using biological or chemical agents, mostly in other parts of the world.

Utility vulnerabilities have existed since they were built and disruptions to services are not uncommon. Water and wastewater utilities have always had to deal with the impacts of extreme weather conditions and pipeline or equipment failures that cause service interruptions. Water systems are frequently tested by natural disasters. Earthquakes, severe weather conditions, aging infrastructure, and the interdependencies among other systems are traditional threats utilities face everyday (Seger, 2003).

Since utility outages are not uncommon, most organizations have a mechanism to deal with smaller scale problems effectively. Utility operators are good at response and routine repairs. They hold a great amount of knowledge about the systems they own and operate. During a planned or manageable system outage, operators can isolate the problem and repair it quickly. System operators can usually reestablish services and return to normal operations within 12 h.

Given the quick recovery time of most service interruptions, customers have become accustomed to immediate restoration of vital water systems. However, water systems can unintentionally contribute to a less resilient community, if their customers are overreliant on immediate recovery. Water systems need to manage customer expectations by reaching out and educating the community to be self-sufficient for at least 72 h. The public will be far more resilient and less panicked if they have an adequate supply of emergency drinking water available for each member of their family during the immediate aftermath of an emergency situation.

The real challenges are preventing more significant system failures. The following list captures the most severe types of water system failures (NDWAC, 2005):

- Loss of pressurized water for a significant part of the system.
- Long-term loss of water supply, treatment, or distribution.
- Catastrophic release or theft of on-site chemicals affecting public health.
- Adverse impacts to public health or confidence resulting from a contamination threat or incident.
- Long-term loss of wastewater treatment or collection capacity.
- Use of the collection system as a means of attack on other key resources or targets.

2.3 Threats to Water Systems

Utilities are adept at maintaining and repairing damage to aging infrastructure related to normal day-to-day operations. However, acquiring the expertise and funding to build the capability to effectively respond to a catastrophic natural disaster or terrorist attack (or to prevent one) that could wipe out an entire water system is a significant challenge. Recognizing threats to the water sector is a critical first step.

There are many types of threats that could harm all or parts of a drinking water or wastewater system. Even though periodic weather emergencies are to be expected, critical infrastructure providers must protect against more sinister threats that include intentional acts and build resiliency to recover from large-scale disasters that could lead to massive system damages. The growing list of threats to water systems is evolving as evident by two homeland security incidents of national significance.

2.3.1 Evolving Threat Environment

The United States has experienced a significant change in the threat environment for utilities during the past decade. Two defining incidents that have changed how our country's leaders think about threats and resiliency are the terrorist attacks of September 11, 2001, and Hurricane Katrina. These catastrophic events resulted in greater awareness of the vulnerabilities of critical infrastructure to intentional acts of terrorism as well as natural disasters. These incidents sparked new security and emergency management regulations through several legislative acts and executive directives.

2.3.1.1 September 11 Terrorist Attacks

The terrorist attacks on the World Trade Center and the Pentagon illustrate that there are people in the world with an expressed interest in harming Americans. They do not limit their targets to military personnel or facilities. Terrorist attacks can happen within our borders, at any time, and without warning. Terrorist actions are beyond criminal. According to terrorism expert Bruce Hoffman, acts of terrorism involve violence or the threat of violence and are specifically designed to spread fear and anxiety through the whole of society and have far-reaching psychological effects (Hoffman, 2006).

Since terrorists are often outnumbered by the military forces of their target country, they most commonly use a strategy of unconventional tactics aimed at nonmilitary, noncombatant targets. This is one of the primary reasons why critical infrastructure and key resources are particularly at risk to a terrorist attack. Critical infrastructure sectors are rich targets, with many vulnerable assets across the country and not enough resources to protect and secure them all.

2.3.1.2 Hurricane Katrina

Local communities throughout the United States and the world are susceptible to natural disasters that create negative impacts on residents, businesses, and visitors. Disasters can also hamper local government agencies' ability to provide essential functions for the welfare of the community. The most frequent and widespread incidents creating hardships for people related to safety of life and economic losses are due to extreme weather events.

In the aftermath of Hurricanes Katrina and Rita in 2005, residents of Louisiana and Mississippi and other areas around the Gulf Coast became painfully aware of how critical water systems are. Many essential functions were severely disrupted. There were massive problems with response and recovery coordination in the region. The lack of resiliency of these infrastructures magnified the hardships that people were experiencing. There was limited continuity of operations or government. Systems were shut down. It was difficult for emergency workers to determine where to begin. Damage assessments were slow, and the prioritization efforts suffered from lack of preplanning. Long-term recovery efforts and return to normal operations took years.

The storms devastated many infrastructure systems. In Louisiana, Mississippi, and Alabama, widespread power outages affected 2.5 million customers. Telecommunication systems collapsed and dangerous hazardous waste chemical facilities were flooded. Katrina destroyed or compromised 170 drinking water facilities and dozens of wastewater treatment facilities (SMGI, 2006). The stark realization of how unprepared the United States was to respond to the disastrous flooding from Hurricane Katrina took centerstage in the media during the aftermath. Government agencies were reactive to the criticism and promptly focused the majority of new homeland security and emergency management initiatives on preparedness and response plans, such as developing Continuity of Operations Plans. These efforts to improve preparedness are vital; however, we need to recognize the full spectrum of threats, not just the ones that have played out most recently.

2.3.2 Threat Assessments

Water and wastewater utilities should conduct a Hazard Identification Vulnerability Analysis (HIVA) to determine which hazards they are most prone to (given their climate and geographic location) and additional threats that could affect the system's operations. To effectively conduct a HIVA, water system operators should first begin by researching all available pre-existing HIVA results for their jurisdiction. For example, relevant HIVAs may be readily available from three different sources:

1. Local City Offices of Emergency Management
2. Local County Offices of Emergency Management
3. State Emergency Management Divisions

Next, water system planners should take inventory of the actual incidents that have caused or led to serious service interruptions during the past 20 years and the frequency of the incidents. This will help identify any particularly troublesome areas of the system that are vulnerable to the most common hazards. The next step involves researching specific threat information germane to your organization's geographic location. This is best achieved by reaching out to law enforcement agencies in your jurisdiction or your state's intelligence/fusion center. A fusion center is an effective and efficient mechanism to exchange information and intelligence, maximize resources, streamline operations, and improve the ability to fight crime and terrorism by merging data from a variety of sources (USDOJ, 2006). It is also important to engage with other stakeholders from within the utility's own organization and partnering agencies, who might be able to add valuable insights and perspectives.

The most common water system threats can be grouped into three different categories: (1) natural disasters; (2) human-caused incidents; and (3) workforce/infrastructure threats. Once you have identified the most likely threats, you can begin to assess the probability and the impact of an occurrence. Figure 2.1 shows the categories of all hazard threats.

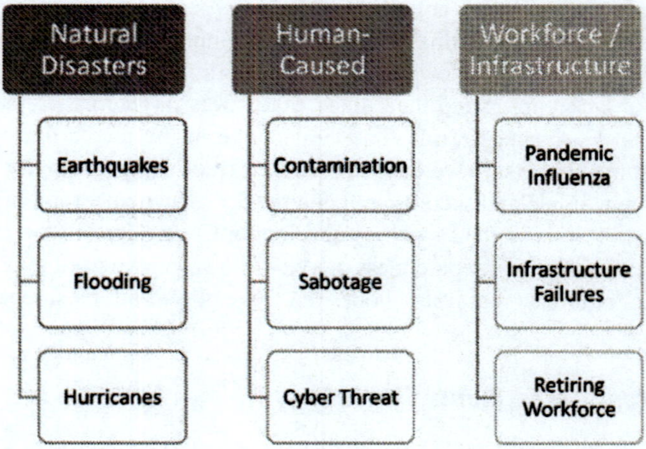

Fig. 2.1 Categories of all hazards and threats

2.3.3 Natural Disasters

Depending on the region, certain threats are more likely. For example, along the Gulf Coast hurricanes may be the most frequent threat to a water or wastewater system. Communities along the West Coast are more prone to earthquakes and California suffers from frequent wildfire conditions resulting from hot, dry weather and high wind conditions. In the Midwest, there have been many problems with historical flooding when rivers rise and jump the banks, threatening homes, businesses, transportation systems, and other critical infrastructure. Other natural disasters that could impact water systems include tornadoes, severe windstorms, snow and ice storms, extended periods of freezing temperatures, lightning strikes, and droughts. Each jurisdiction has a unique set of probable hazards that could wreak havoc on water systems.

It is critical for a utility to have response plans developed for the most probable hazards that could impact their systems. Once plans have been developed, organizations need to train their employees and conduct exercises to ensure that everybody understands what their responsibilities are and what the priorities will be upon such a scenario. The importance of Continuity of Operations Planning (COOP) gained exposure after Hurricane Katrina. The value and benefits of continuity of operations planning is that it will help an organization prioritize tasks and repairs to systems by determining what things can fall off the plate when resources are overwhelmed.

2.3.3.1 Human-Caused Incidents

To fully understand the spectrum of threats facing water utilities, system operators must recognize the types of adversaries, malevolent persons, or groups that may try to prevent utilities from performing one or more of its essential functions. Information gathered about threats is critical to understanding how a potential adversary could carry out an attack on utility assets. This knowledge about potential threats helps utilities form the foundation of a targeted security program to protect critical assets. A water system's comprehensive threat profile is an important factor in risk calculations and methodologies.

There are three broad classes of intentional threats that water utilities should evaluate. They include physical threats, chemical contamination threats, and cyber threats. Each water utility must define the threats that will be used in their risk equation to calculate current and future risks and then propose security upgrades required to reduce those risks.

Information necessary to define threats includes but is not limited to the following:

- Incident reports, suspicious circumstance reports, criminal reports, intelligence reports and any historical data associated with a water utility. Sources for gathering this information include local, state, and federal law enforcement agencies and local/state offices of emergency management, fusion centers, and various other counterterrorism and federal agencies (i.e., FBI).

- Employee data on union disputes, employee conflicts/violence, expressed threats, etc.
- Internet, industry associations, WaterISAC (information sharing, analytic center), professional publications, etc.

Not all human-caused incidents are intentional or caused by outsiders, which is why we will explore external threats deeper by the type of individual or group who might perpetrate an incident. Since an intentional plot to contaminate the potable drinking water for an entire community or to wipe out a water system's ability to meet the service level needs of citizens would be a catastrophic incident, we will first focus on external threats and the possibility of an insider/outsider collusion attack. Figure 2.2 illustrates possible threat sources.

2.3.3.2 External Threats

External threats to a utility include everything from low-level vandals to very high-level terrorist threats. While there has not yet been a catastrophic attack on a drinking water system, there have been attempts demonstrating that the potential, means, motive, and opportunity exists. History has proven that terrorists have considered and carried out attacks on drinking water systems in the United States and other countries.

Critical infrastructure is an attractive target for terrorists due to the potential consequences and ripple effects of a successful attack. Drinking water systems have long been recognized as being vulnerable to an intentional chemical or biological contamination attack even though the probability of such an attack is uncertain (CRS, 2005). The tactic of poisoning an urban drinking water system supports an objective to commit indiscriminate harm on all parts of society. Such an attack could affect entire communities, especially vulnerable populations (i.e., infants, elderly, and immune compromised), businesses, industries, and health-care facilities. The distribution portion of a water system is especially at risk due to the ease of exploiting its vulnerabilities and the potentially large number of deaths and illness that

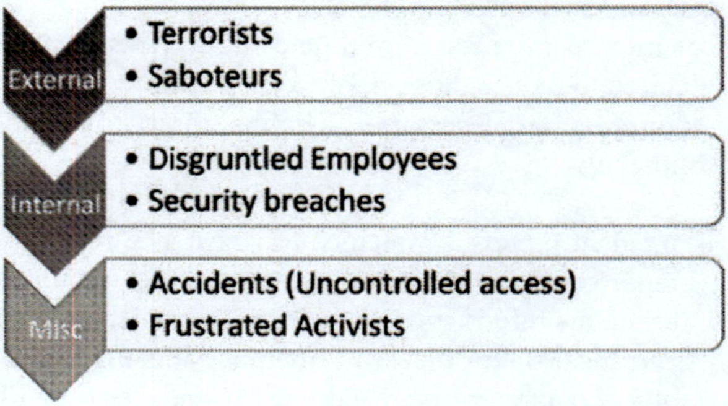

Fig. 2.2 Possible threat sources

could result. The number of casualties from this type of attack, if successful, could surpass the death toll of 9/11.

The DHS has issued advisories to water utilities indicating that al-Qaeda has shown interest in using cyanide, *Botulinum toxin* (Botox), *Salmonella typhi* (the causative agent of typhoid fever), and *Bacillus anthracis* (the causative agent of Anthrax) to attack US water systems (USDHS, 2003). Terrorist organizations such as al-Qaeda are not the only external sources with motives to use chemical or biological weapons to attack a water system. The following list describes others who pose a threat:

- Vandals with no specific agenda, but possessing an interest in chemical and biological weapons and a propensity for violence.
- Anarchists seeking attention and independence.
- Ecoterrorists protesting the use of dams to manage water supply or other perceived environmental impacts related to chemicals or discharges into waterways.

There is empirical evidence that chemical and biological weapons have already been used or considered in plots to contaminate drinking water systems. The chronology of intentional water contamination events with chemical or biological hazards dates back thousands of years. While the actors, motives, tactics, and outcomes of chemical or biological attacks vary, the following is a partial list (Tucker, 2000; Kroll, 2010) of historical incidents that confirm the interest in such attacks:

- R.I.S.E., a neo-Nazi terrorist group, plotted to poison urban water supplies to incapacitate populations and gain attention for their cause. They were arrested and at the time had possession of several biological agents that had been produced in a college laboratory (1972).
- In North Carolina a water reservoir was intentionally contaminated resulting in denial of water to customers. Water had to be trucked in for residents (1977).
- New York City received anonymous threats of plutonium poisoning to the city's water supply. Subsequent testing for plutonium revealed 200 times the normal concentration levels, but not enough to warrant public health concerns (1985).
- Al-Qaeda members were arrested in Rome, in the process of attacking a water distribution system with cyanide near the US Embassy. They had detailed plans and equipment, but were thwarted at the last minute. While the compound turned out to be a benign cyanide derivative, it could have been a pre-event effort to trace the compound and the flow in the water system (2002).
- Two al-Qaeda members in possession of documents about how to poison US water supplies were arrested in Denver (2002).

Saboteurs can come in many forms. A saboteur may not be a terrorist but rather somebody who wants to cause the agency itself harm. They may be an adjacent property owner who is frustrated, but not attempting to make any political statement and not seeking to harm people. They may not be aware of the impact their actions could cause, but they are still a threat. Saboteurs target equipment and assets not people. There are other threats to utilities that can be accidental such as a facility without adequate access control or an employee who does not secure a gate. If anybody can just walk in from the street, unintentional damage may occur. The capabilities and types of attacks that could be carried out by an external threat range from low-level all the way up to very high-level threats.

Low-Level Threat

A threat in this category could include one or two outsiders with no authorized access or inside information with the intent to cause physical damage to the water utility facility or theft of property or equipment.

Medium-Level Threat

A medium-level threat could include a small group of one to three outsiders who possess a limited amount of knowledge about the water system's assets, processes, and security systems. This level of threat may involve equipment or tools that are portable and easy to obtain.

High-Level Threats

A high-level threat could include an organized, highly motivated group of up to five outsiders with intent on sabotage or some type of major disruption to the system. They may be equipped with sophisticated tools, explosives, or weapons. The perpetrators of this type of attack would have extensive knowledge about water system assets, processes, and the security system. They also may have sophisticated cyber capabilities with a moderate level of resources. It is quite possible a planned attack. This category would include a combination of physical and cyber attacks on the water system assets for the purpose of a denial of water attack.

Very High-Level Threat

This group of adversaries possess all of the capabilities listed under the high-level threats, along with access and intent to use weapons of mass destruction, including chemical, biological, radiological, nuclear, or explosive substances. Tactics might include larger than backpack quantities of explosives such as truck bombs and chemical and biological substances with the intent to cause a significant number of deaths and unleash psychological terror on society.

2.3.3.3 Internal Threats

Other threats to a water infrastructure are those that are tied to internal threats. This includes a disgruntled employee who may or may not be currently employed at the organization. There have been attacks on water systems as a direct result of a disgruntled employee. In Pittsburgh, a disgruntled employee deliberately contaminated water mains by injecting weed killer into fire hydrants (Tucker, 2000). Other deliberate actions by insiders include a scenario where pipelines from drinking water distribution system were cross-connected with a wastewater collection pipeline.

Insiders, who include employees, former employees, contractors, and vendors, pose a particularly dangerous threat to utilities. They have specific knowledge of how the systems function. They know where the systems' weaknesses are. They already have access or may know how to circumvent existing security systems. They are trusted partners and can cover up their actions with minimal scrutiny. The capabilities and risks to utilities from insiders can also be categorized from the low-level threat to a high-level threat.

Low-Level Threat

One individual with access to hand or simple power tools, whose intent is to physically damage the water utility or to profit from theft of materials for monetary gain.

Medium-Level Threat

A single motivated insider (employee or contractor) working unaccompanied with authorized access and who possesses extensive knowledge of the utility's systems, processes, procedures, security systems, and emergency response protocols. They also may have knowledge about cyber systems including SCADA systems. This insider has access to hand and power tools and the ability to access on-site chemicals. The intent of this insider is to prevent the delivery of water by damaging or manipulating components of the water system or to introduce substances of concern into the water supply to damage the utility's reputation.

High-Level Threat

A single, disgruntled individual, with motive and intent on harming the utility and/or personnel. This insider has all of the same capabilities as the medium-level threat, in addition to more extensive knowledge about the system, facilities, staffing rotations, and schedules. This individual may have recently undergone disciplinary actions or may have been terminated from employment and might hold other utility personnel or management responsible for their undesirable employment status. This high level of threat adversary may use handguns, explosives, or other violent acts to intimidate or harm people.

2.3.3.4 Cyber Threats

Cyber threats to water systems include the intent of individuals or groups to electronically corrupt or seize control of data or information essential to system operations. Adversaries attempting an attack via cyber mechanisms may seek out information that contains highly sensitive knowledge about a system's vulnerabilities. This includes supervisory control and data acquisition (SCADA) networks, which contain computers and applications that perform remote control functions within the system.

SCADA systems have contributed greatly to water system efficiencies, by allowing the collection and analysis of data and control of equipment such as pumps and valves from remote locations. However, they present a significant security risk (USDOE, 2010). Similar to the vulnerabilities we see due to aging water infrastructure that was not designed with security in mind, SCADA systems were designed primarily to maximize functionality, not security. This leads to some SCADA networks that could be vulnerable to disruption of service, process redirection, or manipulation of operational system components that could result in asset failures and public safety concerns. All water system owners/operators should be cognizant of SCADA vulnerabilities and take actions to secure their SCADA networks (USDOE, 2010).

Additional cyber vulnerabilities are related to the need to secure sensitive information stored on data servers and in paper files. Examples of sensitive utility information includes vulnerability assessments, site security plans, response and recovery plans, water system and asset plans and specifications, descriptions of chemical processes and storage capacity, detailed maps and drawings, customer records, and financial data, all stored in electronic formats on information technology data servers.

The threat of a cyber attack carried out by a hacker can range greatly in sophistication. It can include low-level access via the Internet only or an individual or group with access to the information technology structure within an organization. A hacker may have direct access via modem or PC and may have use of sophisticated hacker tools for the purpose of compromising the system. They also may have access to administrator functions and may coordinate cyber attack with a physical attack. Perpetrators may use sophisticated network gear or other hacker tools. Results of a cyber attack may include denial of service, disruption of business functions, or the ultimate destruction of data and systems.

While the motivations of any of these groups may be unknown, effective security is critical to protect the assets and systems regardless of who might act out the threat or what their tactics might be. To get a better handle on which level of external threat to focus on, an organization should go through a process of determining their design basis threat.

2.3.4 Design Basis Threat

Given the wide variety of potential threats and the various capabilities of the actors involved with carrying out each threat, water systems need to carefully examine their entire threat spectrum. Once all of the potential threats have been collected, utilities need to evaluate and make a determination of what level of threats they are prepared to protect against. This predetermined level of adversary to which the utility must be protected from is called the Design Basis Threat (DBT). Determining the DBT requires consideration of the threat type, tactics, mode of operations, capabilities, threat level, and likelihood of occurrence (ASIS, 2010).

The factors to consider include the adversary's ability to gain access to an asset, the history of any previous attempts on the asset, the type of damage the asset has sustained in the past, the motivation of the adversary, the tactics used, and whether or not the individual or group still exists in the geographic area. Other factors to include are the capabilities, history or intention, or specific targeting. For example, if neighboring utilities have been hit by vandals and criminals who have targeted water tanks, the likelihood that other nearby water tanks will be targeted is also high.

Determining the probability or likelihood of high-level threats is inherently difficult. There is currently a lack of industry-wide information on the probability of threats to water utilities. The most concrete data to pull from are historical events; however, that approach does not fully account for an evolving threat environment. Just because there is no history of a particular type of attack it does not mean an organization should dismiss the possibility of such an attack. We know the risk is greater than zero and therefore utilities must make an assumption of the likelihood of an attack. Water utilities in larger, urban areas will have a higher likelihood of a terrorist attack than rural community water systems. Utilities should attempt to find a ratio or probability factor that can satisfy a reasonable person's test. There are many different risk assessment methodologies available on how to calculate risk, so each utility should find a risk assessment system that meets their needs and will enable them to update their risk profile on an annual basis.

2.3.5 Continuity Threats to Workforce and Infrastructure

It is important to recognize other types of threats to a water system's ability to deliver essential functions. Any type of circumstance that could lead to a significant reduction in workforce or a significant increase in needed resources, should be considered in an organization's threat spectrum. This could be an aging workforce that may result in a spike in retiring employees with substantial system knowledge; aging infrastructure that could lead to a large-scale infrastructure failure; or a public health emergency in which many critical field or office employees are not able to report to work due to illness or dependent care needs.

2.3.5.1 The Dual Threat: Aging Infrastructure and Aging Workforce

Water system employees and infrastructure are both showing signs of the aging process. The risk of losing institutional knowledge about utility systems can dramatically affect the proficiency of maintenance activities of existing infrastructure. The average field employee age in utility industries ranges from 45 to 54. With more than 25 years of experience under their belts, these employees hold a considerable amount of expertise and familiarity with the assets they helped develop, install, and maintain through the years. As baby boomers draw nearer to retirement, a large percentage of lead technicians and crew chiefs will take their individual knowledge base with them (Radice, 2010). With the extreme budget constraints that most utilities are facing, replacing these lost positions is not a guarantee. Many utilities are simply required to do more with less and may sacrifice expertise for quick fix contractors who will not aid in creating a sustainable in-house knowledge base for the next generation.

2.3.5.2 Aging Infrastructure

At the same time as water system employees are aging and preparing to retire, the assets and infrastructure they have cared for through the years are also aging. In every community around the country there are examples of spectacular infrastructure failures that have led to large-scale service interruptions, significant property damage, and human injury as a result of the failure. Much of the nation's critical infrastructure still in service has exceeded its planned operating life and requires major renovations or replacement. The wear and tear of above- and belowground system components is evident as assets are exceeding their life cycle expectancy, many times without plans to replace them prior to failure. Large transmission water lines that are old may provide no warning at all that they are reaching a failure point. While smaller infrastructure breaks can be managed effectively, large-scale infrastructure failures can send a utility into crisis mode.

2.3.5.3 Interdependent Infrastructure Failures

Other types of critical infrastructure are also aging and susceptible to failures that can have an impact on our water system operations. For example, in an urban location, a wide transportation system outage could greatly impact an organization's workforce. A good example of this is the 1-35 Bridge failure in Minneapolis that occurred in 2007 (USA Today, 2007). Just after 6 p.m. on the evening of August 1, 2007, the 40-year-old bridge collapsed into the river and its banks without warning, killing 13 and injuring 121 others. At the time, there were approximately 120 vehicles, carrying 160 people, on the bridge. Transportation infrastructure, especially bridges, can have a significant impact on the mobility of water system employees. A bridge collapse could also wipe out water infrastructure such as transmission pipelines if they are attached to the structure.

2.3.5.4 Workforce Illness

Another threat is that of a public health crisis or a pandemic influenza. In 2009, Mexico, the United States, and many other countries around the world became enthralled in the growing possibility of a worldwide pandemic. The swine flu (H1N1) sent shockwaves through public health and emergency management communities as they scrambled to dust off their pandemic influenza (or bird flu) emergency plans. While the name of the virus may change, planning for a pandemic outbreak that could take away 40% or more of an organization's employees is critically important.

But, the largest threat to water systems may be that of complacency in which the low probability of occurrence outweighs the desire to reduce the risk in advance. The enhancement of security and the abilities of water systems to respond to all types of hazards are key to maintaining reliable supply and delivery of essential functions. Once the owners and operators of a water system better understand the threats they face, they need to become aware of their vulnerabilities.

2.4 Water System Vulnerabilities

Water systems are complex with many intricate connection points and interactive networks. Water system vulnerabilities can be general or specific. An example of a general vulnerability is that most water or wastewater infrastructure in service today was built many years ago. Aging infrastructure has inherent vulnerabilities because the materials used during the initial manufacturing or construction may not be as resilient as current day materials. The wear and tear on system components through the years also contributes to weakened structures. Another generalized vulnerability of water systems is heavy reliance on other critical sectors that are also subject to significant system failures such as electricity and telecommunications. Since older water infrastructure was not built with security as an objective, assets and facilities often were built with an excessive amount of access points (doors, hatches, vaults, etc.), contributing to increased vulnerabilities. Examples of specific infrastructure vulnerabilities might include a treatment facility with inadequate perimeter controls; a pump station with faulty locking mechanisms on a roll-up door; or an elevated water tank co-located with telecommunication towers and antennas that will require frequent access by contractors.

Vulnerabilities can be described as elements that are susceptible to accidents, failures, or attacks that are difficult to defend. Vulnerability assessments are an important step to take prior to identifying and implementing security countermeasures. The components of a water system that should be considered in a comprehensive vulnerability assessment include the following:

- Distribution systems including pipes and constructed conveyances
- Physical barriers
- Water collection, pretreatment, and treatment facilities
- Use, storage, and handling of various chemicals
- Storage and distribution facilities
- Electronic, computer, or other automated or cyber systems

Out of the above-listed system components, distribution systems, chemical treatment facilities, and cyber systems are generally considered the most vulnerable type of assets. The next section will explore vulnerabilities by grouping the system components into three categories: above-ground structures, below-ground structures, and cyber systems.

2.4.1 Above-Ground Structures

Above-ground structures are water system components that are clearly visible, either by passersby or from aerial views. The popularity of satellite photographs and software available for free on the Internet (via GoogleEarth and other geospatial mapping tools) have made it easier for Joe and Jane Public to know exactly where above-ground critical infrastructure assets are located, even if the assets are situated in remote locations. As information sources become more advanced and accessible, water system operators will no longer be able to rationalize lack of security based on the obscurity of an asset's location.

Above-ground water structures include dams, intake structures, wells, water and wastewater treatment plants, pumping stations, reservoirs, tanks and other water storage facilities, exposed conveyance or transmission pipes, open channels, tunnels or support facilities, command and control facilities, and administrative offices. All of these structures are vulnerable to threats, although some have higher level of consequences and risk.

Buildings or complexes that store chemicals such as chlorine, fluorosilicic acid, sodium hypochlorite, oxidizers, propane, diesel, and fluoride can multiply the risks for workers and neighboring communities. Gaseous chlorine is a particularly hazardous chemical of concern that increases risks to communities from the time it leaves the manufacturing facility during transit, to the storage of it on-site, until it is fully utilized in processes at the water facility.

Table 2.1 provides some guidance on how to evaluate the vulnerabilities of above-ground structures. In general, this exercise should provide information about how easily a villain could gain entry to a critical facility.

Table 2.1 Evaluation of above-ground structures

Feature	Quantity/capacity	Quality/construction	Security measures
Perimeter controls	Exterior fences, interior fences, gates, bollards, and vehicle barriers	Height, material, anti-climb, set backs, clear zones, and lighting	Access control, motion detection, and CCTV
Doors, hatches, and vaults	Number of access points	Hollow, steel, reinforced, etc.	Locked hatches, ladder locks, and intrusion detection
Locks and keys	Double entry systems, physical or electronic locks and keys, and padlocks	Automated locking, and tamper-resistant hinges	Door strikes and alarm contacts

2.4.2 Below-Ground Structures

Drinking water and wastewater owner/operators are incorporating an increasing number of underground water infrastructures into their systems. This includes efforts to bury water storage reservoirs that used to be above ground, construction of underground water and wastewater pump stations and overflow storage containment, and various vaults that provide access to electrical panels, equipment, and large transmission and conveyance pipelines. The mere fact that the structures are below ground provides a good barrier to certain types of threats. Below-ground structures are inconspicuous since they are not readily visible to a passerby. Some buried drinking water reservoirs have been turned into parks or open spaces for the enjoyment of neighboring communities. This can be a good activity generator to deter criminal behavior during daylight hours; however, entry points need appropriate security to deter and prevent unauthorized access. Below-ground infrastructure may have more protection against low-/mid-level threats than above-ground structures; however, they may be at an increased risk of other threats such as earthquakes and flooding.

Drinking water distribution systems are incredibly vulnerable due to the thousands of cross-connections and entry points into the system and difficulty detecting an intrusion. A motivated terrorist could facilitate a simple backflow contamination event with pumps and a number of chemical or biological agents. The introduction point into the distribution system could be from a fire hydrant, a residential home or apartment, or a commercial building. An example of how effective this tactic could be is the fact that accidental backflow occurrences have resulted in many incidents of waterborne illness and even death. According to the EPA, backflow events caused 57 disease outbreaks and 9,734 cases of waterborne disease from 1981 to 1998 (USEPA, 2001). If the system is vulnerable to accidents, it is just as vulnerable to a deliberate attack.

An intentional dissemination of a chemical or biological agent or contaminant through a backflow event is a significant concern to the drinking water industry. The detection of such an incident would most likely occur after people become ill and hospitals begin observing a trend. Currently, there are several studies and evaluations of new technology to enable early water contamination warning systems via online water quality monitoring stations throughout a distribution system. The downsides to these systems are that they are very costly to implement, administer, and maintain, and there is a natural resistance to trusting a positive reading (for fear of overreacting to a false-positive reading). All that being said, improvements in this detection area as the science and technology progresses are promising.

2.4.3 SCADA and Cyber Systems

SCADA system vulnerabilities are diverse depending on how each system has developed and deployed the technology. Some systems may have multiple subsystems that are networked together. These systems allow personnel to activate and deactivate pumps and valves from a remote computer system or they can be designed to accommodate local intelligent valve control.

While best practices models tout the importance of physically separated systems on standalone networks, SCADA systems are occasionally linked (even unknowingly) to general utility business computer networks. Utilities that link SCADA networks to its technicians, engineers, or operational decision-makers for convenience sake create vulnerabilities. When any bridge between the two networks occurs, the entire SCADA system becomes only as secure as the weakest point of the business network. Even when the networks are truly separated, many SCADA systems are only protected by simple passwords.

Cyber security vulnerabilities are also related to an organization's posture on public information policies. Revealing too much information about critical utility systems, processes, treatment facilities, and other assets in public forums creates unnecessary vulnerabilities. Utility web sites, as well as those of utility consultants and contractors, frequently provide a goldmine of information that could be used to gain access to additional information or to plot an attack against a water system. For example, some utility web sites might list employee names and e-mail addresses, thus providing a window of opportunity to solicit or seize sensitive data and information, while a consultant web site might boast photographs, drawings, and detailed descriptions of large capital projects involving critical infrastructure.

2.4.4 Vulnerability Assessments

Vulnerability or risk assessments are intended to provide a roadmap for lowering risks. Vulnerability assessments are the best way for an organization to take inventory of their system's critical components and determine what security risks owners/operators should focus on first.

One way to assess vulnerabilities is by pairing up individual assets or system components with a particular threat. This matching up of assets/threat pairs provides an opportunity to evaluate how successful an intentional act to disrupt the system could be. The various types of threats that might be matched up with one individual asset (a pump station) are illustrated below and summarized in Table 2.2.

There are many different formats and categories that can be used when developing a vulnerability assessment. The technical components of a comprehensive vulnerability assessment include the following:

- Characterization of the facility or system
- Inventory of significant assets and areas
- Threat assessment (including DBT and asset/threat pairs)
- Consequence assessment
- SCADA assessment
- Organizational security policies and procedures
- Local, state, and federal interactions

Table 2.2 Potential asset/threat combinations

Asset	Threat	Tactics	Likelihood of success
Pump station	External, sabotage	Explosives, mechanical tampering, arson	*Medium* – Depending on access control and detection capabilities
Pump station	External, cyber	Control of SCADA system, manipulation of valves and equipment	*High* – Could gain control and proceed undetected
Pump station	External, vandal, or criminal	Graffiti, property damage, theft of equipment or wire	*Medium* – Depending on fences and access control
Pump station	Internal, disgruntled employee	Mechanical tampering or electronic panels	*High* – Employees have access, knowledge, and opportunities

- Physical security components
- Risk analysis
- Risk reduction options and recommendations

Vulnerability assessments should be updated after every significant security incident and annually with new information about system facilities, assets, processes, and updated threat analyses. Once the owner/operator of a water utility has completed or updated their vulnerability assessment, they need to make determinations about which recommendations to implement and how to fund the security improvements. It is almost certain that the list of recommendations from a system-wide vulnerability assessment will far outweigh the funding available to address all of the security risks. Resource allocation decisions about how to proceed are not made lightly. Chapter 22 will address the drivers for security improvements, types of physical security measures, and the need for a multilayered security program approach.

References

(ASIS) American Society of Industrial Security. (2010). International, Protection of Assets Manual. http://www.asisonline.org/library/glossary/d.pdf

(CRS) Congressional Research Service. (2005). Report for Congress, Terrorism and Security Measures Facing the Water Infrastructure Sector, Jan 2005, p. 4.

Hoffman, B. (2006). *Inside Terrorism*. Columbia University Press, New York, NY, p. 40.

(HSC) Homeland Security Council. (2007). *National Strategy for Homeland Security*. The White House, Washington, DC, Oct 2007, pp. 1–25.

Kroll, D. (2010). *Securing Our Water Supply: Protecting a Vulnerable Resource*, PennWell Publishers, Tulsa, OK, pp. 19–27.

(NDWAC) National Drinking Water Advisory Council. (2005). Water Security Group Findings, May 18, 2005, p. vii.

Radice, S. (2010). The Dual Threat: Aging Infrastructure and Aging Workforce Call for Integrated Asset and Workforce Management, Electric Energy Online, http://www.electricenergyonline.com/?page=show_article%26;mag=47%26;article=351

Seger, K.A. (2003). *Utility Security: The New Paradigm*, PennWell Publishers, Tulsa, OK, Penwell Corporation, p. 35.

(SMGI) Security Management Group International. (2006). Overview – Hurricane Katrina Crisis, Aug 15, 2006. http://www.smgicorp.com/resources/documents/SMGI-KatrinaCS.pdf.

Tucker, J.B. (2000). "Lessons from Case Studies". *Toxic Terror: Assessing Terrorist Use of Chemical and Biological Weapons*. Edited by J.B. Tucker, Cambridge, MA, MIT Press, pp. 250–251.

USA Today. (2007) On Deadline Blog. Latest on Deadly Minneapolis Bridge Collapse. Retrieved Feb 20, 2008, from message posted to http://blogs.usatoday.com/ondeadline/2007/08/latest-on-deadl.html

(USDHS) U.S. Department of Homeland Security. (2003). Advisory: Potential AI Qaeda Threats to US Water Supply, June 23, 2003.

(USDHS) U.S. Department of Homeland Security. (2006). *National Infrastructure Protection Plan*. Department of Homeland Security, Washington, DC, p. 3.

(USDHS) Department of Homeland Security. (2007a). Homeland Security Threat Assessment: Executive Summary, Aug 2007, p. 8.

(USDHS) U.S. Department of Homeland Security. (2007b). National Strategy for Homeland Security, Oct 2007, p. 28.

(USDHS) U.S. Department of Homeland Security. (2010). National Infrastructure Protection Plan Water Sector Snapshot, http://www.google.com/url?sa=t&source=web&ct=res&cd=4&ved=0CCkQFjAD&url=http%3A%2F%2Fwww.dhs.gov2Fxlibrary2Fassets%2Fnipp_snapshot_water.pdf&rct=j&q=epa+water+sector+security&ei=tkP7S9OtO4KwMomUqb0B&usg=AFQjCNF-6XMn3r4GtVTX3FwqnVyhbAEzcQ

(USDHS & USEPA) U.S. Department of Homeland Security and the U.S. Environmental Protection Agency. (2007). Water Sector Specific Plan as Input to the National Infrastructure Protection Plan. (Office of Ground Water and Drinking Water, EPA 8l7-R-07- OOlA) May 2007, p. 3.

(USDOE) U.S. Department of Energy. (2010). 21 Steps to Improve Cyber Security of Data Networks, http://www.oe.netl.doe.gov/docs/prepare/21stepsbooklet.pdf 3

(USDOJ) U.S. Department of Justice. (2006). Fusion Center Guidelines: Developing and Sharing Information and Intelligence in a New Era, Aug 2006. http://it.ojp.gov/documents/fusion_center_guidelines.pdf

(USEPA) US Environmental Protection Agency. (2001) Potential Contamination Due to Cross-Connections and Backflow and the Associated Health Risks: An Issues Paper, Sept 27, 2001. http://www.epa.gov/cgi-bin/epalink?logname=allsearch&referrer=potential contamination due to cross-connections an issue paper|1|All&target=http://www.epa.gov/safewater/disinfection/tcr/pdfs/issuepaper_tcr_crossconnection-backflow.pdf

Chapter 3
EPA Drinking Water Security Research Program

Hiba S. Ernst, K. Scott Minamyer, and Kim R. Fox

3.1 Background

Following the terrorist attacks of September 11, 2001, and the mailing of letters containing *Bacillus anthracis* spores, the US Environmental Protection Agency (EPA) developed a *Homeland Security Strategy* (USEPA, 2004a) for enhancing national security and protecting human health and the environment. The *Homeland Security Strategy* and the *Public Health Security and Bioterrorism Preparedness and Response Act of 2002* (US Congress, 2002) provided the basis for the water security research that is conducted at the National Homeland Security Research Center (NHSRC) within EPA's Office of Research and Development (ORD). In addition, several Homeland Security Presidential Directives (HSPDs) were issued in 2003 and 2004 and are described on the Department of Homeland Security (DHS) web site (US DHS, 2010b). These directives assigned new responsibilities to the Agency for establishing a strong water security science and research program.

In 2002, the ORD and EPA's Office of Ground Water and Drinking Water (OGWDW) collaborated to identify research needs to better protect the Nation's water and wastewater systems. The *Water Security Research and Technical Support Action Plan* (Action Plan) (USEPA, 2004b) was developed with the help of stakeholders and other federal and state agencies to ensure that research conducted by the EPA is responsive to the needs of the water industry and end-users. The NHSRC, OGWDW, and the Water Environment Federation jointly conducted a

---

Disclaimer The research descriptions herein have been reviewed by the US Environmental Protection Agency and approved for publication. Note that approval does not signify that the contents necessarily reflect the views of the Agency. Mention of trade names or commercial products does not constitute endorsement of recommendation for use.

H.S. Ernst (✉)
US Environmental Protection Agency, National Homeland Security Research Center, Cincinnati, OH, USA
e-mail: ernst.hiba@epa.gov

R.M. Clark et al. (eds.), *Handbook of Water and Wastewater Systems Protection*, Protecting Critical Infrastructure, DOI 10.1007/978-1-4614-0189-6_3,
© Springer Science+Business Media, LLC 2011

series of stakeholder meetings during 2005 to further inform strategic planning and supplement the Action Plan. The National Research Council (NRC), National Academies, reviewed the Action Plan and published a report in 2007 to advise EPA on future research opportunities (NRC, 2004, 2007). EPA's water security research planning is currently described in a 3 year cycle in the Homeland Security Research Program Multi-Year Plan (MYP). The MYP (USEPA, 2010a) describes the research underway and planned by the NHSRC in two main areas: (1) improved protection from and the capability to respond to terrorist attacks on the Nation's water and wastewater infrastructure and (2) wide-area decontamination following a chemical, biological, or radiological (CBR) attack (only the water infrastructure decontamination is described in this chapter).

3.2 Research Drivers

Safe water is critical for multiple uses, such as drinking, sanitation, firefighting, public health, recreation, agriculture, and industry. To protect the public and the Nation's critical infrastructure, the President and Congress assigned responsibilities to various federal agencies. In addition to the Bioterrorism Act, EPA responsibilities were defined in HSPDs 7, 9, 10, 19, and 22 and the National Response Framework published by the DHS (DHS, 2008). The responsibilities that influence the research described in this chapter can be summarized as follows and are detailed in Table 3.1:

1. EPA is the federal government Sector-Specific Agency lead for water infrastructure and has the responsibility to protect water systems from attacks and to detect and recover from successful attacks on water systems.
2. EPA is charged with the development of a nationwide laboratory network with the capability and capacity to analyze for CBR agents for routine monitoring and in response to terrorist attacks.
3. EPA is charged with decontamination and recovery from all hazards.

In addition to the drivers discussed above, research planning is influenced by needs and priorities identified by various advisory councils established by the DHS. These councils facilitate coordination between water utility owners/operators and government agencies. The advisory councils include the Water Sector Coordinating Council, Water Sector Government Coordinating Council, and Critical Infrastructure Protection Advisory Council (CIPAC) (US DHS, 2010a). For example, the Water Sector CIPAC has published recommendations to inform planning for decontamination research (CIPAC, 2008). The EPA also coordinates with the regions, states, local and tribal groups, as well as academia.

Table 3.1 Chronology of EPA roles and responsibilities in water security research

Event chronology	Water infrastructure
Protect against attacks ↓	EPA is responsible for developing strategies and tools to assist utilities with protecting infrastructure against CBR and explosive attacks
Monitor, detect, and confirm CBR attack ↓	EPA is responsible for developing contamination warning systems for monitoring and detection and for verifying an attack with confirmatory analysis
Minimize exposure of the public to the contamination ↓	EPA is responsible for developing strategies and tools for utilities that will minimize spreading of and exposure to CBR-contaminated water
Characterize the nature and extent of contamination ↓	EPA is responsible for determining spatial extent and levels of CBR contamination of water and its infrastructure and establishing the laboratory capability and capacity to analyze for contamination
Assess the risk to human health and develop clean-up goals ↓	EPA is responsible for developing risk assessment methodologies, conducting assessments, and developing clean-up goals for CBR-contaminated water and water infrastructure
Clean-up of the site	EPA is responsible for the treatment of CBR-contaminated water and the decontamination of water infrastructure. Clean-up requires efficacious decontamination and treatment methods, analytical methods to monitor achievement of infrastructure reuse and water treatment criteria, and effective methods to manage contaminated residuals

CBR = Chemical, biological, and radiological
Reproduced from the Homeland Security MYP (2010a)

3.3 Objectives and Desired Outcomes

This chapter describes water security research conducted at the EPA's NHSRC, drivers that guide the research program, and future research planned to support the water sector. The results are intended to support EPA's mission of providing drinking water treatment utilities with tools and methodologies to protect their systems from potential attacks and rapidly detect any contamination in drinking water distribution systems, thus minimizing health, economic, and social consequences. Results also support water systems in establishing consequence management planning and recovery from terrorist attacks and contamination events as rapidly as possible. The goal of the research is the development of tools that will have multiple benefits to ensure the security of water systems, improve drinking water quality, and protect public health and the environment. The EPA's OGWDW is responsible for providing guidance to the water utilities to improve water security.

3.4 Water Security Research

3.4.1 Protection and Prevention

In 2002, methodologies were developed to help drinking water and wastewater systems identify and prioritize threats, determine critical assets, evaluate vulnerabilities, plan for countermeasures to reduce the risk of physical and cyber attacks, and plan for mitigating the consequences of such attacks. These tools are continually being refined to comply with the DHS Risk Analysis and Management for Critical Asset Protection (RAMCAP), for critical infrastructure protection. RAMCAP was developed as a framework to identify and compare risks within a utility, across the water sector, and with other sectors. The RAMCAP process also provides methods to evaluate options for reducing the risks. Examples of NHSRC protection and prevention research are discussed in the following sections.

3.4.1.1 Blast Vulnerability Assessment Tool

The Blast Vulnerability Assessment (BVA) tool was developed to give water utility operators a means to assess the vulnerability of water supply systems to terrorist threats with explosives. Developed for EPA by the US Army Engineer Research and Development Center, it is designed for use with a risk assessment methodology. It provides estimates of damage that can occur in the event of an attack by explosives (Clark et al., 2008). The BVA tool is a PC-based, graphically driven program designed to be used by security professionals and/or engineers with minimal training. Requirements include an overhead site image or map that can be scaled and used to input possible threats and plant critical components. An analysis is run by selecting the threat (e.g., single or multiple improvised explosive devices, using vehicle-based or water-based explosives) and one or many selected critical infrastructure components. Results show potential damage levels and protective standoff distances.

Critical infrastructure currently analyzed by the tool includes various types of elevated water towers, ground tanks, basins, 1-ton chlorine tanks, well heads, earthen and gravity dams, building structures, and general equipment. The BVA tool, user manuals, and training are available to water utilities on the Water Information Sharing and Analysis Centers (WaterISAC) (WaterISAC, 2010), a secure information sharing platform for water utilities and federal and state partners. Improvements to the tool are currently under development to include underground storage tanks and assessments of dams.

3.4.1.2 Design and Renovation of Drinking Water Systems

The NHSRC is investigating methods to incorporate security features into the design of drinking water systems and renovations of existing ones. These studies focus on design features that can potentially reduce the impacts of intentional or accidental contamination incidents. Including these practical and innovative concepts up front in the design approach can potentially minimize public exposure to contaminants, maximize effectiveness of utility response actions, and significantly reduce costs associated with social and economic recovery from such events. The goal of this research is to optimize the ability to control flow patterns using fire hydrants and valves to isolate and contain a contamination incident. This research also investigates optimal distribution system design that would permit the restriction of water flow to individual portions of the distribution system. Additionally, this allows utilities to maximize security features while satisfying other utility design requirements (Grayman et al., 2009).

3.4.1.3 Contingency Planning Following Disasters and Disruption of Service

Contingency planning for water systems includes both alternative water/wastewater treatment options and alternative water sources and supplies. Before an intentional or accidental disruption of normal operations occurs, utilities need to have a plan in place for addressing possible interruption of services. This is critical for continuity of operations while the utility and other authorities investigate the cause of the disruption, and until the system can be restored to normal operation. The NHSRC is collaborating with the American Water Works Association (AWWA) to develop recommendations to water systems on providing alternate water in the event of a large-scale intentional or accidental disaster. The research builds on case studies from natural disasters and international experiences. A report detailing these recommendations was published in 2011 (USEPA, 2011).

3.4.2 Detection

HSPD 9, issued in 2004, charged EPA with developing robust, comprehensive, and fully coordinated surveillance and monitoring systems to quickly detect and respond to CBR contamination. In response, the EPA led the development of a detection system under the Water Security (WS) initiative. The WS initiative is an EPA program that addresses the risk of intentional contamination of drinking water distribution systems. The implementation of this program is the responsibility of the OGWDW but the research that supports implementation decisions is done by the NHSRC. The EPA's goal is to provide water utilities with tools to develop a contamination warning system (CWS) built on the integration of data from multiple sources, including

- Online monitoring of distribution system water quality
- Regular water quality sampling and analysis in distribution systems

- Enhanced security monitoring of the utilities' physical infrastructure
- Surveillance of consumer complaints about their water
- Surveillance of public health (syndromic surveillance)

The NHSRC is conducting research to fill science gaps in several of the five WS initiative components listed above. The detection research focuses on challenging existing commercially available water quality detectors with CBR contaminants to improve continuous monitoring of water quality in the distribution system. These monitoring strategies are followed by sampling and analytical techniques to confirm whether contamination has occurred and, when possible, identify and quantify the contaminants. Research also includes development of models to improve detection, in real time when possible, of contaminants introduced into a drinking water distribution system. This research has been instrumental to the implementation of the WS initiative. Water utilities in Cincinnati, New York, Dallas, Philadelphia, and San Francisco are implementing CWSs using the results of the research described in the following section.

3.4.2.1 Commercially Available Sensors/Detectors

Water quality sensors are historically used by the water industry to monitor treatment effectiveness for public health protection and to optimize their treatment performance. Research on contaminant-specific detectors is scarce and applicability is typically neither cost-effective nor protective against all CBR contaminants. Therefore, the NHSRC initiated an investigation of the response of commercial off-the-shelf water quality sensors (e.g., pH, free/total chlorine residual, total organic carbon (TOC), and oxidation reduction potential (ORP)) to challenges with contaminants such as herbicides, pesticides, *Escherichia coli*, *Bacillus globigii, Bacteriophage MS2*, inorganic contaminants, and warfare agents. The research used pilot-scale, single pass, and recirculating drinking water distribution system simulators (using chlorine and chloramine residual disinfectants to mimic US distribution systems) to investigate the change in baseline water quality parameters due to the introduction of various concentrations of chemical and biological contaminants. In collaboration with other ORD laboratories, the NHSRC also investigated the effectiveness of online toxicity monitors, which use aquatic organisms and an integrated dechlorinating system for the residual disinfectant, to detect changes in water quality. These monitors were highly responsive to copper, cyanide, diazinon, malathion, and toluene (Allen et al., 2008a, b, 2009). The biosensors were also found to be very sensitive to treatment chemicals (e.g., disinfectants) which affected their performance; but the addition of thiosulfate for dechlorination did not have a negative effect on their sensitivity.

Overall, the sensors that responded best to contaminants in chlorinated water were chlorine residual, TOC, ORP, specific conductance, and chloride (Hall et al., 2007). Sensors were also tested in chloraminated water, and results show that total chlorine sensors were either non-responsive or slow (Szabo et al., 2008). No single detector can respond to all contaminants and the background water quality stability in the distribution system can have a significant effect on interpreting the results generated by these sensors. A comprehensive EPA report summarizing these studies and describing the water quality detectors, contaminants, experimental approach, and the results has been published (Hall et al., 2009).

This sensor testing is informational and does not constitute an endorsement by the EPA. For optimum performance of online sensors to serve as contaminant warning systems, the baseline water quality parameters for the system need to be established in order to minimize false alarms from typical water quality fluctuations.

3.4.2.2 Development of New Detectors

In addition to their use to improve water security monitoring, newly developed CBR detectors may provide dual benefits to water utilities to also optimize treatment and distribution system operations (e.g., meet water quality goals and regulatory challenges). A sensor manufacturer is developing a cost-effective TOC online detector and the beta version is undergoing bench and pilot testing by the NHSRC. Additionally, testing is underway for an alpha–beta radiation detector developed by the Department of Energy's Savanna River National Laboratory. The goal is to develop a monitor that can provide online affordable, accurate, and automatic detection of these radiological parameters in water. Existing online alpha–beta detectors generally detect at levels that are several orders of magnitude higher than the maximum contaminant levels and/or protection action guidelines. This type of detector may also be useful in the field for continuous monitoring during remediation following the detonation of nuclear or radiological devices.

3.4.2.3 Threat Ensemble Vulnerability Assessment Research

In response to the need of water utilities to evaluate various CWSs and optimize the placement of water quality sensors in the drinking water distribution system, the NHSRC developed Threat Ensemble Vulnerability Assessment (TEVA) modeling tools under collaborative interagency agreements with the Department of Energy's Argonne National Laboratory, Sandia National Laboratories, and in partnership with the University of Cincinnati. In the detection area, TEVA uses systems analysis and modeling simulation to develop software tools and methodologies to help water utilities prepare for and respond to contamination events. TEVA builds on the EPANET model (Rossman, 2000, 2008), an EPA software tool that is available to the public and models dynamic flow in the distribution system.

Multi-parameter water quality sensor stations monitor pH, free chlorine, TOC, electrical conductivity, turbidity, and other parameters. As shown by Hall et al. (2007, 2009), these parameters may change in the presence of some contaminants. TEVA researchers developed a Sensor Placement Optimization Tool (SPOT) to optimize the physical placement of multiple sensors within the distribution system based on utility-specific distribution system hydraulic models and selected performance objectives (e.g., time to detection and public health measures). This tool

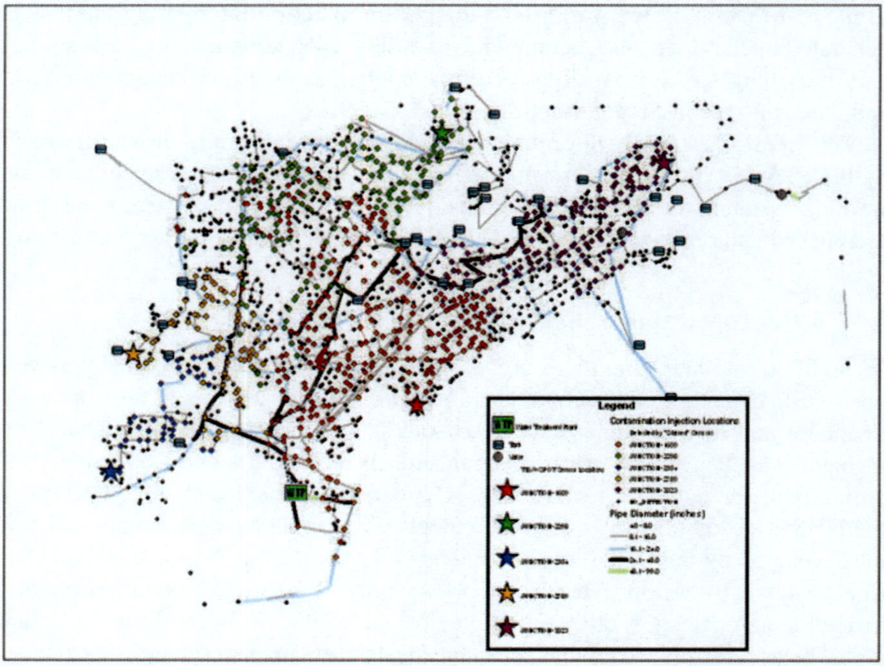

Fig. 3.1 TEVA-SPOT can be used to determine the optimal number and placement of sensors within a drinking water distribution system network (indicated by stars) to support a CWS

allows the system to economically monitor for contamination while maximizing public health protection (USEPA, 2010b). An example of the sensor placement (using TEVA-SPOT) in a typical distribution system network is presented in Fig. 3.1.

The NHSRC also released CANARY, an event detection system that consists of data analysis tools that can analyze water quality data streams from sensor stations to rapidly and accurately identify anomalous conditions in distribution systems. CANARY (name is analogous to the canary in the coal mine) reads data from system sensors in real time and returns an alarm signal to a utility computer system when an anomaly occurs to trigger further investigation (McKenna et al., 2008; Hart and McKenna, 2009). Research continues on improving approach accuracy while reducing the false alarm rate.

In addition to use by WS initiative pilot cities, TEVA-SPOT and CANARY are currently being used by other water utilities (through partnership with the AWWA) to pilot-test the tools and recommend improvements, including algorithm improvements for real-time operation. The NHSRC is developing an extension to EPANET that will incorporate real-time sensor data on water quality, tank levels, pressures, and flows to model characteristics. The model is expected to be able to estimate conditions in the distribution system at locations that lack real-time data.

3.4.2.4 Public Health Surveillance

Another component of a CWS is monitoring public health surveillance data (such as 911 calls, over-the-counter drug sales, and emergency room visits) as a warning of possible water distribution system contamination (Burkom et al., 2011; Babin et al., 2008). The Electronic Surveillance System for the Early Notification of Community-based Epidemics (ESSENCE) syndromic surveillance system integrates water quality data with health indicator data for the early detection of a drinking water contamination event. ESSENCE is a web-based system designed for the early detection of disease outbreaks, illness patterns, and public health emergencies. The water quality data stream algorithms were adapted from CANARY (McKenna et al., 2008; Hart and McKenna, 2009). These tools are being tested by water utilities to improve their effectiveness and decrease the potential for false alarms. One of the most promising public health surveillance data is the use of calls to poison control centers and integrating this with the water quality data collected by the individual utilities to detect possible contamination events.

3.4.2.5 Sampling and Analytical Methods

Following the detection of a suspected contamination incident, analytical methods are needed to confirm the event, identify, and when possible quantify the contaminant. The NHSRC is developing analytical methods and protocols for use by the Nation's laboratories in the EPA Environmental Response Laboratory Network (ERLN) that support the response community in the event of a CBR contamination. The NHSRC developed and published *The Standardized Analytical Methods for Environmental Restoration following Homeland Security Events* (SAM) Rev 6.0 (USEPA, 2010c). This document is a compilation of analytical methods for identifying chemical, biological, and radiological contaminants in many media including water. Sample collection protocols are also developed for each of the analytes included in the SAM document.

3.4.2.6 Improving Microbial Contaminant Detection – Sample Concentration

To improve the detection and identification of microbial pathogens, the NHSRC developed an ultrafiltration device to concentrate bacterial spores and protozoan oocysts from large volumes of water (100 L to yield a retentate of 200–400 mL). In the absence of available standardized analytical methods for biothreat agents, existing microbial techniques are used, but are generally limited by the dilution effect in the water matrix. The ultrafiltration device improves the detection of microbial pathogens through concentration using hollow-fiber filters. The technique was tested under different protocols and at different concentrations for a number of pathogens, including *B. anthracis*, *Bacillus atrophaeus* subsp. *globigii*, and *Cryptosporidium parvum* (Lindquist et al., 2007; Francy et al., 2009). The concentrator was licensed in 2009 to Teledyne Isco for commercialization. The performance of the device is undergoing testing, side by side with a concentrator developed at the Centers for

Disease Control (CDC), which uses the same ultrafiltration technology without the automation. Preliminary results show similar performance (recovery and accuracy) (Gallardo 2010, US Environmental Protection Agency, NHSRC, Cincinnati, OH, personal communication) for both devices. Prototypes of the NHSRC concentrator device are also undergoing testing through the Water Laboratory Alliance (WLA). The WLA is the water part of the ERLN and provides the water sector with an integrated nationwide network of laboratories with analytical capability and capacity for surveillance, response, and remediation. Research is underway to improve field application of the concentrator and to allow for automatic sampling when triggered by a monitoring system.

3.4.3 Containment and Mitigation

The NHSRC is also developing software tools to optimize flushing procedures and the ability of utilities to isolate a contaminant within the distribution system. Following detection of a potential contamination incident, utilities may decide to flush the contaminant from the system or isolate the contamination in place until a decision is made regarding water treatment, infrastructure decontamination, and waste disposal. Optimization models, used in conjunction with EPANET flow models, can assist in identifying the best locations for flushing or isolation and the optimal duration of the flushing program (Baranowski et al., 2008; Haxton and Walski, 2009). Other software tools are being developed that would enable utilities to manage a contamination incident in real time. Such tools include a back-tracking tool to identify the source of a contamination incident following a positive sensor reading downstream (Laird et al., 2006; De Sanctis et al., 2009; Haxton and Uber, 2010), a sampling tool to identify points where samples could be taken to confirm the presence of a contaminant, and a population at risk tool to identify the people who may need to receive medical treatment following exposure to a contaminant.

3.4.4 Water Treatment and Infrastructure Decontamination

Following September 11, 2001, the initial EPA water security research primarily focused on protection, prevention, and detection in order to address identified vulnerabilities in water systems as quickly as possible. While research in these areas continues in NHSRC, more focus is now directed to water treatment and the decontamination of infrastructure contaminated by CBR agents. Research prioritization is done through collaboration and input from the water sector CIPAC and other stakeholders. The NHSRC identified five treatment and decontamination research areas to focus on, including (1) comparative efficacies of various decontamination and treatment protocols and technologies; (2) contaminant fate and transport including modeling; (3) persistence of target contaminants in pipes and infrastructure; (4) appropriate clean-up and verification methodologies; and (5) treatment of contaminated wash water generated during decontamination activities. Examples of NHSRC treatment and decontamination research are provided in the following sections.

3.4.4.1 Inactivation of Biothreat Agents

Research in this area compared inactivation efficacy for a number of bioterrorism organism surrogates, studied by NHSRC, to the respective actual organisms studied by other agencies. Disinfection was done with both free chlorine (Rose et al., 2005) and monochloramine (Rose et al., 2007) for the following agents: *B. anthracis, Brucella melitensis, Burkholderia mallei, Burkholderia pseudomallei, Francisella tularensis,* and *Yersinia pestis.* This research is described in detail in Chapter 9 (Rice, this volume).

3.4.4.2 Persistence of Contaminants in Pipes and Other Water System Infrastructure

Understanding the persistence of CBR contaminants in drinking water distribution systems is important in planning for effective decontamination approaches. Many contaminants of concern can adhere to or become embedded in distribution system pipe corrosion and biofilm. A pilot-scale study tested the level of adherence of arsenic, mercury, *Bacillus subtilis*, diesel fuel, and chlordane to different pipe surfaces (cement-lined ductile iron and PVC pipe surfaces). Figure 3.2 depicts the pilot-scale recirculating loop used to investigate the adhesion of target contaminants.

The study also evaluated different flow regimes (laminar and turbulent) on the fate of the contaminants, impact of decontamination conditions of concentration, pH, and flow, and effectiveness of several decontamination methods (flushing with acidified potassium permanganate or surfactants and shock chlorination). All

Fig. 3.2 Researchers use pilot-scale recirculating pipe loops that simulate drinking water distribution systems to investigate the adhesion of target contaminants to pipes

tested contaminants were found to adhere to cement-lined ductile iron pipe surfaces, indicating the difficulty in decontamination of such pipes by flushing alone. Inorganic contaminants adhered to cement-lined pipes at both flow regimes and removal was improved by the addition of oxidants and surfactants. Chlordane and diesel fuel adhered more to the cement-lined pipes than to the PVC and effective decontamination was achieved with the use of the surfactants. *B. subtilis* showed strong adherence to both types of pipes and shock chlorination at CT values of 30,000 mg/L min demonstrated improved removals over flushing alone (USEPA, 2007). Additional studies evaluating the persistence of *B. globigii* in corroded iron pipes indicated that these spores persisted on corroded iron and adhered to pipe materials and biofilm (Szabo et al., 2007, 2009b). Spores were found to survive even in the presence of a typical chlorine residual in the distribution system, so decontamination with alternative disinfectants and physical removal of corrosion may be necessary (Szabo et al., 2007). *Klebsiella pneumoniae* also showed persistence on corroded iron pipes in the presence of a chlorine disinfectant demonstrating that the biofilm can play an important role in competing for the disinfectant. Additionally, limitations of transport of the disinfectant to the pipe surface were found to affect efficacy (Szabo et al., 2006).

In another NHSRC study, non-radioactive isotopes of cesium and cobalt were studied in annular reactors with biofilm grown using regular tap water with an average of 1 mg/L chlorine residual. Corroded iron coupons were used and the annular reactor was operated with a flow to produce shear on the coupons. Cesium was not detected on corroded iron, likely due to its high solubility in water and the competition of the iron for the other ions, such as calcium and magnesium. Reaction of cobalt with chlorine resulted in a persistent insoluble precipitate on the iron. While flushing alone with chlorine was ineffective, acidification removed greater than 92% of the contaminant (Szabo et al., 2009a). X-ray adsorption spectroscopy is underway to evaluate isotopes and oxidation states.

3.4.4.3 Inactivation of Anthrax Spores in Decontamination Wash/Waste Water

It is anticipated that during decontamination of buildings and other facilities, large amounts of contaminated wash/waste water may be generated. This water needs to be collected, sampled, stored, and in some situations treated before disposal or discharge to a wastewater treatment system. NHSRC is conducting bench-scale studies to determine the effectiveness of chlorine to inactivate anthrax spore surrogates in wash water generated during a decontamination event. The study will test an inactivation technique recommended by the National Response Team (NRT) in its *Environmental Response Technical Assistance Document for Bacillus anthracis Intentional Releases*. Based on studies done in distilled water, the NRT recommends disinfection using bleach and vinegar at doses that result in a 1% hypochlorite solution at pH 7. The NHSRC researchers are conducting bench-scale studies to verify the effectiveness of the recommended approach in actual wash water matrices. A multi-agency study in May 2011, involving decontamination of a building contaminated with anthrax surrogate spores, provided the opportunity to test the application of the bench-scale results and offer additional recommendations for future decontamination needs. The field study will also investigate on-site filtration and inactivation techniques for the wash water generated following the washing of contaminated personal protective equipment and building decontamination.

3.4.4.4 Quantitative Structure Property Relationship Prediction of Chlorine Reactivity with Chemical Contaminants of Concern

This quantitative structure property relationship (QSPR) approach utilizes computational chemistry techniques to quickly, economically, and accurately estimate properties of contaminants related to their fate and transport during water treatment or infrastructure decontamination (Magnuson and Speth, 2005). Many contaminants of concern for homeland security are difficult to study experimentally due to safety concerns or cost. Therefore, estimates of contaminant properties can inform decisions regarding whether or not the contaminant is of concern. For example, a plausible estimation of chlorine reactivity of CBR agents can inform planning for the response to CBR contamination events and provide valuable information to support decontamination approaches. QSPR-generated data not only affect decision making during decontamination activities, but can inform future research plans, particularly for emerging contaminant threats for which little data are available.

3.4.5 Technology Testing and Evaluation Program

The NHSRC's Technology Testing and Evaluation Program (TTEP) focuses on advancing technologies that can be used by water utilities to monitor, detect, and treat contaminants introduced into water systems. Existing or newly developed technologies are tested and evaluated under this program. Test results and recommendations are disseminated to the water community using the WaterISAC when the data are sensitive. Because the testing is quite rigorous and is done using actual CBR contaminants, technology vendors can reference the testing for the commercialization of their equipment. TTEP is currently testing TOC monitors that are under development by manufacturers or already on the market. Other areas TTEP is planning to move forward in are the testing of technologies that have the ability to identify particles in water and multi-parameter probes that measure a variety of water quality parameters in one location. TTEP also tests mobile treatment systems and decontamination systems and technologies for water and wastewater.

3.5 Research Outcomes

The research done by EPA's NHSRC is intended to provide the drinking water community with tools and methodologies that not only benefit water security but also improve water quality monitoring for public health protection and can support regulatory compliance. The impact of a contamination event is often only recognized when emergency room incidences are reported. This is unacceptable given that once emergency room incidences reach a level to be noted, additional exposures will have already occurred.

Proactive tools and methodologies developed by the NHSRC target rapid monitoring and response to contamination events. Thus, they have the potential to minimize public health concerns and possible exposures as well as reduce social and economic impacts, whether the contamination arises from an intentional or accidental incident. Once an event has occurred, rapid response and recovery are critical to returning the drinking water system to normal operations. Any delay in recovering from a contamination event, for example, will not only have economic impacts but also affect public confidence in their water systems.

The research done in NHSRC supports EPA's responsibilities as the Sector-Specific Agency lead. The results are often used immediately by the drinking water utilities, as evidenced by the use of some tools even as improvements are made to them. The data provide the states and the regions with sound science and engineering needed for making informed decisions to help protect and secure the Nation's water and infrastructure systems. In addition, data generated by NHSRC on CBR fate and transport complement the work of other federal agencies (e.g., DHS and CDC). Results from the NHSRC program can be leveraged by other water sector research organizations and academia to build a combined understanding of improving water security and protecting public health.

NHSRC research results are disseminated on its web site at www.epa.gov/nhsrc, through peer-reviewed journals and presentations at national and international conferences. When the results are deemed sensitive, they are distributed to the water utilities through the WaterISAC. Sensitive data can also be shared within the federal government through official channels.

3.6 Future Direction

As its water security research program matures, NHSRC, in partnership with the OGWDW, plans to continue research to address remaining gaps in the areas of prevention, protection, and detection. The research program in the area of water treatment and infrastructure decontamination is expanding and the prioritization will closely align with the Agency's and Center's mission and recommendations of the CIPAC, water utilities, and other key water sector stakeholders.

NHSRC research is targeting the development of tools and models that can be used for multiple benefits of security, resiliency, and sustainability of water systems. Drinking water systems are under increasing financial pressures to meet new water quality regulations, replace aging infrastructure, and prepare for potential natural disasters and terrorist attacks. In addition, long-term trends in the declining availability of high-quality source waters may lead to very different modes of future treatment and operation. NHSRC research explores the connections between security improvements and these competing programs to identify potential synergies. Frameworks on the sustainability of drinking water systems in general, and water distribution systems in particular, are evaluated for their potential to cost-effectively address security issues.

Future research areas that this program plans to address include the following: (1) research to improve response (containment, treatment, and decontamination) to CBR contamination; (2) analytical method development and improvement; (3) detection capabilities for additional classes of contaminants; and (4) expansion of the research in the area of social sciences to improve crisis communication. The research will focus on developing tools and methodologies that provide multiple benefits to the water industry.

Water quality modeling research is moving in the direction of developing real-time models that can use Supervisory Control and Data Acquisition data in real time for more effective detection of contamination and subsequent consequence management decisions. Such models will enhance the ability of utilities to make effective decisions regarding the containment of the contaminated water or flushing the system through effective valve closure. These models can also allow the utility to trace the contaminant to the point of intentional or accidental intrusion. Adsorption and desorption models in the distribution system, models for biofilm attachment, and models for the reaction with chlorine will help inform decontamination decisions.

Acronyms

Action Plan	*Water Security Research and Technical Support Action Plan*
AWWA	American Water Works Association
BVA	Blast Vulnerability Assessment Tool
CBR	Chemical, Biological, and Radiological
CDC	Centers for Disease Control
CIPAC	Critical Infrastructure Protection Advisory Council
CWS	Contamination Warning System
DHS	Department of Homeland Security
EPA	US Environmental Protection Agency
ERLN	Environmental Response Laboratory Network
ESSENCE	Electronic Surveillance System for the Early Notification of Community-based Epidemics
HSPD	Homeland Security Presidential Directives
MYP	Homeland Security Multi-Year Plan
NHSRC	EPA's National Homeland Security Research Center
NRC	National Research Council
NRT	National Response Team
OGWDW	EPA's Office of Ground Water and Drinking Water
ORD	EPA's Office of Research and Development
ORP	Oxidation Reduction Potential

QSPR	Quantitative Structure Property Relationship
RAMCAP	Risk Analysis and Management for Critical Asset Protection
SAM	*The Standardized Analytical Methods for Environmental Restoration following Homeland Security Events, Version 5.0*
SPOT	Sensor Placement Optimization Tool
TEVA	Threat Ensemble Vulnerability Assessment
TOC	Total Organic Carbon
TTEP	Technology Testing and Evaluation Program
WaterISAC	Water Information Sharing and Analysis Centers
WLA	Water Lab Alliance
WS	Water Security

References

Allen HJ, Haught RC, and Macke DA (2008a) Online toxicity monitors and their use in distribution system and watershed early warning systems. Presented at the 2008, Collective Responsibility, Water Contamination Emergencies Conference, London, England

Allen HJ, Muhammed N, and Macke DA (2008b) Effect of sodium thiosulfate on toxicity of malathion, copper, and cyanide to Vibrio fischeri using the ToxControl Online Toxicity Monitor. Presented at the Society of Environmental Toxicology and Chemistry Annual Meeting, Tampa Bay, FL

Allen HJ, Muhammed N, and Macke DA (2009) Effect of sodium thiosulfate on toxicity of cadmium, diazinon, and toluene to Vibrio fischeri using the ToxControl Online Toxicity Monitor. Presented at the Society of Environmental Toxicology and Chemistry Annual Meeting, New Orleans, LA

Babin SM, Burkom HS, Mnatsakanyan Z, Ramac-Thomas L, Thompson MW, Wojcik R, Happel-Lewis S, and Yund C (2008) Drinking water security and public health disease outbreak surveillance. Johns Hopkins APL Technical Digest Public Health Informatics 27(4):403–411

Baranowski T, Janke R, Murray R, Bahl S, Sanford L, Steglitz B, and Skadsen J (2008) Case study analysis to identify and evaluate potential response initiatives in a drinking water distribution system following a contamination event. Presented at the 2008 Borchardt conference, a seminar on advancements in water and wastewater, The University of Michigan, Ann Arbor, MI

Burkom H, Ramac-Thomas L, Babin S, Holtry R, Mnatsakanyan Z, and Yund C (2011) An integrated approach for fusion of environmental and human health data for disease surveillance. Statistics in Medicine 30:470–479. DOI:10.1002/sim.3976

Clark, S, McMahon W, and Taylor V (2008) Blast vulnerability assessment tool. Presented at the 2008 AWWA Water Security Congress, Cincinnati, OH

Critical Infrastructure Partnership Advisory Council's (CIPAC) Water Sector Decontamination Working Group (2008) Recommendations and proposed strategic plan: water sector decontamination priorities final report. http://www.amwa.net/galleries/securityinfo/CIPACDeconReportFinal.pdf. Accessed 03 March 2010

De Sanctis AE, Hachett S, Uber JG, Boccelli DL, and Shang F (2009) Real-time implementation of contamination source identification method for water distribution systems. Presented at the World Environmental and Water Resources Congress, ASCE, Kansas City, MO

Francy DS, Bushon RN, Brady AM, Kephart CM, Likirdopulos CA, Mailot BE, Schaefer FW III, and Lindquist HDA (2009) Comparison of traditional and molecular analytical methods for detecting biological agents in raw and drinking water following ultrafiltration. Journal of Applied Microbiology 107(5):1479–1491

Grayman WM, Murray R, et al. (2009) Effects of redesign of water systems for security and water quality factors. Presented at the World Environmental and Water Resources Congress, ASCE, Kansas City, MO

Hall J, Zaffiro AD, Marx RB, Kefauver PC, Krishnan ER, Haught RC, and Herrmann JG (2007) On-Line water quality parameters as indicators of distribution system contamination. Journal of the American Water Works Association 99(1):66–77

Hall JS, Szabo JG, Panguluri S, and Meiners G (2009) Distribution system water quality monitoring: sensor technology evaluation methodology and results. Publication number EPA/600/R-09/076

Hart DB and McKenna SA (2009) CANARY user's manual, version 4.2. http://www.epa.gov/NHSRC/pubs/600r08040a.pdf. Accessed 09 March 2010

Haxton T and Walski TM (2009) Modeling a hydraulic response to a contamination event. Presented at the World Environmental and Water Resources Congress, ASCE, Kansas City, MO

Haxton T and Uber J (2010) Flushing under source uncertainties. Presented at the 12th Annual Water Distribution Systems Analysis Conference, University of Arizona, Tucson, AZ

Laird CD, Biegler LT, and van Bloemen Waanders BG (2006) Mixed-integer approach for obtaining unique solutions in source inversion of water networks. Journal of Water Resources Planning and Management 132(4):242–251

Lindquist HDA, Harris S, Lucas S, Hartzel M, Riner D, Rochele P, and DeLeon R (2007) Using ultrafiltration to concentrate and detect *Bacillus anthracis*, *Bacillus atrophaeus* subspecies *globigii*, and *Cryptosporidium parvum* in 100-liter water samples. Journal of Microbiological Methods 70(3):484–492

Magnuson ML and Speth T (2005) Quantitative structure–property relationships for enhancing predictions of synthetic organic chemical removal from drinking water by granular activated carbon. Environmental Science and Technology 39:7706–7711

McKenna SA, Wilson M, et al (2008) Detecting changes in water quality data. Journal of the American Water Works Association 100(1):74–85

National Research Council (2004) A review of the EPA water security research and technical support action plan: Parts I and II. National Academies Press, Washington, DC

National Research Council (2007) Improving the nation's water security, opportunities for research. National Academies Press, Washington, DC

Rose LJ, Rice EW, Jensen B, Murga R, Peterson A, Donlan RM, and Arduino MJ (2005) Chlorine inactivation of bacterial bioterrorism agents. Applied Environmental Microbiology 71:566–568

Rose LJ, Rice EW, Hodges L, Peterson A, and Arduino MJ (2007) Monochloramine inactivation of bacterial select agents. Applied Environmental Microbiology 73:3437–3439

Rossman LA (2000) EPANET 2 users manual. http://www.epa.gov/nrmrl/wswrd/dw/epanet/EN2manual.PDF. Accessed 09 March 2010

Rossman LA (2008) EPANET 2.00.12. http://www.epa.gov/nrmrl/wswrd/dw/epanet.html#downloads. Accessed 09 March 2010

Szabo JG, Rice EW, and Bishop PL (2006) Persistence of *Klebsiella pneumonia* on simulated biofilm in a model drinking water system. Environmental Science and Technology 40(16):4996–5002

Szabo JG, Rice EW, and Bishop PL (2007) Persistence and decontamination of *Bacillus atrophaeus* subsp. *globigii* spores on corroded iron in a model drinking water system. Applied Environmental Microbiology 73(8):2451–2457

Szabo JG, Hall JS, and Meiners G. (2008) Sensor response to contamination in chloraminated water. Journal of the American Water Works Association 100(4):33–40.

Szabo JG, Impellitteri CA, Govindaswamy S, et al (2009a) Persistence and decontamination of surrogate radioisotopes in a model drinking water distribution system. Water Research 43:5005–5014

Szabo JG, Muhammad N, Packard B, et al (2009b) *Bacillus* spore uptake onto heavily corroded iron pipe in a drinking water distribution system simulator technical note. Canadian Journal of Civil Engineering 36:1867–1871

U.S. Congress (2002) Public Health Security and Bioterrorism Preparedness and Response Act of 2002: Public Law 107-188. http://thomas.loc.gov/cgi-bin/query/z?c107:H.R.3448.ENR:. Accessed 03 March 2010

U.S. Department of Homeland Security (2008) National Response Framework. http://www.fema.gov/pdf/emergency/nrf/nrf-core.pdf. Accessed 03 March 2010

U.S. Department of Homeland Security (2010a) Critical Infrastructure Sector Partnerships http://www.dhs.gov/files/partnerships/editorial_0206.shtm. Accessed 03 March 2010

U.S. Department of Homeland Security (2010b) Homeland Security Presidential Directives. http://www.dhs.gov/xabout/laws/editorial_0607.shtm. Accessed 09 March 2010

U.S. Environmental Protection Agency (2004a) Homeland Security Strategy, http://www.epa.gov/OHS/pdfs/EPA-HS-Strategy.pdf. Accessed 02 March 2010

U.S. Environmental Protection Agency (2004b) Water Security Research and Technical Support Action Plan. Publication number EPA/600/R-04/063. http://www.epa.gov/nhsrc/pubs/600r04063.pdf. Accessed 02 March 2010

U.S. Environmental Protection Agency (2007) Pilot-scale tests and systems evaluation for the containment, treatment, and decontamination of selected material from T&E building pipe loop equipment. http://www.epa.gov/NHSRC/pubs/600r08016.pdf. Accessed 09 March 2010

U.S. Environmental Protection Agency (2010a) National Homeland Security Research Center water research multi-year plan. Internal document

U.S. Environmental Protection Agency (2010b) Threat Ensemble Vulnerability Assessment Research Program. http://www.epa.gov/nhsrc/water/teva.html. Accessed 09 March 2010

U.S. Environmental Protection Agency (2010c) Standardized analytical methods for environmental restoration following homeland security events, revision 6.0. http://www.epa.gov/sam/. Accessed 27 June 2011

U.S. Environmental Protection Agency (2011) Planning for an emergency drinking water supply. http://cfpub.epa.gov/si/si_public_record_report.cfm?address=nhsrc/&dirEntryId=235197. Accessed 27 June 2011

WaterISAC (2010) Water Security Network (controlled access portal). https://portal.waterisac.org/clearspace_community/pages/page.jspa. Accessed 02 March 2010

Chapter 4
Drinking Water Critical Infrastructure and Its Protection

Rakesh Bahadur and William B. Samuels

4.1 Introduction

Water infrastructure systems include surface and ground water sources for municipal, industrial, agricultural, and household needs; dams, reservoirs, aqueducts, and pipes that contain and transport raw water; treatment facilities that remove contaminants from raw water; finished water reservoirs; systems that distribute water to users; and wastewater collection and treatment facilities. Across the country, these systems comprise approximately 77,000 dams and reservoirs; thousands of miles of pipes, aqueducts, water distribution, and sewer lines; 168,000 public drinking water facilities (many serving as few as 25 customers). The following statistics summarize the drinking water sector (DHS, 2007):

- Of the 168,000 Public Drinking Water Systems (PWS) nearly 140,000 of the public water systems serve 500 people or fewer. Another 360 systems serve more than 100,000 people and provide water to nearly half of the total population served.
- PWS produce 51 billion gallons per day (bgd) of drinking water, out of which 67 percent goes to residential customers and 33 percent to nonresidential customers.
- PWS obtain 63 percent of their source water from surface sources and 37 percent from groundwater.
- There are about 2.3 million miles of distribution system pipes in the United States.

A fairly small number of large drinking water utilities (about 15% of the systems) provide water services to more than 75% of the US population. These systems represent the greatest targets of opportunity for terrorist attacks, while the large number of small systems (serve fewer than 10,000 persons) is less likely to be perceived as

R. Bahadur (✉)
Science Applications International Corporation Center for Water Science and Engineering, McLean, VA 22102, USA
e-mail: bahadurr@saic.com

R.M. Clark et al. (eds.), *Handbook of Water and Wastewater Systems Protection*, Protecting Critical Infrastructure, DOI 10.1007/978-1-4614-0189-6_4,
© Springer Science+Business Media, LLC 2011

key targets by terrorists. However, the smaller systems also tend to be less protected and, thus, are potentially more vulnerable to attack. A successful attack on even a small system could cause widespread panic, economic impacts, and a loss of public confidence in water supply systems.

Drinking water in the United States has long been recognized as among the safest in the world. But the aftermath of September 11, 2001 brought water security to the forefront as a priority. The events of September 11 raised concerns about the security of the nation's drinking water supplies and their vulnerability to attack. Issues include the readiness of water utilities to prevent and respond to attacks on water systems, steps that can be taken to improve preparedness and response capabilities, and the availability of resources to help utilities enhance drinking water security. The basic human need for water and the concern for maintaining a safe water supply are driving factors for water infrastructure protection. In general, there are four areas of primary concentration (NSPP, 2003):

- Physical damage or destruction of critical assets, including intentional release of toxic chemicals;
- Actual or threatened contamination of the water supply;
- Cyber attack on information management systems or other electronic systems; and
- Interruption of services from another infrastructure.

This national infrastructure has been determined to be at risk to accidental and deliberate contamination events, and its security has been made a top priority of the Environmental Protection Agency (EPA, 2005a). Because water supplies directly affect many activities (from drinking water to fighting fires), their disruption could have significant impacts. Ensuring the security of the nations' drinking water supplies poses a substantial challenge, partly because the number of water systems is very large and also because the responsibility for protecting drinking water safety is shared among federal, state, and local governments and utilities. There are no federal standards or agreed upon industry best practices within the water infrastructure sector to govern readiness, response to security incidents, and recovery. Water infrastructure system designers, managers, and operators have long made preparing for extreme events a standard practice. Historically, their focus has been on natural events – major storms, blizzards, and earthquakes – some of which could be predicted hours or longer before they occurred. When considering the risk of man-made threats, operators generally focused on purposeful acts such as vandalism or theft by disgruntled employees or customers, rather than broader malevolent threats by terrorists, domestic or foreign. The events of September 11, 2001, changed this focus. Since September 11, 2001, EPA has increased efforts to help utilities safeguard facilities and supplies from terrorist or other threats in numerous ways (e.g., by providing technical and financial assistance for vulnerability assessments and by supporting research and the establishment of an information sharing center). These efforts have accelerated since September 11, 2011 mainly due to regulations (e.g., Presidential Decision Directive 63 and the Public Health Security and Bioterrorism Preparedness and Response Act of 2002). To improve the flow of information among water sector organizations, the industry has developed its sector-ISAC (Information Sharing Analysis Sector). The WaterISAC provides a secure forum for gathering, analyzing, and sharing security-related information. Additionally, several federal agencies are working together to improve the warehousing of information regarding contamination threats, such as the release of biological, chemical, and radiological substances into the water supply and how to respond to their presence in drinking water.

4.2 Water Security Issues

The potential for terrorism is not new. In 1941, the Federal Bureau of Investigation Director J. Edgar Hoover wrote, "It has long been recognized that among public utilities, water supply facilities offer a particularly vulnerable point of attack to the foreign agent, due to the strategic position they occupy in keeping the wheels of industry turning and in preserving the health and morale of the American populace." (Hoover, 1941). Security in the water sector falls into the following categories:

- Attacks resulting in *physical* destruction to any of these systems could include disruption of operating or distribution system components, power or telecommunications systems, electronic control systems, and actual damage to reservoirs and pumping stations. A loss of flow and pressure would cause problems for customers and would hinder firefighting efforts. Further, destruction of a large dam could result in catastrophic flooding and loss of life.
- Bioterrorism or *chemical* attacks could deliver widespread contamination with small amounts of microbiological agents or toxic chemicals and could endanger the public health of thousands. While some experts believe that risks to water systems actually are small, because it would be difficult to introduce sufficient quantities of agents to cause widespread harm, concern and heightened awareness of potential problems are apparent.
- Factors that are relevant to a *biological* agent's potential as a weapon include its stability in a drinking water system, virulence, and resistance to detection and treatment.
- *Cyber* attacks on computer operations can affect an entire infrastructure network, and hacking could result in theft or corruption of information or denial and disruption of service.

4.2.1 Water Sector Dependencies

The critical infrastructure represents a true "system of systems." Failure in one asset or infrastructure can cascade to disruption or failure in others, and the combined effect could have far-reaching consequences affecting government, the economy, public health and safety, national security, and public confidence.

The efficient functioning of these systems reflects great technological achievements of the past century, but interconnectedness within and across systems also

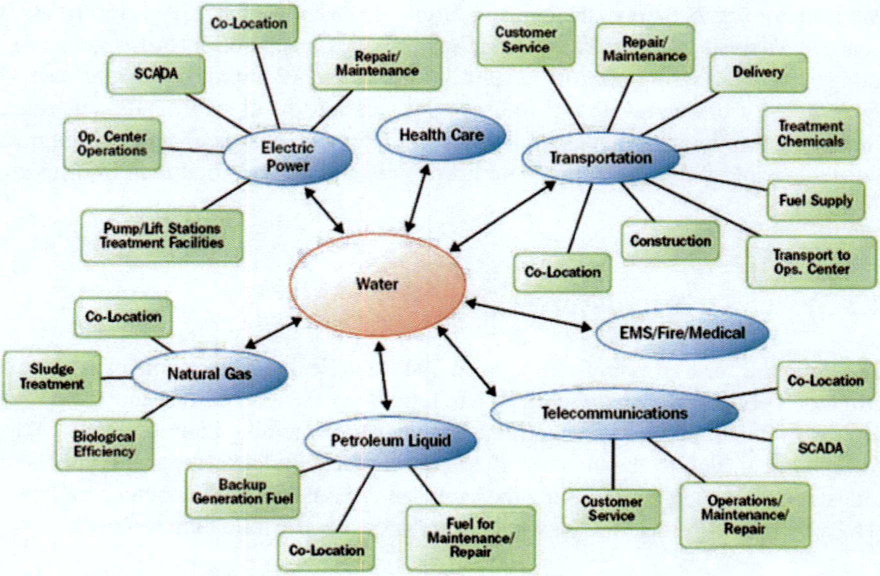

Fig. 4.1 Interdependencies with the water sector (DHS, 2007)

means that infrastructures are vulnerable to local disruptions, which could lead to widespread or catastrophic failures (NAP, 2002). The operations of the water sector depend extensively on other sectors as shown in Fig. 4.1 (DHS, 2007). The heaviest dependence is on the energy sector. For example, running pumps to move water and wastewater and operating drinking water and wastewater treatment plants require large amounts of electricity. To a lesser extent, the water sector also depends on the transportation system for supplies of water treatment chemicals, on natural gas pipelines for the energy used in some operational activities, and on the telecommunications sector. Water systems are increasingly automated and controlled from remote locations for efficiency. The water taken in by a supplier may have been treated and discharged by a user upstream. This situation creates a unique interdependency among individual water or wastewater utilities.

4.2.2 Critical Infrastructure Definition

What is considered to be infrastructure depends heavily upon the context in which the term is used. In US public policy, the definition of "infrastructure" has been evolutionary and often ambiguous. Twenty years ago, "infrastructure" was defined primarily in debates about the adequacy of the nation's public works – which were viewed by many as deteriorating, obsolete, and of insufficient capacity. The US government's definition of "critical infrastructure" has evolved over the years and at any given time has left considerable room for interpretation. Furthermore, since the 1980s, the number of sectors included under that definition has generally expanded

Table 4.1 Evolution of critical infrastructure and key assets over time

Infrastructure	CBO (1983)	NCPWI (1983)	E.O. 13010 (1996)	PDD-63 (1998)	E.O. 13228 (2001)	NSHS (2002)	NSPP (2003)	HSPD-7 (2003)
Transportation	X	X	X	X	X	X	X	X
Water supply/ wastewater treatment	X	X	X	X	X	X	X	X
Education	X							
Public health	X			X		X	X	X
Prisons	X							
Industrial capacity	X							
Waste services		X						
Telecommunications			X	X	X	X	X	X
Energy			X	X	X	X	X	X
Banking and finance			X	X		X	X	X
Emergency services			X	X		X	X	X
Government continuity			X	X		X	X	
Information systems				X	X	X	X	X
Nuclear facilities					X			
Special events					X			
Agriculture/food supply					X	X	X	X
Defense industrial base						X	X	X
Chemical industry						X	X	X
Postal/shipping services						X	X	X
Monuments and icons							X	X
Key industry/tech. sites							X	
Large gathering sites							X	

Moteff and Parfomak (2004)

from the most basic public works to a much broader set of economic, defense, government, social, and institutional facilities, as illustrated in Table 4.1 (Moteff and Parfomak, 2004). The list may continue to evolve and grow as economic changes or geopolitical developments influence homeland security policy.

4.2.3 Contaminants of Concern

Water is particularly vulnerable to chemical or biological attack. One of the greatest concerns facing the United States and other nations is the deliberate use of chemical, biological, or radiological (CBR) weapons by terrorist organizations, including the intentional introduction of CBR contaminants into the nation's drinking water supplies. The CBR agents have been developed and weaponized almost exclusively for airborne assaults, and the effect on a water supply system is secondary. Although less effective as potable water threats, many are potentially capable of inflicting heavy casualties when ingested. For a contaminant to be effective in the drinking water, it must meet the following criteria (Deininger, 2000):

- High toxicity – deadly effect in small amounts.
- No taste or odor.
- Chemical and physical stability.
- Delayed action to protect the sabotage agent.
- Difficult recognition of poisoning – no specific pathologic changes in the organism.
- Difficulties with the detection of the poison with normal analytical methods.
- Unusual effects of poison; no known antidotes.

Biological agents have been used as instruments of warfare for thousands of years. However, the definition and list of agents that could cause mass causalities was not defined. Various researchers (Federal Register, 2005; Burrows and Renner, 1999; Hickman, 1999; Munro et al., 1999; Clark and Deininger, 2000; Hawley and Eitzen, 2000; Deininger, 2000; NRC, 1995; Cristopher, 1997) have defined what is an effective agent. There are many lists of potential water contaminants. These lists have many common entries but are not completely congruent. Some have been prepared by professional or trade organizations, some by academic researchers, some by government researchers, and some by industry. The following section presents an annotated list of databases of contaminants of concern.

4.2.3.1 Agency for Toxic Substances and Disease Registry (ATSDR)

- Maintains a large database on toxic chemicals (http://www.atsdr.cdc.gov)

4.2.3.2 Center for Disease Control (CDC)

- Maintains a list of biological contaminants, many of which pose a water contamination threat (http://www.bt.cdc.gov/agent/agentlist.asp; http://www.bt.cdc.gov/Agent/AgentlistChem.asp)

4.2.3.3 US Environmental Protection Agency (USEPA)

- Water Contaminant Information Tool (WCIT) – EPA has an extensive list of contaminants whose maximum allowable concentration is regulated (https://cdx.epa.gov/SSL/cdx/login.asp)
- The Drinking Water Treatability Database (TDB) presents referenced information on the control of contaminants in drinking water (http://iaspub.epa.gov/tdb/pages/general/home.do)

4.2.3.4 Food and Drug Administration (FDA)

- Publishes the Bad Bug Book on pathogens and biotoxins (http://www.cfsan.fda.gov/~mow/intro.html)

4.2.3.5 World Health Organization (WHO) Database

- Public health response to biological and chemical weapons: WHO guidance (http://www.who.int/csr/delibepidemics/biochemguide/en/index.html)

4.2.3.6 Water Information Sharing and Analysis Center (WaterISAC)

- A proprietary database of contaminants of concern

4.2.3.7 Chemical Weapons Convention

- The list of chemicals found on the three schedules of the Convention (http://www.opcw.org/html/db/cwc/eng/cwc_annex_on_chemicals.html)

4.2.3.8 North Atlantic Treaty Organization (NATO)

- Handbook on the medical aspects of Nuclear Biological and Chemical (NBC) Defensive Operations AmedP-6(B), Departments of the Army, the Navy, and the Air Force, 1996 (http://www.fas.org/irp/doddir/army/fm8-9.pdf)

4.2.3.9 US Department of Defense

- Field manual: Treatment of chemical agent casualties and conventional military chemical injuries, Departments of the Army, the Navy, and the Air Force, and Commandant, Marine Corps, 2000 (http://www.globalsecurity.org/wmd/library/policy/army/fm/8-285/)

4.2.3.10 National Institute of Justice (NIJ)

- Guide for the Selection of Chemical and Biological Decontamination Equipment for Emergency First Responders, Guide 103–00, 2001 (http://www.ncjrs.org/pdffiles1/nij/189724.pdf)

4.2.3.11 Congressional Research Service

- Small-scale terrorist attacks using chemical and biological agents: An Assessment Framework and Preliminary Comparisons, Library of Congress 2004 (http://www.nacwa.org/content/view/58/94/)

4.3 Water Quality Security

The water sector has taken great strides to protect its critical facilities and systems. For instance, government and industry have developed vulnerability assessment methodologies for both drinking water and wastewater facilities and trained thousands of utility operators to conduct them. In response to the *Public Health Security and Bioterrorism Preparedness and Response Act of 2002*, the Environmental Protection Agency (EPA) has developed baseline threat information to use in conjunction with vulnerability assessments. Models and procedures were developed under federal funding to provide physical cyber and water quality security. Efforts to develop voluntary protocols and tools are ongoing since the 2001 terrorist attacks.

4.3.1 Water Security Initiative (WSI)

The Water Security Initiative (EPA, 2007a) is an EPA program that addresses the risk of intentional contamination of drinking water distribution systems. EPA established this initiative in response to Homeland Security Presidential Directive 9, under which the Agency must develop robust, comprehensive, and fully coordinated surveillance and monitoring systems, including international information, for water quality that provides early detection and awareness of disease, pest, or poisonous agents. WSI is a multi-faceted monitoring and surveillance approach for timely detection of water contamination events. It includes water quality monitoring at optimal locations throughout the water distribution system, enhanced security monitoring at key water utility infrastructure assets, consumer complaint surveillance, and innovative uses of public health surveillance data streams. EPA is implementing the WSI in three phases:

- Phase I: develop the conceptual design of a system for timely detection and appropriate response to drinking water contamination incidents to mitigate public health and economic impacts.
- Phase II: test and demonstrate contamination warning systems through pilots at drinking water utilities and municipalities and make refinements to the design as needed based upon pilot results.
- Phase III: develop practical guidance and outreach to promote voluntary national adoption of effective and sustainable drinking water contamination warning systems.

4.3.2 Environmental Technology Verification (ETV) and Technology Testing and Evaluation (TTEP)

Though EPA has taken a limited role in the actual development or financial support of new monitoring technologies, it has been active in testing and verifying the technologies. This has included the following:

- ETV – The ETV program develops testing protocols and verifies the performance of innovative technologies that have the potential to improve protection of human health and the environment. ETV accelerates the entrance of new environmental technologies into the domestic and international marketplaces. The ETV-verified systems include multi-parameter monitors, sensors, and probes and analyzers for specific contaminants (http://www.epa.gov/etv/).

- TTEP – The TTEP program is an outgrowth of ETV. TTEP researchers test, evaluate, and report on the performance of homeland security-related technologies that are designed to detect, contain, decontaminate, or manage chemical, biological, or radiological materials purposefully introduced into structures, drinking water, or the environment. After testing is complete, researchers evaluate the data and compile performance results into individual summary reports. These reports include comparisons of technologies (http://www.epa.gov/NHSRC/pubs/posterTTEP032106.pdf).

4.3.3 The Standardized Analytical Methods for Environmental Restoration Following Homeland Security Events (SAM)

In the aftermath of the terrorist attacks of September 11, 2001, the EPA identified several areas where the country could better prepare itself in the event of future terrorist incidents. One of the most important areas identified was the need to improve the nation's laboratory capacity and capability to analyze environmental samples following a homeland security event. SAM is a compilation of methods for the analysis of chemical, biological, and radiological contaminants in water (EPA, 2007b). SAM identifies a single method per sample type to ensure a consistent analytical approach across multiple laboratories when analyzing environmental samples following an event. The methods include detailed laboratory procedures for confirming the identification of samples and determining their concentrations. The methods are not designed to be used for rapid/real-time response or for conducting an initial evaluation (triage or screening) of suspected material.

4.3.4 Response Protocol Tool Box (RPTB)

The threat management process involves two parallel and interrelated activities: evaluating the threat and making decisions regarding appropriate actions to take in response to the threat. The primary focus of the threat evaluation is public health (i.e., has the water been contaminated at levels of public health concern?). However, the threat evaluation should also consider other potential consequences of a contamination incident such as infrastructure damage, adverse impacts on the aesthetic qualities of the drinking water, and reduced consumer confidence. A key component of any discussion on water security and contamination warning systems is how to respond when a potential contamination event is detected. EPA has developed a Response Protocol Tool Box (RPTB) that is focused on different aspects of planning a response to contamination threats and incidents (EPA, 2003, 2006). RPTB is composed of the following six interrelated modules:

- Module 1 – Water Utility Planning Guide: Includes a brief discussion of the nature of the contamination threat to the public water supply and describes the planning activities that a utility may undertake to prepare for response to contamination threats and incidents.
- Module 2 – Contamination Threat Management Guide: Presents the framework for management (evaluation and actions) of contamination threats to the drinking water.
- Module 3 – Site Characterization and Sampling Guide: Describes the site characterization process in which information is gathered from a contamination incident at a drinking water system.
- Module 4 – Analytical Guide: Presents an approach to the analysis of samples collected from the site of a suspected contamination incident.
- Module 5 – Public Health Response Guide: The public health response measures that would potentially be used to minimize public exposure to potentially contaminated water.
- Module 6 – Remediation and Recovery Guide: Describes the planning and implementation of remediation and recovery activities that would be necessary following a confirmed incident.

4.3.5 Water Contaminant Information Tool (WCIT)

The WCIT is a powerful new information resource for water utilities, public health officials, and agencies responsible for the safety of water supplies. Along with other EPA efforts such as early warning detection systems, WCIT is a major element of EPA's support of water utility incident prevention and detection capability. The WCIT (http://www.epa.gov/wcit/) is a secure, online database that provides information on chemical, biological, and radiological contaminants of concern for water security. Contaminants of concern for water security are those contaminants that may or may not be regulated, but could pose a significant threat to public health if accidentally or intentionally introduced into drinking water. The WCIT database includes some contaminants that are not regulated because they are not typically found in drinking water, but could cause harm if intentionally introduced into a drinking water system. The WCIT database assists in planning for and responding to drinking water and wastewater contamination threats and incidents. For each contaminant, WCIT contains the detailed information for the following parameters:

- Name, Chemical Abstracts Service (CAS) ID
- Physical or pathogen properties
- Fate and transport
- Medical and toxicity information

- Early warning indicators
- Field detection and analysis information
- Drinking water and wastewater treatment
- Environmental impacts
- Infrastructure decontamination

4.4 Water Quality Security Models

Modeling, simulation, and analysis activities help to prioritize critical infrastructure and key asset protection activities and investments. Modeling, simulation, and analysis can also facilitate protection planning and decision support by enabling the mapping of complex interrelationships among the elements that make up the risk environment. Using models and simulations, responsible authorities can evaluate the risks associated with particular vulnerabilities more accurately and subsequently make more informed protection decisions. Modeling and simulation can also be used as a real-time decision support tool to help mitigate the effects of an attack or avert a secondary attack altogether.

Models have a wide application in water sector even when it is not possible to accurately model the behavior of a chemical or a microbe. Model output regarding the water and contaminant movement in the distribution system can determine which portions of the system are exposed to water/contaminants from particular sources, tanks, or pipe breaks. One of the current applications of models is to help utilities to respond due to either accidental or intentional contamination. Hydraulic models have evolved since their first use over the years. This evolution was mainly influenced to meet water system requirements. The first models were based on steady-state hydraulic models, evolved into extended period simulation (EPS) and gradually evolved into GIS-based water quality (W.Q.) security models (Fig. 4.2).

There are numerous models available (EPANET,[1] H2OMAP/H2ONET,[2] Helix delta-Q,[3] InfoWater™ Protector,[4] MIKE NET,[5] OptiDesigner,[6] PIPE2000/KYPIPE,[7] PipelineNet,[8] STANET®,[9] WADISO SA,[10] and WaterCAD[11]) that can be used not only for water quality but also for water security modeling. Each of

[1] http://www.epa.gov/nrmrl/wswrd/dw/epanet.html
[2] http://www.mwhsoft.com/
[3] http://www.helixtech.com.au/
[4] http://www.mwhsoft.com/
[5] http://www.bossintl.com/
[6] http://www.optiwater.com
[7] http://www.kypipe.com/
[8] http://www.tswg.gov/tswg/ip/PipelineNetTB.htm
[9] http://www.stafu.de
[10] www.wadiso.com
[11] www.haestad.com

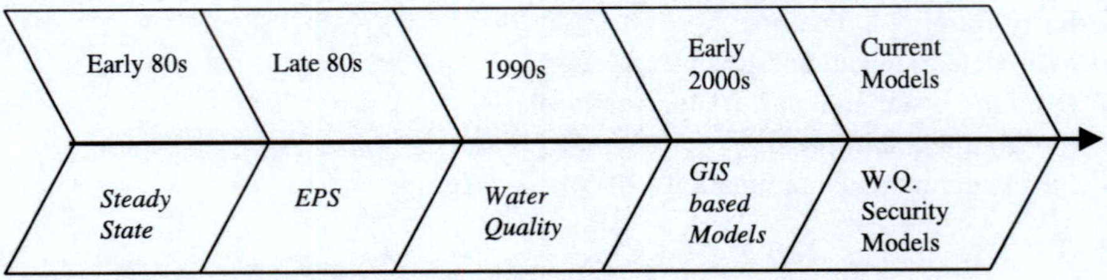

Fig. 4.2 Evolution of modeling in water distribution systems

these models has unique hydraulic and water security capabilities and applications in water security. A more in-depth description of these water models is provided in ASCE (2004). Water quality security applications of water distribution models include the following:

- *Sensor monitoring/instrument placement* – Models used to help determine the optimum sensor placement.
- *Pre-event response scenarios* – Models can help to simulate what-if scenarios before an event occurs to facilitate response planning.
- *Design/upgrade of water systems* – Model simulations can identify the weak points in the water distribution system.
- *Identifying location of contamination* – Models can be used to predict location and concentration of the contaminants during an actual event.
- *Confirmation of positive event* – Models can be used to verify the contaminant concentration at a location and compare it with the actual samples taken in the distribution system.

Hydraulic/water quality network models are also the most practical tools for identifying the candidate locations for monitoring instruments and sensors. The use of intuitive methods for locating sampling sites may not be effective in meeting all these objectives. Mathematical programming (optimization) methods together with hydraulic/water quality network models are used to address and tackle this problem. Using this methodology, optimal sampling sites can be ascertained in the distribution system. Model sophistication needed is a major disadvantage using this methodology.

4.4.1 EPANET Family of Codes

EPANET models water distribution piping systems and performs extended period simulation of the hydraulic and water quality behavior within pressurized pipe networks. EPANET tracks the flow of water in each pipe, the pressure at each node, the height of the water in each tank, and the concentration of a chemical species throughout the network during a simulation period. As shown below, EPANET has been extended to support homeland security research efforts.

- EPANET-MSX (http://www.epa.gov/nhsrc/news/news073007a.html) is an extension to EPANET that allows the user to define and simulate chemical reactions that involve multiple species.
- EPANET-DPX (Distributed Processor eXtension) is a Java application that schedules execution of multiple EPANET input files for execution on distributed processors.
- EPANET-MCX (Monte Carlo eXtension) is a stand-alone C application that integrates EPANET and EPANET-MSX with the general Monte Carlo simulation package MCSIM.
- EPANET-RTX is an extension to EPANET that allows for real-time estimation of water demands using real-time hydraulic measurements of pressure and flow rate.
- EPANET-PBX is an API extension to EPANET for particle backtracking analysis using an EPANET hydraulic solution.

4.4.2 TEVA's Sensor Placement Optimization Tool (SPOT)

SPOT enables water utilities to determine and evaluate sensor placement. TEVA-SPOT (http://www.epa.gov/nhsrc/news/news112607.html) requires specific information from the utility and allows users to select design objectives and compare and contrast the benefits of different sensor placements (Murray et al., 2004, 2008; Skasden et al., 2008). A major focus of the TEVA research program has been on developing software tools to assist water utilities in determining the best location for sensors (e.g., water quality monitors) within water distribution systems. There are many considerations in locating sensors, including physical requirements (e.g., ease of access for maintenance, power, and sewage), design objectives (e.g., public health protection, spatial coverage, and detection time), and costs.

4.4.3 CANARY

CANARY (http://www.epa.gov/nhsrc/news/news122007.html) software evaluates standard water quality data (e.g., free chlorine, pH, and total organic carbon) over time and uses mathematical and statistical techniques to identify the onset of anomalous water quality incidents. Before using CANARY for the first time, historical utility data must be used to determine the natural variation of these water quality parameters. This allows the water utility to adapt CANARY to work accurately at multiple locations within the water distribution system and helps utility operators to understand the expected false alarm rates associated with CANARY and contamination incident detection (Hart et al., 2007).

4.5 Physical Security

The attributes that contribute to a water utility's security may include components like intakes, collection systems, treatment plants, storage, chemicals, and distribution systems. Physical security is important to guard against unauthorized access to these components. The physical security measures make the water treatment plant and distribution system more secure and reduce the probability of contamination. Effective physical protection is based on the following four elements (ASCE, 2006):

- *Deterrence* – Deterrence is not generally considered a part of a physical protection system with a predictable level of effectiveness; however, it can reduce the occurrence of crime or low-level vandal attacks. Security measures such as lighting, the presence of closed circuit television (CCTV), a clearly visible facility with no visual obstructions, or people in the area may deter an adversary from attacking a facility.
- *Detection* – A detection system includes electronic features such as sensors as well as cameras. Security measures (e.g., sensors) are intended to detect the presence of an intruder. Depending on the types of sensors, a detection system may include lighting systems, motion detectors, monitoring cameras, access control equipment, or other devices.
- *Delay* – Physical barriers are designed to delay an adversary until a response force can interrupt the adversary's actions. Delay features consist primarily of physical hardening devices often employed in multiple layers to provide protection in-depth. Delay features are only effective when placed within a layer of detection.
- *Response* – Response refers to actions taken to interrupt the adversary's task. Utility staff, the utility's security response force, or law enforcement may carry out the response with the appropriate responder depending on the threat and policy of the utility.

4.5.1 Physical Security Tools

Water sector risk assessment tools enable drinking water utilities to identify, inventory, and assess the criticality of utility-specific components in much greater detail. EPA has supported development of risk assessment tools for drinking water and wastewater systems of all sizes. These tools address unique and fundamental security concerns. Risk assessment tools developed for the water sector and supported by EPA funding or by others include the following:

- *Risk Assessment Methodology for Water Utilities (RAM-W)* – This comprehensive security risk assessment methodology was designed for large drinking water utilities. Developed by Sandia National Laboratories (Sandia, 2002), the methodology contains sensitive information and is subject to strict nondisclosure requirements. It covers all aspects of water utility operations. Extensive fault trees throughout the analysis help the utility systematically assess its vulnerabilities to attack. The results provide a prioritized list of relative risks to be considered for system or security upgrades. RAM-W is available only to relevant stakeholders in the water supply community.

- *Risk Assessment Methodology for small and medium utilities* – This is a simplified version of RAM-W for use by small and medium water utilities and provides step-by-step instructions.
  - o Determining the important utility components to protect;
  - o Determining the consequences of losing key components;
  - o Defining the types of threats and likelihood of their occurrence;
  - o Defining safeguards to protect the utility from sabotage;
  - o Analyzing the utility to determine constraints; and
  - o Developing an ERP to counter or minimize risks.
  - o Interdependencies with other sectors, employee screening, security policies, and contingency plans are also addressed.
- *Vulnerability Self-Assessment Tool (VSAT)* – Three VSAT tools (http://www.vsatusers.net/), developed by the Association of Metropolitan Sewerage Agencies (AMSA), are available for drinking water, wastewater, and combination drinking water/wastewater systems. The tools cover the full range of utility components, including physical plant, employees, IT, communications, and customers. Its threat library contains information on man-made disruptions and natural disasters that utilities can apply to determine their potential consequences to each system component. VSAT output provides the user with a vulnerability assessment report and updated Emergency Response Plan (ERP). The VSAT software tool enables utilities to
  - o Assess their vulnerability;
  - o Determine potential solutions for the prioritized vulnerabilities;
  - o Develop priorities for security improvements; and
  - o Plan for emergency responses.
- *Security Vulnerability Self-Assessment Guide for Very Small Drinking Water Systems Serving Populations of Fewer Than 3,300* – This guide was developed by the Association of State Drinking Water Administrators (ASDWA) and the National Rural Water Association (NRWA) in consultation with EPA. The goal for this guide is to help these systems (fewer than 3,000) understand the basics of water system security that includes Vulnerability Assessments (VAs), Emergency Response Plans (ERPs), and practical actions to improve system security. The guide also helps these systems assess their critical components and identify security measures that should be implemented (http://www.epa.gov/safewater/watersecurity/pubs/very_small_systems_guide.pdf).
- *Security Vulnerability Self-Assessment Guide for Small Drinking Water Systems Serving Populations of 3,300–10,000* – Developed by the Association of State Drinking Water Administrators (ASDWA) and the National Rural Water

Association (NRWA), this guide was designed to help drinking water systems serving populations of between 3,300 and 10,000 to identify critical components of vulnerability assessments, complete assessments required under the Bioterrorism Act, and identify security measures to be implemented. This vulnerability assessment guide provides a streamlined tool for small drinking water utilities as they inventory their critical components, conduct self-assessments, and prioritize needed actions (http://www.epa.gov/OGWDW/dwa/pdfs/vulnerability3300-10000.pdf).

- *Voluntary Water Infrastructure Security Enhancement (WISE) Initiative* – EPA grant funding facilitated the development of guidance, training, and voluntary standards that cover the design of online contaminant monitoring systems and physical security enhancements of drinking water, wastewater, and storm water infrastructure systems. The interim voluntary guidance documents will assist drinking water and wastewater utilities in mitigating the vulnerabilities of their systems to man-made threats through the design, construction, operation, and maintenance of both new and existing systems of all sizes. The effort has been carried out in the following three stages:
  - The first stage resulted in three interim voluntary security guidance documents that cover the design of online contaminant monitoring systems and physical security enhancements of drinking water, wastewater, and storm water infrastructure systems.
  - The second phase resulted in training modules to provide instruction on the guidance documents.
  - The third phase resulted in the development of draft standards for trial use to advance physical security measures at drinking water, wastewater, and storm water utilities (http://www.asce.org/static/1/wise.cfm).

- *Security and Emergency Management Systems* – This software tool for drinking water utilities serving between 3,300 and 10,000 persons is based on the Self-Assessment Guide for Small Drinking Water Systems Serving Populations of Between 3,300 and 10,000. The CD-ROM provides a step-by-step process for evaluating water utility and developing a vulnerability assessment. Upon completion of the assessment, the software provides the user with a vulnerability assessment report and updated ERP (http://www.vulnerabilityassessment.org/).

4.6 Cyber Security

Supervisory Control and Data Acquisition (SCADA) networks contain computers and applications that perform key functions in providing essential services and commodities (e.g., electricity, natural gas, gasoline, water, waste treatment, and transportation) to all Americans. They are part of the nation's critical infrastructure and require protection from a variety of threats that exist in cyber space today. SCADA systems allow a water utility to collect data (water level, temperature, pressure, contaminant concentration, and pipeline flow rates) from sensors and control equipment located at remote sites. Advances in information technology and the necessity of improved efficiency have resulted in increasingly automated and interlinked infrastructures and created new vulnerabilities due to equipment failure, human error, weather and other natural causes, and physical and cyber attacks. Some areas and examples of possible SCADA vulnerabilities include (EPA, 2005b) the following:

- *Human* – People can be tricked or corrupted and may commit errors.
- *Communications* – Message can be fabricated, intercepted, changed, deleted, or blocked.
- *Hardware* – Security features are not easily adapted to small units with limited power supplies.
- *Physical* – Intruders can break into a facility to steal or damage SCADA equipment.
- *Natural* – Tornados, floods, earthquakes, and other natural disasters can damage equipment.
- *Software* – Programs can be poorly written.

Federal Directives highlighted the need to secure cyber space, including SCADA, from terrorists and other malicious actors and stated that securing SCADA is a national priority. SCADA networks were initially designed to maximize functionality, with little attention paid to security. As a result, performance, reliability, flexibility, and safety of distributed control/SCADA systems are robust, while the security of these systems is often weak. This makes some SCADA networks potentially vulnerable to disruption of service, process redirection, or manipulation of operational data that could result in public safety concerns and/or serious disruptions to the nation's critical infrastructure. Sandia National Laboratory (Stamp et al., 2003) described the common problems identified with the SCADA system vulnerability and categorized them in the following five categories:

- *System Data* – Important data attributes for security include availability, authenticity, integrity, and confidentiality. Data should be categorized according to its sensitivity, and ownership and responsibility must be assigned.
- *Security Administration* – Vulnerabilities emerge because many systems lack a properly structured security policy, equipment and system implementation guides, configuration management, training, and enforcement and compliance auditing.
- *Architecture* – Many common practices negatively affect SCADA security. For example, while it is convenient to use SCADA capabilities for other purposes such as fire and security systems, these practices create single points of failure. Also, the connection of SCADA networks to other automation systems and business networks introduces multiple entry points for potential adversaries.
- *Network* (including communication links) – Legacy systems' hardware and software have very limited security capabilities, and the vulnerabilities of contemporary systems (based on modern information technology) are

publicized. Wireless and shared links are susceptible to eavesdropping and data manipulation.

- *Platforms* – Many platform vulnerabilities exist, including default configurations retained, poor password practices, shared accounts, inadequate protection for hardware, and nonexistent security monitoring controls. In most cases, important security patches are not installed, often due to concern about negatively impacting system operation; in some cases technicians are contractually forbidden from updating systems by their vendor agreements.

The President's Critical Infrastructure Protection Board[12] and the Department of Energy have developed 21 steps to help improve the security of SCADA networks. These steps address essential actions to be taken to improve the protection of SCADA networks and are divided into the following two categories.

4.6.1 Specific Actions to Improve Implementation

- Identify all connections to SCADA networks.
- Disconnect unnecessary connections to the SCADA network.
- Evaluate and strengthen the security of any remaining connections to the SCADA network.
- Harden SCADA networks by removing or disabling unnecessary services.
- Do not rely on proprietary protocols to protect your system.
- Implement the security features provided by device and system vendors.
- Establish strong controls over any medium that is used as a backdoor into the SCADA network.
- Implement internal and external intrusion detection systems and establish 24-h-a-day incident monitoring.
- Perform technical audits of SCADA devices and networks and any other connected networks, to identify security concerns.
- Conduct physical security surveys and assess all remote sites connected to the SCADA network to evaluate their security.
- Establish SCADA "Red Teams" to identify and evaluate possible attack scenarios.

4.6.2 Actions to Establish Essential Underlying Management Processes

- Clearly define cyber security roles, responsibilities, and authorities for managers, system administrators, and users.

[12] http://www.oe.netl.doe.gov/docs/prepare/21stepsbooklet.pdf

- Document network architecture and identify systems that serve critical functions or contain sensitive information that require additional levels of protection.
- Establish a rigorous, ongoing risk management process.
- Establish a network protection strategy based on the principle of defense-in-depth.
- Clearly identify cyber security requirements.
- Establish effective configuration management processes.
- Conduct routine self-assessments.
- Establish system backups and disaster recovery plans.
- Senior organizational leadership should establish expectations for cyber security and performance and hold individuals accountable for their performance
- Establish policies and conduct training to minimize the likelihood that organizational personnel will inadvertently disclose sensitive information regarding SCADA system design, operations, or security controls.

The American Water Works Association (AWWA, 2008) developed a roadmap that aims to provide a framework to address the full range of needs for mitigating cyber security risk of Industrial Control Systems (ICS) across the water sector. By implementing this roadmap, water sector industry leaders believe that within 10 years, ICS throughout the water sector will be able to operate with no loss of critical function in vital applications during and after a cyber event. This roadmap considers all variables for mitigating vulnerabilities and reducing the risk of industrial control systems in the water sector, including water and wastewater stakeholders and infrastructures; partnerships; critical functions and applications; near-, mid-, and long-term cyber security activities; and 10-year time frame.

References

ASCE. (2004). "Interim Voluntary Guidelines for Designing an Online Contaminant Monitoring System." Publication No. 817R08007, http://www.asce.org/files/pdf/wise/3.pdf

ASCE. (2006). "Guidelines for the Physical Security of Water Utilities." Publication No. 817R08009, http://www.asce.org/files/pdf/wise/4.pdf

AWWA. (2008). "Roadmap to Secure Control Systems in the Water Sector," http://www.awwa.org/files/GovtPublicAffairs/PDF/WaterSecurityRoadmap031908.pdf

Burrows, W. D., and Renner, S. E. (1999). "Biological Warfare Agents as Threats to Potable Water." *Environmental Health Perspectives*, Vol. 107, No. 12, pp. 975–984.

CBO. (1983). "Public Works Infrastructure: Policy Considerations for the 1980s." U.S. Congressional Budget Office.

Clark, R. M., and Deininger, R. A. (2000). "Protecting the Nation's Critical Infrastructure: The Vulnerability of U.S. Water Supply Systems." *Journal of Contingencies and Crisis Management*, Vol. 8, No. 2, pp.73–80.

Cristopher. G. W., Cieslak, T. J., Pavlin, J. A., and Eitzen, E. M. Jr. (1997). "Biological Warfare: A Historical Perspective." *JAMA*, Vol. 278, No. 5, pp. 412–417.

Deininger, R. A. (2000). "The Threat of Chemical and Biological Agents to Public Water Supply Systems." Science Applications International Corporation, McLean, VA.

DHS. (2007). "Water: Critical Infrastructure and Key Resources; Sector-Specific Plan as input to the National Infrastructure Protection Plan." Dept. of Homeland Security, http://www.dhs.gov/xlibrary/assets/nipp-ssp-water.pdf

E.O. (1996). "Executive Order 13010 – Critical Infrastructure Protection." *Federal Register*, Vol. 61, No. 138, pp. 37347–37350, July 17, 1996.

E.O. (2001). "Executive Order *13228* – Establishing the Office of Homeland Security and the Homeland Security Council." *Federal Register*, Vol. 66, No. 196, pp. 51812–51817, October 8, 2001.

EPA. (2003). "Response Protocol Toolbox: Planning for and Responding to Drinking Water Contamination Threats and Incidents," http://www.epa.gov/safewater/watersecurity/pubs/rptb_response_guidelines.pdf

EPA. (2005a). "The Water Security Research and Technical Support Action Plan, Progress Report for 2005." EPA National Homeland Security Research Center and Water Security Division, http://www.epa.gov/nhsrc/pubs/600r05104.pdf

EPA. (2005b). "EPA Needs to Determine What Barriers Prevent Water Systems from Securing Known Supervisory Control and Data Acquisition (SCADA) Vulnerabilities." Report No. 2005-P-00002, January 6, 2005.

EPA. (2006). "A Water Security Handbook: Planning for and Responding to Drinking Water Contamination Threats and Incidents," EPA 817-B-06-001.

EPA. (2007a). "Contamination Warning System Deployment." Office of Water, EPA 817-R-07-002, 2007, http://www.epa.gov/safewater/watersecurity/pubs/guide_watersecurity_securityinitiative _interimplanningpdf.pdf

EPA. (2007b). "Standardized Analytical Methods for Environmental Restoration following Homeland Security Events." EPA/600/R-07/136, http://www.epa.gov/NHSRC/pubs/600r07136.pdf

Federal Register. (2005). Title 42- Public Health, Part 72 – Interstate Shipment of Etiologic Agents. http://www.access.gpo.gov/nara/cfr/waisidx_06/42cfr72_06.html

Hart, D., McKenna, S., Klise, K., Cruz, V., and Wilson, M. (2007). "CANARY: A water quality event detection algorithm development tool." Proceedings of the 2007 *World Water and Environmental Resources Congress*, May 15–19, 2007, Tampa, Florida, sponsored by the Environmental and Water Resources Institute of the American Society of Civil Engineers.

Hawley, R. J., and Eitzen, E. M. (2000). "Bioterrorism and Biological Safety." Diane O. Fleming and Debra L. Hunt (eds.) "Biological Safety: Principles and Practices," 3rd edition. American Society for Microbiology (ASM) Press, Washington, DC, p. 784.

Hickman, D. (1999). "A Chemical and Biological Warfare Threat: USAF Water Systems at Risk". *The Counter proliferation Papers*, Future Warfare Series No. 3, http://www.au.af.mil/au/awc/awcgate/cpc-pubs/hickman.htm

Hoover, J. E. (1941). "Water Supply Facilities and National Defense." *Journal of the American Water Works Association*, Vol. 33, No. 11, pp. 1861–1865.

HSPD-7. (2003). "Homeland Security Presidential Directive: Critical Infrastructure Identification, Prioritization, and Protection," http://www.whitehouse.gov/news/releases/2003/12/20031217-5.html

Moteff, J., and Parfomak, P. (2004). "Critical Infrastructure and Key Assets: Definition and Identification." October 1, 2004, Resources, Science, and Industry Division, *Congressional Research Service*, RL32631.

Munro, N. B., Talmage, S. S., Griffin, G. D., Waters, L. C., Watson, A. P., King, J. F., and Hauschild, V. (1999). "The Source, Fate, and Toxicity of Chemical Warfare Agent Degradation Products." *Environmental Health Perspectives*, Vol. 107, No. 12, pp. 933–974.

Murray, R., Janke, R., and Uber, J. (2004). "The Threat Ensemble Vulnerability Assessment (TEVA) Program for Drinking Water Distribution System Security." Proceedings of the 2004 *World Water and Environmental Resources Congress*, June 27–July 1, 2004, Salt Lake City, Utah, sponsored by the Environmental and Water Resources Institute and the American Society of Civil Engineers.

Murray, R., Janke, R., Hart, W. E., Berry, W., Taxon, T., and Uber, J. (2008). "Sensor Network Design of Contamination Warning Systems: A Decision Framework." *Journal of the American Water Works Association*, Vol. 100, No. 711, pp. 97–109.

NAP. (2002). "Making the Nation Safer the Role of Science and Technology in Countering Terrorism." National Academy Press, Washington, DC.

NCPWI. (1983). "Fragile Foundations: A Report on America's Public Works." National Council on Public Works Improvement. Final Report to the President and Congress. Washington DC, February, 1988, p. 33.

NRC. (1995). "National Research Council, Guidelines for Chemical Warfare Agents in Military Drinking Water." PB95-267142.

NSHS. (2002). "The National Strategy for Homeland Security." U.S. Office of Homeland Security. July 16, 2002, http://www.whitehouse.gov/homeland/book/nat_strat_hls.pdf

NSPP. (2003). "The National Strategy for the Physical Protection of Critical Infrastructure and Key Assets." Office of the President. February, 2003. p. 71, http://www.whitehouse.gov/pcipb/physical_strategy.pdf

PDD-63. (1998). "The Clinton Administration's Policy on Critical Infrastructure Protection: Presidential Decision Directive No. 63." White Paper, May 22, 1998.

Sandia. (2002). "Risk Assessment Methodology for Water (RAM^{TM})

Skasden, J., Janke, R., Grayman, W., Samuels, W., TenBroek, M., Steglitz, B., and Bahl, S. (2008). "Distribution system monitoring for detecting contamination events and water quality changes." *Journal of the American Water Works Association*, Vol. 100, No. 7, pp. 81–94.

Stamp, J., Dillinger, J., Young, W., and DePoy, J. (2003). "Common Vulnerabilities in Critical Infrastructure Control Systems." (2nd edition, 22 May 2003; revised 11 November 2003)," Sandia National Laboratories, http://www.oe.netl.doe.gov/docs/prepare/vulnerabilities.pdf

Chapter 5
Wastewater Critical Infrastructure Security and Protection

Rakesh Bahadur and William B. Samuels

5.1 Introduction

Properly treated wastewater is critical to modern life. Wastewater utilities comply with regulatory requirements to prevent harmful pollutants from being released into the nation's waters. Significant damage to the nation's wastewater facilities or collection systems could result in loss of life, catastrophic environmental damage to rivers, lakes, and wetlands, contamination of drinking water supplies, long-term public health impacts, destruction of fish and shellfish production, and disruption to commerce, economy, and our nation's way of life.

Wastewater systems vary by size and other factors, but all include a collection system, pumping system, and treatment facility. Collection systems are geographically dispersed and have multiple access points, including drains, catch basins, and manholes. The majority of the access points in a collection system are not monitored. Wastewater treatment facilities use a series of physical, biological, and chemical processes to treat wastewater. Chemicals used in this process are typically stored on site. Wastewater systems have become increasingly computerized and rely on the use of automated controls to monitor and operate them.

The wastewater collection, treatment, and discharge facilities in the United States consist of more than 800,000 miles of conduit and are valued at more than $2 trillion. There are approximately 2.3 million miles of collection system pipes and approximately 16,255 publicly owned treatment works (POTWs) and 100,000 major pumping stations in the United States. Approximately 75% of the total US population is served by POTWs, while the remainder is served by decentralized or private septic systems. POTWs treat 32 billion gallons of wastewater every day. Though 80% of the systems treat less than 1 million gallons per day (MGD), these systems serve only 11% of the population (Fig. 5.1). In contrast, systems that treat more than 1 MGD provide wastewater treatment to 89% (http://cfpub.epa.gov/safewater/watersecurity/basicinformation.cfm) of the population served by POTWs. Only 3%

R. Bahadur (✉)
Science Applications International Corporation Center for Water Science and Engineering, McLean, VA 22102, USA
e-mail: bahadurr@saic.com

R.M. Clark et al. (eds.), *Handbook of Water and Wastewater Systems Protection*, Protecting Critical Infrastructure, DOI 10.1007/978-1-4614-0189-6_5,
© Springer Science+Business Media, LLC 2011

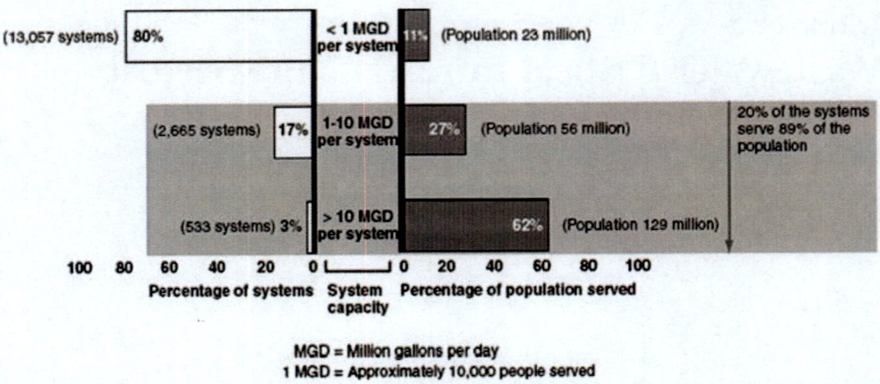

Fig. 5.1 POTW by system size and population served (GAO, 2005)

of the nation's wastewater systems (approximately 500 systems) provide service to 62% of the populations served by POTWs. Each of these systems treats more than 10 MGD of wastewater (GAO, 2005).

5.2 Wastewater Security Regulations

Wastewater systems have been determined to be at risk to accidental and deliberate contamination events, and their security has been made a top priority of the Environmental Protection Agency, EPA (US EPA, 2005). Authority for EPA's broadened responsibilities is derived from the President's National Strategy for Homeland Security (July 2002) and Homeland Security Presidential Directive-7 (HSPD-7). In 2004, the EPA developed a Water Security Research and Technical Support Action Plan (US EPA, 2004) which addresses potential threats to wastewater collection and treatment infrastructures as a part of the overall water infrastructure, and a Vulnerability Self Assessment Tool (VSAT^{TM}) was developed through a cooperative agreement with the Association of Metropolitan Sewerage Agencies (AMSA) (WERF, 2004a). A summary of laws pertaining to wastewater are listed:

- Since 1972, the Federal Water Pollution Control Act (known as the Clean Water Act) has established EPA as the primary regulatory authority over publicly owned and privately owned treatment works in the United States. EPA is ultimately responsible for implementation and achievement of the goals of the Clean Water Act "to restore the chemical, physical, and biological integrity of the Nation's waters." However, 46 States and the US Virgin Islands manage the day-to-day operations of their National Pollution Discharge Elimination System (NPDES) program under EPA approval. The Clean Water Act NPDES programs for the following States and territories are managed by the Environmental Protection Agency: Idaho, Massachusetts, New Hampshire, New Mexico, District of Columbia, American Samoa, Guam, Johnson Atoll, Midway/Wake Islands, Northern Marianas, and Puerto Rico.

- The Presidential Decision Directive No. 63 in 1998 has established EPA as the primary Federal agency responsible for security in the water sector.
- After 9/11, the HSPD-7 established EPA as the lead Federal agency to oversee the security of both drinking water and wastewater facilities.
- The Department of Homeland Security (DHS) established new chemical facility security standards in 2007. This act specifically exempted wastewater treatment works from DHS. The 2007 Act preserved EPA as the lead Federal agency responsible for the security of wastewater facilities. This statutory exemption from DHS regulation for wastewater treatment works expires on October 4, 2010. The "Wastewater Treatment Works Security Act of 2009" amends the Clean Water Act to enhance the security of operations at wastewater treatment works/sewage treatment facilities from intentional acts that may substantially disrupt the ability of the facility to safely and reliably operate or that may have a substantial adverse impact on critical infrastructure, public health or safety, or the environment. This title preserves EPA's historical regulatory oversight of sewage treatment facilities (http://www.govtrack.us/congress/billtext.xpd?bill=h111-2868).

5.3 Early Warning System

An Early Warning System (EWS) is an integrated system for monitoring, analyzing, interpreting, and communicating data, which can then be used to make decisions that are protective of public health and minimize unnecessary concern and inconvenience to the public. The goal of an early warning monitoring system is to reliably identify low-probability high-impact contamination events (chemical, biological, and radiological). It is essential to be able to accurately identify and measure in real time a wide range of chemical and biological agents, at levels much lower than toxic, in vapor and on surfaces, preferably from a distant position. It should also be able to predict the dispersal of chemical and biological agents in the environment following an attack (NSF, 2002).

5.3.1 Selection Criteria for an Early Warning System

An EWS should meet criteria listed below to become effective and reliable for wastewater systems (adapted from US EPA, 2005; Alai et al., 2005; WERF, 1994, 2009a):

- *Detection*
    - Provide an accurate, continuous, in situ, and rapid response.
    - Detect a sufficiently wide range of potential contaminants.
    - Permit minimal false positives/false negatives.

- *Sensitivity*
  - The sensitivity must exceed the baseline wastewater quality parameters.
  - Demonstrate sufficient sensitivity to detect contaminants.
- *Cost*
  - Inexpensive enough to be utilized in great numbers.
  - Cost-effective relative to off-line laboratory analysis.
  - Acquisition, maintenance, and upgrades at an affordable cost.
- *Maintenance and Operation*
  - Minimum maintenance interval (a longer interval is highly desirable).
  - Allow remote operation, adjustment, and third-party testing and verification.
  - Robust and rugged for continual operation in a wastewater environment.

As discussed below, the Water Environment Research Foundation (WERF, 2009a) developed a four-step hierarchical system of selection sensor classification for a deployable EWS. This tiered approach aids in the analysis and more easily conveys the current state of the technology.

- *Tier 4* sensors are all the sensors available for water sector without any distinction for its applicability; whether for wastewater, drinking water, or an air matrix.
- *Tier 3* sensors are subset of Tier 4 sensors that are suitable for wastewater only.
- *Tier 2* sensors are wastewater sensors which can measure some physical parameter (e.g., pH, ORP, conductivity, turbidity, TOC, temperature, etc.) in the wastewater stream and communicate with the SCADA system in a real time and can be deployed in the wastewater as an online instrument.
- *Tier 1* sensors are those that pass Tier 2 and determine a specific chemical, biological, or radiological agent in the wastewater stream.

5.3.2 Wastewater Contaminants of Concern Database

Unlike drinking water, the contaminants in the wastewater are not known. There are more than 85,000 chemicals commercially available in the United States with new chemicals and technologies introduced each year. The contaminants in wastewater constantly change and increase with the introduction of new chemicals. The specific chemical, biological, and radiological (CBR) contaminants that could potentially be found in the wastewater are enormous; thus a wide variety of contaminants can find their way into wastewater collection and treatment systems. There are multiple lists of potential contaminants (prepared by professionals, trade organizations, researchers, government researchers, and industry) which have many common entries but are not completely congruent (Table 5.1).

Table 5.1 List of CBR databases

Name of CBR database	Internet availability
Agency for Toxic Substances and Disease Registry (ATSDR)	www.atsdr.cdc.gov
Centers for Disease Control and Prevention	images/stories/public/2006dwlcdcct.pdf
CDC. "List of Select Biological Agents"	http://www.selectagents.gov/agentToxinList.htm
Registry of Toxic Effects of Chemical Substances	http://www.bt.cdc.gov/Agent/AgentlistChem.asp http://www.cdc.gov/niosh/npg/default.html
Congressional Research Service	http://www.nacwa.org/images/stories/public/2006dwlcrssmscale.pdf
EPA Drinking Water Treatability Database	http://iaspub.epa.gov/tdb/pages/general/home.do
US EPA's Maximum Contaminant Levels	http://www.epa.gov/safewater/mcl.html#mcls
Water Contaminant Information Tool (WCIT)	http://www.epa.gov/wcit/pdfs/fs_watersecurity_wcit-2007.pdf
ACToR Aggregated Computational Toxicology Resource	http://epa.gov/ncct/databases.html
Army, Navy, Air Force, and Commandant, Marine Corps Field Manual	http://www.vnh.org/
National Institute of Justice Database	http://www.ncjrs.org/pdffiles1/nij/189724.pdf
Query Database for individual agents	http://www.chrismanual.com
Target Compounds: USGS	http://toxics.usgs.gov/regional/contaminants.html
TOXicology Data NETwork Toxnet	http://www.nlm.nih.gov/pubs/factsheets/toxnetfshtml
Physician Preparedness for Acts of Water Terrorism	www.waterhealthconnection.org/bt/index.asp
US Coast Guard. 2001. "Chemical Hazards Response Information System"	Not available on Internet
WaterISAC/UKWIR: Secured web site for members only	www.waterisac.org
WERF 92-OPW-1: Online Monitoring to Control Transients in Wastewater Treatment. WERF 03-CTS-2S: Identify, Screen and Treat Contaminants to Ensure Wastewater Security	http://www.werf.org/
WHO Database	www.who.int/csr/delibepidemics/biochemguide/en/

5.3.3 Wastewater Sensors

An Early Warning System (EWS) includes sensors to detect the contaminant, systems to acquire and analyze data, links for communication and notification, and protocols for decision making and emergency response. There are a limited number of sensors that can be used to detect CBRs in wastewater. The rest are in various stages of research and development or are not suitable to be used in wastewater matrices. Currently, it is not possible to measure individual contaminants with one sensor. It is impractical to have a separate detection technology for each contaminant based on characteristics. Several of the commercially available sensor systems (Table 5.2) measure surrogate parameters (e.g., physical parameters such as

Table 5.2 List of wastewater sensor manufacturers

Company name	Measured parameters
Applied Spectrometry Associates – www.chemscan.com	Nutrients and chemicals
Analytical Systems International – www.asiwebpage.com	Oil in H_2O, VOCs, H_2S
Analytical Technology, Inc. – http://analyticaltechnology.com/cms/	Turbidity, pH, ORP, conductivity, and DO
Biochem Technology – Sentry C2	NH_4, NO_3
Challenge Environmental Systems, Inc. – http://challenge-sys.com	Anaerobic gas production
Data Rangers LLC – UniTech – www.tech-associates.com	pH, ORP, conductivity, turbidity, lead, Cl, TOC, temp., α-, β-, γ-rays, flow, and press
GE – www.ge.com	TC, IC, and TOC
HACH LANGE LTD – www.hach-lange.co.uk/	Multiple parameters
KELMA, Niel, Belgium – www.kelma.com/	NO_3, BOD in activated sludge
microLAN – TOXcontrol – www.microlan.nl/index.php	Photosynthetic activity of the algae
Myron L company – www.myronl.com/	Conductivity, TDS
N-CON Systems Co. – www.n-con.com/nconsys.html	Oxygen consumption
OAKTON Instruments – www.4oakton.com/toc.asp	Cyanide
Oxymax ER – www.respirometer.com	Changes in gas composition
REAL Tech Inc. – www.realtech.ca	UV 254 nm organic monitoring
Respirometry Plus, LLC. – www.respirometryplus.com	Influent toxicity and organic loadings
S-CAN – www.s-can.at/index.php?id=35	TS, COD, BOD, NO_3, NO_2, H_2S, NH_4, K, pH, ORP, conductivity, temperature, saxitoxin, benzene, and carbendazim
Siemens – Strantrol ORP analyzers – www.water.siemens.com/en/	ORP, PPM
SECOMAM – Model STAC – www.secomam.fr/index.php	COD, BOD, and TSS
Severn Trent Services	pH, ORP, DO, CL, ClO_2, bromine, O_3, conductivity, and temperature
SYSTEA Srl – MICROMAC C – www.systea.it/	Alkalinity, aluminum ammonia, boron, calcium, chloride, Cl, chromium 6+,
TriOS GmbH – www.trios.de/__industrial/uk/index.html	Nitrate, solids, organic carbon, TOC, DOC, COD, BOD, benzene
US Endress+Hauser – www.us.endress.com	Toxicity, DO in activated sludge
Wallace & Tiernan–Siemens, US – www.siemens.com/wallace-tiernan	Cl, SO_2, Cl, Res. ClO_2, $KMNO_4$, and turbidity
WALCHEM – www.walchem.com	pH, ORP, DO_3, conductivity, Cl, Res. ClO_2, and turbidity
Wedgewood Analytical, Inc. (on 1/1/08 became Endress + Hauser) – electrochemical analyzer	pH, ORP, conductivity, DO, Cl, ClO_2, UV, absorbance, cell growth, bubble detection, and color
WTW – www.wtw.com	pH, ORP, oxygen, turbidity/TSS, ammonium, nitrate, COD, temperature, NH_4, and NO_3
YSI – www.ysi.com	DO, conductivity, TDS, pH, ORP, turbidity, temperature, flow, and rhodamine

temperature, turbidity, conductivity, pH, and total organic carbon) rather than measuring a specific contaminant. By using surrogate parameters, the presence, identity, and concentrations of contaminants are inferred from measurements of other properties in the wastewater. In the case of surrogate measures, it is the change in the system that indicates a potential problem (ASCE, 2004). While the data from the surrogate measures may be reliable, the connection between the measured surrogate parameters and the identity and concentrations of a specific contaminant is not established.

5.3.4 Sensor Locations

Online sensors can provide an early warning of impending problems. Early warning allows operational staff an opportunity to prepare for a change in wastewater characteristics. In response to this information, controls can be designed to take corrective action (shutting the plant influent, diversion of pollutants to holding basins, and/or chemical treatment to neutralize contaminants) to minimize the impact of the contamination. Sensor placement is of concern to wastewater utilities as it involves planning and analysis and has costs associated with the purchase, maintenance, and operation of individual sensors. While many utilities are able to identify locations to help monitor daily water quality, it is more difficult to identify locations for monitoring wastewater security. Each treatment plant's NPDES permit specifies sampling locations. Therefore, the locations for grab and composite samples are known and fixed within the treatment plant. But the same is not true for online samples. The current practice appears to be locating them in pump station wet wells, manholes, and at key facilities, but no guidance or explanatory information was found on the optimal selection of these locations. Moreover, this practice is not based on expert guidance. American Society of Civil Engineers Interim Volunteer Guidelines base sensor location on the following two factors (ASCE, 2004).

5.3.4.1 Local Conditions

- Easy access to the instrument site
- Available space for the instruments and auxiliary equipment
- Suitability of candidate instruments or sample collection method for the sampling site
- Physical security of the instrument site
- Hydraulic conditions at sampling sites

5.3.4.2 System-Wide and Topological Factors

- Potential areas or entry points of contamination
- Likely contaminants and contaminant transport time and concentration
- Instrument accuracy and detection limits

- Vulnerable populations
- Frequency of sampling, i.e., periodic vs. continuous sampling

Currently, research is going on in the drinking water sector for sensor location, but very little research is being funded for wastewater (WERF, 2009a). Unlike the drinking water sector, in wastewater, there are no mathematical models or methodology for optimal location of sensors. As a rule of thumb, sensor location in a wastewater collection and treatment system should be early in the treatment process to allow time for mitigation steps. It also should be positioned to minimize fouling, downtime, and maintenance issues.

5.3.5 Anomaly Detection

Currently, a gap exists between the contaminants and what is being measured by the probes/sensors. This gap is filled by using Event Detection Algorithms (Kumar et al., 2007). There are a few contaminant-specific probes which can continuously monitor for constituents of concern. Even when contaminant-specific probes are used, the detection may not always yield a continuous concentration value. The current technologies are more suitable to support the detection (WERF, 2008) of the presence of a contaminant in the network via changes in commonly monitored water quality parameters, such as pH, conductance, and chlorine.

In an EWS, a large amount of data are generated and collected by the sensors and the probes. These data are compared with stored cases to determine the presence or absence of an appropriate match in the stored database. If there is no match, then the data are declared as an anomaly. Water sector security depends upon several factors including the capability of recognizing anomalous behavior in water quality parameters. The main challenge is to distinguish between the benign anomalies that arise in the operation of the system and potential contamination events. This requires a thorough understanding of these benign changes and what characteristics can be used to differentiate them from contaminant events. In order to maximize detection likelihoods, EWSs can incorporate multiple detection technologies. Currently, anomaly detection technologies or event detection systems for wastewater are in the very early stages of prototyping and development. One of the reasons for this is paucity of CBR-specific sensors or probes for wastewater. The detection of anomalous process behavior relies largely on computational methods with experience-based discriminatory ability. Various anomaly detection methodologies are described:

- *CANARY* – The goal of CANARY (Hart et al., 2007) was to take standard water quality data and use statistical and mathematical algorithms to identify the onset of periods of anomalous water quality, while at the same time, limiting the number of false alarms that occur. CANARY can be run on historical data to help set the configuration parameters in order to provide the desired balance between event detection sensitivity and false alarm rates. The CANARY program can test multiple event detection algorithms simultaneously on multiple data sets.

- *Artificial Intelligence (AI)/Artificial Neural Network (ANN)* – ANNs are mathematical modeling tools that are especially useful in the field of prediction and forecasting in complex settings. Historically, they were meant to operate through simulating the activity of the human brain. The ANN accomplishes this through a large number of highly interconnected processing elements (neurons), working in unison to solve specific problems, such as forecasting and pattern recognition. Each neuron is connected to certain of its neighbors with varying coefficients or weights that represent the relative influence of the different neuron inputs to other neurons. The different forms of AI are Genetic Algorithms, Ant Colony Optimization, Particle Swarm Optimization, Neural Networks, etc. There are many kinds of ANNs (Hamed et al., 2004) and they can be classified based on the following attributes: applications (classification, clustering, function approximation, and prediction); connection type (static (feedforward) and dynamic (feedback)); topology (single layer, multilayer, recurrent, and self-organized); learning methods (supervised and unsupervised). A typical artificial neural network (Chen et al., 2003) model generally consists of three independent layers: input, hidden, and output layers. Each layer is comprised of several processing neurons. While input (arriving signals) and output layers perform as a boundary between the neural network and the environment, the hidden layer and input/output layers may fully interconnect with each other through the information flow channels between the neurons.
- *Data Mining Techniques* – Utilities spend considerable resources collecting data of all kinds over long periods of time. This data is seldom used and analyzed to its full potential. Data mining techniques can be applied to identify the time points at which the changes occur, thus triggering the presence of potential contamination events. Data mining is the science of extracting valuable knowledge from large databases. Data mining is recognized as an iterative process in which the understanding derived from the reports of the data mining models is fed back into data cleaning, data enrichment, data selection, and data transformation. Data preparation includes statistical methods, signal processing, chaos, and machine learning. These methods are applied to maximize the information content in raw data and reduce the influence of poor or missing measurements. These methods include clustering, filtering, spectral decomposition, estimation of non-linear invariants and time delays, and synthesizing missing data (WERF, 2009a).
- *Decision Support Systems* – A decision support system (DSS) is both a process and a tool for solving problems that are too complex for manual solution, but too qualitative for computers. A DSS is capable of aggregating all competing objectives to identify the best optimal strategy. DSS is an amalgamation of functions such as monitoring, anomaly detection, machine learning, situational awareness, and remediation. A DSS collects, organizes, and processes information and then translates the results into management plans that are comprehensive and justifiable.

Though there are some similarities between the private sector and wastewater utilities, yet differences between them remain significant to the point that lesson learned from private sector cannot be directly transferred to the wastewater. These lessons need to be considered with caution. Reference business process models (SCOR, TOM, etc.) are not available for wastewater utilities (WERF, 2004c). WERF (2009b) developed and applied a DSS prototype for the prediction of contaminant events at wastewater treatment plants.

5.4 Security Incidents: Consequences and Response

CBR contamination can occur directly in wastewater collection or due to flushing operations in a drinking water distribution system. Currently, wastewater treatment utilities lack a systematic framework of tested response guidelines and corrective action protocols to be implemented during and after toxic shock events. There are emergency response plan guidances (WERF, 2004b) to help wastewater utilities but they do not discuss the upsets and response. There are three kinds of responses to an upset event as follows:

- Response to upset event – decontamination
- Response to upset even in wastewater treatment from CBR
- Response to upset event in wastewater – normal conditions

The *Wastewater Response Protocol Toolbox* (WWRPTB) addresses the preparedness and response needs for threats and contamination events in wastewater systems. These events can include contamination with toxicants as well as infectious, flammable, explosive, or radioactive substances (US EPA, 2008). The WWRPTB discusses the response to accidental and negligent contamination events in addition to its primary focus on intentional contamination. The WWRPTB is designed to be a planning tool. It is not intended to be a reference document for use during an actual emergency when decisions need to be made rapidly. Furthermore, the WWRPTB is not based on any statutory authority and, therefore, contains no mandatory requirements. It is merely provided as guidance to aid utilities in planning for contamination threats and events. The guidance document is composed of six interrelated modules:

- Wastewater Utility Planning Guide
- Contamination Threat Management Guide
- Site Characterization and Sampling Guide
- Analytical Guide
- Public Health and Environmental Impact Response Guide
- Remediation and Recovery Guide

5.4.1 Decontamination

Decontamination is an agent-specific task and involves both cleaning the infrastructure and the contaminated water. As shown, decontamination information is available from multiple sources:

- *US EPA* – The US EPA's Water Contaminant Information Tool (WCT) and National Environmental Methods Index for Chemical, Biological, and Radiological methods (NEMI-CBR) can be accessed under WCIT (https://cdx.epa.gov/SSL/cdx/login.asp)
- *Department for Environment, Food and Rural Affairs. UK* – Strategic national guidance: The decontamination of the open environment exposed to chemical, biological, radiological, or nuclear (CBRN) substances or material. Crown Publishing, March 2004 (images/stories/public/2006dwlcbrnsn.pdf)
- *US Department of the Army* – Edgewood Chemical and Biological Center, US Army Soldier and Biological Chemical Command. Guidelines for Responding to a Chemical Weapons Incident. Domestic Preparedness Program, Rev 1. Aug, 2003 (images/stories/public/2006dwguisechem.pdf)
- *US Department of Energy* – Interagency Steering Committee on Radiation Standards, ISCORS. Assessment of Radioactivity in Sewage Sludge: Recommendations on Management of Radioactive Materials in Sewage Sludge and Ash at Publicly Owned Treatment Works, ISCORS 2004-04, EPA 832-R-03-002B, DOE/EH-0668 (images/stories/public/2006dwliscors.pdf)
- *US Army Soldier and Biological Chemical Command (SBCCOM)* – Lake, William A., Fedele, Paul D., and Marshall, Stephen M., Guidelines for Mass Casualty Decontamination During a Terrorist Chemical Agent Incident. US Army Soldier and Biological Chemical Command (SBCCOM) publication, January 2000 (images/stories/public/2006dwlgmcas.pdf)
- *Naval Facilities Engineering Service Center and US Army ERDC* – Potable Water CBR Contamination and Countermeasures (images/stories/public/2006dwlpotwcbr.pdf)
- *Water UK* – Protocol for the Disposal of Contaminated Water. v 2.1, September 2003 (images/stories/public/2006dwlwateruk.pdf)

5.4.2 Response to Wastewater Treatment from CBR

Decontaminated water, if not contained and treated properly, can move into wastewater treatment facilities through the collection system. It is important to minimize the impact of CBR substances on the sewer system, treatment processes, biosolids, air emissions, facility personnel, and the environment. Currently, not much information is available for response to wastewater treatment from CBR. Water Contaminant Information Tool (US EPA, 2007) contains information for select contaminants that can be used for treating CBR in a wastewater treatment facility. The National

Environmental Methods Index for Chemical, Biological, and Radiological (NEMI-CBR) can be accessed under WCIT and provides a mechanism to compare and contrast the performance, speed, and relative cost of analytical methods for response to both intentional (i.e., terrorist attacks) and accidental (i.e., spills) contamination events from CBR agents. The CBR Advisor, an expert system companion to NEMI-CBR, provides information and advice for which method to select (or not to select) based on the situation at hand. National Association of Clean Water Agencies (NACWA, 2005) also developed general guidelines for mitigating the impacts of wastewater contaminated with CBR.

5.4.3 Response to Upset Event in Wastewater – Normal Conditions

Wastewater treatment operations can encounter upsets attributed to the presence of inhibitory compounds and slug loading conditions that occur without warning. These upsets can cause nuisances such as poor biosolids settleability or, in the worst case, compromise treatment, leading to effluent that does not meet quality targets. Biological wastewater treatment systems are susceptible to toxic shock loads of industrial chemicals, which can adversely affect the efficiency of the treatment process. Studies have shown that chemical toxins can negatively impact the essential processes within an activated sludge system (Henriques et al., 2007). The cause and effects for upsets is usually unknown and cannot be linked to a single influent constituent, industrial event, or operating event. WERF (2005) measured water quality assays and parameters to determine whether a treatment process effect has occurred. WERF (2005) performed a comprehensive source–effect relationship study for activated sludge exposed to shock loads of six chemical toxins. This study covered six different classes of industrially relevant chemicals as sources and the effects of varying shock concentrations of those toxins on activated sludge COD removal ability, flocculation ability, biomass growth, respiration rates, settleability, and dewaterability.

5.5 EPA-WERF Wastewater Security Research

A cooperative agreement between WERF and EPA commenced in October 2002 and extended to September 2004 with supplemental funding, for a total of $3.6 million. The overarching objective of this grant-funded research is to protect public health and the nation's wastewater infrastructure from multiple hazards (natural and human-induced disasters). This cooperative agreement is vital to ensuring the continued safety of our facilities and the communities they serve. In wastewater systems the three issues of the greatest concern are the deliberate use of chemical, biological, or radiological (CBR) weapons by terrorist organizations, including the intentional introduction of CBR contaminants into wastewater collection systems; physical attack on wastewater infrastructure; and attack on the SCADA system of wastewater utilities. The cooperative agreement enabled the development of an array of robust security products and tool kits such as emergency communication protocols, guidance documents, fact sheets, technical libraries and databases, expert software systems, and GIS-based simulation models for contaminant assessments in wastewater collection and treatment systems. In total, 12 projects (see Table 5.3) were funded by this grant and covered the following broad thematic areas of security research as follows:

- Emergency Response Plan (ERP) and Contingency Planning
- C/B/R (Chemical/Biological/Radiological) Contamination Events
- Wastewater Sensors and Early Warning Systems
- Cyber Security (Process Control Systems)
- Risk Communications

Table 5.3 WERF security projects

Project number/key deliverable	Project title
03-CTS-1S Key Deliverable: Workshop Document	Securing Wastewater Infrastructure and Protecting Public Health; Mar 03–Dec 03
03-CTS-2S Key Deliverable: Priority List of CBR contaminants of concern	Identify, Screen, and Treat Contaminants to Ensure Wastewater Security; May 04–Aug 09
03-CTS-3SCO Key Deliverable: Control Systems Cyber Security Self Assessment Tool "CS2SAT"	Security Measures for Computerized and Automated Systems at Water and Wastewater Facilities; Dec 04–Jun 09
03-CTS-4S Key Deliverable: Guidance Document	Emergency Response Plan Guidance for Wastewater Systems; Apr 04–Apr 05
03-CTS-5SCO Key Deliverable: Guidance Document	Emergency Communications with Your Local Government and Community; Aug 04–Jun 08
03-CTS-6S Key Deliverable: VSAT^{TM} tool	Software to Prioritize Wastewater Asset Failure and Security Risks; Jun 04–Sep 05
03-CTS-7S Key Deliverable: Expert System Tool	Feasibility Testing of Expert Software Support Systems to Prevent Treatment Upsets; Jun 04–Feb 09
04-CTS-9S Key Deliverable: Scoping Study/Guidance Document	Chemical, Biological, and Radiological Sensors for Early Warning Systems in Wastewater Utilities; Feb 08–Sep 09
04-CTS-9SW Key Deliverable: Scoping Study	Scoping Study for -04-CTS-9S; Apr 05–May 07
04-CTS-10S Key Deliverable: Simulation Tool – SewerNet	Integrated, GIS-based Consequence Assessment Model for Sewer and Stormwater; Nov 05–Sep 09
04-CTS-11S Key Deliverable: Guidance Document	Detailed Protocols for Treatment Process, Standard Response, and Collection System Disruptions; Feb 06–Aug 09
04-CTS-12S Key Deliverable: Guidance Document	Security Strategies for Small Wastewater Systems for Safeguarding Assets, Emergency Response Planning and Risk Communications; Mar 06–Jun 09

5.6 Summary

The wastewater infrastructure of the United States is vulnerable to accidental impacts as well as purposeful attack. Wastewater utilities vary a lot in their configuration and operation. This implies a need for extensive adaptation of any general guidance provided to the specific circumstances of the utility in question. Utilities may want to establish an EWS for the protection of infrastructure and worker safety. Proper location of sensors also helps manage the complex risks that are related to contamination. Sensor selection for wastewater is a complex problem because much of the basic scientific and engineering knowledge and the instrumentation needed are not yet available. The current technology for the detection of CBR contaminants in wastewater is insufficiently developed to provide a comprehensive and secure EWS to protect wastewater collection and treatment systems from accidental or deliberate contamination. Therefore, a limited number of sensors can be used to detect CBRs in wastewater. The rest are in various stages of research and development or are not suitable for wastewater matrix use.

Several of the current commercially available sensor systems measure surrogate parameters (e.g., physical parameters such as temperature, turbidity, conductivity, pH, and total organic carbon) rather than measuring a specific contaminant. By using surrogate parameters, the presence, identity, and concentrations of contaminants are inferred from measurements of other properties in the wastewater. While the data from the surrogate measures may be reliable and accurate, the connection between the measured surrogate parameters and the identity and concentrations of a specific contaminant is not established. Currently, it is not possible to measure individual contaminants with one sensor. It is impractical to have a separate detection technology for each contaminant based on characteristics.

The Internet, communications, information technologies, and nanotechnologies are revolutionizing sensor technology. Emerging technologies can decrease the size, weight, and cost of sensors and sensor arrays by orders of magnitude and can increase their spatial and temporal resolution and accuracy. Communications networks provide rapid access to information and computing, eliminating the barriers of distance and time for detecting toxic agents. Currently, EWS development for wastewater is at an early stage of development but this is changing rapidly. It is envisioned that in the coming years, a wastewater treatment EWS will be able to perform the following:

- Move beyond sensor technologies that merely warn of system contamination.
- Include sensors capable of accurate, continuous, in situ measurement of a specific CBR agent without the need for sample transport, sample conditioning, or reagent addition.

References

Alai, M., Glascoe, L., Love, A., and Johnson, M. (2005). *Sensor Acquisition for Water Utilities: A Survey and Technology List*, 2005. https://e-reports-ext.llnl.gov/pdf/317385.pdf.

ASCE. (2004). *Interim Voluntary Guidelines for Designing an Online Contaminant Monitoring System*, 2004. http://www.asce.org/static/1/wise.cfm.

Chen, J. C., Chang, N. B., and Shieh, W. K. (2003). *Assessing wastewater reclamation potential by neural network model*. Engineering Applications of Artificial Intelligence. Special Issue on Applications of Artificial Intelligence for Management and Control of Pollution Minimization and Mitigation Processes, Volume 16, Issue 2, March 2003, Pages 149–157.

GAO. (2005). *Wastewater facilities: Experts' Views on How Federal Funds Should Be Spent to Improve Security*. www.gao.gov/cgi-bin/getrpt?GAO-05-165.

Hamed, M. M., Mona G. Khalafallah, and Ezzat A. Hassanien. (2004). *Prediction of wastewater treatment plant performance using artificial neural networks*. Environmental Modelling & Software, Volume 19, Issue 10, Pages 919–928.

Hart, D., McKenna Sean, A., Klise, K., Criz, V., and Wilson, M. (2007). *CANARY: A Water Quality Event Detection Algorithm Development Tool*. World Environmental and Water Resources Congress 2007: Restoring Our Natural Habitat, 2007.

Henriques, I. D. S., Kelly, R. T. II, Dauphinais, J. L., and Love, N. G. (2007). *Activated sludge inhibition by chemical stressors–a comprehensive study*. Water Environment Research, Volume 79, Issue 9, Pages 940–951.

Kumar, J., Zechman, E. M, Brill, E. D., Mahinthakumar, G., Ranjithan, S., and Uber, J. (2007). *Evaluation of Non-Uniqueness in Contaminant Source Characterization Based on Sensors with Event Detection Methods*. World Environmental and Water Resources Congress 2007: Restoring Our Natural Habitat, 2007.

NACWA. (2005). *Vulnerability Self Assessment Tool*^{TM} *for Water & Wastewater Utilities* (Version 3.2 Update). February 2005. http://www.nacwa.org/pugs/index.cfm.

NSF. (2002). *The New Challenges of Chemical and Biological Sensing*; National Science Foundation Workshop, January 9–10, 2002, Arlington, VA.

US EPA. (2004). *The Water Security Research and Technical Support Action Plan*, EPA/600/R-04/063 March 2004. http://www.epa.gov/safewater/watersecurity/pubs/action_plan_final.pdf.

US EPA. (2005). *The Water Security Research and Technical Support Action Plan, Progress Report for 2005*, published by EPA National Homeland Security Research Center and Water Security Division. http://www.epa.gov/nhsrc/pubs/600r05104.pdf.

US EPA. (2007). Water Contaminant Information Tool (WCIT); 2007. http://www.epa.gov/wcit/pdfs/fs_watersecurity_wcit-2007.pdf.

US EPA. (2008). *Wastewater Response Protocol Toolbox: Planning for and Responding to Wastewater Contamination Threats and Incidents*. Interim Final, 2008.

WERF. (1994). *Online Monitoring to Control Transients in Wastewater Treatment – Sensor Technology*, Project No. 92-OPW-1, 1994.

WERF. (2004a). *Enhancing Security in the Wastewater Sector: A Prioritized Research Agenda, Security Symposium Proceedings*, August 7–8, 2003, published by Water Environment Research Foundation January, 2004.

WERF. (2004b). *Emergency Response Plan Guidance for Wastewater Systems*, Project No. 03-CTS-4S, Water Environment Research Foundation, VA, 2004.

WERF. (2004c). *Decision Support Systems for Wastewater Facilities Management*, Project No. 00-CTS-7, Water Environment Research Foundation, VA, 2004.

WERF. (2005). *Upset Early Warning Systems for Biological Treatment Processes: Source and Effect Relationships,* Stock No. 1CTS2a; Water Environment Research Foundation, Alexandria, VA, 2005.

WERF. (2008). *Strategy, Guidance, and Decision Support Systems For Deployment and Development of Upset Early Warning Sensor Systems for Wastewater Collection And Treatment Operations*. WERF 04-CTS-9S, 2008.

WERF. (2009a). *Chemical, Biological and Radiological Sensors for Early Warning Systems in Wastewater Utilities*, Project No. 04-CTS-9S.

WERF. (2009b). *Feasibility Testing of Support Systems to Prevent Upsets*. Project No. 03-CTS-7S.

Chapter 6
Protecting Water and Wastewater Systems

Randy G. Fischer

6.1 Introduction

After the events of September 11, 2001, public water systems began to realize their vulnerability to disaster. Prior to September 11, public water systems were not thought of as a likely terrorist target.

In 2003, the Environmental Protection Agency, because of the President's signing of Public Law 107–188 on June 12, 2002 (Public Health Security and Bioterrorism Preparedness and Response Act 2002), gave awards to states to take measures to assure that water systems within their borders prepare for a terrorist attack or other emergency. In Nebraska, the new Drinking Water Security Program set the following goals: (1) to encourage public water systems to secure their facilities to the greatest extent possible, (2) to train public water system personnel to develop an effective emergency response plan, (3) to develop a sense of cooperation and teamwork among all emergency responders that ensures effective action in the wake of a disaster, and (4) to meet with and produce a video for law enforcement personnel to educate them in the particulars of crime scene evidence related to public water systems.

6.2 Nebraska's Response to PL 107–188

To comply with PL 107–188, the Nebraska Department of Health and Human Services (NE DHHS) developed an Emergency Response Plan outlining the actions that the Department and its employees would take in response to any type of water emergency. For the most part, Department response was defined as an advisory role. However, if needed, on-site assistance and response could be given.

R.G. Fischer (✉)
Division of Public Health, Nebraska Department of Health and Human Services (NE DHHS), Lincoln, NE 68509-5026, USA
e-mail: randy.fischer@nebraska.gov

R.M. Clark et al. (eds.), *Handbook of Water and Wastewater Systems Protection*, Protecting Critical Infrastructure, DOI 10.1007/978-1-4614-0189-6_6,
© Springer Science+Business Media, LLC 2011

Probably the best example of on-site response is when Department staff responded to flooding that had adversely impacted the wells of a water system. Water samples were taken and activities were coordinated by the Department staff because the water system had limited experience responding to an experience and positive action was needed to correct the problems in a timely manner. The Department provides assistance for various water emergencies including floods, tornadoes, earthquakes, fires, chemical spills, and unspecified system contamination events.

6.2.1 Vulnerability Assessments

PL 107–188 required community water systems with a population of greater than 3,300 to conduct a vulnerability assessment of their systems and submit a copy of that assessment to the USEPA. In Nebraska, approximately 40-plus systems had to meet this requirement. Systems had to identify what might be vulnerable and how a terrorist could impact their systems.

While public water systems across the state with a population of less than 3,301 did not have to do this USEPA-required vulnerability assessment, the Nebraska Health and Human Services System (now the Nebraska Department of Health and Human Services, or DHHS) decided that all community water systems regardless of size should be encouraged to do a vulnerability assessment as well. Unlike large systems, small systems could not be compelled to do so.

Vulnerability assessments are a detailed inventory of all physical water system facilities, including personnel and all water system controls. The assessments were to determine the vulnerability of the facilities/personnel/controls to terrorist threats and other potential threats to the water system, including both man-made and natural disasters.

In planning, the vulnerability of a water system is the likelihood or degree to which the service of an adequate supply of safe water would be adversely affected by a disaster. An effective vulnerability assessment will provide systems with a view of potential events that could disrupt normal operations.

The vulnerability of a water system involves more than structural consideration. It is also concerned with supplies and equipment required for emergency operations and to expedite recovery efforts and the availability of trained, experienced personnel. Vulnerability studies will point out the various operations which may be affected from which general estimates of outage time, estimated cost of repairs, and risk to public health can be determined.

This assessment of vulnerability provides the procedural framework for local governments to request needed emergency assistance. Requests should be timely in order to shorten repair delay and to reduce possible effects to public health.

In addition to pointing out the direct response needs of a water system, a comprehensive vulnerability assessment can be used to establish priorities for disaster mitigation, preparation, and recovery, the other phases of emergency planning. As a result and with the cooperation of all concerned parties, definite facility goals can be formulated using this tool. An assessment should evaluate the following components of a water system: source of drinking water, including wells and intake structures; treatment facilities; transmission; storage; distribution; structures; power; communication; equipment; materials and supplies; manpower; emergency procedures; and security. When accomplishing an assessment of a public water system, the components noted above should be reviewed to determine vulnerability to either natural (tornado, windstorm, snow or ice storm, flood, earthquake, drought, or loss of water sources) or man-made (fire, riot or civil disorder, contamination, failure of equipment, nuclear attack or accident, terrorism, and sabotage) hazards to public water supply systems.

"Failure of Equipment" was included because many of the smaller systems in Nebraska do not typically have auxiliary or standby equipment readily available in the event of failure of any of the system components. The impact on the water system due to equipment failure can be the same as if a disaster in the usual sense has occurred.

The primary components of a water system are source, treatment, storage, distribution, and power. Additional components that are not a physical part of the system, but are necessary to its operation are personnel, transportation, and communication. Failure or impairment of any of these components has a serious effect on the ability of the system to deliver safe drinking water to the consumer.

6.2.2 Progress in Completing Vulnerability Assessments

As of today (42), all systems required by PL 107–188 in Nebraska to do a vulnerability assessment had done one. Fifteen percent of systems were not by law required to do one but voluntarily completed one (http://cfpub.epa.gov/safewater/watersecurity/index.cfm).

Training on how to conduct a vulnerability assessment was essential to meeting the requirements. For training, Nebraska's response to PL 107–188 was to work with their 2% Technical Assistance Team. This team was developed as Nebraska's response to the capacity development requirements that were mandated in USEPA's 1996 reauthorization of the Federal Safe Drinking Water Act.

The 2% Technical Assistance Team works under contract for the Department. These contracts are paid for by funds from the 2% set aside from the Drinking Water State Revolving Fund, which was also made possible by the referenced reauthorization in 1996.

The 2% Technical Assistance Team is made up of the Midwest Assistance Program, Nebraska Rural Water Association, the League of Nebraska Municipalities, and the Nebraska American Water Works Association. These organizations provided, and continue to provide, training and technical assistance to water operators and public water systems. This assistance can range from emergency planning and conducting vulnerability assessments to educating new operators as to the day-to-day operations of a public water system and regulatory requirements.

In 2003 and 2004, the Department relied on the 2% Technical Assistance Team to provide the training that was mandated under PL 107–188. Funding was, and continues to be, provided for this purpose by the federal government in the form of security grants to the Department through USEPA. In August 2004, the Department provided training to water operators on how to effectively complete emergency response plans that cover the unique needs of each system.

After the signing of PL 107–188 and subsequent federal actions, the Department determined that a full-time person was needed to fulfill the state's obligations to provide necessary assistance, planning, and training to the state's public water systems. In 2004, the Department hired the author as the water system security coordinator to conduct training and to continue with ongoing emergency planning. With 20 years of experience in law enforcement, 10 years of experience with fire and rescue, and several years of experience operating a public water system, it was expected that this background would prepare him to do the planning, training, and technical assistance necessary to meet the requirements of PL 107–188.

To prepare, the author attended the terrorism and emergency preparedness courses, as well as a railroad tank car training. Many courses are put on across the nation on emergency planning for public water systems. The training is general in scope, meaning that they covered such things as securing a facility, determining contamination issues, and a variety of possible disaster scenarios. Online courses to improve skills in Developing and Managing Volunteers, Leadership and Influence, and the National Incident Management System (NIMS) introduction are available.

The author spent time establishing a working relationship with various groups such as the county health emergency response coordinators, county emergency management coordinators, and law enforcement officials. While water operators were the primary target population for training, the program could not be effective without the participation of county emergency management coordinators, county health emergency response coordinators, county/city board members, law enforcement personnel, and state's drinking water program field personnel.

Public water systems operating within the State of Nebraska are charged with the responsibility of providing a continuous supply of safe drinking water to all customers situated within their service areas. Many community water systems have the additional responsibility of furnishing water in sufficient quantities and at proper pressures to assist in the fire defense of real property located within their sphere of operation. Because of the generally excellent record of past performance by public water systems, Nebraskans have come to expect their supply of water to be furnished on an uninterrupted basis.

Each separate system is faced with potential disruptions in service due to recurring and normal operational problems including water main breaks or leaks; valve failures; equipment breakdowns; temporary losses of electric power; and fluctuations in either quantity or quality of the water supply. Factors such as system size, preparation, and frequency of system disruptions dictate the water system's perceived urgency of such occurrences. These perceptions range from everyday operation and maintenance problems to an emergency condition.

Both natural and man-made disasters may create emergency conditions that disrupt water service or damage system components. Natural disasters typically include events such as floods, high winds and tornadoes, earthquakes, extended periods of weather extremes, freezing, or drought. Man-made causes include inadvertent or deliberate contamination of water supplies, vandalism, sabotage, war, or civil disorder. Finally, labor walkouts and strikes involving water works or allied services including other utilities, industries, and communications or transportation services can create emergency conditions.

6.2.3 Response to Emergencies

Emergency situations demand prompt and appropriate response on the part of each public water system in order to quickly and efficiently reduce the negative impact of the particular circumstance on the populace served. The first and immediate responsibility to react to an emergency rests with the local public water system involved; however, the Governor of Nebraska may be requested to issue a state of emergency proclamation if the emergency exceeds the local government's capability to adequately respond.

Adequate pre-emergency planning might well shorten the term of water system outage for a given emergency. It may well maximize service to all classes of users and provide the highest degree of protection of safety and health to the residents, businesses, and the general public.

6.3 Nebraska's Public Water System Emergency Response Plan

The Department's Public Water System Emergency Response Plan establishes an organizational structure and procedural guidelines to be utilized by the Department in response to typical water supply and water quality emergencies. This plan also recommends the components which should be considered by each public water system in the development of their own local emergency plan. Finally, this plan is intended as a supplement to, and to coordinate with, the state's Emergency Operations Plan.

Title 179 NAC 22-004 item 6 requires Emergency Response Plans for community and non-transient non-community water systems. "Maintain an emergency plan of operations for safeguarding the water supply, protecting the drinking water, and, if necessary, providing for an alternate drinking water supply in the event of natural or manmade disasters." The plan must include a list of individuals who may be called for help in times of disaster, their titles, and their phone numbers. This list must be updated annually with a copy provided to the Department. The plan must state the basic domestic water needs and usage under normal conditions. Any special institutional, commercial, or industrial users must be shown. Any special backup or standby equipment or auxiliary power supply must be included as well as alternate sources of supply or bottled water sources. All available chemicals and equipment for the purpose of disinfection must be listed. The emergency plan must outline all emergency operations and must be updated at least every 3 years with copies provided to the Department for inclusion in the state Drinking Water Emergency Plan.

The emergency plan must be placed at key locations, clearly marked and readily accessible to utility personnel.

Public water system means a system for providing the public with water for human consumption, if such system has at least 15 service connections or regularly serves an average of at least 25 individuals daily at least 60 days per year. Community water system means a public water system that (a) serves at least 15 service connections used by year-round residents of the area served by the system or (b) regularly serves at least 25 year-round residents. Non-transient, non-community water system means a public water system that is not a community water system and that regularly serves at least 25 of the same individuals over 6 months per year.

The Department's Public Water System Emergency Response Plan contains guidance and actions in response to public water system emergencies and is intended to assure the provision of safe drinking water to the public during emergencies. The plan is statewide in scope and is applicable to all types of disasters which may affect a public water system. This plan was developed as a guide for public water systems and is not designed to directly address private supplies. Private water supply emergencies are addressed elsewhere in the Department. All individuals within the state, however, may be addressed by this plan through local governmental response activities to disaster victims via mass rescue, shelter, food, and water services supported and coordinated herein.

The plan outlines responsibilities and provides for coordination of Department responses to local water system emergencies together with support from other services and agencies. The plan will supplement local emergency plans, if required, and provide guidance in the development of local emergency safe drinking water plans. This plan also addresses response during regional, national, and statewide emergencies.

6.3.1 Implementing the Plan

The author's first task was to revise the Department's Emergency Response Plan (ERP). The plan needed to be revised to inform public water systems on exactly how the state would respond to a water emergency.

After all management-level personnel agreed to the plan, it was put in place and training started. The Department conducted training across the state with its partners on the plan itself. The partners were public health, EPA, and Nebraska Emergency Management Agency and the target audiences were water operators, local health officials, and County Emergency Management personnel.

The next task of the author was to create an Emergency Plan of Operations template for systems to use. The template was required to include a contact list for emergencies and lay out what actions a system might take to recover from an incident, such as any man-made or natural disaster. After the template was completed, an internal review was conducted to make sure the template contained all items that systems needed to complete their plan. The evaluation asked the following questions: Did the plan contain all information required by Title 179? Did it cover most, if not all, types of potential emergencies that may overwhelm a water system?

After the development of the template, the author traveled across Nebraska, holding seminars, meeting water operators, and sharing the template with them. Local health departments, law enforcement officials, and emergency management planners also participated in these training sessions. The importance of partnerships was stressed: that they need to know each other, they need to understand who is responsible for various activities during an emergency, and they need to work together to be effective.

The regulations require an emergency plan to be detailed, to include items such as how the water system plans to provide drinking water to its customers if something should happen to the system, where can repair parts be obtained, how the system will be disinfected if needed, and many other varied items and necessary components. The goal was to develop a plan that met the needs of the drinking water industry. This plan had to lay out actions or steps that public water systems needed to take. This plan needed to be more than just a phone list of people to call to assist. The plan had to contain action steps that the public water system was going to take immediately to recover from emergency event(s) such as floods, ice storms, tornadoes, or man-made disasters. Systems needed to understand that all incidents are a local incident first. Assistance might come later, but the action of recovery needs to start immediately.

6.3.1.1 Training Activities

Training is the first part of developing effective emergency response plans. While the major thrust of training was over by June 30, 2005, ongoing training and education is necessary to foster continuing cooperation and teamwork and to teach new concepts as they emerge. The Department conducted training sessions for public water system personnel and other emergency responders regarding emergency plan development across the state in a joint effort with the USEPA and the Nebraska Emergency Management Agency.

By the end of the first quarter of 2005, a total of 767 people had been trained. The majority of them were water operators, but there were 17 county emergency management coordinators, 16 county health emergency responders, 2 law enforcement personnel, and many others that were not categorized. Participant evaluation forms indicate an overwhelming positive response. Between August and December of 2004, only 4% of the water operators indicated they would *not* use the Department's template to update their emergency plan.

A testament to the effectiveness of the Water System Security Program is the number of public water systems represented at the training sessions. Over 80% of community and non-transient non-community public water systems participated in this training between August 2004 and March 2005.

Nebraska's approach to the Water System Security Program includes training that might be considered outside-the-box, placing a great emphasis on developing relationships among emergency responders and getting out across the state to meet and train emergency responders. Other states have developed templates for water system emergency response plans, as the author has, but they have not been as involved in developing relationships and organizing and training emergency responders.

The wide variety of personnel attending these sessions indicates that the Department's goal of encouraging a sense of cooperation and teamwork among all persons involved in responding to an emergency is being accomplished. The Department sees this as an indication of effectiveness and a positive sign that all persons who are involved in responding to an emergency want to learn how to be as effective as possible.

One of the key things in Nebraska is how everyone works together. Nebraska is primarily a rural state, with a population of 1.78 million. Seventy percent of the state's 93 counties have less than 10,000 in population. Small communities lack many resources and have few funds to do projects to make their systems more secure. The role of a water system security coordinator for the state is to make sure that the state could provide water systems with the understanding of what was needed and what could be done at a low cost to be able to secure even the smallest system.

There was a need to build partnerships and to bring water operators into the picture of emergency planning. Water personnel have never been thought about in the law enforcement arena or worked closely with local health departments. Developing relationships with the key players in the event a real water emergency happens. In all the training provided, building relationships was stressed. Building relationships with all local emergency response personnel today would save valuable time when faced with an emergency. Communication and getting to know other players and how they would react to a water emergency is critical to bringing the system back online as soon as possible. Water operators also need to learn what other partners might bring to the table to assist in the event of a water emergency. Putting a face with a name in today's world is essential. It not only provides understanding about whom the other person(s) is but also gives both parties the understanding of what each other's role is or will be in an emergency response situation.

Training across all areas needs to continue and build. Partnerships are not developed overnight or by meeting a person one time. This needs to continue on at least an annual basis. Community planning or community partnerships are like a good marriage. For a marriage to survive, communication needs to be of the utmost importance. The same goes for emergency planning partnerships. It needs to continue to grow; understanding who, what, and how a person reacts to a water emergency is important. Remembering people's names and faces is key to emergency planning. The door to communicating needs to remain open, and there must be a continuity of flow of information. A key component to this continuity is the ongoing understanding and dissemination of new developments that are occurring in water system emergency planning. Just as things change in society, so it is in the water industry. New ideas and new partnerships develop.

After doing all of the training on plan development, systems needed to test their plans. Many tabletop exercises were conducted across the state on how systems would react to different scenarios. It gave the systems and their partners (local health departments, police, fire and rescue, etc.) a chance to make changes to their plan if needed, after the scenario was completed. It gave all participants a chance to see what their roles might be in a real event dealing with a water emergency.

An example of a tabletop exercise is one dealing with the derailment of a car carrying a toxic substance, creating a major emergency. Railroads in Nebraska pass through many communities. If a derailment occurred, knowledge was needed regarding the hazards of the product(s) that may have been spilled. In rural Nebraska, many water wells are located in close proximity to rail lines.

Once emergency plans of operation are prepared, more training, in the form of tabletop exercises related to disaster preparedness, was offered. These exercises help water operators determine if they need to revise their plans. They also allow the participants to determine the responsibilities of disaster responders and the situations in which each is to be contacted. These tabletop exercises brought together the various players and allowed them to interact and learn about each other's roles.

Law enforcement personnel expressed an interest in having a video produced relating to public water system disasters for training purposes. In October 2005, the Department produced a video that was shot at the Waverly, Nebraska, water plant. The video was produced using employees from the Waverly Water Department and the Lancaster County Sheriff's Department. The video includes a scenario where the plant was broken into and the role of the water operator and of law enforcement is focused on. Law enforcement entities in Nebraska thought this scenario was not possible.

Partnerships with law enforcement personnel are critical and need to be developed. After the video was completed, it was sent to all 93 County Sheriffs in Nebraska and the 200-plus police departments across the state.

6.4 Potential for Replication

The training program that Nebraska offered has great potential for replication. One of the potential problems in replicating this project could be coordinating all the different disaster responders and getting them to think in terms of a team. Turf wars are not easily overcome. Many water operators are independent and believe in doing their own work. They may think they can handle anything. Pointing out some of the devastation that can occur with a tornado or hurricane may help them to get a different perspective. Natural and man-made disasters require very similar if not identical preparations and/or responses.

The training part would not have to be replicated in exactly the same way. Each state would need to determine what works best for them. Teleconferencing holds great potential for reaching a lot of people at the same time. Other groups may also be trained to go out and do some of the training. *Benefits: flexibility, adaptability, etc.*

The choice of resource allocation might be another limitation other states may find. Travel is labor, cost, and time intensive; however, having one's presence known throughout the state and indicating a willingness to work with everyone makes the program more approachable.

Nebraska found county health departments and emergency management coordinators very willing to cooperate and be an integral part of the training. Nebraska is largely rural and law enforcement personnel in rural areas need to be aware of what types of emergencies or bioterroristic acts they might have to address one day.

For national implementation, each state would need to make a commitment to work with emergency responders to make their program effective. The program in Nebraska will continue indefinitely, as long as funding can be obtained. In order to effectively protect public health, there is definitely a need for public water systems to respond effectively and efficiently to disasters that affect their systems.

6.5 Summary and Conclusions

The efforts carried out by the Office of Public Health Emergency Response seem to be effective. Table 6.1 summarizes responses to emergency response training conducted during 2004 and Table 6.2 summarizes responses for 2005. The Appendix provides a summary of the security needs assessment summary based on data from community and non-transient non-community public water supplies serving populations less than 3,301.

Table 6.1 Summary of responses to the emergency response planning training conducted during year 2004. Cities where the evaluation was used were Lincoln, Grand Island, Hartington, Creighton, Ainsworth, Kearney, Lexington, Holdrege, and Minden

1. Did you obtain new information during this training that will be of use to you in developing an Emergency Plan of Operations for your water system? (Check one)

Responses:	YES: 120	NO: 27

Other: 3 – N/A

2. After this training session, do you believe that your current Emergency Plan of Operations is adequate to address possible emergency situations affecting your water system? (Check one)

Responses:	YES: 57	NO: 54

Other: 2 – unknown

Comments: 1 – But it can be improved

1 – Maybe needs to be double checked

1 – Don't know

3. If you answered "NO" to #2 above, what area(s) of emergency response might be in need of improvement in your current plan?

Responses ranged from "we really don't have anything in place" to "always room for improvement" to "response resources" to "securing wells better."

Table 6.1 (continued)

4. When was the last time your Emergency Plan of Operations was reviewed and updated?

Responses to this question ranged from "don't know" and "never" to recently. Some expressed plans to do it soon or in the future. The responses from most indicated the importance of training on the need for an emergency plan of operation, and perhaps a need for . . . a requirement or expectation from DHHS that the emergency plan be reviewed on a set basis or time frequency.

5. If revision of your Emergency Plan of Operations is made, will the template presented at this training session be used? (Check One)

Responses:	YES: 105	NO: 5		
Other: 8	1 – Not sure 2 – Maybe	4 – N/A	1-?	1 – unknown

Comments: 1 – At least sections of it
1 – Maybe already used it/probably

6. Were there any areas presented in this training session that you believe should be covered in more depth? (Check one)

Responses:	YES: 7	NO: 105
Other: 4	3 – N/A	1 – Not sure

7. If you answered "YES" to #6 above, what areas should be covered in more depth?

Responses:

1. "Help in doing emergency plan"
2. "More information on generator rental and sizes"
3. "If the plan has been submitted and approved this training can be helpful in updates"
4. "I think it was covered well/it just needs to be implemented"
5. "Mutual Aid with other communities and bottle water sources"

Other comments included:

- "will be a useful tool to be used in the future"
- "presentation did bring up other possible scenarios that I had not thought about in our present emergency plan"
- "as board chairman of our village I found this presentation extremely helpful to help us understand how important it is to have an emergency plan"
- "template will be very helpful in completing this information and I really like the idea of all systems being set up on the same template for an emergency response plan"
- "very good for the health department to know what is required of water operators and for us to meet. . .we are looking forward to conducting a tabletop in the future"
- "plan (in template) is more comprehensive than we have"
- "(need) smaller format for the NTNC or CWS under 100 connections"
- "can see the need of a plan"
- "very informative"
- "a lot of very good information"
- "good class"

Table 6.2 Summary of responses to the emergency response planning training conducted during 2005 (January–March). Cities where the evaluation was used were North Platte, Hastings, Beatrice, Dakota City, Omaha, Lyons, Norfolk, and Weeping Water

1. Did you obtain new information during this training that will be of use to you in developing an Emergency Plan of Operations for your water system? (Check one)

Responses:	YES: 86	NO: 1
Other:		

2. After this training session, do you believe that your current Emergency Plan of Operations is adequate to address possible emergency situations affecting your water system? (Check one)

Responses:	YES: 52	NO: 30
Other:		
Comments: 1 –	Don't know	
	1 – N/A	

3. If you answered "NO" to #2 above, what area(s) of emergency response might be in need of improvement in your current plan?

Responses included:

- "(need to) overhaul and update plan"
- "this will be our initial plan"
- "do not have one in place"
- "where to find needed water, where to find generator to run current system"
- "different types of emergency protection, equipment lists"
- "bulk water carriers, video security system, inventory of spare piping"
- "contact numbers, repair numbers, possible backup"
- "more water haulers, disinfection"
- "what resources are available"
- "we need a total plan"
- "get names and numbers posted"
- "acquire spare chlorinator"
- "long-term bulk water supply/transportation"
- "communications"
- "needs to be more detailed"
- "a complete update"
- "organization, water use restriction, emergency response action"
- "generator need, power supply hookup for generator at valve control"
- "water supply by truck"

4. When was the last time your Emergency Plan of Operations was reviewed and updated?

Responses ranged from "this year" to "unknown," "not updated" and "never had one."

Table 6.2 (continued)

5. If revision of your Emergency Plan of Operations is made, will the template presented at this training session be used? (Check One)

Responses:	YES: 69	NO: 3

Out of eight comments, responses ranged from "don't know" to "I think it would at least be looked at."

6. Were there any areas presented in this training session that you believe should be covered in more depth? (Check one)

Responses:	YES: 4	NO: 75

7. If you answered "YES" to #6 above, what areas should be covered in more depth?

Responses ranged from "unknown" to "grant or money information or resources at the state level and the (state???) players."

Other comments included:

- "A very informational seminar and a lot of participation by the group."
- "We use most of this information in our plan but we will update any new information and add it to our plan as well"
- "I am a waste treatment operator. I am glad to know there is a program like this, I do not know if we have a plan"
- "The training session was presented well and also the template will be very useful"
- "Good Introduction"
- "I thought the session was excellent and I gained more insight into what information communities need to assemble. It's a good tool for other things as well as for emergencies"
- "This training brought me together with other operators and agencies for information about their resources"
- "Helpful information"
- "This was a very good class with much needed information"
- "This plan is very detailed and will give operators better preparation in case of an emergency"
- "I think this was a very beneficial class"
- "Good presentation, good discussion"

Appendix: Security Needs Assessment Summary

This is a summary of the information gleaned from the Water System Security Needs Survey. The survey was mailed to Community and Non-Transient Non-Community PWSs which serve a population of less than 3,301. The survey was mailed out in mid-May 2005 with a return date of July 15, 2005.

The numbers in parenthesis in the first table represent the number of systems that completed and returned the survey. Surveys were mailed to *564 CWS*. A total of 316 CWS surveys were returned for a 56% response rate. Surveys were mailed to *183 NTNC*. A total of 68 NTNC surveys were returned for a 38% response rate.

Population	CWS	NTNC	% Reporting
0–100	83 (45)	112 (48)	47.6
101–500	290 (156)	56 (18)	50.2
501–1,000	102 (53)	7 (2)	50.4
1,001–3,300	89 (62)	8 (2)	65.9

The numbers in parenthesis for questions 2 through 9 represent the percentage of responding systems that do or do not have these components in place for their PWS. The numbers in ( ) for question #10 represent the percentage of responding systems that feel they need these improvements at their facilities.

The categories that do not total 100% indicate that the question was not answered on each survey.

Question #1: How many wells and storage facilities does your water system have? The numbers in ( ) represent the average number of facilities per responding PWS

	Wells	Storage
Community	706 (2.2)	350 (1.1)
NTNC	105 (1.5)	45 (.6)

Question #2: Are facilities fenced, including well houses and storage facilities?

	Yes	No
Community	62 (20%)	256 (80%)
NTNC	22 (32%)	47 (68%)

Question #3: Are doors, windows, and other points of entry such as tank and roof hatches and vents kept closed and locked?

	Yes	No
Community	293 (93%)	41 (13%)
NTNC	51(75%)	17 (25%)

Question #4: Is there external lighting around critical components of your water system?

	Yes	No
Community	151(47%)	166 (53%)
NTNC	42 (60%)	29 (40%)

Question #5: Are warning signs (no tampering, unauthorized access, etc.) posted on all critical components of your water system?

	Yes	No
Community	115 (36%)	201 (64%)
NTNC	14 (20%)	54 (80%)

Question #6: Do you have an alarm system that will detect unauthorized entry or attempted entry at critical components?

	Yes	No
Community	19 (6%)	297 (94%)
NTNC	4 (5%)	64 (95%)

Question #7: Do you have a key control and accountability policy?

	Yes	No
Community	202 (64%)	101 (32%)
NTNC	26 (38%)	42 (62%)

Question #8: Are entry codes and keys limited to water system personnel?

	Yes	No
Community	226 (72%)	76 (24%)
NTNC	25 (36%)	43 (64%)

(continued)

Question #9: Do you have law enforcement check your critical components while on their patrol?

	Yes	No
Community	128 (41%)	175 (56%)
NTNC	14 (21%)	53 (78%)

Question #10: List possible security enhancements that you would like to see done at your facility

	Fencing	Lights	Alarms/Cameras	Signs/Locks	Blank/None
Community	110(35%)	92 (29%)	78 (27%)	19 (6%)	124 (39%)
NTNC	9 (13%)	9 (13%)	3 (4%)	3 (4%)	41 (60%)

Chapter 7
Spatial Distributed Risk Assessment for Urban Water Infrastructure

Michael Möderl and W. Rauch

7.1 Introduction

The approach presented and tested serves for managing water infrastructure which is categorized, e.g., according to Reid (2009) as critical infrastructure taking into account abnormal, critical, and future conditions of the entire system. Terrorist attacks may damage critical infrastructure components or contaminate water resources. Interruptions of water supply are the consequence (Gleick, 2006). Pipe bursts caused by deterioration result in supply deficits and generate financial penalties (Jayaram and Srinivasan, 2008). Excavators potentially damage pipes and other infrastructure components during road reconstructions. If climate change results in an increase of rainfall intensities, urban flooding also increases. Further, land use change varies demand and runoff. Infrastructure and managers of it have to be prepared for such abnormal, critical, and future scenarios. This is a matter of risk assessment. Generally, risk is defined as a product of consequence and likelihood. In this chapter a different definition is used. For that the terms hazard and vulnerability maps are described in the following.

7.1.1 Hazard Maps

If components of a WSS are located in hazard zones, the likelihood of a system failure caused by a detrimental event is high. However, hazard maps (e.g., avalanche hazard (Gruber and Bartelt, 2007); seismic hazard (Torres-Vera and Antonio Canas, 2003); landslide hazard (van Westen et al., 2008); and land use change hazard (Chen et al., 2009)) express only the likelihood of the hazardous events, but neither the likelihood of a subsequent failure nor the consequences of the latter. For instance an avalanche damages a certain part of the system (e.g., due to a pipe breakage) but

M. Möderl (✉)
Institute of Infrastructure, University of Innsbruck, 6020 Innsbruck, Austria
e-mail: Michael.Moederl@uibk.ac.at

R.M. Clark et al. (eds.), *Handbook of Water and Wastewater Systems Protection*, Protecting Critical Infrastructure, DOI 10.1007/978-1-4614-0189-6_7,
© Springer Science+Business Media, LLC 2011

additional information is required to estimate how much the water system is actually hampered under this critical condition. To understand the true relevance of hazard events, an investigation of the vulnerability of the system is necessary.

7.1.2 Vulnerability Maps

Chosen from Ezell (2007), where different definitions for vulnerability are collected, vulnerability assessment is defined as "*the identification of weaknesses* in a system, *focusing on defined threats that could compromise its ability to provide a service*." Vulnerability maps for water systems are commonly used, but traditionally they refer only to component vulnerability and not to intrinsic vulnerability (see below). Component vulnerability maps report characteristics of the individual elements, e.g., material or age, and thus represent a valuable help for the planning of system maintenance. But the failure of a component does not automatically constitute a critical system failure. For instance, a failure of a water reservoir or the main combined sewer overflow facility is more harmful to the entire system than a pipe burst located in a cul-de-sac or a failure in a pumping station in an upper section of the sewer system, respectively. However, to quantify the consequence of each component failure to the system performance, hydraulic simulations are necessary.

7.1.3 Intrinsic Vulnerability Maps

An example for a system-wide vulnerability assessment approach is shown in Ezell (2007). Therein a value model was used to measure the vulnerability of infrastructure systems (Infrastructure Vulnerability Assessment Model, I-VAM). They did not apply a hydraulic solver to measure the system-wide performance. Vulnerability assessment tools used in the water security sector (e.g., RAM-W (Risk Assessment Methodology for Water), RAMCAP (Risk Analysis and Management for Critical Asset Protection), and VSAT (Vulnerability Self Assessment Tool); for more details see Brashear and Stenzler (2007)) aid in describing critical facilities and assets to protect by identifying system vulnerabilities and determining the level of protection to which the security system should be designed. None of these tools utilize hydraulic or water quality simulations. In Nilsson et al. (2005) intrusion into water distribution systems is simulated. In Vreeburg et al. (1994) a quantitative method to determine reliability of water supply systems (WSSs) is introduced. Similarly to Mark et al. (1998) the assessment is based on hydraulic simulations. These papers apply the principles of intrinsic vulnerability, but do not exploit the results spatially by using GIS methodology. In Khanal et al. (2006) the influence of contamination events on water distribution system performance is investigated. Therein a zone of influence map (i.e., vulnerability map) was constructed by superimposing the set of population exposure values onto their respective nodes.

Following the principal idea of Khanal et al. (2006) the aim of the presented method is to improve the risk assessment procedure by generating intrinsic vulnerability maps for water systems that take into account the effect of a potential failure and modification of each component (pumps, reservoirs, pipes, combined sewer overflows, etc.) to the performance of the entire system. In order to generate this spatial information the software tools VulNetWS (Vulnerability of Networks, Water Supply; Möderl et al., 2008) and VulNetUD (Vulnerability of Networks, Urban Drainage; Möderl et al., 2009) are used. With these tools, it is possible to identify weak points in a network by the assessment of the spatially referenced intrinsic vulnerability of urban water systems based on hydraulic and hydrodynamic simulations. Thereby the impacts of component failures and system changes are evaluated holistically.

7.1.4 Risk Maps

The risk of a system is calculated by multiplying its vulnerability with its belonging hazard (UN DHA, 1992, Fig. 7.1). The risk analysis presented is only shown for the case of water supply and focuses on four alpine natural hazards, i.e., flooding, landslides, debris flow, and avalanches in an alpine region including five municipalities. The natural hazards listed above are the most prevalent in the alpine region of Tyrol (BMLFUW, 2007). Global warming and changes in precipitation will further influence these hazard types in future (Rauch and De Toffol, 2006; Soldati et al., 2004; Höller, 2007; Stoffel and Beniston, 2006). The novelty of the methodology is the combination of vulnerability maps (calculated based on hydraulic simulations) and hazard maps based on Geographical Information System (GIS) analysis. Using the presented methodology, the vulnerabilities of infrastructure are identified and (if necessary) eliminated by technical measures, resulting in a higher security. Furthermore potential hazard zones can be located to construct protection measures at proper sites considering the available budget.

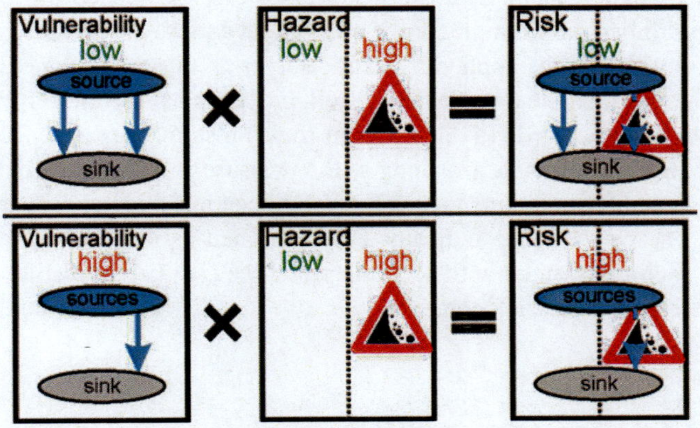

Fig. 7.1 Merging vulnerability with hazard maps to calculate risk

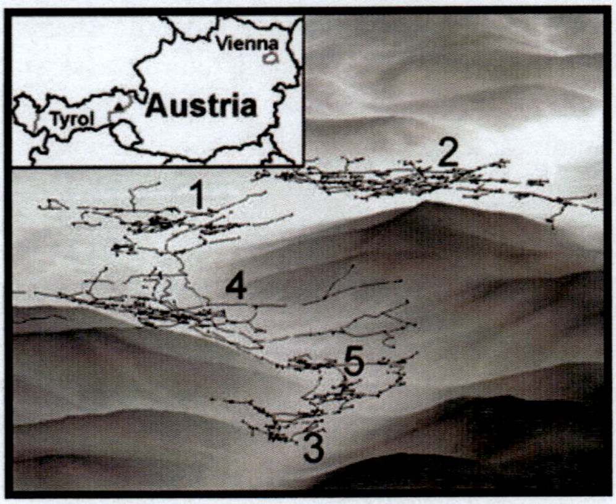

Fig. 7.2 Location of the alpine WSS (number of WSS corresponds with that of Table 7.3)

7.2 Materials and Methods

First, performance evaluation for water systems is defined as the basis for vulnerability assessment. Second, the spatially distributed sensitivity analysis for vulnerability assessment is described. Third, examples for a hazard evaluation are shown. Finally, it is demonstrated how risk is assessed by merging vulnerability and hazard.

7.2.1 Performance Evaluation for Water Supply

The performance decrease of a WSS that is caused by system impacts is assessed by means of performance indicators (PI) which estimate the vulnerability of a component. High values of the PIs indicate a low vulnerability of the system or, in other words, a redundant system. The following three PIs, introduced by Möderl et al. (2007), serve only as an example for the assessment and can be replaced at ease. The only restriction is that the applied hydraulic solver, in our case Epanet2 (Rossman, 2000), is able to calculate the relevant system properties of the chosen PI.

Performance indicator 1 (PI1 in %) refers to water supply pressure. The system is assumed to operate sufficiently as long as the pressure is within a predefined range. The actual pressure requirements are defined by setting a lower (p_l) and an upper (p_u) limit. The performance indicator PI1 is defined as the sum of the delivered demand at each junction (j) with accurate pressures $Q_{del}(p)$ divided by the sum of water required (Q_{req}) at each node:

$$\text{PI1} = \frac{\sum_{j=1}^{J} Q_{\text{del},j}(p)}{\sum_{j=1}^{J} Q_{\text{req},j}} \cdot 100\ (\%) \qquad Q_{\text{del}}(p) : p \in (p_l, p_u) \qquad (7.1)$$

The second indicator (PI2 in %) refers to water quality. Water quality is affected when the residence time of the water in the system is longer than a maximum acceptable quality limit in terms of time (t_u) (Engelhardt et al., 2000). EPANET2 computes the mean residence time of the water in the system at each node. Indicator PI2 is defined as the relation of the demand delivered with sufficiently low residence time at each junction (j) (i.e., the demand that meets the quality requirement) divided by the sum of water delivered (Q_{req}) at each junction:

$$\mathrm{PI2} = \frac{\sum_{j=1}^{J} Q_{\mathrm{del},j}\,(\mathrm{Age})}{\sum_{j=1}^{J} Q_{\mathrm{req},j}} \cdot 100\ (\%) \qquad Q_{\mathrm{del}}\,(\mathrm{Age}) : \mathrm{Age} \in (0, t_u) \tag{7.2}$$

For convenience, also a third indicator (PI3 in %) is used for the evaluation which is a measure of the combination of both hydraulic and water quality. This PI does not give any extra information on the vulnerability of the system. However, it is potentially useful to summarize both hydraulic and quality performance for GIS post processing.

$$\mathrm{PI3} = \min\,(\mathrm{PI1}, \mathrm{PI2}) \tag{7.3}$$

Summarizing, a value of 25% of a PI indicates that 25% of the total demand is supplied complying the limits. Consequently 75% of the total demand is delivered inadequate, i.e., either with too small pressure or too long retention time in the system.

7.2.2 Performance Evaluation for Urban Drainage

The change in the performance of an urban drainage system (UDS) is assessed by means of a comparison of PIs which estimates the intrinsic vulnerability of a component taking into account the entire system performance. In total seven predefined PIs for surcharging, flooding, CSO efficiency, and pollutant emissions are calculated using SWMM (Rossman, 2004) for each impact simulation. A description of all PIs can be found in Möderl (2009). The most interesting PIs are described in detail below.

The indicator PI1 refers to the Combined Sewer Overflow (CSO) efficiency of a combined sewer system. The greater the volume of surface runoff which is transported to the wastewater treatment plant (WWTP), the lesser the volume of wastewater that is discharged into receiving water bodies. PI1 is calculated by a mass balance defined as that part of surface runoff (V_{R}) treated at the WWTP ($V_{\mathrm{R}} - V_{\mathrm{CSO}}$):

$$\mathrm{PI1} = 1 - \frac{\left(\sum_{i=1}^{N} V_{\mathrm{CSO},i}\right)}{\left(\sum_{j=1}^{C} V_{\mathrm{R},j}\right)} \tag{7.4}$$

In the formula i denotes the CSO facility and j indicates the subcatchment. This performance indicator is derived from the Austrian guideline ÖWAV-RB 19 (2007) and described by De Toffol (2006).

The second indicator (PI2) refers to water quality. Again, a mass balance is used to quantify the emission-based performance. The mass (M_{Po}) of a specific pollutant that is discharged into the receiving water body is divided by the mass of the pollutant that enters the UDS, either from rainfall (M_{R}) or from dry weather flow (M_{DWF}):

$$\mathrm{PI7} = 1 - \frac{\sum_{i=1}^{N} M_{\mathrm{Po},i}}{\sum_{j=1}^{C} M_{\mathrm{R},j} + \sum_{i=1}^{N} M_{\mathrm{DWF},i}} \tag{7.5}$$

VulNetUD also contains PIs for sewer flooding. For example, PI4 is the sum of maximum pond volume (V_{P}) over each node divided by the total rainfall runoff volume:

$$\mathrm{PI4} = 1 - \frac{\sum_{i=1}^{N} \max\left(V_{\mathrm{P},i}\right)}{\sum_{j=1}^{C} V_{\mathrm{R},j}} \tag{7.6}$$

The absolute values of the PIs vary, if different rainfall events are simulated. However, relative values between UDS components are assumed to be independent of rainfall events. For the vulnerability assessment in this study the relative performance $\left(\mathrm{PI}_{\text{changed system}}/\mathrm{PI}_{\text{base system}}\right)$ is analyzed and calculated with SWMM using a design storm event with a return period of 5 years. In general the application is possible using any user-defined rainfall input.

7.2.3 Vulnerability Assessment

Several types of hazards impact urban water infrastructure, e.g., natural hazards (such as landslides and river flooding) and anthropogenic hazards (abnormal traffic load, construction incidents, land use change, etc.). These incidents may result in damaged pumps, broken pipes, contaminated water intakes, blocked CSO facilities among other system changes. In terms of simulation, the software VulNet mimics the impact of such incidences by adjusting component parameters so that all relevant hazard impacts can be considered. For instance blockage of CSO facilities is mimicked by setting the geometry of the weir opening to zero or a broken pipe is mimicked by closing it. VulNet evaluates impacts for each spatially distributed component by calculation of performance indicators based on hydraulic simulation results. Then a spatially referenced sensitivity map is created by mapping indicator values according to corresponding component locations. These sensitivity maps are regarded as vulnerability map due to the parameter variation mimics of a specific hazard impact.

7.2.4 Hazard Assessment

A lot of different hazard types impact water systems. As an example for a hazard analysis, the impact of natural hazards on WSS is described, but the methodology is applicable to every hazard type and also to UDS. In Table 7.1 discussed natural hazards are listed. All hazards impact the water sources (springs and ground water wells). Flooding impacts the Water Distribution System (WDS) hydraulically, if the pressure outside the pipe is higher than inside the pipe. It also impacts the water sources by polluting them and disabling electrical pumping equipment. It is assumed that landslides and debris flows impact a water distribution system mechanically by sweeping pipes along with soil material and water sources by submerging the latter. Avalanches impact only the water source, by assuming that they do not whirl up soils. The hazard zones are categorized based on existing hazard zone maps (HZM) (regulated by BGBl 436, 1976) and if not available on a GIS-based spatial assessment. For the latter different geo-data sets are used, such as a digital elevation model (DEM) and land use data (e.g., the European CORINE database). The resolution for the GIS-based rasterized analysis is 125 m. All hazards were categorized in the subsets as low, moderate, and high.

7.2.4.1 Flooding Hazard (h1)

A study to determine the potential flooding zones in Austria (Blöschl et al., 2006) is used for analyzing flooding hazard. This study assumes a breakdown of existing flooding protection measures and provides flooding zones with a return period of 30, 100, and 200 years. In this work these zones are characterized as high, moderate, and low flooding hazard zones, respectively.

7.2.4.2 Landslide Hazard (h2)

The methodology of Perotto-Baldiviezo et al. (2004) is modified to determine the hazard of landslides, taking into account slope and land use. Potential landslide hazard increases with higher slopes. Forests offer a protection against potential

Table 7.1 Table of natural hazards

Natural hazard	Impact on	Areas covered by HZM	Other areas
h1 – flooding	Water source and WDS	HORA[a]	–
h2 – landslide	Water source and WDS	HZM	GIS-based spatial assessment
h3 – avalanche	Water source		
h4 – debris flow	Water source and WDS		

[a]"Flood Risk Zoning" provided by the "Federal Ministry of Agriculture, Forestry, Environment and Water Management"

landslides, while pastures and grassland are potentially hazardous. Bare rock – as found regularly in alpine environments – poses no threat. The combination of these two factors (slope and land use) resulted in low, moderate, and high landslide hazard.

7.2.4.3 Avalanche Hazard (h3)

The method used for this analysis gives information on the theoretical (potential) starting zone of an avalanche. The results are summarized in an avalanche hazard map. Information on the existing protection measures is not considered. Nowadays there is a large variety of different software tools for the calculation of avalanche hazard (Christen et al., 2002; Gruber and Bartelt, 2007). The method used in this study is based on a qualitative estimation for regional analysis. The main factors influencing avalanche release are multiplied with each other without weighting. GIS data for slope, sun exposition, and land cover are used. The values attributed for each category are chosen on the basis of the following consideration.

Avalanches are most frequent for slope inclinations between 35° and 50°. Higher inclinations are less dangerous because of little possibility for snow to accumulate and to build an avalanche of higher weight. At inclinations between 20° and 35° there is the possibility of avalanche formation, but only in the presence of high amounts of snow. For downslope hillsides with inclinations smaller than 20° the probability of avalanches is very low. The probability of release of an avalanche depends also on the orientation of a slope face with respect to the sun. The exposition with higher hazard is in the directions south and southeast in late spring. In such hillsides, snow melts faster. Thus the snow is heavier. Further melting and icing processes will build ice crusts. Thirty two percent of all avalanches in Austria occur on south- to southeast-faced slopes (Höller, 2007). In winter, the north and northeast zone is more unstable because the snow cover is colder and tends to build weak layers with formation of dry snow avalanches. Another important factor for the avalanche release is land cover. Vegetation plays an important role in avalanche protection. The forest prevents accumulation of snow, intercepts the snow fall, reduces the formation of slab avalanches, and interrupts the fraction line. If a snow avalanche is already moving it can be stopped from trees, but only if the already gained velocity and weight are small. The potential hazard map obtained with the above-described method is then integrated with HZM provided from the Austrian environmental protection agency (regulated by BGBl 436, 1976). This map takes into account the protection measures, but includes only residential areas generally located in the bottom of alpine valleys.

7.2.4.4 Debris Flow (h4)

Debris flow is a considerable process in alpine streams. It consists of inhomogeneous loose geological material mixed with water and may be generated when hillside colluvium or landslide material becomes saturated with water and flows into a channel. Debris flow hazards are the outcome of natural process of erosion and sediment motion interacting with human systems (Davies, 1997). There are many detailed GIS-based models for debris flow hazard mapping; a literature review can be found in Yu et al. (2006). However, geological and soil data with high resolution are required. As for the catchment analyzed in this study such data are not available. Thus a simplified method based on the Austrian hazard zone map is used. For areas where the Austrian hazard zone map is available it is rasterized; for other areas the hazard zone is modeled along known torrents available as a geo-data set.

7.2.5 Risk Assessment

According to UN DHA (1992) risk (*R*) is calculated by hazard (*H*) times vulnerability (*V*). Not only risk maps are calculated to evaluate the risk by merging vulnerability and hazard maps but also a risk matrix is defined. The values in the matrix are the number of cells (125×125 m resolution) of related categories. These values are calculated separately for each hazard type (Table 7.2). Cells referred to critical categories (marked in gray) are recommended for further investigation.

7.2.6 Case Studies for Water Supply

The application for water supply is described by means of five systems which are located close to each other in the Tyrolean Alps, Austria. The total population supplied by the five WSSs amounts to 21,200. Each WSS is an independent network. Figure 7.2 gives an overview of the region and the layout of the networks. A more detailed description of the study area is given in Vanham et al. (2008).

Data of the infrastructure and water resources are collected. The demand is computed as annual average by a population density raster with a resolution of 150×150 m provided by STATISTIK AUSTRIA. In the region tourism is by far the largest industry, with 1.6 million tourist overnight stays recorded in the year 2001. Thus, water demand of tourism is considered as equivalent of overnight stays. No daily demand pattern was applied. The tanks are modeled as fixed pressure points, but the method can also be applied to a network model including tanks with dynamic behavior. The initial water age was set equally to the water age at steady-state conditions. The process of model building and calibration is not in the focus of

Table 7.2 Schema risk matrix (h1 – flooding, h2 – landslide, h3 – avalanches, h4 – debris flow)

Risk matrix	Low *H*	Moderate *H*	High *H*	Sum
Low *V*	h1 – h2 – h3 – h4	h1 – h2 – h3 – h4	h1 – h2 – h3 – h4	h1 – h2 – h3 – h4
Moderate *V*	h1 – h2 – h3 – h4	h1 – h2 – h3 – h4	h1 – h2 – h3 – h4	h1 – h2 – h3 – h4
High *V*	h1 – h2 – h3 – h4	h1 – h2 – h3 – h4	h1 – h2 – h3 – h4	h1 – h2 – h3 – h4
Sum	h1 – h2 – h3 – h4	h1 – h2 – h3 – h4	h1 – h2 – h3 – h4	h1 – h2 – h3 – h4

Table 7.3 Basic features of WSSs

Case study	Reservoirs	Tanks	Junctions	Pumps	Pipes	Components
1 – Oberndorf	4	0	144	3	157	308
2 – St. Johann	2	2	253	1	281	539
3 – Jochberg	1	3	84	1	93	182
4 – Kitzbuehel	6	3	132	6	156	303
5 – Aurach	3	2	100	1	102	208
Total	16	10	713	12	789	1,540

this chapter, but has received sufficient attention in order to ensure the quality of the result. Table 7.3 outlines basic features of the WSSs.

7.2.7 Case Studies for Urban Drainage

One case study is used to demonstrate the potential of the methodology for urban drainage management. This case study is based on a sewer system which drains a city and several surrounding municipalities to a WWTP. CSOs are discharged into two rivers, a large River1 and smaller River2. The system is divided into six main sub-catchments based on the main collectors. The main sewer collectors have huge storage capacities due to the size of the sewers. The combined sewer system serves an impervious area of approximately 3,000 ha. In total 300,000 people equivalents are connected. Furthermore, approximately 40 CSO facilities, 20 pumping stations, and 400 nodes are included in the system.

7.3 Results and Discussion

The results section is structured into three parts. First, vulnerability maps for WSSs are shown. Correspondingly, a vulnerability map for a UDS is shown. As an example of a risk assessment, risk maps are discussed based on the WSS of Aurach.

7.3.1 Vulnerability Maps

The innovative feature of the presented methodology is the spatial explicit information on the intrinsic vulnerability of the system. As vulnerability is expressed in terms of indicator values a desired indicator for spatial information has to be chosen. For better visualization all vulnerability maps presented in the following have been exported to a geographical information system. Yellow colors constitute low vulnerability (high performance under critical conditions) while red color indicates high vulnerability (low performance under critical conditions).

7.3.2 Vulnerability Maps for WSSs

The WSS of Aurach (Fig. 7.3, left) reveals a fairly redundant behavior. One hundred sixty-eight of the total 208 components of the system exhibit values higher than 90% for the pressure indicator PI1. A weak point in the system is in the northern part where the yellow and orange colors indicate higher vulnerability. The problem arises as the northern part of the supply system is mainly connected by means of these vulnerable pipes. A failure of these components would leave a significant part of the consumers without water. To improve the reliability of the system the water source in the northwestern should be connected more efficiently. In order to demonstrate the capability of the method for improvement, the right-hand side of Fig. 7.3 shows the result for a system improvement. The connection of the northwestern water source to the WSS can be improved by only optimizing the valve parameter settings in the tank. However, due to this slight change the overall vulnerability of the system is decreased significantly.

7.3.3 Vulnerability Maps for UDSs

Figure 7.4 maps the effects of blocked, i.e., fluvial surcharged, CSO facilities. This is represented by setting the geometry of the weir opening to zero. Such a blockage can be caused by river flooding. Blocked CSO1 and CSO2 decrease performance significantly. This is because the northern part of the UDS is drained through a culvert with reduced capacity (nearby CSO1) to the main sewer and the sewer along River2 is connected to the main sewer by a conduit with limited capacity too. In total, 7 of the 40 CSO facilities were identified as vulnerable. Emergency pumps which have sufficient capacity to pump the wastewater when high water levels occur would reduce the vulnerability.

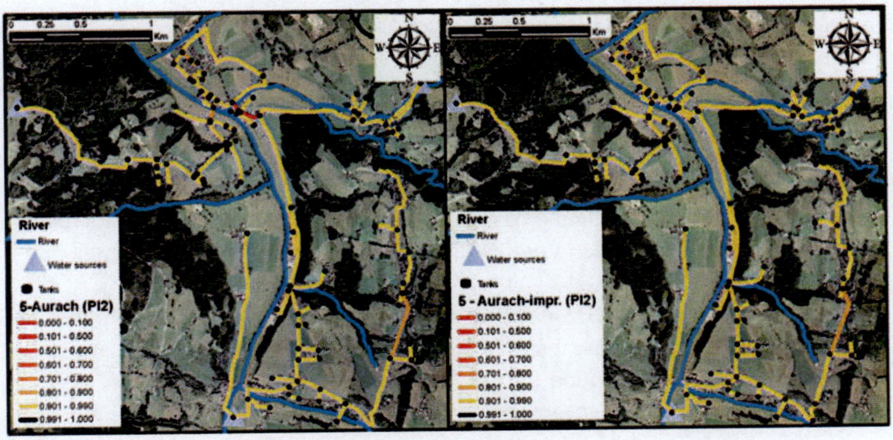

Fig. 7.3 *left*: Mapping the results of PI2 – Aurach; *right*: mapping PI2 of improved WSS

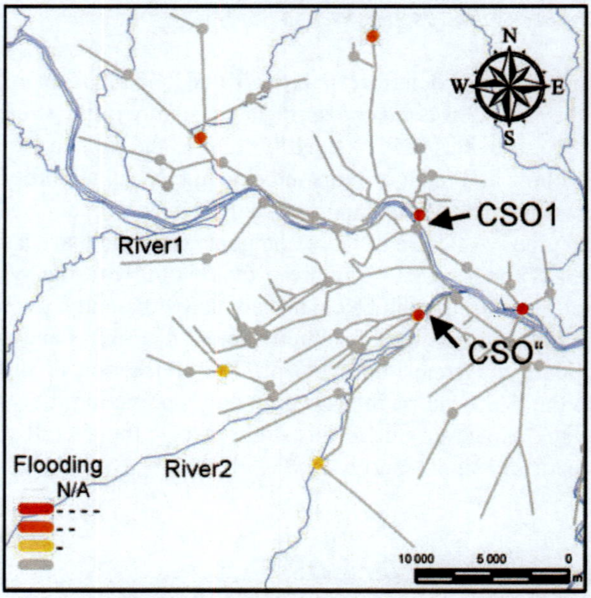

Fig. 7.4 Impact of blocked CSOs on flooding efficiency

7.3.4 Risk Maps

In Fig. 7.5 a risk map for the WSS of Aurach is shown. The vulnerability map for isolation of a component failure is based on PI3 which quantifies pressure quality and water age. The WSS of Aurach is highly vulnerable compared to urban municipalities and moderately vulnerable compared to other rural municipalities. Most of the components are moderately vulnerable and one part of the system is highly vulnerable. In the background the debris flow hazard zones are plotted. Red cells indicate high and yellow color moderate debris flow hazard. Debris flow hazard is located along torrents. The vulnerable components which connect the upper part of the entire system are located in a high hazard zone (framed in gray). Due to high vulnerability and high hazard this site of the WSS is at high risk. Preventive measures are suggested to reduce it. There are two options: On the one hand, the redundancy of the network can be increased so that the upper part of the system is better connected. In the planning of rehabilitation this fact has to be considered. On the other hand, it is possible to protect these vulnerable system parts against debris flow impact.

A risk assessment for UDS can be worked out in a similar manner. For instance a vulnerability map for fluvial surcharged CSOs can be merged with the data set "Flood Risk Zoning" provided by the "Federal Ministry of Agriculture, Forestry, Environment and Water Management."

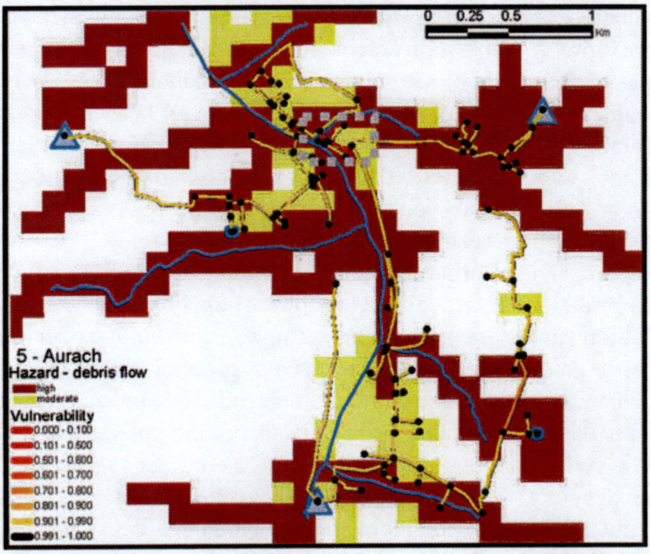

Fig. 7.5 Vulnerability map for isolated component failures and debris flow hazard map

7.3.5 Summary of Risk Analysis

In Table 7.4 the numbers of cells with corresponding hazard and vulnerability categories for each evaluated natural hazard and for all five WSSs are listed. In the majority of cases the raster cells are categorized as low vulnerable. Based on the matrix, only 27 cells out of total of 1,368 cells (respectively, 0.42 km^2) are categorized as zones with moderate and high vulnerability or hazard (marked gray in Table 7.4). These risk zones are recommended for an in-depth analysis and – if required – for preventive measures.

7.4 Conclusions

For the analysis of the vulnerability for water systems against hazards it is a common undertaking to map the characteristics and the individual vulnerability of the components of the system. However, this neglects that failures and functional changes

Table 7.4 Resulting risk matrix (h1 – flooding, h2 – landslide, h3 – avalanches, h4 – debris flow). The numbers indicate the number of cells in each category

Risk matrix	Low *H*	Moderate *H*	High *H*	Sum
Low *V*	16 – 272 – 0 – 0	17 – 156 – 2 – 203	183 – 53 – 3 – 329	216 – 480 –5 –532
Moderate *V*	0 – 4 – 0 – 0	0 – 4 – 0 – 2	1 – 0– 0 – 11	1 –8 –0 –13
High *V*	0 – 4 – 0 – 0	2 – 0 – 0 – 3	1 – 0 – 0 – 3	3 –4 –0 –6
Sum	16 – 280 – 00 – 00	19 – 160 – 08 – 15	185 –53 – 12 – 25	220 – 493 –5 –551

of individual components in the system have a very distinct influence on the performance of the entire system. In this chapter a methodology is developed where the effect of functional changes of a component are computed by means of a hydraulic simulation and expressed in terms of indicator values. When this is done for each individual component of the entire system, spatial information on the intrinsic vulnerability of the system is generated. VulNet is a software tool that performs these computations and also the subsequent assessment of the vulnerability.

The methodology has been tested for five WSSs and one UDS. It was demonstrated that the spatial information of the intrinsic vulnerability of WSSs offers significant information on critical sections of the supply system and indicates also how the situation can be improved. For example, vulnerabilities occur if different demand areas (e.g., separated by a river) are not properly connected. By strengthening these connections, vulnerabilities are reduced. The application of the method using VulNet is seen as a valuable tool for managers and operators of water utilities to improve the performance of their system and to consider system vulnerability in rehabilitation planning.

Additionally, an alpine region including five municipalities is chosen to evaluate the public drinking water supply security. A methodology is developed to identify, on a regional basis, zones with high risk by merging information on vulnerability and four potential natural hazards. The methodology aids water management to make decisions on which sites of the WSS should be chosen for preventive measures.

Acknowledgments The work reported was funded by project "KIRAS PL 3: Achilles," project no. 824682 under the "Sicherheitsforschungs – Förderprogramm KIRAS" of the Austrian Federal Ministry for Transport, Innovation and Technology (BMVIT) and the Austrian Research Promotion Agency (FFG).

References

BGBl 436 (1976). "Verordnung des Bundesministers für Land- und Forstwirtschaft vom 30.07.1976 über die Gefahrenzonenpläne." P. b. b. Verlagspostamt, Wien.

Blöschl, G.; Merz, R.; Humer, G.; Hofer, M.; Hochold, A. and Wührer, W. (2006). HORA – Hydrologische Arbeiten. Endbericht an das BMLFUW, Sektion VII, Institut für Wasserbau und Ingenieurhydrologie, TU Wien (in German), Wien.

BMLFUW (2007). "Zahlen und Fakten 2007 (Statistics of "The Federal Ministry of Agriculture, Forestry, Environment and Water Management") (in German)."

Brashear, J. and Stenzler, J. (2007). "Water and Wastewater Specific RAMCAP Guidance." AWWA/WEF Joint Management Conference, Portland, OR, USA, Feb 25–28.

Chen, Y.; Xu, Y.P. and Yin, Y.X. (2009). "Impacts of land use change scenarios on storm-runoff generation in Xitiaoxi basin, China." Quaternary International, 208, 121–128.

Christen, M.; Bartelt, P. and Gruber, U. (2002). "AVAL-1D: An avalanche dynamics program for the practice." International Congress INTERPRAEVENT 2002 in the Pacific Rim – Matsumoto/Japan Congress publication, vol. 2, pp. 715–725.

Davies, T.R.H. (1997). "Using hydroscience and hydrotechnical engineering to reduce debris flow hazards." Debris-Flow Hazards Mitigation: Mechanics, Prediction, and Assessment. Proceedings, First International Conference on Debris Flow Hazards Mitigation; American Society of Civil Engineers, San Francisco, CA, pp. 787–810.

De Toffol, S. (2006). "Sewer system performance assessment – an indicator based methodology." PhD thesis, Unit of Environmental Engineering, University of Innsbruck.

Engelhardt, M.O.; Skipworth, P.J.; Savic, D.A.; Saul, A.J. and Walters, G.A. (2000). "Rehabilitation strategies for water distribution networks: a literature review with a UK perspective." Urban Water, 2(2), 153.

Ezell, B.C. (2007). "Infrastructure vulnerability assessment model (I-VAM)." Risk Analysis, 27(3), 571–583.

Gleick, P.H. (2006). "Water and terrorism." Water Policy, 8(6), 481–503.

Gruber, U. and Bartelt, P. (2007). "Snow avalanche hazard modelling of large areas using shallow water numerical methods and GIS." Environmental Modelling & Software, 22(10), 1472–1481.

Höller, P. (2007). "Avalanche hazards and mitigation in Austria: a review." Natural Hazards, 43(1), 81–101.

Jayaram, N. and Srinivasan, K. (2008). "Performance-based optimal design and rehabilitation of water distribution networks using life cycle costing." Water Resources Research, 44(1), 15.

Khanal, N.; Buchberger, S.G. and McKenna, S.A. (2006). "Distribution system contamination events: exposure, influence, and sensitivity." Journal of Water Resources Planning and Management-Asce, 132(4), 283–292.

Mark, O.; Wennberg, C.; van Kalken, T.; Rabbi, F. and Albinsson, B. (1998). "Risk analyses for sewer systems based on numerical modelling and GIS." Safety Science, 30(1–2), 99–106.

Möderl, M. (2009). "Modelltechnische Analyse von Netzwerksystemen der Siedlungswasserwirtschaft." PhD thesis, Institute of Infrastructure, University of Innbruck.

Möderl, M.; Fetz, T. and Rauch, W. (2007). "Stochastic approach for performance evaluation regarding water distribution systems." Water Science and Technology, 56(9), 29–36.

Möderl, M.; Vanham, D.; De Toffol, S. and Rauch, W. (2008). "Potential impact of natural hazards on water supply systems in Alpine regions." Water Practice and Technology, 3(3). doi: 10.2166/wpt.2008.060.

Möderl, M.; Kleidorfer, M.; Sitzenfrei, R. and Rauch, W. (2009). "Identifying weak points of urban drainage systems by means of VulNetUD." Water Science & Technology, 60(10), 2507–2513.

Nilsson, K.A.; Buchberger, S.G. and Clark, R.M. (2005). "Simulating exposures to deliberate intrusions into water distribution systems." Journal of Water Resources Planning and Management-Asce, 131(3), 228–236.

ÖWAV-RB 19 (2007). "Richtlinie für die Bemessung von Mischwasserentlastungen." Österreichischer Wasser- und Abfallwirtschaftsverband, Wien.

Perotto-Baldiviezo, H.L.; Thurow, T.L.; Smith, C.T.; Fisher, R.F. and Wu, X.B. (2004). "GIS-based spatial analysis and modeling for landslide hazard assessment in steeplands, southern Honduras." Agriculture, Ecosystems & Environment, 103(1), 165–176.

Rauch, W. and De Toffol, S. (2006). "Climate change induced trends in high resolution rainfall." 7th International Workshop on Precipitation in Urban Areas, St. Moritz, Swizerland.

Reid, R.L. (2009). "Guiding critical infrastructure." Civil Engineering, 79(2), 50–55.

Rossman, L.A. (2000). EPANET 2 user manual. National Risk Management Research Laboratory – U.S. EPA, Cincinnati, OH.

Rossman, L.A. (2004). Storm water management model – User's manual Version 5.0. National Risk Management Research Laboratory, U.S. Environmental Protection Agency, Cincinnati, OH.

Soldati, M.; Corsini, A. and Pasuto, A. (2004). "Landslides and climate change in the Italian Dolomites since the Late glacial." CATENA, 55(2), 141–161.

Stoffel, M. and Beniston, M. (2006). "On the incidence of debris flows from the early Little Ice Age to a future greenhouse climate: A case study from the Swiss Alps." Geophysical Research Letters, 33, L16404, doi:10.1029/2006GL026805.

Torres-Vera, M.A. and Antonio Canas, J. (2003). "A lifeline vulnerability study in Barcelona, Spain." Reliability Engineering & System Safety, 80(2), 205–210.

UN DHA (1992). Internationally Agreed Glossary of Basic Terms Related to Disaster Management. UN DHA (United Nations Department of Humanitarian Affairs), Geneva.

Vanham, D.; Fleischhacker, E. and Rauch, W. (2008). "Technical Note: Seasonality in alpine water resources management – a regional assessment." Hydrology and Earth System Sciences, 12(1), 91–100.

van Westen, C.J.; Castellanos, E. and Kuriakose, S.L. (2008). "Spatial data for landslide susceptibility, hazard, and vulnerability assessment: An overview." Engineering Geology, 102(3–4), 112–131.

Vreeburg, J.H.G.; Hoven, T.J.J.v.d. and Hoogsteen, K.J. (1994). "A quantitative method to determine reliability of water supply systems." Water Supply, 12(1–2), 7.9–7.13.

Yu, F.-C.; Chen, C.-Y.; Chen, T.-C.; Hung, F.-Y. and Lin, S.-C. (2006). "A GIS process for delimitating areas potentially endangered by debris flow." Natural Hazards, 37, 167–189.

Chapter 8
US Water and Wastewater Critical Infrastructure

Robert M. Clark

8.1 Introduction

Concern over the possibility of attacks against targets within the United States by domestic and foreign terrorists resulted in the formation of the President's Commission on Critical Infrastructure to evaluate the vulnerability of the following infrastructure categories to internal and external terrorism (President's Commission on Critical Infrastructure Protection, 1996):

- Information and communication
- Physical distribution
- Banking and finance, energy
- Vital human services

The rapid proliferation of telecommunication and computer systems, which connect infrastructures to one another in a complex network, compounds this vulnerability.

Vital human services include community water supply systems on local and state levels. In terms of public administration, water supply systems are generally governmental in nature. However, each supply system tends to be highly localized. And it had been assumed that failures in one community would have little direct impact on other communities.

Following the terrorist attacks of September 11, 2001, and the mailing of letters containing *Bacillus anthracis* spores, the US Congress enacted the *Homeland Security Strategy* and the *Public Health Security and Bioterrorism Preparedness and Response Act of 2002* (US Congress, 2002). Subsequently the US Environmental Protection Agency (EPA) developed a *Homeland Security Strategy* (USEPA, 2004) for enhancing national security and protecting human health and the environment. In addition, several Homeland Security Presidential Directives (HSPDs) were issued in 2003 and 2004. These directives are described on the Department of Homeland

R.M. Clark (✉)
9627 Lansford Drive, Cincinnati, OH 45242, USA
e-mail: rmclark@fuse.net

R.M. Clark et al. (eds.), *Handbook of Water and Wastewater Systems Protection*, Protecting Critical Infrastructure, DOI 10.1007/978-1-4614-0189-6_8,
© Springer Science+Business Media, LLC 2011

Security (DHS) web site (US DHS, 2010). A more recent concern is the issue of cyber security and the interconnectedness of various vital functions such as the electrical grid upon that support water and wastewater systems (USGAO). This concern seems justified give the substantial water supply and wastewater infrastructure that has been built in the United States including extensive storage and distribution facilities in the West and Southwest.

The first municipal water utility in the United States was established in Boston in 1652 to provide domestic water and fire protection (Hanke, 1972). The Boston system emulated ancient Roman water supply systems in that it was multipurpose in nature. Subsequently, although many water supplies in the United States were primarily introduced in cities for the prevention of fires, most have been adapted to serve commercial and residential properties with water. The first water treatment plant in the United States was constructed in Richmond, Virginia, in 1832 and the second municipal treatment plant was constructed in 1855. It consisted of a small charcoal sand and gravel filter located at Elizabeth, New Jersey (EPA, 2002).

Wastewater treatment prior to 1900 consisted of physically separating solids and floating debris from wastewater before discharging it into receiving waters (EPA, 2001). The first wastewater treatment plant was built in Gloversville, NY, and in 1916 Chicago, IL, constructed an activated sludge treatment plant (EPA, 2002).

Major reductions in waterborne disease outbreaks were brought about by the use of sand filtration, disinfection via chlorination, and the application of drinking water standards (Clark et al., 1985) and the construction of wastewater treatment systems. The investment in water and wastewater treatment in the United States has proven to be a major contributor to ensuring the nation's public health.

The two primary laws that deal with Safe Drinking Water and Wastewater in the United States are the Safe Drinking Water Act of 1974 and the Clean Water Act of 1948. Consequently these laws will have a major impact on the ability of drinking water and wastewater utilities to adapt to global climate changes.

8.2 Safe Drinking Water Act

Since the late 1890s, concern over waterborne disease and uncontrolled water pollution has regularly translated into water quality legislation at the federal level. The first water-related regulation was promulgated in 1912 under the Interstate Quarantine Act of 1893. At that time interstate railroads made a common cup available for train passengers to share drinking water while on board – a practice that was prohibited by the Act. Several sets of federal drinking water standards were issued prior to 1962, but they too applied only to interstate carriers (Grindler, 1967; Clark, 1978). By the 1960s, each of the states and trust territories had established their own drinking water regulations, although there were many inconsistencies among them. As a consequence, reported waterborne disease outbreaks declined from 45 per 100,000 in 1938–1940 to 15 per 100,000 in 1966–1970. Unfortunately, the annual number of waterborne disease outbreaks ceased to fall around 1951 and may have increased slightly after that time, leading, in part, to the passage of the Safe Drinking Water Act (SDWA) of 1974 (Clark, 1978).

Prior to the passage of the SDWA, most drinking water utilities concentrated on meeting drinking water standards at the treatment plant, even though it had long been recognized that water quality could deteriorate in the distribution system – the vast infrastructure downstream of the treatment plant that delivers water to consumers. After its passage, the SDWA was interpreted by the US Environmental Protection Agency (EPA) as meaning that some federal water quality standards should be met at various points within the distribution system rather than at the water treatment plant discharge. This interpretation forced water utilities to include the entire distribution system when considering compliance with the SDWA. Consequently water quality in the distribution system became a focus of regulatory action and a major interest to drinking water utilities. Maintaining a high level of water quality downstream of the treatment plant can be a challenge because water is transported through extensive pipe networks, many of which are corroded and decaying (NRC, 2006).

In the United States drinking water quality is defined as a measure of the suitability of water for human consumption, based on selected physical, chemical, and biological characteristics. EPA has promulgated many rules and regulations as a result of the SDWA that requires drinking water utilities to meet specific guidelines and two types of numeric standards for water quality. One set of standards is enforceable and is collectively referred to as maximum contaminant levels (MCLs). The other set of standards is non-enforceable and referred to as maximum contaminant level goals (MCLGs). The MCLGs are set at a level at which no known or anticipated adverse human health effects occur. Where it is not economically or technologically feasible to ascertain the level of a contaminant, a treatment technique is prescribed by EPA in lieu of establishing an MCL. For example, because the protozoan *Giardia lamblia* is very difficult to measure, it has been established that if water is treated at a given pH, temperature, and chlorine concentration for a specified length of time (all of which must be verified by the water utility), a fixed level of *Giardia* inactivation will take place. In summary, the EPA guidelines and standards are designed to ensure that drinking water is adequately treated and managed by water utilities to protect public health (Clark and Feige, 1993).

8.3 Clean Water Act

The Federal Water Pollution Control Act or Clean Water Act is the principal law that is concerned with the control of pollution in the nation's streams, lakes, and estuaries. It was enacted in 1948 and completely revised by amendment in 1972. These amendments were the basis for the current form of the Act and specified ambitious programs for water quality improvements that have been or are being put into place by industries and cities. Additional amendments were added to the CWA in 1977, 1981, with a set of comprehensive amendments in 1987 (Copeland, 2006). The Act consists of two major parts:

- Regulatory provisions that impose progressively more stringent requirements on industries and cities in order to meet the statutory goal of zero discharge of pollutants.
- Provisions that authorize federal assistance for municipal wastewater treatment construction.

The 1987 Amendments added a new section to the Act under which states were required to develop and implement programs to control nonpoint sources of pollution or rainfall runoff from farms and urban areas. States are required to identify pollutant-impaired water segments and develop "total maximum daily loads (TMDLs)" that set the maximum amount of pollution that a water body can receive without violating water quality standards (Copeland, 2006). Other issues that affect the CWA are

- Storm water discharges
- Combined and separate sewer overflows
- Wetlands
- Development of strategies concerning animal feeding operations

In 1972 Congress mandated that all publicly owned treatment works (POTWs) provide secondary treatment of wastewater. By 1996 fewer than 200 systems out of 16,204 nationwide failed to met that standard (EPA, 2002).

8.4 Water and Wastewater Infrastructure

8.4.1 Drinking Water Infrastructure

Distribution system infrastructure is generally the major asset of a water utility. This infrastructure includes the pipes, pumps, valves, storage tanks, reservoirs, meters, fittings, and other hydraulic appurtenances that connect treatment plants to consumers' taps. The American Water Works Association (AWWA, 1974) defines the water distribution system as one "including all water utility components for the distribution of finished or potable water by means of gravity storage feed or pumps though distribution pumping networks to customers or other users, including distribution equalizing storage." As mentioned previously, these systems must also be able to provide water for nonpotable uses, such as fire suppression, street watering, and irrigation of landscaping. They span almost 1 million miles in the United States (Kirmeyer et al., 1994) and include an estimated 154,000 finished water storage facilities (AWWA, 2003). As the US population grows and communities expand, 13,200 miles (21,239 km) of new pipes are installed each year (Kirmeyer et al., 1994).

Because distribution systems represent the vast majority of physical infrastructure for water supplies, they constitute the primary management challenge from both an operational and public health standpoint. Furthermore, their repair and replacement represent an enormous financial liability; EPA estimates the 20-year water transmission and distribution needs of the country to be $183.6 billion, with storage facility infrastructure needs estimated at $24.8 billion (EPA, 2005a).

Water utilities in the United States vary greatly in size, ownership, and type of operation. The SDWA defines public water systems as consisting of community water supply systems, transient, non-community water supply systems, and non-transient, non-community water supply systems. A community water supply system serves year-round residents and ranges in size from those that serve as few as 25 people to those that serve several million. A transient, non-community water supply system serves areas such as campgrounds or gas stations where people do not remain for long periods of time. A non-transient, non-community water supply system serves primarily non-residential customers but must serve at least 25 of the same people for at least 6 months of the year (such as schools, hospitals, and factories that have their own water supply). There are over 162,000 water systems in the United States that meet the federal definition of a public water system (EPA, 2005b). Thirty-three percent (52,838) of these systems are categorized as community water supply systems, 55% are categorized as transient, non-community water supplies, and 12% (19,375) are non-transient, non-community water systems (EPA, 2005b). Overall, public water systems serve 297 million residential and commercial customers. Although the vast majority (98%) of systems serves less than 10,000 people, almost three-quarters of all Americans get their water from community water supplies serving more than 10,000 people (EPA, 2005b). Not all water supplies deliver water directly to consumers, but rather deliver water to other supplies. Community water supply systems are defined as "consecutive systems" if receiving their water from another community water supply through one or more interconnections (Fujiwara et al., 1995).

Some utilities rely primarily on surface water supplies while others rely primarily on groundwater. Surface water is the primary source of 22% of the community water supply systems, while groundwater is used by 78% of community water supply systems. Of the non-community water supply systems (both transient and non-transient), 97% are served by groundwater. Many systems serve communities using multiple sources of supply such as a combination of groundwater and/or surface water sources. In a grid/looped system, the mixing of water from different sources can have a detrimental influence on water quality, including taste and odor, in the distribution system (Clark et al., 1988, 1991a, b).

Some utilities, like the one operating in New York City, own large areas of the watersheds from which their water source is derived, while other utilities depend on water pumped directly from major rivers like the Mississippi River or the Ohio River and therefore own little if any watershed land. The SDWA was amended in 1986 and again in 1996 to emphasize source water protection in order to prevent microbial contaminants from entering drinking water supplies (Borst et al., 2001). Owning or controlling its watershed provides an opportunity for a drinking water utility to exercise increased control of its source water quality (Peckenham et al., 2005).

The water supply industry in the United States has a long history of local government control over operation and financial management, with varying degrees of oversight and regulation by state and federal governments. Water supply systems serving cities and towns are generally administered by departments of municipalities or counties (public systems) or by investor-owned companies (private systems). Public systems are predominately owned by local municipal governments, and they serve approximately 78% of the total population that uses community water supplies. Approximately 82% of urban water systems (those serving more than 50,000 persons) are publicly owned. There are about 33,000 privately owned water systems that serve the remaining 22% of people served by community water systems. Private systems are usually investor owned in the larger population size categories but can include many small systems as part of one large organization. These investor-owned utilities are in business to generate profits for their shareholders. In the small- and medium-sized categories, the privately owned systems tend to be owned by homeowners, associations, or developers and they simply provide water to their clientele with little regard for profit. Finally, there are several classifications of state chartered public corporations, quasi-governmental units, and municipally owned systems that operate differently than traditional public and private systems. These systems include special districts, independent non-political boards, and state chartered corporations (NRC, 2006).

Table 8.1 provides a snapshot of the size and the population served for public water systems in the United States.

The extent of water distribution pipes in the United States is estimated to be 980,000 miles (1.6×10^6 km) in length, with pipes being replaced at an estimated rate of once every 200 years (Grigg, 2005). Rates of repair and rehabilitation have not been estimated. There is a large range in the type and age of the pipes that make up water distribution systems. The oldest cast iron pipes from the late 19th century are typically described as having an average useful life span of about 120 years because of the pipe wall thickness (AWWA, 2001; AWWSC, 2002). In the 1920s the manufacture of iron pipes changed to improve pipe strength, but the changes also produced a thinner wall. These pipes have an average life of about 100 years. Pipe manufacturing continued to evolve in the 1950s and 1960s with the introduction of ductile iron pipe that is stronger than cast iron and more resistant to corrosion. Polyvinyl chloride (PVC) pipes were introduced in the 1970s and high-density polyethylene in the 1990s. Both of these are very resistant to corrosion but they do not have the strength of ductile iron. Post-World War II pipes tend to have an average life of 75 years (AWWA, 2001; AWWSC, 2002).

In the 20th century, most of the water systems and distribution pipes were relatively new and well within their expected life span. However, as is obvious from the above paragraph and recent reports (AWWA, 2001; AWWSC, 2002), these different types of pipes, installed during different time periods, will all be reaching the end of

Table 8.1 Public water system inventory data

System size by population served		Very small 500 or less	Small 501–3,300	Medium 3,301–10,000	Large 10,001–100,000	Very large >100,000	Total
CWS	**# systems**	**30,417**	**14,394**	**4,686**	**3,505**	**361**	53,363
	Pop. served	**5,010,834**	**20,261,508**	**27,201,137**	**98,706,485**	**122,149,436**	273,329,400
	% of systems	**57%**	**27%**	**9%**	**7%**	**1%**	100%
	% of pop	**2%**	**7%**	**10%**	**36%**	**45%**	**100%**
NTNCWS	**# systems**	16,785	2,786	97	16	2	19,686
	Pop. served	2,327,575	2,772,334	506,124	412,463	279,846	6,298,342
	% of systems	85%	14%	0%	0%	0%	100%
	% of pop	37%	44%	8%	7%	4%	100%
TNCWS	**# systems**	85,366	2,657	96	29	4	88,152
	Pop. served	7,315,647	2,602,706	528,624	619,248	12,269,000	23,335,225
	% of systems	97%	3%	0%	0%	0%	100%
	% of pop	31%	11%	2%	3%	53%	100%
	Total # of systems	132,568	19,837	4,879	3,550	367	161,201

Column 1 in Table 8.1 shows the various categories of water supplies: community water supplies (CWS), non-transient non-community water supplies (NTNCWS), transient non-community water supplies (TNCWS). Column 2 shows the number of systems, the population served, the number of systems, and the percent of population served in each catagory. Columns 3 through 8 show the categories of water supply considered. For example, very small systems serve 500 or less people.

Source: EPA (2003)

their expected life spans in the next 30 years. An estimated 26% of the distribution pipe in the country is unlined and in poor condition. Analysis of main breaks at one large Midwestern water utility that kept careful records of distribution system management documented a sharp increase in the annual number of main breaks from 1970 (approximately 250 breaks per year) to 1989 (approximately 2,200 breaks per year) (AWWSC, 2002). Thus, the water industry is entering an era where it must make substantial investments in pipe repair and replacement. An EPA report on water infrastructure needs (EPA, 2002) predicted that transmission and distribution replacement rates will need to be around 0.3% per year in 2005 and will rise to 2.0% per year by 2040 in order to adequately maintain the water infrastructure. This is about four times the current replacement rate.

Utilities vary in their approach to infrastructure maintenance and in their commitment to its replacement. Some utilities do not rehabilitate or replace infrastructure until a serious failure is imminent or after failure has occurred (Hughes, 2002) while others use advanced methods for planning and prioritization. Given the aging of the nation's infrastructure, there is a concern that in the near future many failures may occur over a narrow time period, overwhelming the water industry's capability to react effectively (Beecher, 2002). The needs of the aging infrastructure are compounded by conservative design philosophies, increasingly stringent standards and regulations, negligence in maintenance and repair, and the concern of the public over the ability of distribution systems to maintain water quality to the consumer (Clark et al., 1999). Good water utility leadership will be needed in order to ensure a reliable supply of high water quality at the tap, to meet regulations, and to respond to customer needs, all the while controlling costs (Westerhoff et al., 2005).

8.4.2 Wastewater Infrastructure

Of the 16,024 publicly owned treatment plants in the United States less than 200 do not provide secondary treatment. Ninety-eight of US wastewater treatment works are publicly owned and provide service to 190 million people or 73% of the US population. Seventy-one percent serve less than 10,000 people and 25% of the population is not connected to centralized treatment and use some form of on-site treatment system. There are approximately 600,000 miles of publicly owned pipes (EPA, 2002).

In many older cities sanitary sewage and storm water runoff are collected in a single sewage system. This type of combined system is referred to as a combined sewer system (CSS). This type of sewer system provides partially separated channels for sanitary sewage and storm water runoff. It provides backup capacity for the runoff sewer when runoff volumes are unusually high. However, it is considered to be antiquated and is vulnerable to sanitary sewer overflow during peak rainfall events. A combined sewer overflow is an apparatus built into a combined sewer system that allows a certain amount of untreated flow to discharge into a water course to keep the systems from becoming surcharged in storm conditions. The combined sewer overflow often contains a screen which may be a mechanical or static arrangement

Table 8.2 US publicly owned treatment and collection systems

Treatment facilities	% of population served	Collection systems
16,024	71.8	20,670

Source: Seidenstat (2003)

depending on the frequency of spills per year. During heavy rainfall when the storm water exceeds the sanitary flow the sewage from homes would be diluted. However, combined sewage can be a major environmental problem and municipalities have begun to look for ways to mitigate the environmental effects of such overflow locations. One solution is to build a combined sewer overflow (CSO) facility, which consists of some low-level treatment, storage, and return of the sewage to the normal system (EPA, 2002).

CSOs, however, can create major public health and water quality problems. These overflows have contributed to closing of beaches and shellfish bed and contamination of public water supplies. The cost of providing adequate storm water infrastructure is one of the major expenditures in many urban areas. EPA estimated (1996) that the development of adequate storm water infrastructure would require an expenditure of $44.7 billion. State Revolving Fund (SRF) loans accounted for $2.08 billion from 1998 to 2000. In 2000, CSO loans accounted for $411 million or over 12% of the total SRF activity for that year (EPA, 2001). Table 8.2 provides an overview of publicly owned treatment and collection systems (Seidenstat, 2003).

8.4.2.1 Other US Water Infrastructure

This chapter has focused on water and wastewater. However, there is extensive infrastructure throughout the United States devoted to transfer and storage of water for a wide variety of purposes. There are more then 80,000 dams and reservoirs and millions of miles of canals, pipes, and tunnels in the contiguous 48 states. These facilities are devoted to support agriculture, including irrigation, power, and energy and industrial, commercial, and mining activities. For example, there is an extensive water transfer system from the Colorado River and Northern CA to Southern CA. Throughout the Southwestern part of the United States as well as in the Southeastern United States there are major water storage and water storage facilities which are critical to the survival and economic viability of the communities they serve (Jacobs et al., 2001). This report, with few exceptions, will not deal with those types of infrastructure issues.

8.5 Drinking Water Infrastructure Needs

As indicated previously there are a large number of community-owned water utilities in the United States. However, many of these utilities will need major investments for a variety of purposes. For example, the USEPA's 2003 Needs Assessment found that the nation's water systems will need to invest $276.8 billion over the next 20 years in order to continue to provide safe drinking water to their consumers (EPA, 2005a). These investments will be required for the installation of new infrastructure as well as rehabilitation or replacement of deteriorated or undersized infrastructure. In addition, there will be a need to replace aging infrastructure that may be adequate now but will require replacement or significant rehabilitation within the next 20 years. The biggest percentage of these investments will be directed toward water system's needs to continue to deliver water to their customers and not related to violations of SDWA regulations.

EPA's 1995 and 1999 Needs Assessments estimated the total national drinking water investment needs at $167.4 and $165.5 billion, respectively (EPA, 2005a). The current assessment estimates a need of $276.8 billion which exceeds previous assessments by more than 60%. The methods used to collect both the current and the previous estimates were essentially the same as in the previous Needs Surveys. The 2003 estimate does emphasize improved management of assets, including collection of better data on infrastructure condition, and long-term planning for rehabilitation and replacement. These needs are summarized in Table 8.3.

Although the 2003 Needs Assessment estimate represents a substantial increase over previous assessments, it is still within the range identified in other reports. EPA's "Clean Water and Drinking Water Infrastructure Gap Analysis" estimated drinking water systems' 20-year capital needs between $170 and $493 billion, with a point estimate of $303 billion (EPA, 2002). The Congressional Budget Office (CBO) report "Future Investment in Drinking Water and Wastewater Infrastructure" estimates annual water system needs of $12.2–21.2 billion, which would extrapolate to a 20-year total need in the range of $245–424 billion (CBO, 2002). The Water Infrastructure Network's (WIN) "Clean and Safe Water for the 21st Century – A Renewed National Commitment to Water and Wastewater Infrastructure" estimates water system needs of $21 billion annually, which extrapolates to $420 billion over 20 years (WIN, undated).

The nation's 1,041 largest community water systems (serving more than 50,000 people) account for $122.9 billion, or 44%, of the total national need. Medium and small community water systems also have substantial needs of $103.0 billion and $34.2 billion, respectively.

The 2003 Needs Assessment differentiates "current" from "future" needs; the definitions of these two types of needs, as well as examples, are described below (EPA, 2005a). About 60% of the total needs, $165.0 billion, are identified as current needs. Current needs are defined as projects that a system considers a high priority for near-term implementation to enable a water system to continue to deliver safe drinking water. Future needs are projects that water systems do not currently need, but would expect to address in the next 20 years as part of routine rehabilitation or replacement of infrastructure because of predictable events.

Growth-related needs are not consistent with the eligibility requirements for the DWSRF. The 2003 Needs Assessment did not include projects that would be undertaken solely to accommodate future growth (e.g., extension of service lines to new housing developments).

Table 8.3 Total need by project type (in millions of January 2003 dollars)

System size and type	Distribution and transmission	Treatment	Storage	Source	Other	Total need
Large community water systems (serving over 50,000 people)	$89, 779.9	$20, 091.3	$6, 994.5	$4, 715.8	$1, 270.2	$122, 851.7
Medium community water systems (serving 3,301 to 50,000 people)	$73, 454.4	$14, 906.2	$9, 473.3	$4, 392.8	$790.9	$103, 017.4
Small community water systems (serving 3,300 and fewer people)	$18, 624.3	$6, 164.1	$6, 263.8	$2, 871.0	$248.3	$34, 171.5
Costs associated with the recently promulgated arsenic rule		$947.4				$947.4
Not-for-profit non-community water systems	$425.3	$670.2	$1, 620.3	$681.0	$0.8	$3, 397.5
American Indian and Alaska Native village water systems	$1, 347.3	$462.2	$490.3	$135.1	$13.6	$2, 448.5
Subtotal national need	$183, 631.1	$43, 241.4	$24, 842.2	$12, 795.6	$2, 323, 7	$266, 834.1
Costs associated with proposed and recently promulgated regulations		$9, 927.4				$9, 927
Total national need	$183, 631.1	$53, 168.8	$24, 842.2	$12, 795.6	$2, 323.7	$276, 761.5

Source: EPA (2005a)

Treatment projects represent the second largest category of need, $53.2 billion, nearly one-fifth of total need, over the next 20 years (EPA, 2005a). This category consists of projects needed to reduce contaminants through treatment processes such as filtration, disinfection, corrosion control, and aeration. The installation, upgrade, or rehabilitation of treatment infrastructure also enables removal of contaminants that can cause chronic health effects or taste, odor, and other aesthetic problems.

The total 20-year need for storage projects is $24.8 billion (EPA, 2005a). This category includes projects to construct new or rehabilitate existing finished water storage tanks.

Construction of new tanks is necessary if the system cannot provide adequate flows and pressure during peak demand periods. Many projects in this category involve rehabilitating existing tanks to prevent structural failures or sanitary defects that can allow microbiological contamination.

The source category includes projects that are necessary to obtain safe supplies of surface water or ground water (EPA, 2005a). The infrastructure needs in this category include the installation and rehabilitation of drilled wells and surface water intakes. The total 20-year needs for source water projects are $12.8 billion.

Transmission and distribution projects constitute the largest category of need, accounting for almost two-thirds of the total need with $183.6 billion needed over the next 20 years (EPA, 2005a). Little of this need is related to any federal mandate. Instead, utilities need to install and maintain distribution systems to provide potable water to their customers while preventing contamination of that water prior to delivery. Although treatment plants or elevated storage tanks are usually the most visible components of a water system, most of a system's infrastructure is underground in the form of transmission and distribution mains. Failure of transmission and distribution mains can interrupt the delivery of water leading to a loss of pressure, possibly allowing a backflow of contaminated water into the system. Broken transmission lines can also disrupt the treatment process. The transmission and distribution category also comprised the largest proportion of the total need in the 1995 and 1999 Needs Assessments.

The other category accounts for an estimated $2.3 billion (EPA, 2005a). This category captures needs that cannot be assigned to one of the other categories. Examples include emergency power generators not associated with a specific system component, computer and automation equipment, and projects for system security.

The SDWA requires that public water systems meet national standards to protect consumers from the harmful effects of contaminated drinking water, and so some of the needs are directly attributable to specific SDWA regulations (EPA, 2005a). These needs are subdivided into the needs associated with existing SDWA regulations and the needs associated with recently promulgated and proposed regulations. The total regulatory need is estimated as $45.1 billion, or 16% of the total national need.

While most of the total need is not driven by compliance with a particular regulation, properly maintaining a system's infrastructure is not only economical in the long run but also protective of public health. These nonregulatory costs include routine installation, upgrade, and replacement of basic infrastructure and are borne by the system regardless of regulations.

The estimated needs directly associated with existing SDWA regulations are $35.2 billion. Microbial Contaminants, projects that address microbiological contamination comprise 86%, or $30.2 billion, of the total existing regulatory need. Chemical Contaminants or projects designed to protect the public health from chemical contaminants comprise $5.0 billion, or 14%, of the total existing regulatory need. As shown in Table 8.3 the total need associated with proposed and recently promulgated regulations is $9.9 billion.

8.6 Wastewater Infrastructure Needs

Table 8.4 reports the total needs for water and wastewater in the United States based on the Clean Watershed Needs Survey (CWNS) 2004 Report to Congress (EPA, 2008). These needs as reported (in January 2004 dollars) for the wastewater treatment and collection categories (Categories I through V) increased from $180.2 billion in the CWNS 2000 to $189.2 billion in this report. This is a $9.0 billion (or

Table 8.4 Comparison of total needs for 1996–2004 (billions of dollars)

Need category	1996	2000	2004	$B	% change between 1996 and 2004
I – Secondary wastewater treatment	32.8	41	44.6	3.6	8.8
II – Advanced wastewater treatment	21.6	22.7	24.5	1.8	7.9
III-A – Infiltration/ inflow correction	4,1	9.1	10.3	1.2	13.2
III-B – Sewer replace-ment/rehabilitation	8.6	18.7	21.0	2.3	12.3
IVA – New collector sewers and appurtenances	13.3	15.9	16.8	0.9	5.7
IVB – New interceptor sewers and appurtenances	13.3	16.5	16.2	0.7	4.2
V – Combined sewer overflow correction	55.2	56.3	54.8	–1.5	–2.7
VI – Storm water management programs	9.1	6.2	9.0	2.8	45.2
X – Recycled water distribution	–	–	4.3	4.3	NA
Total	158	186.4	202.5	16.1	8.6

Needs for 1996 and 2000 were adjusted to 2004 for comparison

5.0%) increase. Most (94%) of this increase can be attributed to needs increases of more than $100 million each in only 92 of the 10,152 facilities with reported needs. An additional 78 facilities had needs that decreased by at least $100 million each. The most significant increases in needs related to wastewater treatment and collection are the following: Category I (secondary wastewater treatment), increased by $3.6 billion; Category III-A and III-B (infiltration/inflow correction and sewer replacement/rehabilitation), by $3.5 billion; and Category II (advanced wastewater treatment), by $1.8 billion. Increases in Categories I and II could be due to a variety of issues including rehabilitation of aging infrastructure, facility improvements to meet more protective water quality standards, and in some cases providing additional treatment capacity for handling wet-weather flows.

New needs (needs reported for the first time) account for $10.0 billion of the Category I needs, $7.6 billion of the Category II needs, and $5.6 billion of the Category III-B needs. The amounts for projected facilities are $2.1 billion in Category I and $3.6 billion in Category II needs. By definition, Category III-B needs would be entered only for existing facilities.

Category III-A and III-B needs are for inflow/infiltration (I/I) correction and sewer replacement or rehabilitation. I/I occurs when flow from wet-weather conditions enters collection systems through various means, such as pipe cracks and broken joints.

Wet-weather events are known to cause a variety of water quality problems throughout the nation. Under various circumstances, precipitation in the form of snow or rain generates runoff that can be contaminated by a number of different pollutant sources (e.g., industrial operations, roadways, and land use practices). Where combined sewer systems are in use, wet-weather contributes to CSOs. CSOs contain not only storm water but also untreated human and industrial waste, toxic materials, and debris. These materials can be a major water pollution concern for cities with combined sewer systems.

8.7 Summary and Conclusions

It is clear that drinking water and wastewater systems are part of the critical infrastructure in the United States. There are over 162,000 community water supplies in the United States and over 50,000 community water supplies. The USEPA's 2003 Needs Assessment found that the nation's water systems will need to invest $276.8 billion over the next 20 years in order to continue to provide safe drinking water to their consumers (EPA, 2005a). These investments will be required for the installation of new infrastructure as well as rehabilitation or replacement of deteriorated or undersized infrastructure. In addition, there will be a need to replace aging infrastructure that may be adequate now but will require replacement or significant rehabilitation within the next 20 years. The biggest percentage of these investments will be directed toward water system's needs to continue to deliver water to their customers and not related to violations of SDWA regulations.

There are over 16,000 publicly owned wastewater treatment facilities in the United States. Based on the Clean Watershed Needs Survey (CWNS) 2004 Report to Congress (EPA, 2008) the new wastewater investment needed (2004) is $202.5 billion. These needs include wastewater treatment and collection infiltration/inflow correction, sewer replacement/rehabilitation), and storm water management.

References

American Water Works Association (AWWA). (1974). Water distribution research and applied development needs. Journal of the American Water Works Association 6:385–390.

American Water Works Association (AWWA). (2001). Reinvesting in Drinking Water Structure: Dawn of the Replacement Era. Denver, CO: AWWA.

American Water Works Association (AWWA). (2003). Water Stats 2002 Distribution Survey CD-ROM. Denver, CO: AWWA.

American Water Works Service Co., Inc. (AWWSC). (2002). Deteriorating buried infrastructure management challenges and strategies. Available online at http://www.epa.gov/safewater/tcr/pdf/infrastructure.pdf. Accessed March 16, 2005.

Beecher, J.A. (2002). The infrastructure gap: myth, reality, and strategies. In: Assessing the Future: Water Utility Infrastructure Management. D.M. Hughes (ed.). Denver, CO: AWWA, pp. 1–15.

Borst, M., M. Krudner, L. O'Shea, J.M. Perdek, D. Reasoner, and M.D. Royer. (2001). Source water protection: its role in controlling disinfection by-products (DBPs) and microbial contaminants. In: Controlling Disinfection By-Products and Microbial Contaminants in Drinking Water. R.M. Clark and B.K. Boutin (eds.). EPA/600/R-01/110. Washington, DC: EPA Office of Research and Development, pp. 4-1–4-25.

Clark, R.M. (1978). The Safe Drinking Water Act: implications for planning. In: Municipal Water Systems – The Challenge for Urban Resources Management. D. Holtz and S. Sebastian (eds.). Bloomington, IN: Indiana University Press, pp. 117–137.

Clark, R.M. and W.A. Feige. (1993). Meeting the requirements of the safe drinking water act. In: Strategies and Technologies for Meeting the Requirements of the SDWA. R.M. Clark and R.S. Summers (eds.). Lancaster, PA: Technomic Publishing Company.

Clark, R.M., J.A. Goodrich, and J.C. Ireland. (1985). Costs and benefits of drinking water treatment. Journal of Environmental Systems 14(1):1–30.

Clark, R.M., W.M. Grayman, and R.M. Males. (1988). Contaminant propagation in distribution systems. Journal of Environmental Engineering, ASCE 114(2):929–943.

Clark, R.M., W.M. Grayman, and J.A. Goodrich. (1991a). Water quality modeling: its regulatory implications. Proceedings of the AWWARF/EPA Conf. on Water Quality Modeling in Dist. Systems, Cincinnati, OH.

Clark, R.M., W.M. Grayman, J.A. Goodrich, R.A. Deininger, and A.F. Hess. (1991b). Field testing of distribution water quality models. Journal of the American Water Works Association 83(7):67–75.

Clark, R.M., G.S. Rizzo, J.A. Belknap, and C. Cochrane. (1999). Water quality and the replacement of drinking water infrastructure: the Washington, DC case study. Journal of Water Supply Research and Technology – Aqua 48(3):106–114.

Congressional Budget Office (CBO). (2002). "Future Investment in Drinking Water and Wastewater Infrastructure," (November 2002), p. ix.

Copeland, C. (2006). Water Quality: Implementing the Clean Water Act, CRS Report for Congress, The Library of Congress, 2006.

Fujiwara, M., J.M. Manwaring, and R.M. Clark. (1995). Drinking water in Japan and the United States: conference objectives. In: Drinking Water Quality Management. R.M. Clark and D.A. Clark (eds.). Lancaster, PA: Technomic Publishing Company, pp. 1–20.

Grigg, N.S. (2005). Assessment and renewal of water distribution systems. Journal of the American Water Works Association 97(2):58–68.

Grindler, B.J. (1967). Water and Water Rights: A Treatise on the Laws of Water and Allied Problems: Eastern, Western, Federal. Vol 3. Indianapolis, IN: The Allan Smith Company.

Hanke, S.H. (1972). Pricing urban water. In: Public Prices for Public Products, Selma Mushkin (ed.). Washington, DC: The Urban Institute, pp. 283–306.

Hughes, D.M. (ed.). (2002). Assessing the Future: Water Utility Infrastructure Management. Denver, CO: AWWA.

Jacobs, K., D.B. Adams, and P. Gleick. (2001). Potential consequences of climate variability and change for the water resources of the United States. In: Climate change impacts on the United States – the potential consequences of climate variability and change. Jerry Meilillo, Anthony Janetos, and Thomas Karl (eds.). New York, NY: Cambridge University Press. (Chapter 14)

Kirmeyer, G., W. Richards, and C.D. Smith. (1994). An assessment of water distribution systems and associated research needs. Denver, CO: AWWARF.

National Research Council. (2006). Drinking Water Distribution Systems: Assessing and Reducing Risks, National Academy of Sciences. Washington DC: National Academy Press. pp. 15–46.

Peckenham, J.M., C.V. Schmitt, J.L. McNelly, and A.L. Tolman. (2005). Linking water quality to the watershed: developing tools for source water protection. Journal of the American Water Works Association 97(9):62–69.

President's Commission on Critical Infrastructure Protection. (1996). Critical Infrastructure: Protecting Americans Critical Infrastructure. Available at http//www.pccip.gov

Seidenstat, P. (2003). Organizing water & wastewater industries to meet the challenges of the 21st century. Public Administration & Management, An Interactive Journal 8(2):69–99.

United States Government Accountability Office. (USGAO). Internet Infrastructure: DHS Faces Challenges in Developing a Joint Public/Private Recovery Plan. GAO-06-672, June, 2006.

US Congress. (2002). Public Health Security and Bioterrorism Preparedness and Response Act of 2002: Public Law 107-188. http://thomas.loc.gov/cgi-bin/query/z?c107:H.R.3448.ENR:. Accessed 03 March 2010.

US Department of Homeland Security. (2010). Homeland Security Presidential Directives. http://www.dhs.gov/xabout/laws/editorial_0607.shtm. Accessed 09 March 2010.

USEPA. (2001). Report to Congress: Implementation and Enforcement of the Combined Sewer Overflow Control Policy. EPA 833-R01-003, Washington DC.

USEPA. (2002). The Clean Water and Drinking Water Infrastructure Gap Analysis. EPA-816-R-02020. (September 2002) Washington DC.

USEPA. (2003). Factoids: Drinking Water and Ground Water Statistics for 2003.

USEPA. (2004). Homeland Security Strategy, http://www.epa.gov/OHS/pdfs/EPA-HS-Strategy.pdf. Accessed 02 March 2010.

USEPA. (2005a). Drinking Water Infrastructure Needs Survey and Assessment Third Report to Congress, June 2005, U.S. Environmental Protection Agency, Office of Water, EPA 816-R-05-001, Washington DC 20460.

USEPA. (2005b). Factoids: Drinking Water and Ground Water Statistics for 2003. EPA 816-K-05-01. Washington, DC: EPA Office of Water.

USEPA. (2008). Clean Watersheds Needs Survey 2004 Report To Congress, January 2008, U.S. Environmental Protection Agency, Office of Water. Washington DC 20460.

Water Infrastructure Network (WIN). (undated). "Clean and Safe Water for the 21st Century – A Renewed National Commitment to Water and Wastewater Infrastructure," p. 3-1.

Westerhoff, G., H. Pomerance, and S. Robinson. (2005). It's all about leadership. Underground Infrastructure Management Jan/Feb:22–25.

Chapter 9
Microbial Issues in Drinking Water Security

Eugene W. Rice

9.1 Introduction

A fundamental issue following both natural and man-made disasters is the need for an adequate supply of safe drinking water. The presence of microbial pathogens in a water supply following a disaster poses an urgent threat to public health. While there is an extensive amount of literature available on the classical waterborne pathogens, there is by contrast a rather limited amount of information on the overt bio-threat or bio-warfare agents which could be introduced into a water system. Studies on the persistence, detection, and treatment for these agents have not received wide attention, even though the feasibility of such threats has been recognized for many years (Berger and Stevenson, 1955). Most studies regarding bio-threat agents in water have been centered on their role in military situations as opposed to domestic terrorist attacks. A bioterrorism incident in a municipal drinking water system would have the potential for causing widespread disease and disruptions of vital public services which could affect large segments of the population (Nuzzo, 2006).

In a review of water terrorism events, Gleick (2006) noted three reports within the last 50 years where biological agents were implicated as potential threats:

1. In 1972 a group known as the "Order of the Rising Sun," who when they were arrested in Chicago, Illinois, had in their possession the causative agent of typhoid fever (*Salmonella enterica* serovar Typhi). The organism was allegedly

---

Disclaimer: The material presented in this chapter has been subjected to review and has been approved for publication by the US Environmental Protection Agency through its Office of Research and Development. Note that approval does not signify that the contents necessarily reflect the views of the Agency.

E.W. Rice (✉)
National Homeland Security Research Center, U.S. Environmental Protection Agency, Cincinnati, OH 45268, USA
e-mail: rice.gene@epa.gov

R.M. Clark et al. (eds.), *Handbook of Water and Wastewater Systems Protection*, Protecting Critical Infrastructure, DOI 10.1007/978-1-4614-0189-6_9,
© Springer Science+Business Media, LLC 2011

to be used to contaminate water systems in several US cities including Chicago, Illinois, and St. Louis, Missouri.

2. In 1973 a German biologist seeking to obtain a ransom threatened to contaminate water supplies with *Bacillus anthracis* and a biotoxin (a biologically produced chemical agent).
3. In 1984 the Rajneeshee religious cult planned to contaminate the Dalles, Oregon, municipal water supply with the bacterial pathogen *Salmonella enterica* serovar Typhimurium. While thought to be unsuccessful in their attempt to contaminate the water system, the group was later implicated in using this organism to contaminate salad bars in local restaurants, resulting in a community-wide outbreak of salmonellosis and one of the largest bioterrorism events ever reported in the United States.

This chapter reviews recent developments for assessing the role of microbial pathogens which have the potential for being used as bio-threat agents when intentionally introduced into a water system.

9.2 Bio-threat Agents

Pathogens associated with naturally occurring waterborne outbreaks consist of several groups of microorganisms including bacteria, viruses, and protozoa. It should be noted that these traditional waterborne pathogens also could be used to intentionally contaminate a water system. Pathogenic bacteria, including several members of the family Enterobacteriaceae, can be transmitted by water. Bacteria are the etiological agents of many of the well-known waterborne diseases, such as typhoid fever (*S. enterica* serovar Typhi) and cholera (*Vibrio cholerae*). In most instances bacteria can persist in water for several days to weeks and are readily removed or inactivated by conventional water treatment processes. Viral agents have also been implicated as common waterborne pathogens. Among these are members of the viral taxonomic families Adenoviridae, Astroviridae, Caliciviridae, Picornaviridae, and Reoviridae. Depending upon various abiotic and biotic conditions, viruses have the ability to persist in the aquatic environment and compared to bacteria are more resistant to commonly used water disinfectants. The encysted protozoa, most notably *Crytposporidium* spp. and *Giardia* spp., have been linked to numerous waterborne disease outbreaks. In their encysted form these protozoa can exist for extended periods of time in water and exhibit high levels of resistance to disinfection processes. Numerous studies have been published regarding the occurrence, detection, and efficacy of water treatment processes for these classical waterborne pathogens (American Water Works Association, 2006; American Public Health Association, 2005).

Bio-threat organisms have been classified by the US Centers for Disease Control and Prevention (CDC) as Select Agents (http://www.bt.cdc.gov/agent/agentlist.asp). Work with these agents requires special governmental permission and is limited to laboratories with enhanced bio-safety and bio-security capabilities. Many of these agents would not traditionally be considered waterborne pathogens, but some have the potential to be disseminated via water upon intentional introduction in a water system. Included in these Select Agents are viral taxonomic groups such as Arenaviridae, Bunyaviridae, Filoviridae, Flaviviridae, Orthomyxoviridae, Poxviridae, and Togaviridae. These viral families are responsible for a wide range of diseases including hemorrhagic fevers, small pox, respiratory conditions, and arthropod-borne encephalitides. Several of these viral groups are known to possess an outer lipid envelope, viz., Bunyaviridae, Flaviviridae, Orthomyxoviridae, and Poxviridae, which makes them more sensitive to chemical disinfection. These Select Agent viruses are generally not considered to be high-probability threats for water. This assumption is based upon both the lack of availability of many of these agents and the inability to produce these pathogens in large quantities and their sensitivity to chlorination.

Table 9.1 provides a list of potential waterborne bacterial bio-threat agents. Most of these Select Agents are primarily zoonotic pathogens. These organisms are self-replicating and can be propagated in large numbers in the microbiology laboratory. Of particular interest has been the bacterium *B. anthracis*, the etiological agent of anthrax. *B. anthracis* can form dormant structures known as spores which exhibit increased resistance to various environmental stresses. In the bioterrorist attacks in the United States in October, 2001, spores of *B. anthracis* were disseminated in letters which consequently contaminated private buildings, US Postal Service facilities, and the US Capital. This organism can be transmitted by direct contact, ingestion, and inhalation. Water could potentially provide a vehicle for all three modes of transmission.

The other listed bacterial Select Agents do not form spores and exist in the less resistant vegetative cell stage. While being considered bio-threat agents, some are also responsible for naturally occurring waterborne outbreaks. *Francisella tularensis* is the causative organism of the disease tularemia, sometimes referred to as rabbit fever. Tularemia can also be acquired by contact, ingestion, and inhalation. This organism, like *B. anthracis*, has previously been directly linked with state-supported biological warfare programs. *F. tularensis* subsp. *holarctica* is a known

Table 9.1 Potential waterborne bacterial bio-threat agents

Agent	Disease
Bacillus anthracis[a]	Anthrax
Brucella melitensis	Brucellosis
Brucella suis	
Burkholderia mallei	Glanders
Burkholderia pseudomallei	Melioidosis
Francisella tularensis	Tularemia
Yersinia pestis	Plague

[a]Spore-forming organism

naturally occurring waterborne pathogen which has been responsible for numerous outbreaks particularly in northern Europe. *Burkholderia mallei* is the etiological agent of the highly communicable animal disease known as glanders. This disease has primarily been associated with equine species, but is considered a potential waterborne bio-threat agent (Burrows and Renner, 1999). *Burkholderia pseudomallei*, the causative agent of melioidosis, is associated with waterborne transmission through skin wounds coming into contact with contaminated water. *B. pseudomallei* is endemic to areas of Southeast Asia and northern Australia.

Other bacteria which have been proposed as potential water bio-threat agents include the causative agents of brucellosis (*Brucella* spp.) and the plague bacillus *Yersinia pestis.* Neither of these organisms would be classified as naturally occurring waterborne pathogens. Brucellosis, sometimes referred to as undulant fever, occurs on a worldwide basis, with particular emphasis in Mediterranean countries. The disease is caused by several different species, viz., *Brucella abortus, Brucella melitensis, Brucella suis* and *Brucella canis*. Primary reservoirs are cattle, swine, sheep and goats, and canine species. *Y. pestis*, the etiological agent of bubonic plague, is cosmopolitan in occurrence. It is transmitted by arthropod vectors with various wild rodents and lagomorphs being the primary natural reservoirs. Like *B. anthracis* and *F. tularensis*, the plague bacillus has been implicated in state-sponsored programs for biological warfare.

Along with the above-mentioned bacterial species the rickettsial or rickettsia-like organism *Coxiella burnetii*, the causative agent of Q fever, has also been proposed as a potential agent for infection through drinking water (Burrows and Renner, 1999). Q fever is an acute febrile illness. Natural reservoirs include cattle, sheep, goats, and arthropod vectors. The disease can be transmitted by direct contact, inhalation, or in the case of contaminated milk, by ingestion. While considered a potential water threat, there is little information regarding *C. burnetii* in water.

9.3 Persistence

The survival of bio-threat organisms in water is not a well-studied area. Many of these organisms are zoonotic pathogens and have been isolated from various bodily secretions (e.g., urine, feces, and saliva) as well as from sewage and thus could serve as sources for water contamination. It is difficult to make generalizations regarding the persistence or survival in water for a given organism. Variations reported in the literature from studies using a wide array of experimental conditions (e.g., temperature, pH, water type, detection methodology, and inoculum preparation) all contribute to the fairly wide variability cited in published reports. Differences in survival characteristics have also been reported between strains of the same species. It is, however, widely accepted that spores of *B. anthracis* can persist for a longer time in the environment than vegetative cells of the other bacterial agents. Further information on environmental stability of various pathogens is available in several references (Mitscherlich and Marth, 1984; Sinclair et al., 2008).

9.4 Detection

The recognition that a microbiological agent has been introduced into a water system can be a formidable task. Routine monitoring or special case monitoring may detect changes in water quality parameters such as presence of fecal contamination, increases in turbidity, or depletion of disinfectant residual. Recent advances in the use of various sensor devices, placed at different points in a water system, reportedly show promise for assisting in these monitoring efforts. Yet it remains difficult to equate these observed changes in water quality with specific threats. Unfortunately, as in the case with most naturally occurring waterborne outbreaks, an incident may not be apparent until illness is detected in a community. It is important to have established procedures in place regarding the response to a suspected bioterrorism incident. These procedures should include specific protocols covering the various aspects associated with proscribed sampling and analysis plans for detection and analysis of bio-threat agents in environmental samples. The protocols should encompass such factors as choice of appropriate laboratory for analysis, sample transport conditions, chain of custody, bio-safety, and security concerns. Coordination of activities using a unified chain of command associated with environmental response teams is essential for insuring that samples comply with the various requirements both from a law enforcement and public health perspective. These protocols may vary during the course of an investigation of a bioterrorist incident as the perspective changes from the initial characterization phase to a remediation response.

The detection of a microbiological contaminant in a water supply, which in most instances will occur at low levels, will generally require some form of sample concentration. This initial concentration step in a sampling protocol has often been a major limiting factor in the ability to detect microorganisms. Recent advances in this area have involved the use of ultrafiltration devices, which allow for the concentration of target organisms from large volumes of water (Lindquist et al., 2007; Polaczyk et al., 2008).

Analysis procedures are dependent upon the level of information provided for a given sample type. This information may vary from a sample with completely unknown constituents to one where the analysis would be aimed at detecting a specific target organism. Microbiological analysis of samples can be divided into three basic approaches: cultural procedures involving in vitro cultivation and biochemical characterization, molecular procedures employing nucleic acid amplification and identification based upon unique genetic signatures, and immunological procedures based upon antigen–antibody reactions. All three approaches can be utilized for both initial isolation and identification. In practice often all three are used in laboratory analysis. Successful isolation by cultural methods addresses the issue of agent viability and provides individual isolates which can be further characterized by biochemical, immunological, and molecular genetic procedures. Antibiotic sensitivities can also be determined for these isolates. Amplification of genetic material using procedures such as real-time polymerase chain reaction (PCR) assays can provide valuable information in a timely fashion. Sequencing of genetic material provides a means for very specific identification of an organism and also allows for detecting strain variations. In the United States definitive identification of bio-agents is reserved for those laboratories, many of which are state public health laboratories, which are members of the CDC Laboratory Response Network (LRN). These LRN facilities are specialized laboratories which maintain high levels of bio-safety and bio-security and have access to specific diagnostic reagents not available in other microbiological laboratories.

A recent comprehensive study (Francy et al., 2009) compared the use of traditional procedures, viz., cultural and microscopic, with quantitative polymerase chain reaction (qPCR) for detecting biological agents in large volume drinking water samples which had been concentrated by ultrafiltration. The study encompassed two classical bacterial waterborne pathogens (*S. enterica* serovar Typhi and *V. cholerae*), the cysts of the protozoan parasite *Cryptosporidium parvum*, and surrogate organisms for two bacterial Select Agents: spores of *B. anthracis* Sterne and the attenuated *F. tularensis* LVS. It was concluded that qPCR, as well as traditional methods, could be used to rapidly detect these agents from large volumes of drinking water following concentration using ultrafiltration procedures. While it was noted that qPCR would require additional improvement for some of the assays, the procedure did hold promise as a method for detecting bio-threat agents in water.

9.5 Treatment

The issuance of "boil water advisories" is often the first public health measure undertaken in response to a known contamination event or the recognition of a waterborne disease outbreak. Bringing water to a rolling boil and maintaining those conditions for 2–3 min is normally deemed sufficient to inactivate the majority of waterborne pathogens, including encysted forms of protozoa, vegetative bacterial cells, and enteric viruses. However, depending on conditions, viz., covered vessel and altitude, boiling for this time period may not be sufficient to inactivate bacterial spores (Rice et al., 2004). Prior to the issuance of a "boil water advisory," water utility and public health authorities need to ascertain if this is the correct course of action since in some instances a "do not use advisory" maybe a more appropriate response.

Disinfection is a primary means for inactivating microbes in water. Conditions for the inactivation of microorganisms using chemical disinfectants are often described using *Ct* values, which are the product of the disinfectant concentration (*C*, mg/L) and the exposure time (*t*, min). *Ct* values are derived from experimental data and are used to determine required conditions to achieve a desired order of magnitude inactivation for a given microorganism under specific conditions of water temperature and pH. This concept has been utilized by regulatory agencies to insure adequate disinfection of potable water. As a general rule, inactivation proceeds at a faster rate as temperature increases. Free chlorine inactivation is more rapid at pH 7 than at pH 8 where the concentration of the more biocidal hypochlorous acid

Table 9.2 Chloride and monochloramine *Ct* values for inactivation of bacterial bio-threat agents

Agent	Temperature (°C)	Ct (mg-min/L) for 3 $\log_{10}$ inactivation: Chlorine pH 7	Monochloramine pH 8
Bacillus anthracis[a]	5	339	15,164
	25	102	1,847
Brucella melitensis	5	0.5	580
	25	0.2	117
Brucella suis	5	0.4	157
	25	0.2	56
Burkholderia mallei	5	0.2	194
	25	0.2	65
Burkholderia pseudomallei	5	0.7	156
	25	0.6	46
Francisella tularensis	5	10[b]	116
	25	4	37
Yersinia pestis	5	0.7	116
	25	0.6	33

Ct = *C*, disinfectant concentration (mg/L); *t*, time (minutes) – adapted from Rose et al. (2005, 2007)
[a]Spores
[b]Extrapolated value

species predominates over hypochlorite ion. Monochloramine is the predominate form of chloramine used in drinking water treatment and this form predominates at pH 8. *Ct* values for free chlorine and monochloramine inactivation (Table 9.2) have now been determined for many of the bacterial bioterrorism agents which have been proposed as potential water threats (Rose et al., 2005, 2007). As can be seen in Table 9.2, the spores of *B. anthracis* are much more resistant than the vegetative forms of the other bacterial agents.

Limited studies have been conducted on the use of alternative disinfectants for inactivating bio-threat agents. A recent report, where *Bacillus atrophaeus* subsp. *globigii* was used as a surrogate for *B. anthracis*, evaluated the effectiveness of chlorine dioxide for spore inactivation (Hosni et al., 2009). It was reported that both time and disinfectant concentration were equally important in determining inactivation rates for both free chlorine and chlorine dioxide, but that under the conditions studied (pH 8, 20°C), chlorine dioxide was more efficacious than free chlorine. This difference was attributed to fewer mass transfer limitations and the greater oxidizing power of chlorine dioxide.

The use of ultraviolet (UV) irradiation in the treatment of drinking water has gained in acceptance owing to the ability of this unit process to inactivate encysted protozoa, especially *Cryptosporidium* spp. For inactivation of microorganisms the fluence or UV dose is measured in units of energy (e.g., milli-Joules, mJ) per area (e.g., square centimeter, cm^2). It has been reported that vegetative bacterial bio-threat agents were susceptible to UV light (254 nm) with a four order of magnitude inactivation being achieved at a fluence of <12 mJ/cm^2. However, spores of

B. anthracis exhibited increased resistance to UV exposure, and specific fluences for given levels of inactivation could not be determined owing to a tailing effect seen in the inactivation curves. Higher UV fluences were effective for inactivating encysted protozoa and viruses (Rose and O'Connell, 2009).

Differences in resistance to disinfection have been observed in species of the same bacterial genus and even between strains of the same species. Studies comparing the inactivation of spores of *Bacillus* spp. have noted differences in chlorine resistance between attenuated and virulent strains of *B. anthracis* and between these strains and other *Bacillus* species. As seen in Table 9.3, *B. atrophaeus* subsp. *globigii*, an often used surrogate organism, was more resistant to chlorination than the virulent *B. anthracis* Ames strain, which in turn was more resistant than the attenuated *B. anthracis* Sterne strain. This increased resistance suggests that spores of *B. atrophaeus* subsp. *globigii* might serve as a conservative indicator of chlorine inactivation (Sivaganesan et al., 2006). Spores of *Bacillus thuringiensis* subsp. *israelensis* yielded *Ct* values which were closest to the values obtained for the virulent *B. anthracis* Ames strain (Rice et al., 2005).

These observed differences have not been confined to the spore-forming *Bacillus* species. In a comprehensive study designed to address apparent discrepancies in the published literature concerning the resistance to chlorination of *B. pseudomallei*, O'Connell et al. (2009) reported that differences between strains were correlated with the relative amount of extracellular material produced by each isolate. Mucoid strains were found to have *Ct* values an order of magnitude higher than non-mucoid strains for similar levels of inactivation. Rose et al. (2005) reported a greater than threefold increase in free chlorine *Ct* values for a virulent strain of *F. tularensis* compared to the attenuated live vaccine strain (LVS) of this organism which is commonly used as a surrogate for virulent strains of this agent. The use of surrogates can often obviate the stricter bio-safety and bio-security concerns associated with experimentation with virulent strains of bioterrorism organisms. However, as these

Table 9.3 Mean free chlorine *Ct* values at pH 7 for inactivation of spores of *Bacillus* spp.

		Ct (mg-min/L)	
Organism	Temperature (°C)	2 log_{10} inactivation	3 log_{10} inactivation
Bacillus atrophaeus subsp.	5	372	446
globigii	23	108	136
Bacillus anthracis Ames	5	220	339
	23	79	102
Bacillus anthracis Sterne	5	140	210
	23	45	68
Bacillus cereus	5	117	175
	23	41	62
Bacillus thuringiensis subsp.	5	229	344
israelensis	23	66	99

Adapted from Sivaganesan et al. (2006)

reports suggest care should be taken when evaluating disinfection efficacy based solely upon results obtained from data derived from a single strain of a pathogen or from surrogate organisms.

Physical associations of organisms have also been linked to increased resistance to disinfection. Chlorination becomes much more problematic when organisms are associated with biofilms or corrosion products such as those which might be encountered in a drinking water distribution system. Using a biofilm annular reactor Szabo et al. (2006, 2007) showed that *Klebsiella pneumoniae*, a surrogate for vegetative bacterial bioterrorism agents, and spores of *B. atrophaeus* subsp. *globigii*, a surrogate for *B. anthracis*, were able to persist for extended periods of time in chlorinated water when in association with biofilm and corrosion products.

9.6 Summary

Concerns regarding biological threats to a water system remain an important consideration for protecting the public from acts of terrorism. Several authors (Meinhardt, 2005; Nuzzo, 2006) have noted the need for increased scientific understanding of the true vulnerability of water supplies to intentional contamination with biological threat agents. In the event of such an incident information on stability and treatability of these agents would be a paramount concern for water utilities and public health authorities. This sentiment has been previously expressed by other authors (Clark and Deininger, 2000; Khan et al., 2001) shortly before or at the time of the aforementioned anthrax bioterrorism events in the United States. The ability to distinguish between a natural occurring waterborne outbreak and one which is the result of an intentional contamination event appears to be one area which should be given high priority (Grunow and Finke, 2002). As cited above, much progress has been made in recent years in providing information on bio-threat agents which could be used to contaminate public water supplies. Data on the resistance to disinfection as commonly applied in water systems for many of these organisms have only now been determined for the first time. A recent report by the Commission on the Prevention of Weapons of Mass Destruction Proliferation and Terrorism (http://www.preventwmd.gov/static/docs/report-card.pdf), a group commissioned by the US Congress, noted in their reporting on potential terrorist activity that a critical area of concern centered on the use of biological weapons. The development of specific response plans, coupled with continued research efforts, is clearly warranted to address the issue of protecting water supplies from biological attacks.

References

American Public Health Association. (2005). *Standard Methods for the Examination of Water and Wastewater,* 21st ed. Washington, DC: American Public Health Assoc.

American Water Works Association. (2006). *Waterborne Pathogens, Manual of Water Supply Practices M48,* 2nd ed. Denver, CO: American Water Works Assoc.

Berger, Bernard B., and Stevenson, Albert H. (1955). "Feasibility of Biological Warfare Against Public Water Supplies." *Journal of the American Water Works Association* 47, No. 2 February, pp. 101–110.

Burrows, W. Dickinson, and Renner, Sara E. (1999). "Biological Warfare Agents as Threats to Potable Water." *Environmental Health Perspectives* 107, No. 12 December, pp. 975–984.

Clark, Robert M., and Deininger, Rolf A. (2000). "Protecting the Nation's Critical Infrastructure: The Vulnerability of U.S. Water Supply Systems." *Journal of Contingencies and Crisis Management* 8, No. 2 June, pp.73–80.

Francy, D.S., Bushon, R.N., Brady, A.M.G., Bertke, E.E., Kephart, C.M., Likirdopulos, C.A., Mailot, B.E., Schaefer, F.W., 3rd, and Lindquist, H.D. Alan. (2009). "Comparison of Traditional and Molecular Analytical Methods for Detecting Biological Agents in Raw and Drinking Water Following Ultrafiltration." *Journal of Applied Microbiology* 107, No. 5 November, pp. 1479–1491.

Gleick, Peter H. (2006). "Water and Terrorism." *Water Policy* 8, No. 6, pp. 481–503.

Grunow, R. and Finke, E.-J. (2002). "A Procedure for Differentiating Between the Intentional Release of Biological Warfare Agents and Natural Outbreaks of Disease: Its Use in Analyzing the Tularemia outbreak in Kosovo in 1999 and 2000." *Clinical Microbiology and Infection* 8, No. 8 August, pp. 510–521.

Hosni, A.A., Shane, W.T., Szabo, J.G., and Bishop, P.L. (2009). "The Disinfection Efficacy of Chlorine and Chlorine Dioxide as Disinfectants of *Bacillus globigii*, a Surrogate for *Bacillus anthracis*, in Water Networks: a Comparative Study." *Canadian Journal of Civil Engineering*, 36, No.4 April, pp. 732–737.

Khan, Ali S., Swerdlow, David L., and Juranek, Dennis D. (2001). "Precautions Against Biological and Chemical Terrorism Directed at Food and Water Supplies." *Public Health Reports* 116, January-February, pp. 3–14.

Meinhardt, Patricia L. (2005). "Water and Bioterrorism: Preparing for the Potential Threat to U.S. Water Supplies and Public Health." *Annual Review of Public Health* 26, April, pp. 213–237.

Mitscherlich, Eilhard, and Marth, Elmer H. (1984). *Microbial Survival in the Environment*. Berlin: Springer.

Nuzzo, Jennifer B. (2006). "The Biological Threat to U.S. Water Supplies: Toward a National Water Security Policy." *Biosecurity and Bioterrorism: Biodefense Strategy, Practice and Science* 4, No. 2, pp. 147–159.

Lindquist, H.D.A., Harris, S., Lucas, S., Hartzel, M., Riner, D., Rochele, P., and DeLeon. R. (2007). "Using Ultrafiltration to Concentrate and Detect *Bacillus anthracis, Bacillus atrophaeus* subspecies *globigii*, and *Cryptosporidium parvum* in 100-Liter Water Samples. *Journal of Microbiological Methods* 70, No. 3 September, pp. 484–492.

O'Connell, H.A., Rose, L.J., Shams, A., Bradley, M., Arduino, M.J., and Rice, E.W. (2009). "Variability of *Bukholderia pseudomallei* Strain Sensitivities to Chlorine Disinfection." *Applied Environmental Microbiology* 75, No. 16 August, pp. 5405–5409.

Polaczyk, A.L., Narayanan, J., Cromeans, T.L., Hahn, D., Roberts, J.M., Amburgey, J.E., and Hill, V.R. (2008). "Ultrafiltration-based Techniques for Rapid and Simultaneous Concentration of Multiple Microbe Classes from 100-L Tap Water Samples." *Journal of Microbiological Methods* 73, No. 2 May, pp. 92–99.

Rice, Eugene W., Rose, L.J., Johnson, C.H., Boczek, Laura A., Arduino, Matthew J., and Reasoner, Donald J. (2004). "Boiling and *Bacillus* Spores." *Emerging Infectious Diseases* 10, No.10 October, pp. 1887–1888.

Rice, E.W., Adcock, N.J., Sivaganesan, M., and Rose, L.J. (2005). "Inactivation of Spores of *Bacillus anthracis* Sterne, *Bacillus cereus*, and *Bacillus thuringiensis* subsp. *israelensis* by Chlorination." *Applied Environmental Microbiology* 71, No. 9 September, pp. 5587–5589.

Rose, Laura J., Rice, Eugene W., Jensen, Bette, Murga, Richardo, Peterson, Alicia, Donlan, Rodney M., and Arduino, Matthew J. (2005). "Chlorine Inactivation of Bacterial Bioterrorism Agents." *Applied Environmental Microbiology* 71, No.1 January, pp. 556–568.

Rose, Laura J., Rice, Eugene W., Hodges, Lisa, Peterson, Alicia, and Arduino, Matthew J. (2007). "Monochloramine Inactivation of Bacterial Select Agents." *Applied Environmental Microbiology* 73, No.10 May, pp. 3437–3439.

Rose, L.J., and O'Connell, H. (2009). "UV Light Inactivation of Bacterial Biothreat Agents." *Applied Environmental Microbiology* 75, No. 9 May, pp. 2987–2990.

Sinclair, Ryan, Boone, Stephanie A., Greenberg, David, Keim, Paul, and Gerba, Charles P. (2008). "Persistence of Category A Select Agents in the Environment." *Applied Environmental Microbiology* 74, No. 3 February, pp. 555–563.

Sivaganesan, M., Adcock, N.J., and Rice, E.W. (2006). "Inactivation of *Bacillus globigii* by Chlorination: A Hierarchical Bayesian Model." *Journal of Water Supply: Research and Technology-AQUA* 55.1, pp. 33–43.

Szabo, Jeffrey G., Rice, Eugene W., and Bishop, Paul L. (2006). "Persistence of *Klebsiella pneumoniae* on simulated biofilm in a model drinking water system." *Environmental Science and Technology* 40, No.16 August, pp. 4996–5002.

Szabo, Jeffrey G., Rice, Eugene W., and Bishop, Paul L. (2007). "Persistence and Decontamination of *Bacillus atrophaeus* subsp. *globigii* Spores on Corroded Iron in a Model Drinking Water System." *Applied Environmental Microbiology* 73, No. 8 April, pp. 2451–2457.

Chapter 10
Rapid Detection of Bacteria in Drinking Water and Wastewater Treatment Plants

Rolf A. Deininger, Jiyoung Lee, and Robert M. Clark

10.1 Introduction

Water systems are spatially diverse and are therefore inherently vulnerable to physical, chemical, and biological threats that might compromise a systems' ability to reliably deliver safe water. Community water supplies are designed to deliver water under pressure and generally supply most of the water for fire-fighting purposes. A loss of water or a substantial loss of pressure could, therefore, disable fire-fighting capability, interrupt service, and disrupt public confidence. This loss might result from sabotaging pumps that maintain flow and pressure or disabling electric power sources that might cause long-term disruption. Many of the major pumps and power sources in water systems have custom-designed equipment and could take months or longer to repair and/or replace (Clark and Deininger, 2001).

Major areas of vulnerability include

- Raw water source (surface or groundwater)
- Raw water channels and pipelines
- Raw water reservoirs
- Treatment facilities
- Connections to the distribution systems
- Pump stations and valves and
- Finished water tanks and reservoirs

Each of these system elements presents unique challenges to a water utility in safeguarding water supply (Clark and Deininger, 2000) and water systems are vulnerable to both physical attacks and/or contamination.

The ability of a water supply to provide water to its customers can be compromised by destroying or disrupting key physical elements of the water system.

R.A. Deininger (✉)
School of Public Health, The University of Michigan, Ann Arbor, MI, USA
e-mail: rad@umich.edu

R.M. Clark et al. (eds.), *Handbook of Water and Wastewater Systems Protection*, Protecting Critical Infrastructure, DOI 10.1007/978-1-4614-0189-6_10,
© Springer Science+Business Media, LLC 2011

These elements include raw water facilities (dams, reservoirs, pipes, and channels) treatment facilities, and distribution system elements (transmission lines and pump stations). Physical disruption may result in significant economic cost, inconvenience, and loss of confidence by customers, but has a limited direct threat to human health. Exceptions to this generalization include (1) destruction of a dam that causes loss of life and property in the accompanying flood wave and (2) an explosive release of chlorine gas at a treatment plant. Water utilities should examine their physical assets, determine areas of vulnerability, and increase security accordingly. An example of such an action might be to switch from chlorine gas to liquid hypochlorite, especially in less secure locations, which decreases the risk of exposure to poisonous chlorine gas. Redundant system components would provide backup capability in case of accidental or purposeful damage to facilities.

Contamination is generally viewed as the most serious potential terrorist threat to water systems. Chemical or biological agents could spread throughout a distribution system and result in sickness or death among the consumers and for some agents the presence of the contaminant might not be known until emergency rooms report an increase in patients with a particular set of symptoms. Even without serious health impacts, just the knowledge that a group had breached a water system could seriously undermine consumer confidence in public water supplies (Clark, 2002). Accidental contamination of water systems has resulted in many fatalities. Examples of such outbreaks include cholera contamination in Peru (Clark et al., 1995), *Cryptosporidium* contamination in Milwaukee, Wisconsin (USA) (Fox and Lytle, 1996), and *Salmonella* contamination in Gideon Missouri (USA). In Gideon the likely culprit was identified as pigeons infected with *Salmonella* that had entered a tank's corroded vents and hatches (Clark et al., 1996).

10.2 Waterborne Pathogens

Waterborne pathogens have been recognized as a threat to human public health throughout history but the development of drinking water treatment techniques has controlled this threat since the beginning of the twentieth century. Although modern drinking water treatment has virtually eradicated waterborne disease from developed countries, drinking water treatment systems have been identified as a potential security vulnerability.

Water-related microbial pathogens can be categorized as water-based or waterborne pathogens. Water-based pathogens spend part of their life cycle to reach and infect a potential host. An excellent example of a water-based pathogen is malaria for which mosquitoes are a vector. Since water-based pathogens are not transmitted totally through water they are not potential agents of bio-terrorism.

However waterborne pathogens are those transmitted through ingestion of contaminated water primarily through the fecal-oral route. In this case water acts as a passive carrier of infectious agents. Some waterborne pathogens that can cause problems in drinking water include *Campylobacter jejuni*, pathogenic *Escherichia coli*, *Yersinia enterocolitica*, enteric viruses such as rotavirus, calicivirus, astrovirus, and parasites such as *Giardia lamblia*, *Cryptosporidium parvum*, and Microsporidia. Some species of environmental bacteria have demonstrated the ability to survive in drinking water biofilms and have been identified as opportunistic pathogens including *Legionella* spp., *Aeromonas* spp., *Mycobacterium* spp., and *Pseudomonas aeruginosa* (Abbaszadegan and Alum, 2004; Rice et al., 1999; Geldreich et al., 1992).

Bacterial pathogens can cause gastroenteritis including cramps, diarrhea, nausea, vomiting, chills, and mild fever. Bacterial pathogens are generally sensitive to disinfectants such as chlorine and include (Abbaszadegan and Alum, 2004; Clark and Deininger, 2000, 2001; Field, 2004)

- *Salmonella*
- *Shigella*
- *E. coli* O157:H7
- *Yersinia*
- *Vibrio*
- *Campylobacter*
- *Legionella*

Viral pathogens can pose a 10- to 10,000-fold higher infection risk than bacteria. Important waterborne viral pathogens include

- *Adenovirus*
- *Astroviruses*
- *Hepatitis A*
- *Hepatitis E*
- *Norovirus*
- *Rotaviruses*

Parasitic pathogens are a significant threat to drinking water supplies. Nearly 20,000 protozoan parasites have been identified of which 20 genera are known to cause disease in humans including

- *Acanthamoeba*
- *Cryptosporidium parvum*
- *Entamoeba histolytica*
- *Microsporidia*
- *Naegleria*

In general, the most effective mechanism for controlling these pathogens is disinfection with especially chlorine. In addition to general pathogens of concern in water supplies some pathogens can be categorized as biological warfare agents. Some of the organisms that have potential use in bio-terrorism are discussed below (Burrows and Renner, 1998, 1999; Clark and Deininger, 2000, 2001).

10.3 History of Water System Contamination

As discussed by Clark et al. (Chapter 8, this volume) the recorded history of attacks on water systems dates from 4,500 years ago (Gleick, 2006). Urlama, King of Lagash, and his son Illater cut off the water supply to Girsu, a city in Umma during the period 2450 to 2400 BC. In New York in 1748 an angry mob burned down a ferry house on the Brooklyn shore of the East River. It is reported that this act was a revenge for unfair allocation of East River water rights. Small groups attacked small dams and reservoirs in the 1840s and 1850s in the eastern and central USA due to concerns about threats to health and to local water supplies. In the Owens Valley of California between 1907 and 1913 farmers repeatedly dynamited the aqueduct system being built to divert their water to the growing city of Los Angeles.

Bacillus anthracis, a spore-forming bacterium, which has been weaponized for aerosol application, was used by the Japanese Army during World War II to contaminate food and water supplies of Chinese cities (WHO, 1970; Williams and Wallace, 1989; Burrows and Renner, 1999). Abdominal pain, fever, vomiting, bloody diarrhea, and shock are the principal manifestations of this form of the disease, which has an incubation period of 2–7 days. Anthrax spores are easily removed by any water treatment filter system with pore size <1 μm. The Japanese Army also used a number of other organisms to contaminate food and water including the bacterium *Vibrio cholera*; plague a disease of rodents, both wild and domestic, caused by the bacillus *Yersinia pestis* and transmissible to humans generally considered to be a threat in water as well; *Salmonella typhimurium* often found in outbreaks of food poisoning; and *Shigella dysenteriae.* Mycotoxins: T-2 toxin is one of several trichothecene mycotoxins isolated from cereal grains infected with *Fusarium* and some other genera of fungi. Russian experience with infected agricultural products indicates that ingested trichothecenes could impose a deadly threat (Burrows and Renner, 1999). Unconfirmed and controversial findings suggest the use of trichothecenes as BW agents in Laos, Cambodia, and Afghanistan, and Iraq has investigated the weaponization of trichothecenes. Other trichothecenes, viz., nivalenol, 4-deoxynivalenol, and diacetoxyscirpenol, may be present in crude preparations; their toxicities are probably similar to but no greater than that of T-2.

10.3.1 Recent Water Contamination Experience in the USA

In the fall of 1996 we sampled drinking water in Washington D.C. A sample of high bacterial level is shown in Fig. 10.1. The Capitol is shown in the background. The drinking water supply was overchlorinated but otherwise safe. Tests in other places also showed contamination, but not above the allowable levels in Washington D.C. (see Figs. 10.1 and 10.2).

There have been many examples of contamination and potential water supply contamination in the USA including waterborne outbreaks. For example, in New York City, low levels of plutonium were found in the drinking water (in the order of 20 fCi). The usual background is below 1 fCi. However, a person would have

Fig. 10.1 Fountain in the center of Washington, D.C. with the capitol in the background

Fig. 10.2 Analyzing for bacteria levels in front of the Smithsonian Institute

to drink several million liters of water to acquire a lethal dose estimated at about 100 μCi. A femtocurie is nine orders of magnitude smaller than a microcurie (Clark and Deininger, 2000). Another case was the contamination of salad bars in Dalles, Oregon, by the Rajneeshee religious cult, using vials of *S. typhimurium. S. typhimurium* is a highly toxic bacteria frequently carried by birds. The cult also contaminated a city water supply tank using Salmonella. A community outbreak of salmonellosis resulted in which at least 751 cases were documented in a county that typically reports fewer than 5 cases per year. The cult apparently cultured the organisms in their own laboratories (Clark and Deininger, 2000; Gleick, 2006). Monitoring of chlorine residuals is not a universal practice, but it can be done at minimal cost. On July 10, 1986, the water supply to the presidential rooms of the White House was cut off after a monitor indicated a lack of chlorine. President Reagan got his morning coffee anyhow, using bottled water. The supply to the West Wing was not shut off, but the staff was warned not to drink the water (New York Times, 1986). Strategically placed monitors in a distribution system provide one solution for the protection of a water supply system.

10.3.1.1 Cabool Missouri Case Study

Cabool, located in the Southeastern corner of Missouri, experienced a large outbreak of *E. coli* O157:H7 during the winter of 1989–1990. Cabool, a town of approximately 2,100 people, experienced a waterborne disease outbreak in which 243 cases were reported, with 32 hospitalizations and 4 deaths. This was the largest waterborne outbreak of *E. coli* O157:H7 that had been reported in the USA. At the time of the outbreak, the water source was untreated groundwater. Shortly after the outbreak, the USEPA sent a team to conduct a research study to determine the underlying cause of the outbreak.

Exceptionally cold weather prior to the outbreak contributed to two major water system line breaks and 43 water meter replacements throughout the city area. The sewage collection lines in Cabool were generally located away from the drinking water distribution lines, but did cross or were near water lines in several locations. At the time of the outbreak, storm water drained via open ditches along the sides of the streets and roads. During heavy rainfalls, sewage was observed to overflow manhole covers and flow into streets, parking lots, and residential foundations.

The town's water system (untreated groundwater) was implicated in the outbreak. Two of the town's four wells were operating at the time of the outbreak: one was 305 m deep and the other was 396 m deep. Both wells had protected wellheads, and the monitoring data from the 10 years before the outbreak indicated that no coliforms had been detected in either well. Investigation of the outbreak indicated that the distribution system was not well maintained and was vulnerable to sewage contamination at several points. Approximately 35% of the total flow was lost in the system – suggesting leaks, inaccurate meters, or unmetered connections. The town sewer system was also in poor condition and operating beyond capacity, resulting in regular sewage backups and overflows.

A number of risk factors apparently contributed to this outbreak. In mid-December 1989, unusually cold weather caused 2 large water mains and 45 in-ground water meters to fail. Ten cases of bloody diarrhea were reported to the local health department on January 4, 1990. A boil-water order was issued on January 5, and water chlorination was initiated on January 12. Analyses of the temporal distribution of the cases indicated that the first cases occurred 7 days before the first water main break (December 23), and the last case occurred 3 days after the implementation of water chlorination. The early cases may have been due to leaks and holes that developed prior to the main break. There was a small increase in the incidence of diarrhea after the first main break and a large increase in diarrhea cases about 4 days after the second main break on December 26.

Hydraulic and water quality models were applied to examine the movement of water and contaminants in the system. Steady-state scenarios were examined and a dynamic analysis of the movement of water and contaminants associated with meter replacement and the aforementioned breaks was conducted. Typical demand patterns were developed from available meter usage for each service connection and it was found that the water demand was 65% of the average well production, indicating inaccurate meters, unmetered uses, and a high water loss in the system.

The modeling effort revealed that the pattern of illness occurrence was consistent with water movement patterns in the distribution system assuming two water line breaks. It was concluded, therefore, that some disturbance in the system, possibly the 2 line breaks or 43 meter replacements, allowed contamination to enter the water system. Analysis showed that the simulated contaminant movement covered 85% of the infected population.

Replacement of the failed water meters may have contributed to contamination of the distribution system. During the replacement of the meters and main break repairs, the lines were subjected to "limited flushing" but were not disinfected by super-chlorination, and no water samples were tested for microbial indicators to examine the water quality before bringing the lines back into service. Although sewage overflow into the distribution system via the main breaks and intrusion was believed to be responsible for the outbreak, microbial contamination of the distribution system could not be confirmed because water samples from the distribution system were never collected and analyzed. Hydraulic modeling of the system reinforced the evidence that the second main break had the potential to contaminate a greater portion of the distribution system, including the northern part of the town where 36% of the cases occurred (Geldreich et al., 1992).

10.3.1.2 Gideon, Missouri Case Study

In 1993, the town of Gideon, Missouri, suffered from an outbreak of salmonellosis that affected more than 650 people and caused 7 deaths (Hrudey and Hrudey, 2004). At the time of the outbreak, Gideon was a small town (population 1,100) in a rural, agricultural area with an unemployment rate greater than 11%. Twenty-five percent of the population was living below the poverty level at the time.

The Missouri Department of Health (MDOH) had identified 31 cases of laboratory-confirmed salmonellosis (Clark et al., 1996). The State Public Health Laboratories identified 21 of these isolates as dulcitol negative *Salmonella* serovar *Typhimurium*. *Salmonella* is a pathogenic bacterium that has been classified into several serotypes (common set of antigens). *Salmonella* serovar *Typhimurium* is among the most common *Salmonella* serovars causing salmonellosis in the USA. Fifteen of the 31 laboratory culture-confirmed patients were hospitalized (including 2 patients hospitalized for other causes and who developed diarrhea while in the hospital). These 15 patients were admitted to 10 different hospitals. Seven nursing home residents exhibiting diarrheal illness died; four of these patients were culture confirmed (the other three were not cultured). Two of the patients had positive blood cultures. Interviews conducted by the MDOH during this period suggested that there were no food exposures common to a majority of the patients. However, all of the ill persons, including the culture-confirmed patients, had consumed municipal water which supported the association. The MDOH reported their suspicion to the Missouri Department of Natural Resources (MDNR).

The Gideon municipal water system was originally constructed in the mid-1930s and obtained water from two adjacent, 1,300 ft deep wells. The well waters were not disinfected at the time of the outbreak. After the outbreak emergency, chlorination was initiated, and later a permanent chlorination system was installed. The distribution system consisted primarily of small diameter (2-, 4-, and 6-in.) unlined, steel and cast iron pipe. Tuberculation and corrosion were major problems in the distribution pipes. Raw water temperatures were unusually high for a groundwater supply system (58°F), because the system overlies a geologically active fault. Under low-flow or static conditions, the water pressure was close to 50 psi. However, under high-flow or flushing conditions the pressure dropped dramatically. These sharp pressure drops were evidence of major problems in the Gideon distribution system. The municipal system had two elevated tanks. One tank was a 50,000 gallon (gal) tank (referred to as small tank) and the other was a 100,000 gal tank (referred to as large tank).

In early November, a cold snap caused a thermal inversion in the water storage tanks that resulted in taste and odor problems. In response, the water system was systematically flushed on November 10. The first cases of acute gastroenteritis were reported on November 29 and diagnosed as *S. typhimurium*. However, the outbreak investigation later revealed that diarrhea cases in Gideon started around November 12 with a peak incidence around November 20. By early December, there was a 250% increase in absenteeism in the Gideon schools and a 600% increase in anti-diarrheal medication sales. Over 40% of nursing home residents suffered from diarrhea and seven people died (Angulo et al., 1997). However, the outbreak was not linked to the water system until December 15 when the water system samples were reviewed and investigative water sampling was initiated. A boil-water advisory was issued on December 18. On December 22, emergency chlorination was added to the production well and the two municipal storage tanks were superchlorinated. The last reported cases occurred on December 28.

Water samples collected from the distribution system on December 16, 17, 20, and 21 were positive for total coliforms, and the samples from December 20 and 21 were also positive for fecal coliforms. The outbreak strain of *S. typhimurium* was detected in one large volume sample collected from a fire hydrant. Inspection of the water storage tanks suggested that the outbreak was probably caused by contamination by bird feces in one or more of the tanks. The larger of the municipal tanks was in disrepair and had birds roosting on the roof. The private storage tank had an unscreened overflow pipe and a hole at the top of the tank that was large enough for birds to enter. This private tank had been drained on December 30, but the outbreak strain of *S. typhimurium* was detected in samples of sediment collected on January 5, 1994. The remaining water on the bottom of the tank was described as black and very turbid, with rust, suspended particles, and bird feathers floating on the top. Initially attention was focused on the private tank as the source of the outbreak as reported by Skala (1994). On January 14, 1994, an EPA field team, in conjunction with the CDC and the State of Missouri, initiated a field investigation that included a sanitary survey and microbiological analyses of samples collected on-site. A system evaluation was also conducted in which EPANET was used to develop various scenarios to explain possible contaminant transport in the Gideon system.

Although the private tank was initially suspected as the cause of outbreak an in-depth hydraulic analysis of the Gideon system, conducted as part of the outbreak investigation, raised questions about the possibility of the private tank being the source of the outbreak. A subsequent review of as-built drawings of the Gideon system by MO DNR personnel revealed that the private tank was separated from the municipal system by a backflow prevention valve, but there were conflicting reports about whether this valve was open or closed when the tank was first inspected. In order to test the integrity of the valve a "fire pumper" was brought to the site and it was found that the backflow prevention valve held under a pressure of 100 psi. In a subsequent hydraulic analysis the private tank was eliminated as a contamination source for the outbreak. The analysis demonstrated that elimination of the private tank yielded results that were consistent with the behavior of the system as observed during the outbreak scenario. This analysis also pointed to the largest municipal tank as the most likely source of the outbreak. A visual inspection of the large municipal tank revealed broken and rusted hatches, bird parts and feathers on the top of the tank, and bird parts and feathers floating on the surface of the tank water. Therefore, the subsequent EPA field investigations and modeling efforts focused on the two municipal tanks as the source of contamination.

The key analysis was focused on a flushing program conducted earlier by the utility in response to taste and odor complaints. A sequential flushing program was conducted on November 10, 1993, involving all 50 hydrants in the system. The flushing program was started in the morning and continued through the entire day. Each hydrant was flushed for 15 min at an approximate rate of 750 gallons per minute (gpm). It was observed that the pump at one of the wells was operating at full capacity during the flushing program (approximately 12 h), which would indicate that the municipal tanks were discharging during this period.

During the evaluation, it was hypothesized that the taste and odor problems may have resulted from a thermal inversion that had taken place due to a sharp temperature drop prior to the day of the complaint. If stagnant or contaminated water were floating on the top of a tank, a thermal inversion could have caused this water to be mixed throughout the tank and to be discharged into the system resulting in taste and odor complaints (Fennel et al., 1974). As a consequence, the utility initiated the aforementioned city-wide flushing program. Turbulence in the tank from the flushing program could have stirred up the tank sediments that were subsequently transported into the distribution system. It is likely that the bulk water and/or the sediments were contaminated with *Salmonella* serovar *Typhimurium*. During the EPA field visit, a large number of pigeons (bird droppings are known to contain *Salmonella*) were observed roosting on the roof of the 100,000 gal municipal tank.

The EPA study team evaluated the effects of distribution system design and operation, demand, and hydraulic characteristics on the possible propagation of contaminants in the system. Given the evidence from the lab samples and the results from the valve inspection of the private tank, it was concluded that the most likely contamination source was bird droppings in the large municipal tank. Therefore, the analysis concentrated on propagation of water from the large municipal tank in conjunction with the flushing program. Other possible sources of contamination such as cross-connections were also studied, but ruled out as a source of the outbreak.

The system layout, demand information, pump characteristic curves, tank geometry, flushing program, and other information needed for the modeling effort were obtained from maps and demographic information and numerous discussions with consulting engineers and city and MDNR officials. EPANET was used to conduct the contaminant propagation study (Rossman et al., 1994).

The EPANET network model was calibrated by simulating flushing at the hydrants assuming a discharge of 750 gpm for 15 min. The "C" factors (pipe roughness) were adjusted until the head loss in the model matched head losses observed in the field. After the calibration, the hydraulic model was simulated for 48 h. Thereafter, the flushing program was simulated starting at 8 am on day 3, by sequentially imposing a 750 gpm demand on each hydrant for 15 min. Utilizing the TRACE option in EPANET, the percentages of water from both municipal tanks were calculated at each node over a period of 72 h.

During the simulation of the flushing program, the pump at one of the wells was operated (as previously observed) at full capacity, which was over 800 gpm, and then reverted to cyclic operation. The simulation results showed that the tank elevation fluctuated for both municipal tanks, and both the tanks discharged during the flushing program. At the end of the flushing period, nearly 25% of the water from the large municipal tank passed through the small municipal tank where it was again discharged into the system. The model predicted dramatic pressure drops during the flushing program. Based on the information available, it was felt that these modeling results replicated the conditions that existed during the flushing program closely enough to provide a basis for an analysis of water movement in the system.

Data from the simulation study, the microbiological surveillance data, and the outbreak data were utilized to provide insight into the nature of both general contamination problems in the system and the outbreak itself. The water movement patterns showed the majority of the collected samples that were total coliform and fecal coliform (FC) positive occurred at points within the zone of influence of the small and large tanks. During both the flushing program and for large parts of normal operation, these areas were predominately served by tank water, which confirmed the belief that the tanks are the source of the fecal contamination since there were positive FC samples prior to chlorination.

10.3.1.3 Walkerton, Canada Case Study

The first documented outbreak of *E. coli* O157:H7/*Campylobacter* spp. gastroenteritis associated with a municipal water supply in Canada occurred in the small rural town of Walkerton, Ontario (population 1,261), in May 2000 (Walkerton Herald Times, 2000; Grayman et al., 2004). At the time of the outbreak, the town's drinking water was supplied by three wells (Wells 5, 6, and 7), which fed a common distribution system. A hydro-geological assessment revealed that two of the three wells (Wells 5 and 6) were under the influence of surface water. A microbiological investigation found *E. coli* O157:H7 and *Campylobacter* spp. on a cattle farm adjacent to Well 5 with identical molecular characteristics to those isolated from outbreak victims. The water supply pumped by generator to the local hospital is seen in Fig. 10.3.

Fig. 10.3 The water supply pumped by generator to the hospital in Walkerton

In order to understand the factors that caused the outbreak a water quality model of the Walkerton water distribution system was developed. Using a "cross-sectional" study, it was demonstrated that during the outbreak, residents living in homes connected to the municipal water supply and consuming Walkerton water were 11.7 times more likely to have developed gastroenteritis than those not exposed to Walkerton water.

Modeling of the Walkerton water system involved estimating the following parameters for use in WaterCAD (a proprietary water quality model):

- Pipe diameter and length, location, age, and composition of all water pipes
- Size, storage capacity, and active volumes of the two stand pipes (water towers) in the system
- Well pump specifications (including pump curves)
- Pipe friction

The objective of the water modeling exercise was to recreate the pattern of water flow throughout the town's distribution system immediately before and during the outbreak period. The system was modeled as a network of nodes, which included water sources, tanks, and demand (customer) locations. Individualized temporal water demand patterns were assigned to each commercial node and residential users were assigned to the nearest residential node. As residential users in Walkerton were not metered, hourly demand was estimated using the daily volume of water supplied to the system after accounting for commercial users and fire events and literature-based hourly demand patterns. Well pump controls were added to the model and pump on and off times were set in accordance with the historical pump records. Computerized data from the supervisory control and data acquisition (SCADA) system for the water supply containing 15-min pumpage rates for the three wells were utilized. Exposure scenarios were created by adding a hypothetical inert contaminant to each well at predetermined times and concentrations and then using the model to follow the movement and relative concentrations of contaminants through the distribution system.

Conditional logistic regression was used to quantitatively evaluate the relationship between exposure to potentially contaminated well water, from each of Walkerton's three wells, and the likelihood of experiencing infection. A median incubation period of 2–5 days was assumed for both *E. coli* O157:H7 and *Campylobacter* for water supplied to homes 2–5 days prior to the illness onset date, and potential exposure scenarios were calculated for each case. This was done for each of the six exposure scenarios.

The results of this study clearly supported the hypothesis that Well 5 was the primary, if not the only, well involved in the Walkerton *E. coli*/*Campylobacter* waterborne outbreak. The results also suggest that an extreme rainfall event, which occurred just prior to the peak of the outbreak, may have played a significant role in the propagation of the contaminants.

10.3.1.4 Cryptosporidium Outbreak in Milwaukee

Milwaukee's *Cryptosporidium* outbreak started in March and April of 1993 (Fox and Lytle, 1996). Historically it was the largest US waterborne disease outbreak causing illness in more than 400,000 people. Milwaukee was served by two treatment plants, the Linwood Water Treatment Plant primarily served the northern part of the city and the Howard Avenue Water Treatment Plant (HAWTP) served the southern part of the city. Analysis of the outbreak data indicated that the HAWTP was the source of the outbreak and consequently the plant was closed and a boil-water order was issued. The Linwood plant was able to serve the entire city.

10.4 Need for Development of a Rapid Detection Method

Because of the occurrence of waterborne outbreaks and the potential for intentional contamination in water systems, there is a great deal of interest in development of rapid techniques for the identification of potential water system contamination events (Deininger et al., 1997; Deininger and Lee, 1998; Lee and Deininger, 1999). Although most current microbiological methods focus on a single group of indicator organisms to measure the bacteriological safety of drinking water (Federal Register, 1990) there is important information that can be gained from determining the total number of heterotrophic bacteria in water samples. Many opportunistic pathogens are not in the coliform group and a high HPC has been shown to interfere with coliform determination. For example in the Gideon waterborne outbreak high HPC counts were found in water samples prior to the outbreak and a rapid HPC method would have been helpful.

Figure 10.4 shows the luminometer and equipment, both fit well on a clipboard.

The present HPC method using an R2A agar is known to be the most sensitive test for enumerating the bacteria from treated water. The only disadvantage of the test is that it takes 7 days to complete and when the results are known, the water has been consumed. A test is needed to determine the total bacterial population in a very short time so that corrective actions can be taken in a timely manner.

The ATP bioluminescence assay allows an estimation of bacterial populations within minutes and can be applied using a local platform. The estimation of a bacterial count based on the ATP bioluminescence of the water is not new. Standard methods (APHA-AWWA-WEF, 1995) indicate that the method requires 1 hour, 1l of water, and has a sensitivity of 100,000 cells. This method is also described and specified in ASTM d 4012-81. However, the method discussed here is 100 times more sensitive, requires one-hundredth of the sample volume, and is over 10 times faster than the standard method. In order to validate the method a series of cooperative studies were conducted in collaboration with the Michigan Department of Community Health in Lansing. Several waters were tested and it was found that the method was valid (Michigan Department of Community Health, Jan 29, 2002).

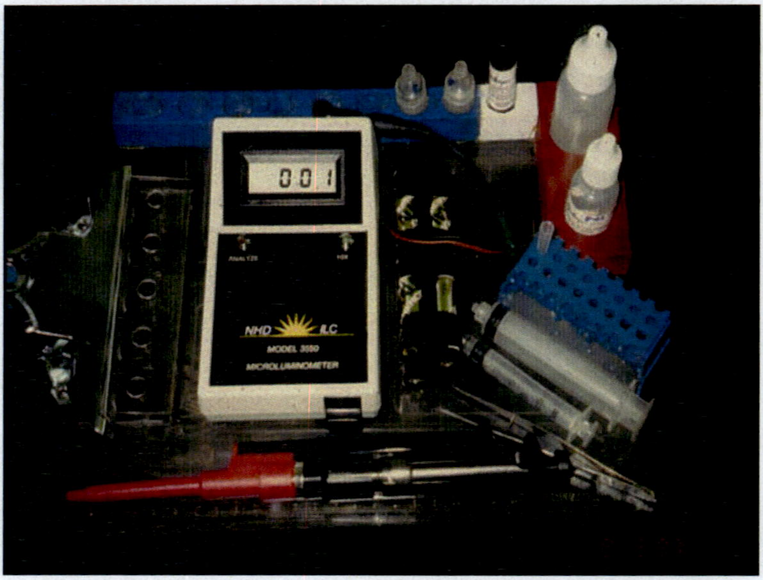

Fig. 10.4 The luminometer and equipment fit well on a clipboard

10.5 Method Development

10.5.1 Sample Filtration

In order to prepare the water samples for testing a Filtravette, which is a combination of a filter and a cuvette with a filtration size of 0.45 μm, was placed in a Swinnex filter holder (13 mm Millipore Corporation, Bedford, MA). A sterile syringe was used to extract the sample. Water extraction volumes varied between 0, 1, and 10 ml, based on the expected number of bacteria in the water sample. The filter holder was screwed onto the syringe and the water sample was forced through the filter. The Filtravette was taken from the filter holder and placed on a sterile plotter lying over a sterile plotting paper. The remaining water in the 3 ml syringe was removed with a specially converted 3 ml syringe by applying gentle positive air pressure.

10.5.2 ATP Bioluminescence

A somatic cell releasing agent (New Horizon Diagnostic Corporation, Columbia, MD) was used to lyse all non-bacterial ATP through the filter. This was performed twice. At this stage the Filtravette contains all the cell membranes intact on its surface after this procedure. The Filtravette was then inserted into a microluminometer. At this stage, the Filtravette retains the bacteria on its surface, and the bacterial ATP remains within the bacterial cell membranes. A bacterial cell releasing agent was then added to lyse the bacterial cells. Released bacterial cell ATP was then mixed with 50 ml of luciferin/luciferase (NHD, Columbia, MD) and the drawer of the micro-luminometer is closed. The light emissions are evaluated after a 10-s period and the light emission is recorded (RLU). The result was expressed as RLU/ml by dividing the water volume.

The detection limit and the sensitivity of the luminometer were tested using a serially diluted ATP solution (NHD, Columbia, MD). Distilled de-ionized water was used for the dilution with a pH of 7.8. The activity of the ATP was checked, and it was found that the RLU are proportional to the amount of ATP, and it is proportionally equivalent to the amount of viable material.

10.5.3 Bacterial Enumeration: AODC, DVC, and HPC

The total (nonviable and viable) bacterial cells were determined from a formaldehyde fixed (2% v/v final concentration) sample with the AODC method. Bacterial cells were stained with acridine orange (0.01% w/v, Fluka, Switzerland) after filtration onto a 0.2 μm pore size black polycarbonate membrane filter (Poretics, Livermore, CA). Cells were enumerated at a magnification of × 1,000 with an Olympic Provis epi-fluorescence microscope (Olympus Optical Co., Japan) equipped with a mercury arc lamp and a 400–490 mm excitation filter. The number of bacteria was counted in 10 microscopic fields using three subsamples and was then averaged. The number of bacteria per milliliter of sample was calculated using the equation in Standard Methods (APHA-WEF-AWWA, 1995).

The viable cells were counted by the DVC method. The samples were then incubated with yeast extract (0.005% w/v, Difco, Detroit, MI) and nalidixic acid (10 mg/l, Sigma, St. Louis, MO) without dilution for 24 h at 20°C. The modifications were used at a lower concentration of yeast extract and no dilution. After incubation, the fixation, counting, and calculations of elongated bacteria were done following the AODC method.

The HPC was determined for each water sample in triplicate using a R2A medium (Difco, Detroit, MI). The bacterial colonies were counted after an incubation period of 7 days at 28°C.

10.6 Method Validation

The intent of the study was to determine if a rapid ATP assay could estimate bacterial populations in real water samples in a practical and timely manner (Deininger and Lee, 2007; Lee and Deininger, 2001). For quality control purposes and to test the accuracy of both the ATP and HPC test, a direct enumeration of the bacteria in a water sample was done using two epi-fluorescence methods. The AODC method was used to enumerate the total number of bacteria, which includes both the number of viable cells and the nonviable cells. The other method was the DVC method that selectively enumerates the viable cells.

10.6.1 Collection of Water Samples

Water samples were taken from drinking waters fountains or distribution systems in the USA and abroad. Samples in the USA were taken from locations in California, Colorado, Florida, Georgia, Illinois, Kentucky, Maryland, Michigan, New York, Ohio, Oregon, Tennessee, Texas, Washington, and Washington, D.C. Some of the samples were taken from airports (California, Illinois, Kentucky, Maryland, New York, Oregon, Tennessee, Texas, and Washington) and others were obtained from cooperating utilities (California, Colorado, Florida, Georgia, Michigan, and Ohio).

Samples were also taken from Argentina, Austria, Australia, Brazil, Egypt, France, Germany, Hungary, Japan, Korea, Lithuania, the Netherlands, Israel, Panama, Peru, Saudi Arabia, Switzerland, Ukraine, and the United Kingdom.

The ATP bioluminometer, Profile 1 was used from New Horizon Diagnostics. Figure 10.4 shows the entire instrument. Everything fits on a small table. Samples from the USA and the European capitals are shown in Figs. 10.5 and 10.6. The US data and the overseas data show that the relationship can be described by

$$\log \text{ HPC (CFU/ml)} = 1.74 \times \log \text{ ATP (RLU/ml)} \tag{10.1}$$

For the countries overseas the relationship is

$$\log \text{ HPC (CFU/ml)} = 1.68 \times \log \text{ ATP (RLU/ml)} \tag{10.2}$$

The two relationships are very similar. This means that the US standard operations procedure can be used in any foreign country.

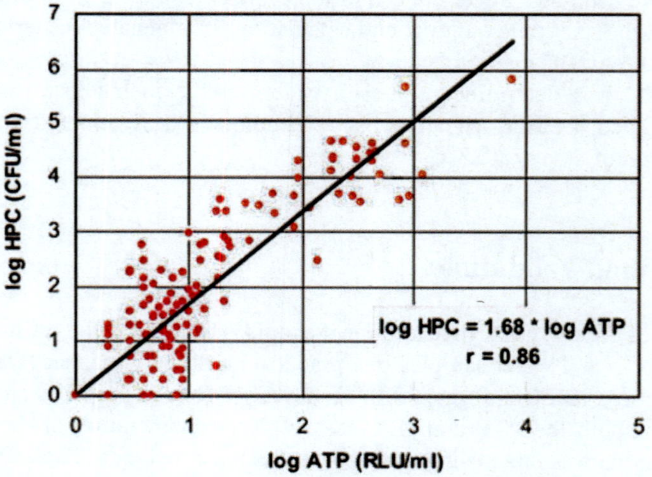

Fig. 10.5 ATP analysis of foreign water supplies

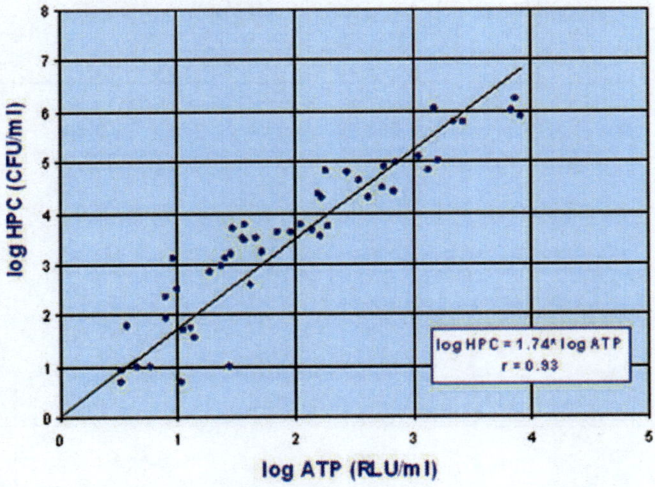

Fig. 10.6 ATP analysis of domestic water supplies

10.6.2 Public Water Supplies in France

To illustrate the application of the technique we report on a number of public drinking supplies taken in Paris, France, in August of 1999 (Lee and Deininger, 1999). Samples were taken at the Charles De Gaulle Airport, the Eiffel Tower, Hospital Pitie, The Louvre, Notre Dame, Palais Chaillot, Sacre Coeur, and the UNESCO buildings. Date were collected on the metal concentrations and the French method of analysis for ATP was applied to the samples. The water samples were analyzed for their ATP and the metal content using ICP/MS. Table 10.1 shows the metal concentrations and Table 10.2 shows bacterial concentrations using three methods. The drinking water samples were analyzed using three methods. We used an ATP assay,

Table 10.1 Concentration of metals in Paris drinking water (micrograms/liter)

Location	Pb	Mg	Al
Palais Chaillot	0.2	2,659	1.1
Eiffel Tower	3.1	1,454	1.5
Hotel St. Andre	5	1,421	1.2
Notre Dame	3.6	1,365	1.2
Parc Viviani	91	1,655	2.6
Near Louvre	2.4	1,647	1.9
Hospital Pitie	0.6	2,400	13
UNESCO	5.5	1,930	1.2
Sacre Coeur	5.5	1,930	1.2
CDG Airport	0.02	5,370	20
Bottled Water	0.02	16,000	0.7
Water Dispenser	0.1	4,781	0.8

Table 10.2 Bacteria in Paris drinking water (CFU/ml)

Location	ATP (RLU/ml)	R2A (CFU/ml)	French method (CFU/ml)
Palais Chaillot	728	1,735	250
Eiffel Tower	435	10,920	2,300
Hotel Andre	952	45,000	18,200
Notre Dame	4	65	10
Parc Viviani	296	16,310	1,000
Near Louvre	10	80	15
Hospital Pitie	539	52,000	3,610
UNESCO	47,600	176,000	2,000
Sacre Coeur	27	33	5
CDG Airport	173	4,670	170
Bottled Water	13	108	2
Water dispenser	813	57,400	4,350

the HPC using an R2A agar for 7 days, and a plate count using the French method. The method uses the following:

- Casein peptone 5 g
- Yeast extracts 2.5 g
- Glucose 1 g
- 15 g/l of distilled water
- Incubation at 22°C for 3 days

Since the water samples were in transit for about 2 days one should not place a great emphasis on the absolute number, but rather the relative relationships. The water quality varied significantly from location to the next location, and visual observation showed different patterns of colony size and colors. Figure 10.7 shows bacterial counts after 7 days and Fig. 10.8 shows bacterial counts elongated after 7 days.

10.6.3 Results and Discussion

The detection limit of ATP was determined with high accuracy. It showed that the micro-luminometer was able to determine ATP concentrations as low as 0.2 pg. It is known that the average ATP concentration in one bacterial cell is about 10^{-15} g (i.e., 1 fg). Thus the 0.2 pg corresponds to about 200 bacterial cells.

As a follow-up about 120 water samples were analyzed with ATP bioluminescence, HPC, DVC, and AODC methods, in triplicate. The correlation coefficient between ATP and HPC was 0.84, and the correlation coefficient between ATP and DVC was 0.8 which was statistically highly significant. The prediction of HPC can be accomplished by the following equation. HPC (CFU/ml) = RLU exp 1.47.

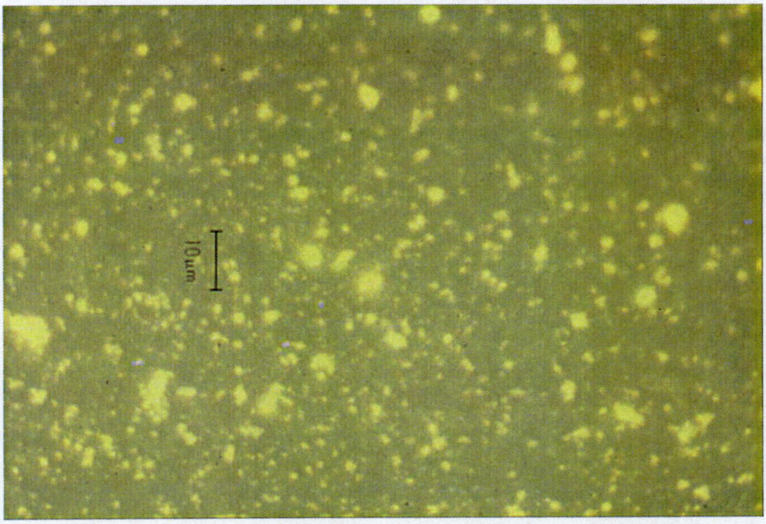

Fig. 10.7 Bacterial count after 7 days

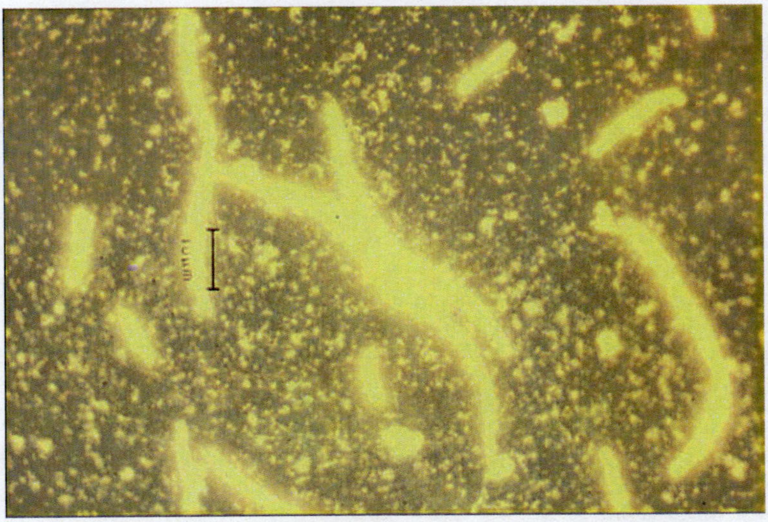

Fig. 10.8 Bacterial count elongated after 7 days

10.7 Summary and Conclusions

It is anticipated that the HPC technique has the potential for determining specific bacterial counts. We believe unique antibodies in specific bacteria will allow analysts to determine the bacterial content of water samples within a short time.

Other publications that support the use of the ATP method are as follows:

- Delahaye et al. (2003)
- Webster et al. (2005)
- Meier et al. (2008)
- Frundzhyan and Ugarova (2007)
- Bushon et al. (2009)
- Weilen and van der Kooij (2010)
- Wall Street Journal (2008)
- Lee et al. (2010)

Results from the ATP test are summarized by Weilen and Kooij as follows: "Overall, the results from our study demonstrate that ATP is a suitable indicator parameter to easily, rapidly and quantitatively determine the total microbial activity in distributed drinking water."

The authors believe that ATP techniques have a strong potential for application to wastewater as well as drinking water (Lee and Deininger, 2010; Bushon et al., 2009).

References

Abbaszadegan, M. and Absar Alum, "Microbiological Contaminants and Threats of Concern," in L.W. Mays (ed) *Water Supply Systems Security*, McGraw-Hill, New York, NY, pp. 2.1–2.12, 2004.

Angulo, F.J., S. Tippen, D.J. Sharp, B.J. Payne, C. Collier, J.E. Hill, T.J. Barrett, R.M. Clark, E.E. Geldreich, H.D. Donnell, and D.L. Swerdlow, A Community Waterborne Outbreak of Salmonellosis and the Effectiveness of a Boil Water Order. *American Journal of Public Health*, 87(4):580–584, 1997.

APHA (American Public Health Association), American Water Works Association (AWWA), Water Environment Federation (WEF). *Standard Methods for the Examination of Water and Wastewater*, 19th Edition, M.A. Franson, Mgn. Ed., American Public Health Association, Washington, DC, 1995.

ASTM d 4012-81, Standard test method for adenosine triphosphate (ATP) contents of microorganisms in water.

Burrows, W.D. and S.E. Renner, Biological Warfare Agents as Potable Water Threats, U.S. Army Combined Arms Support Command, Fort Lee, VA, p. 10, 1998.

Burrows, W.D. and S.E. Renner, Biological Agents as Threats to Potable Water. *Environmental Health Perspective*, 107(12): 975–984, 1999.

Bushon, R.N., A.M. Brady, C.A. Likirdopulos, and J.V. Cireddu, Rapid Detection of *Escherichia coli* and Enterococci in Recreational Water Using an Immunomagnetic Separation/Adenosine Triphosphate Technique. *Journal of Applied Microbiology*, 106(2): 432–441, Feb 2009.

Clark, R.M., "Assessing the Etiology of a Waterborne Outbreak: Public Health Emergency or Covert Attack," in Jennifer Hatchett (ed), *Proceedings of the First Water Security Summit*, Haested, Heasted Methods, Waterbury, CT, pp. 170–179, 2002.

Clark, R.M. and R.A. Deininger, Protecting the Nation's Critical Infrastructure: The Vulnerability of U.S. Water Supply Systems. *Journal of Contingencies and Crisis Management*, 8(2): 73–80, 2000.

Clark, R.M. and R.A. Deininger, "Minimizing the Vulnerability of Water Supplies to Natural and Terrorist Threats" in *Proceedings of the American Water Works Association's IMTech Conference*, Atlanta, GA, April 8–11, pp. 1–20, 2001.

Clark, R.M., L. Rossman, and L. Wymer, Modeling Distribution System Water Quality: Regulatory Implications. *Journal of Water Resources Planning and Management*, ASCE, 121(6): 423–428, Nov/Dec 1995.

Clark, R.M., E.E. Geldreich, K.R. Fox, E.W. Rice, C.H. Johnson, J.A. Goodrich, J.A. Bsarnick, and F. Abdesaken, Tracking a Salmonella Serovar Typhimurium Outbreak in Gideon, Missouri: Role of Contaminant Propagation Modeling, *Journal of Water Supply Research and Technology-Aqua* 45(4): 171–183, 1996.

Deininger, R.A. and J. Lee, "Rapid Determination of Bacteria in Water." in *Water Quality Technology Conference, Proceedings*, American Water Works Association, San Diego, CA, November 1–4, 1998.

Deininger, R.A. and J. Lee, "*Rapid Determination of Bacterial Loads for the Assessment of Water Quality*, Detection Technologies 2007, San Diego, CA, November 2, 2007.

Deininger, R.A., J. Lee, and P. Klangsin, Rapid Onsite Determination of Bacteria in a Water Distribution System, *American Water Works Association*, Norfolk, VA, June 7–8, 1997.

Delahaye, E., B. Welte, Y. Levi, G. Leblon, and A. Montiel, An ATP-Based Method for Monitoring the Microbiological Drinking Water Quality in a Distribution Network. *Water Research*, 37: 3689–3696, 2003, Elsevier Science Ltd.

Federal Register. National Primary Drinking Water Regulations: Analytical Techniques. *Coliform Bacteria*, Federal Budget. U.S.EPA 55: 22752–22756, 1990.

Fennel, H., D.B. James, and J. Morris, Pollution of a Storage Reservoir by Roosting Gulls. *Journal of Society of Water Treatment Exam*, 23:5–24, 1974.

Field, M. S., "Assessing the Risks to Drinking-Water Supplies from Terrorists Attacks," in L.W. Mays (ed) *Water Supply Systems Security*, McGraw-Hill, New York, NY, pp. 6.1–6.26, 2004.

Fox, K.R. and D.A. Lytle, Milwaukee's Crypto Outbreak Investigations and recommendations. *Journal of the American Water Works Association*, 88(9): 87–94, 1996.

Frundzhyan, V. and N. Ugarova, Bioluminescent Assay for Total Bacterial Contamination of Drinking Water. *Luminescence*, 22: 241–244, 2007.

Geldreich, E.E., K.R. Fox, J.A. Goodrich, E.W. Rice, R.M. Clark, and D.L. Swerdlow, Searching for a Water Supply Connection in the Cabool, Missouri Disease Outbreak of Escherichia coli O157:H7. *Water Research*, 26(8): 1127–1137, 1992.

Gleick, P.H., Water and Terrorism. *Water Policy* 8: 481–503, 2006.

Grayman, W.M., R.M. Clark, B.L. Harding, M.L. Maslia, and J. Aramini, "Reconstructing Historical Contamination Events," in L.W. Mays (ed) *Water Supply Systems Security*, McGraw-Hill, New York, NY, Pp. 10.1–10.55, 2004.

Hrudey. S.E. and E.J. Hrudey, *Safe Drinking Water : Lessons from Recent Outbreaks in Affluent Nations*. IWA Publishing, London, 2004.

Lee, J. and R.A. Deininger, A Rapid Method for Detecting Bacteria in Drinking Water. *Journal of Rapid Method and Automation in Microbiology*, 7: 135–145, 1999.

Lee, J. and R.A. Deininger, Rapid Quantification of Viable Bacteria in Water Using an ATP Assay. *American Labaratory News*, 33(21): 21–26, 2001.

Lee, J. and R.A. Deininger, Real Time Determination of the Efficacy of Residual Disinfection to Limit Sewage Contamination in a Water Distribution System Using Filtration Based Luminescence. *Water Environment Research* 82(5): 474–478, 2010.

Lee, C. et al. Cov-IMS/ATP Enables Rapid In-field Detection and Quantification of Escherichia coli and Enterococcus spp. in Freshwater and Marine Environment. *Journal of Applied Microbiology*, 109(1): 324–333, 2010.

Meier, T., J. Lee, and R.A. Deininger, et al., Quantification, Distribution and Possible Source of Bacterial Bio-film in Rodent Automated Watering Systems. *Journal of the American Association of Laboratory Animal Science*, JAALAS, 47(2): 63–70, 2008.

Michigan Department of Community Health, January 29, 2002.

New York Times, White House water cut off temporarily. July 10, p. 16, 1986.

Rice, E.W., R.M. Clark, and C.H. Johnson, Chlorine Inactivation of Escherichia coli O157:H7. *Emerging Infectious Diseases*, 5(3): 461–463, May–June 1999.

Rossman, L.A., R.M. Clark, and W.M. Grayman, Modeling Chlorine Residuals in Drinking Water Distribution Systems. *Journal of Environmental Engineering*, 120(4): 803–820. July/August 1994.

Skala, M.F., Waterborne Salmonella Outbreak in Southeastern Missouri. *Missouri Epidemiologist*, 17(2): 1–2, 1994.

The Walkerton Herald Times, Vol. 141, No. 31, Wednesday, August 2, 2000.

Wall Street Journal , 2008, "China Turns Away Shipment of Evian, Australian Seafood, May 30, 2007.

Webster, A.H., J. Lee, and R.A. Deininger, Rapid Assessment of Microbial Hazards in Metal Working Fluids. *Journal of Occupational and Environmental Hygiene*, 2: 243–218, April 2005.

Weilen, P.W.J.J. and Dick van der Kooij, Adenosine Tri Phosphate (ATP) as a Parameter to Determine Microbiological Activity in Distributed Drinking Water, Elsevier, *Water Research*, 44(17): 4860–4867, 2010.

Williams, P. and D. Wallace, Unit 731, The Japanese Army's Secret of Secrets, Hodder & Stoughton, London, 1989.

World Health Organization, Health aspects of chemical and biological weapons, Appendix 5, 1970.

Chapter 11
Chlorine Residual Management for Water Distribution System Security

Jeanne M. VanBriesen, Shannon L. Isovitsch Parks, Damian E. Helbling, and Stacia T. McCoy

11.1 Introduction and Background

A majority of Americans receive their drinking water from community water systems, which make use of pressurized pipes to deliver potable water from treatment plants to consumers. Compromised water supply systems can endanger consumer health if contaminated drinking water contains pathogenic microorganisms or hazardous substances at concentrations sufficiently high enough to cause illness or injury. The drinking water distribution system's primary defense against accidental or intentional contamination is its disinfectant residual, typically in the form of chlorine (free chlorine, chloramines, or chlorine dioxide). Chlorine-based chemicals are strong oxidants that accomplish two goals in water treatment: primary disinfection to kill microbial pathogens that were not removed through coagulation, settling, and filtration and secondary disinfection to inhibit microbial re-growth in the distribution system. Being non-selective oxidants, chlorine residuals also react with a variety of chemical agents, increasing the value of chlorine as a response to multiple potential intrusions. Further, changes in chlorine residuals are useful as an indicator of biological or chemical contaminants, introduced accidentally or intentionally.

Distribution system water quality may decline with time and distance from the plant due to the chlorine reacting with natural organic matter in the bulk water and/or on pipe surfaces or with pipe materials themselves. Further degradation of water quality is possible via additional reactions between the chlorine residual and introduced materials following events such as the infiltration of sewage-contaminated ground water, cross-connections between water and sewer lines, pressure loss resulting in groundwater intrusion, or the intentional introduction of chemical or biological agents. Systems that routinely have difficulty maintaining a chlorine residual due to long travel times in their systems may employ booster chlorination

J.M. VanBriesen (✉)
Department of Civil and Environmental Engineering, Carnegie Mellon University, Pittsburgh, PA, USA
e-mail: jeanne@andrew.cmu.edu

R.M. Clark et al. (eds.), *Handbook of Water and Wastewater Systems Protection*, Protecting Critical Infrastructure, DOI 10.1007/978-1-4614-0189-6_11,
© Springer Science+Business Media, LLC 2011

stations to inject additional chlorine at critical locations in the distribution system. The use of booster stations has been shown to require less total chlorine to maintain adequate residual and to result in a more uniform level of chlorine residual throughout the distribution system when compared to adding chlorine only at the source of the distribution system (Boccelli et al., 1998). The presence of the residual suggests that the water does not contain significant organic chemicals or pathogenic organisms as these agents would react with and consume the chlorine residual. The chlorine residual, therefore, is a central parameter for water security planning and management.

Free chlorine is the most common form of chlorine used as a disinfectant in drinking water systems in the United States (Haas et al., 1992), and it is monitored in the distribution system along with biological water quality as a requirement of the Safe Drinking Water Act (SDWA). Grab samples from a limited number of sites collected at a relatively low frequency are evaluated for indicator organisms and chlorine residual (ASCE, 2004a). These requirements result in spatially and temporally sparse data sets that do not provide adequate warning of sudden contamination events. Alternatively, there are several commercially available sensors specifically designed for use in water distribution systems that monitor free or total chlorine concentrations and can provide continuous chlorine concentration data (ABB, 2008; Chemtrol, 2008; Hach, 2008; Kuntze, 2008; Teledyne, 2008; Yokogawa, 2008). The use of chlorine sensors has become more prevalent in the drinking water industry; however, sensors are primarily used within the treatment plant to control the chlorine concentration entering the distribution system (States et al., 2003). Online water quality monitoring in the distribution system has not yet been widely implemented; deployments are recent and involve relatively few (one to five) sensors at few distribution system locations (ASCE, 2004b). However, sensor networks for water distribution systems are increasingly being proposed to facilitate enhanced security (Ostfeld et al., 2006; Berry et al., 2006; Propato et al., 2006; Preis and Ostfeld, 2006). When extensive, real-time chlorine data are available, a drinking water utility then can use these data to (1) evaluate system vulnerabilities, (2) detect system anomalies including accidental or intentional intrusions, and (3) manage chlorine booster stations.

11.2 Vulnerability Assessment Using Chlorine Residual Modeling

A vulnerability assessment based on chlorine residual concentrations will identify areas in a distribution system where there is a high probability of insufficient disinfectant residual yielding decreased protection against intrusions of biological or chemical contaminants. Vulnerability assessment begins with an accurate hydraulic and water quality model of the distribution system that enables documentation of the consequences of potential attacks on the drinking water system (Murphy and Kirmeyer, 2005). The first water distribution system models were based on steady-state hydraulic models and became available in the 1980s; fully dynamic water quality models are now widely used and are better able to represent the time-variant nature of a chemical (disinfectant *or* contaminant) in a water distribution system. EPANET is the most commonly used water distribution system model used in the United States. It was developed by the Water Supply and Water Resources Division of the Environmental Protection Agency's (EPA) National Risk Management Research Lab as a research tool for understanding the fate and transport of drinking water constituents in distribution systems. EPANET has been widely used in research and in industry and forms the basis for a number of commercial water distribution system modeling packages (e.g., H2OMAP (MWH, 2007), PipelineNet (WIMMT, 2007), and WaterCAD (Bentley, 2007)). Version 2 of EPANET was released in 2000, and an updated version, EPANET Multi-Species eXtension (EPANET-MSX), was released in 2007. Water distribution system models are most frequently used in industry to predict hydraulic conditions in the system. Water quality models are developed less frequently due to the need for significant calibration of chemical reaction parameters in the system. However, prediction of chlorine residual through a water quality model is critical for vulnerability assessment.

EPANET can be used to generate temporal and spatial chlorine residual profiles for a given distribution system, and sensitivity and uncertainty analyses can be conducted to determine the most critical parameters for adequate modeling of the system. The sample system analyzed here (shown in Fig. 11.1) demonstrates the relationship between water age, chlorine residual, and system vulnerability. Typical values for water flow, demand, and chlorine decay rates were used for simulations in this hypothetical system; real systems would require a calibration step. An initial parametric sensitivity analysis was used to determine which model parameters control chlorine concentrations and, consequently, system vulnerability. System parameters were characterized by distributions rather than point values in order to determine system vulnerability for a range of conditions.

The total chlorine decay rate modeled in the system is based on the sum of the reaction rate coefficients for the bulk flow and on the pipe wall surfaces: $k = k_b + k_w/r_h$, where k is total reaction rate coefficient, k_b is bulk flow reaction rate coefficient, k_w is pipe wall reaction rate coefficient, and r_h is hydraulic radius of the pipe. Chlorine reaction rates and the associated coefficients can be approximated from distribution system tracer studies or bench-scale chlorine demand experiments. In the present work, we consider bulk flow reaction rate coefficient; pipe wall reaction rate coefficient; demand; and roughness coefficient. Uncertainty was characterized by selecting distributions for bulk flow (log-logistic) and for pipe wall (Weibull) reaction rate coefficients (Powell et al., 2000; Hallam et al., 2002). Assuming that values greater or less than the median or typical value (global multiplier = 1.00) are less likely to occur, uncertainty in the demand and roughness coefficient was characterized by a triangular distribution. To simulate the natural variability in parameters, 100 Monte Carlo simulations were run with four random variables, one value drawn independently from each of the parameter distributions, for each 15-day simulation.

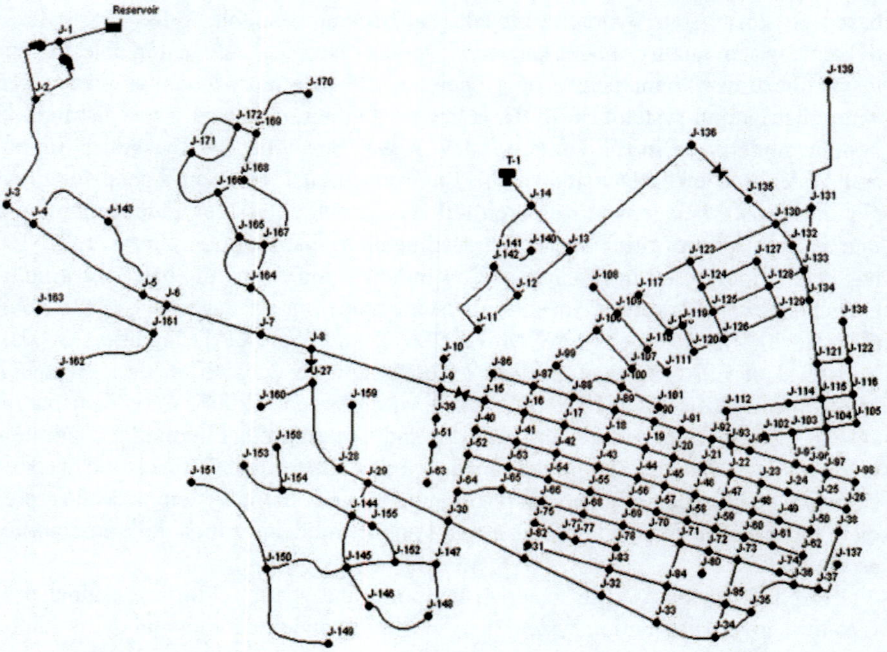

Fig. 11.1 Map of sample distribution system (Haested Methods WaterCAD) showing the clear well-labeled reservoir, 166 nodes labeled J-1 through J-172, a storage tank labeled as T-1, and the pipes connecting all these elements. (*Source*: Thompson et al. (2007), with permission from ASCE.)

For the sensitivity analysis, eight 15-day simulations were run with one parameter at its extreme value, minimum or maximum, and all other parameters held at their nominal or mean values (mean values for each parameter are shown in Table 11.1). For the bulk flow and pipe wall reaction rate coefficients, the minimum and maximum were defined as the 5th and 95th percentiles of their respective distributions, and the lower and upper limits of the triangular distribution served as minimum and maximum values for the demand and roughness coefficient, respectively. Using the EPANET results from the last 10 days in each simulation, overall system measures, such as the mean chlorine concentration and the fraction of the time the chlorine

Table 11.1 Mean parameter values

Input parameter	Mean value
Initial chlorine concentration, C_0	4.00 mg/L
Bulk flow decay constant, k_b	1.64 day^{-1} (Powell et al., 2000)
Pipe wall decay constant, k_w	0.15 ft/day (Hallam et al., 2002)
Base demand	0.00–12.00 gallons/min
Pipe roughness coefficient	100.00

concentration is below a required threshold, were calculated. Current regulations require a detectable disinfectant residual at all points of consumption, using the best available technology (USEPA, 1999). In systems using free chlorine, this minimum residual concentration is often assumed to be 0.2 mg/L (Propato and Uber, 2004), and this disinfectant level is considered to provide a protective barrier against contamination in subsequent analyses.

For this example distribution system, the bulk flow reaction rate coefficient, k_b, has the greatest influence on system measures, such as mean chlorine concentration, as demonstrated by the wide value range in Fig. 11.2. Demand and the pipe wall reaction rate coefficient also have significant influence on system measures while roughness has minimal influence. Based on this preliminary sensitivity test, uncertainty in the reaction rate coefficients and demand constitutes most of the uncertainty in chlorine concentrations and, consequently, vulnerability. This parametric sensitivity analysis requires minimal computational power and would provide insight to public utility operators regarding important factors for control of chlorine residual in their system.

A probabilistic method for vulnerability assessment, which is demonstrated here, expands on the parametric sensitivity analysis by simulating 100 random sets of parameters (bulk flow reaction rate coefficient, pipe wall reaction rate coefficient, and demand) to determine the distribution of the minimum chlorine concentration at all nodes in the system. These minimum chlorine concentration results from all

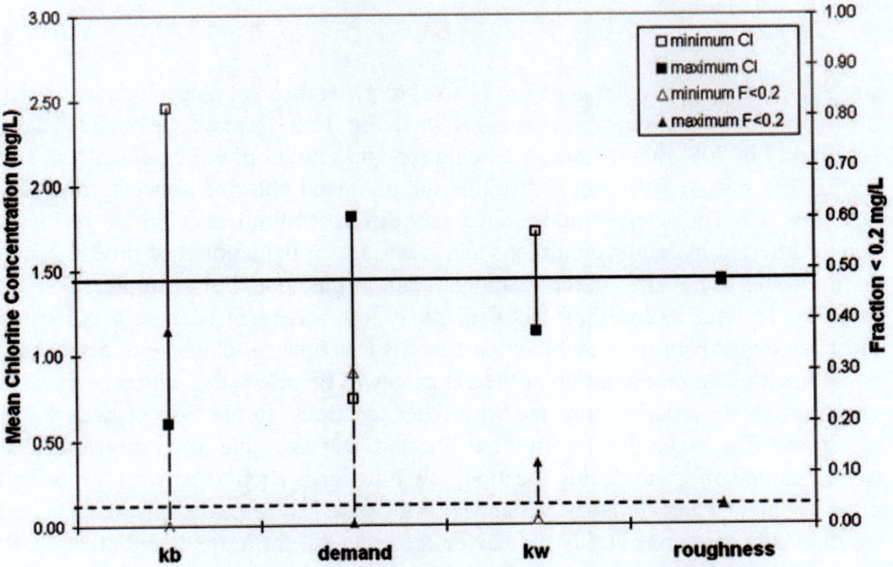

Fig. 11.2 Mean chlorine residual concentration and fraction of nodes, considering all time steps, with a chlorine residual below 0.2 mg/L. Simulation results when a parameter is minimized are shown as *open squares* or *triangles* while maximized parameter results are shown as *solid squares* or *triangles*. The *solid line* represents the mean chlorine concentration for the entire system and the *dashed line* indicates the average fraction of nodes with a residual below the protective threshold

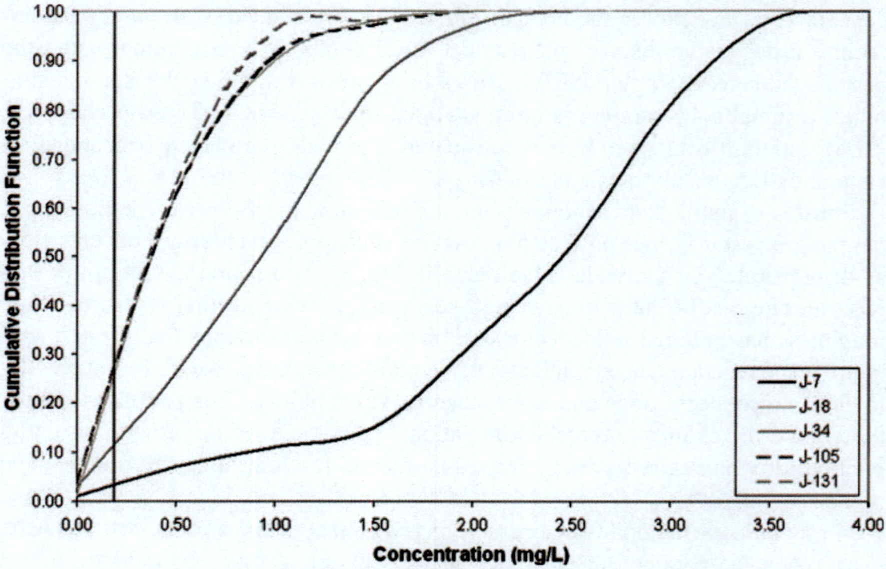

Fig. 11.3 Minimum chlorine concentration cumulative probability density functions for selected nodes generated from the results of Monte Carlo simulations with variable input parameters k_b, k_w, demand, and roughness. *Solid vertical line* indicates the minimum chlorine concentration, 0.2 mg/L, required to protect the distribution system. As a result of different hydraulic patterns and water residences times, the probability that a selected node will not have a protected residual ranges from 0.03 to 0.30

simulations for a particular node can be used to generate a node-specific cumulative distribution function, such as the ones shown in Fig. 11.3. Here we see node J-7 has a probability of 90% that the predicted chlorine concentration will be *above 1.0*, and node J-131 has an 80% probability that the predicted chlorine concentration will be *below 0.5*. The selected nodes have very different minimum chlorine residuals due to different hydraulic conditions and water age or the amount of time required for the water to travel from the treatment plant to the node. For example, node J-7 (see Fig. 11.1 for its location) has a relatively low water age because it is located directly downstream from the plant, so there is less time for chlorine to decay and, consequently, the probability that the chlorine will be below the protective level is very low, close to 0.00, compared with other locations. In the case of node J-131 (again see Fig. 11.1), this location has a high water age, due to an increased distance from disinfectant dosing location, low flow areas such as network grids, and other hydraulic characteristics, which yields increased chlorine decay and decreased residual concentrations at this site. Given the potential for more disinfectant decay and consequently lower concentrations, it is not surprising that the probability that the chlorine is below 0.2 mg/L is much higher in this location, nearly 0.30.

Moving from cumulative distribution functions to probabilities, it is easy to predict nodes that are more vulnerable to chlorine-sensitive contaminants. Nodes having a low probability of the minimum chlorine concentration being below the protective threshold, such as J-7, are less vulnerable than nodes such as J-105 and J-131, where there is a high probability that disinfectant levels will be insufficient. For this example system, the greatest vulnerability risk occurs in system locations further from the chlorine source, which contribute to increased water ages and decreased chlorine concentrations; in industrial areas with low flow regimes outside of working hours; and in network areas with limited flow pathways, which also contribute to increased residence times and greater disinfectant decay. This is shown graphically in Fig. 11.4, where the size of the dots at the node locations indicates the probability that a given node has a concentration of chlorine that is *below* required levels. Thus, nodes with large dots are more vulnerable since a contaminant introduced at these locations will meet with less chlorine to counteract it.

This method for assessing vulnerability in drinking water distribution systems can be used by public water utilities to identify and address high-risk system locations, where there is a high probability that the site will be vulnerable to microbial contamination due to low disinfectant residual. This probabilistic method may also be valuable to public water utilities for identification of spatially or temporally specific vulnerabilities. Due to dynamic hydraulics and water demand patterns, nodal chlorine concentrations fluctuate, yielding temporal variability in the spatial location

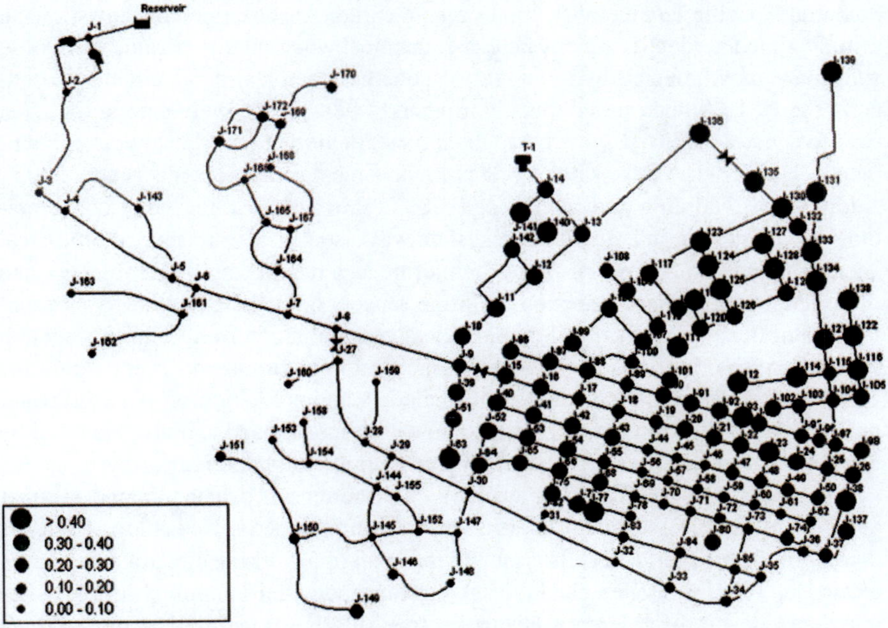

Fig. 11.4 Vulnerability map of the distribution system that shows the probability that the minimum chlorine concentration is less than 0.2 mg/L for all system nodes. The probabilities are indicated by different sized *circles* with the *largest circles* indicating probabilities greater than 0.40, the *second largest circles* having a probability range of 0.30–0.40, the *medium circles* 0.20–0.30, the next to smallest circles 0.10–0.20, and the *smallest circles* 0.00–0.10. (*Source*: Thompson et al. (2007), with permission from ASCE.)

of nodes having increased vulnerability. It would be advantageous for a public utility to examine particular scenarios, such as the time period following overnight demand, when there is increased system vulnerability due to hydraulic changes and low demand resulting in stagnant "old" water, as well as to consider special interest locations, such as government buildings, schools, or hospitals. Using the chlorine residual as a surrogate for contamination risk is an effective approach for assessing system vulnerability.

11.3 Chlorine Sensors for Nonspecific Event Detection

When considering the number of potential chemical or biological agents that could be involved in accidental or intentional contamination of a drinking water distribution system, specific contaminant detection is cost prohibitive and impractical. However, many contaminants of concern are reactive species that may alter physical or chemical water quality parameters whose expected values are easily measured and relatively predictable. For example, contaminants such as the chemical cyanide or the biological agent *Escherichia coli* could significantly change the typical or expected values of easily measured parameters such as pH, oxidation reduction potential, specific conductivity, total organic carbon, or chlorine concentration. In a study aimed at identifying physical and chemical water quality parameters whose values are most affected by a number of potential chemical contaminants of concern, the USEPA determined that chlorine and free chlorine were among the most sensitive water quality parameters to the largest fraction of tested chemical contaminants (Hall et al., 2007; United States Environmental Protection Agency, 2005). Additionally, Helbling and VanBriesen (2007) showed that chlorine at concentrations typically seen in a distribution system was also sensitive to several biological agents, including an agent resistant to chlorine as a disinfectant. Therefore, in lieu of targeted contaminant detection, chlorine sensors offer the potential for nonspecific identification of chemical or biological contamination events. In this section we discuss chlorine sensors and their response to contamination events, chlorine sensors used for validating water quality models and prediction of expected water quality, event detection algorithms, and how chlorine demand signals generated by a contamination event may propagate within a community distribution system.

The majority of commercially available chlorine sensors can be grouped as either amperometric sensors or colorimetric sensors. Amperometric free chlorine sensors consist of electrochemical cells typically made up of a working electrode, a counter electrode, and a reference electrode. A constant potential is applied to the working electrode by the reference electrode, free chlorine is reduced at the working electrode, and the counter electrode is oxidized. The free chlorine concentration is determined as a function of some physical change in the system, e.g., change in potential across the electrode. Colorimetric sensors are often designed to automate the *N*,*N*-diethyl-*p*-phenylenediamine (DPD) colorimetric method (American Public Health Association, 2005) that is used to manually determine free chlorine concentrations in grab samples, making measurements from these sensors more transparent than those from an amperometric sensor. Colorimetric sensors generally consist of an external housing cabinet, a system connection and pressure regulator that allow for small volumes of the drinking water to enter a photometric cell, a pump assembly to pump reagents into the photometric cell, and a light source to make the photometric measurements.

Amperometric chlorine sensors are rapidly responding and therefore chlorine concentrations can be queried and recorded at a rate of approximately once per every few seconds. Further, amperometric sensors are also reagentless and require minimal maintenance, which may include monthly recalibration of the electrode response, electrolyte replacement, or other minor manufacturer-specific maintenance recommendations. The primary disadvantage of using amperometric sensors to measure free chlorine concentrations is a limitation arising from the pH-dependent speciation of free chlorine. Free chlorine is defined as the sum of hypochlorous acid (HOCl, $p\mathrm{Ka} = 7.6$) and its dissociation product, hypochlorite (OCl^-). Hypochlorous acid is a stronger oxidant and thus reacts differently within the amperometric cell. These differences can be accounted for with integral pH measurement within the sensor; however, calibration of the electrode at one pH may introduce errors if the pH changes, especially at pHs greater than 8.0, when hypochlorite is the dominant species.

Colorimetric sensors are generally slower to respond when compared to amperometric sensors as the water must be sampled, sent to the photometric cell, combined with the reagents, mixed, and analyzed before a measurement is available. Measurements for colorimetric sensors can be queried and recorded at a rate of approximately once every few minutes. Colorimetric sensors also require more maintenance with monthly reagent replenishment and periodic repairs to small pumps and internal plumbing. However, there are no pH dependencies or speciation concerns as there are with amperometric sensors, making colorimetric sensors a more suitable option in distribution systems operating at higher pHs.

In order to test the ability of commercially available sensors to respond to contaminant-induced changes in free chlorine concentrations within a distribution system, amperometric and colorimetric free chlorine sensors were installed on a laboratory-scale distribution system (LDS) built from 1-in. PVC pipe and equipped with contaminant injection ports (Helbling and VanBriesen, 2008). To ensure that all of the induced changes in chlorine concentration that were observed were attributable to the contaminant of interest, all contaminants were suspended in a chlorine demand free buffer (pH=7.4). As a control, chlorine demand free buffer was injected into the LDS at varying pHs to ensure that any observed sensor response was attributable to the suspended contaminants and not the buffer itself. The water in the laboratory-scale distribution system had a pH range of 8.1–8.3 during the duration of the control experiment. Figure 11.5 shows that the colorimetric sensor captures the natural variation in the chlorine concentration within the bulk water without responding to the injected demand free buffer. However, the amperometric sensor shows large errors in chlorine measurement at a pH of 7.4 and smaller errors as the pH was increased to the background (and electrode calibration) pH

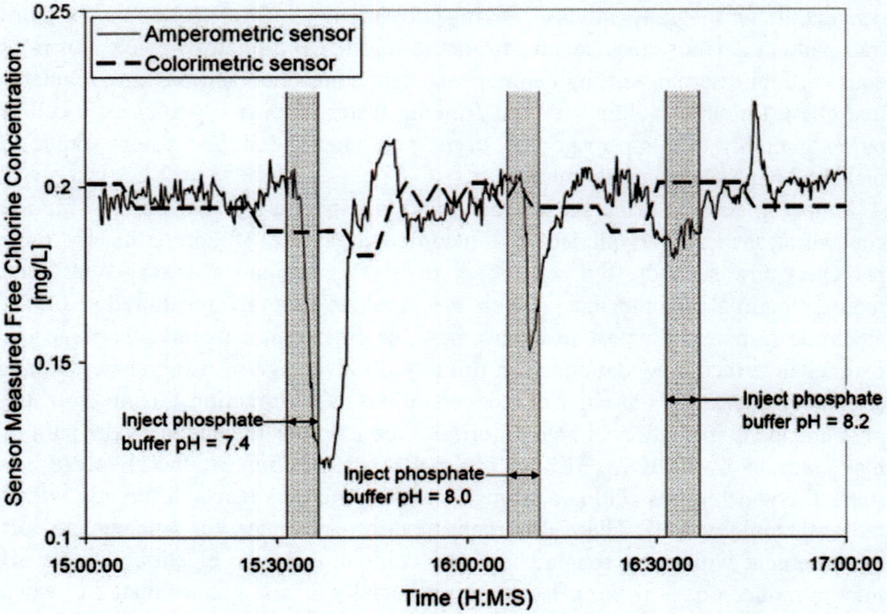

Fig. 11.5 Results of a control experiment where phosphate buffer of varying pH was injected into a continuously running laboratory-scale distribution system. Results show that erroneous free chlorine concentrations were determined with the amperometric sensor at magnitudes proportional to the difference in pH between the distribution system water and the buffer

of approximately 8.2. All measurements were checked with grab samples, which confirmed the measurements of the colorimetric sensor. These results highlight the potential for false alarms a utility may see with amperometric chlorine sensors when distribution system water sees a sudden change of pH. Utilities should check with manufacturers of amperometric sensors to see how the sensor is designed to compensate for these types of errors.

Using the colorimetric sensor integral to the LDS, a panel of experimental contamination scenarios was developed by considering various contaminants, hydraulic conditions, background free chlorine concentrations, and injection locations (Helbling and VanBriesen, 2008). The goals of these experiments were to evaluate the ability of a free chlorine sensor to rapidly respond to contaminant-induced changes to in situ free chlorine concentrations and to determine whether the induced changes were predictable using hydraulic data and batch-determined reaction kinetics. All of the contaminants were injected into the LDS while suspended in chlorine demand free buffer at a pH of 7.4. The results of this panel of experimental contamination scenarios showed that commercially available colorimetric sensors respond rapidly to induced changes in residual concentration and that these changes are directly proportional to the initial background free chlorine concentration, the injected contaminant concentration, and the amount of contact time available between the injection location and sensor, which is a function of flow rate and distance. Further, it was shown that reaction kinetics can be applied within a general advection-reaction transport model to predict the sensor response over a large range of experimental conditions (Helbling and VanBriesen, 2008). These results support the use of chlorine sensors as nonspecific sentinels for contamination events within the distribution system.

As discussed in Section 11.2, several commercial software packages are available to model water quality (particularly chlorine concentrations) in drinking water distribution systems. These models use as inputs the system architecture, known demand patterns, and bulk water and pipe wall reaction rate coefficients to provide expected chlorine concentrations at spatial and temporal locations within a prescribed resolution. Hourly, daily, and seasonal variations in parameters such as demand and temperature result in temporally variable concentrations at a single spatial location. Traditionally, water quality models used for vulnerability assessment or to gain insight into system operational conditions are calibrated and validated by temporally and spatially sparse grab samples. Online sensor placement that can provide continuous and long-term data collection and storage can capture trends not made readily transparent with sparse grab samples and could offer a better validation of water quality models.

Understanding and predicting the expected value of free chlorine concentration at a specific spatial location is imperative in the identification of contaminant-induced chlorine demand signals. In addition to their utility in validating existing water quality models, online sensors could be used to collect continuous chlorine concentration data from vulnerable locations within the distribution system to be used in conjunction with time-series analyses and forecasting tools to predict expected chlorine concentrations at a finer resolution than models seeking entire distribution system coverage. In an effort to study the patterns in long-term chlorine concentration data at a single location, a colorimetric chlorine sensor was installed within the distribution system of a college campus and data were collected continuously for 12 months. Figure 11.6 summarizes the results of this data collection effort as hourly average chlorine concentrations for weekdays and weekends, which clearly shows strong hourly and daily trends. Higher chlorine concentrations are associated with periods of higher demand such as on weekdays between 9 am and 5 pm, while lower concentrations are observed during periods of lower demand such as overnight and on weekends. The simplest method available to predict expected water quality at a future time step based on continuously collected chlorine concentration data is to state that the expected concentration at the next time step (C_{t+1}) is equal to the current concentration (C_t). Using this simple predictor over each data point in the data set summarized in Fig. 11.6 results in an average prediction error (PE $= C_{t+1} - C_t$) of $-3.36 \times 10^{-0.5}$ mg/L with a standard deviation of 0.0246 mg/L. These data suggest that even simple predictors may be suitable for defining expected chlorine concentrations, though more sophisticated tools such as autoregressive moving average models or Kalman filters that can take advantage of larger windows of previous data may be more appropriate.

While prediction of expected chlorine concentrations is enabled by collecting long-term data from online sensors placed within distribution systems, algorithms

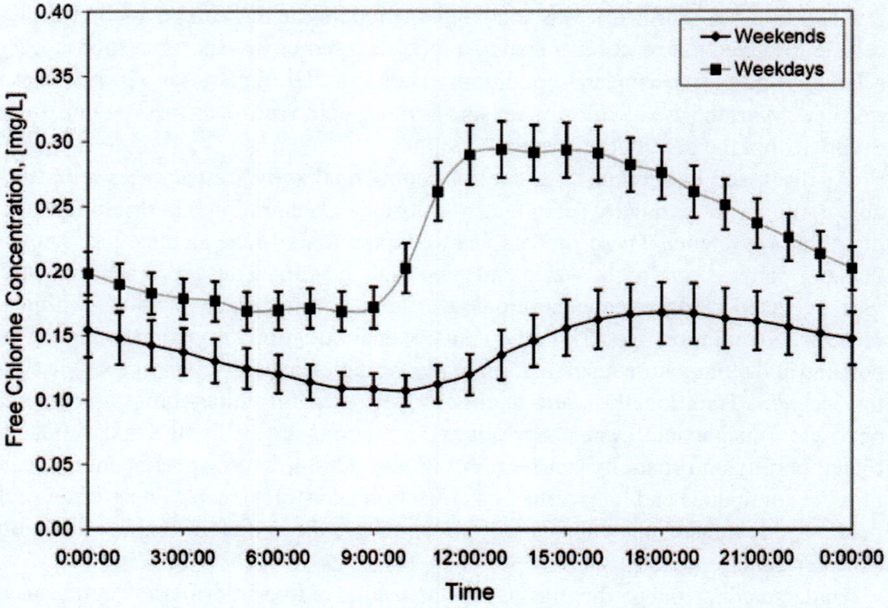

Fig. 11.6 Average hourly weekday and weekend values of free chlorine concentration for a 12 month period at an academic institution with 95% confidence intervals

still must be developed that alarm a utility operator when measured concentrations deviate significantly from expected values. Significant research is ongoing in the field of algorithm development for anomaly detection within continuous streams of water quality data. Several vendors of sensor products have developed software that is designed to continuously analyze the data streams generated by their sensors and provide warning to utility operators of anomalous conditions. Byer and Carlson (2005) used standard deviations to define confidence limits of expected values, and events were defined as measured values lying outside these confidence intervals. Klise and McKenna (2006) and McKenna et al. (2006) described methods for predicting expected values of sensor data based on historic data and defined thresholds within which expected values should fall. While these and other researchers have contributed to the development of event detection algorithms, false alarm rates remain high, and additional research and development is needed.

Successful application of sentinel systems consisting of chlorine sensors is predicated on sensor selection and placement, chlorine response to a given contaminant, adequate prediction of expected water quality, anomaly detection, and the widespread propagation of the chlorine demand signal. A chlorine demand signal induced by a contaminant must be significant in magnitude and thus detectable at locations downstream from the intrusion. If a chlorine demand signal is not significant or detectable downstream from an intrusion – due to the hydraulic regime, slow reaction rates, or dilution due to mixing – then the proposition of using chlorine as a sentinel for contamination is in vain. It is not possible to test the propagation of chlorine demand signals within a real distribution system without contaminating the system. Therefore, models must be developed to simulate the transport of an injected contaminant along with the expected chlorine response. Simultaneous simulations in the absence of a contaminant intrusion can provide the expected chlorine concentrations, and simple algorithms can be used to determine whether propagated chlorine demand signals could be detected at distribution system locations downstream and distant from the origin of contamination.

The recently developed EPANET Multi-Species eXtension (EPANET-MSX; Shang et al., 2007) model allows for the consideration of multiple species of reactants within the bulk water. With this model, reactions can be defined between chlorine and the bulk water, pipe walls, and any number of reactive contaminants. The kinetics of reactive contaminant decay may be determined in batch systems and then incorporated into EPANET-MSX as additional rate expressions. Contaminant decay kinetics are easily estimated for single chemical species; however, biological agents pose complications as they are an aggregate of oxidizable material. Pseudokinetic models are needed like those used for reactions between free chlorine and natural organic matter that make up the bulk water reaction rate coefficient. Development of these models allows for simulation of contaminant intrusion into a system with corresponding chlorine response to the contaminant. Utilities can use these types of models and simulations to identify particularly vulnerable locations within their system or for placement of sensors to maximize the probability of event detection.

A multi-species water quality model was used to simulate contaminant transport and subsequent free chlorine response in a model community drinking water distribution system (Helbling and VanBriesen, 2009). First, the model was run as a simple water quality simulator (as discussed in Section 11.2) to establish expected chlorine concentrations at each node within the distribution system in the absence of contamination. Then, contamination simulations focused on daily variation by selecting a single node in the network and simulating an intrusion of two separate contaminants at 3-h intervals over a 24-h period. The resulting dynamic chlorine response patterns were then compared to the expected values derived from the control simulations. The results of these simulations were positive as significant losses in chlorine residual were observed at locations significantly downstream from the intrusion location when compared to the expected values. Factors that were determined to affect the magnitude of the response were (1) community demand patterns, which are a function of time and system architecture and affect contaminant transport and contaminant-specific reaction rates, (2) initial concentrations of chlorine and contaminant, and (3) the overall transport time.

11.4 Booster Response to Low Chlorine

As discussed in the previous sections, maintenance of a chlorine residual in the distribution system provides the first line of protection of drinking water. However, concerns about disinfection by-products restrict disinfectant loads at the treatment plant, and as a result, booster disinfection is employed in some systems. Booster disinfection is the injection of additional disinfectant (chlorine) at strategic locations throughout a drinking water distribution system that results in a more uniform distribution of chlorine throughout the system. Currently, many water utilities rely on booster disinfection stations within their distribution systems. Although operation, maintenance, and security concerns may deter utilities from adopting booster disinfection (Walski et al., 2003; Boccelli et al., 1998), recent literature discussing its benefits (Munavalli and Kumar, 2003; Prasad et al., 2004; Tryby et al., 2002; Propato and Uber, 2004; Ostfeld et al., 2006) suggests an increase in use and comfort level among drinking water utilities. In addition, it has been shown that booster disinfection used in conjunction with a water quality sensor network can be an effective first response to a contamination event that caused disinfectant residuals to drop to unsafe levels (Parks and VanBriesen, 2009). The extent of protection afforded by booster systems varies according to the attributes of the drinking water distribution system, such as branched versus looped flow paths and the number and location of storage tanks (Propato and Uber, 2004).

The potential for a boost-response system (chlorine boosters triggered by a sensor network) to provide substantial protection and allow for uninterrupted service during an intrusion event was evaluated by Parks and VanBriesen (2009). Random contamination events were simulated in a model water distribution system with an optimized sensor network. The simulations included contaminant injection, contaminant reaction with chlorine residual, and booster response to the contaminant when chlorine level dropped below a defined threshold, and "decontamination" was defined when the boosted chlorine reached contaminated nodes. Figure 11.7a shows the network that was modeled (Ostfeld et al., 2006), with nodes shown as small dots, sensors as stars, the booster as a large dot, the reservoir source as a triangle, and the contaminant injection point as a cross. A disinfection boost was simulated to begin the instant contamination reached a sensor, and a range of reaction rate coefficients were applied to the contaminant to simulate reaction with the disinfectant. Figure 11.7b–d shows the system after the contaminant has been injected but before it has been detected (black nodes are contaminated, panel b), the system the instant the contaminant reaches a sensor (top right star) and the booster has been triggered (gray nodes are decontaminated, panel c), and the system after the booster has been triggered and most nodes have been decontaminated (panel d). After completing many simulations with various reaction rate coefficients and booster locations, cumulative distribution curves of the volume of consumed contaminated water for various response levels illustrated that a boost-response system could be effective in significantly reducing the volume of consumed contaminated water, but only under very specific circumstances. A boost-response system where boosted disinfectant can reach most nodes and can rapidly inactivate the contaminant is still limited by the effectiveness of the sensor network, highlighting the need for cost-effective and reliable sensors easily deployed by water utilities. However, booster disinfection can still provide a practical first response to intrusion mitigation. It is impractical to assume that a water utility would 'shut down' or issue a 'do not consume' order

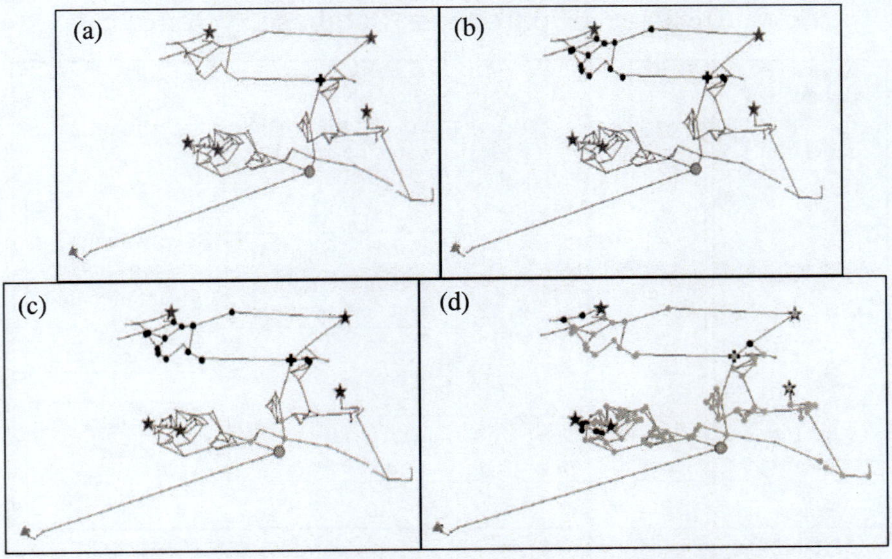

Fig. 11.7 Simulation results of a disinfection boost beginning the instant contamination reaches a sensor: (**a**) the system before contamination and boost, (**b**) the system after the contaminant has been injected but before it has been detected (*black nodes* are contaminated), (**c**) the system the instant the contaminant reaches a sensor (*top right star*) and the booster has been triggered (*gray nodes* are decontaminated), and (**d**) the system after the booster has been triggered and most nodes have been decontaminated. Nodes are shown as *small dots*, sensors as *stars*, the booster as a *large dot*, the reservoir source as a *triangle*, and the contaminant injection point as a *cross*

when a chlorine sensor responds to a nonspecific event, due to the consequential customer insecurity and the possibility of a false positive from the sensor network. As a result, a less drastic approach to contamination mitigation, such as hyperchlorination, is likely to be employed. A second, confirmatory detection may subsequently lead to a 'do not consume' order. This combination of a boost at the first detection and a 'do not consume' order at the second detection can be an effective means of reducing the volume of consumed contaminated water (Parks and VanBriesen, 2009).

Parks and VanBriesen (2009) also illustrated how the use of distribution system modeling and database management can be useful to utilities for efficient disaster response. The study showed that if exhaustive disinfection scenarios are simulated and the results managed in a database, a utility needs to merely simulate the contamination scenario once and query the database to determine the effects of a multitude of responses, such as boost response, issuing a 'do not consume' order, or waiting for a second detection to respond. Figure 11.8 shows the steps a utility would take to design a disaster preparedness system based on the use of chlorine to respond to a sensor network detection.

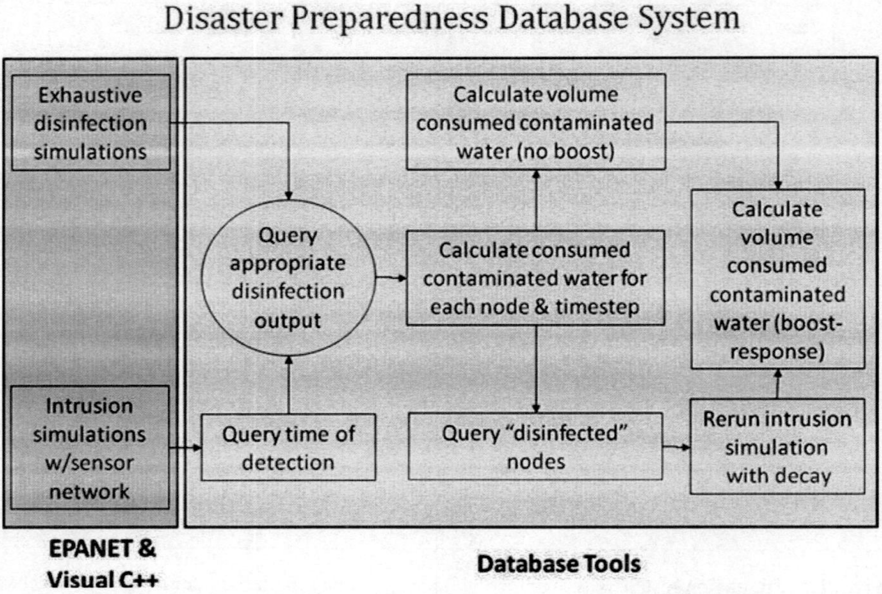

Fig. 11.8 Flowchart illustrating steps to estimate the volume of consumed contaminated water under various boost-response scenarios

11.5 Future Directions in Chlorine Sensing for Drinking Water Distribution System Control

The threat of an accidental or intentional contaminant intrusion into a drinking water distribution system is of significant concern. In this chapter we have summarized the steps a utility can take to understand distribution system vulnerability, evaluated the value of chlorine sensors for detection of intrusions, and evaluated the potential of chlorine boosters to respond to detected intrusions. Application of these methods requires that a utility have a map of their distribution system (with locations of infrastructure and information on water flows) and a calibrated hydraulic and water quality model capable of predicting system flow and chlorine concentrations. Many utilities have not reached this level of sophistication in managing their distribution systems. However, the work presented here provides strong motivation to do so.

Once a distribution system model is calibrated and can accurately simulate chlorine concentrations under normal operating conditions, use of chlorine sensors, anomaly detection algorithms, and chlorine booster stations can enable a utility to respond to low chlorine levels caused by accidental or intentional contamination. While not discussed in the present work, monitoring chlorine levels within the distribution system can also enable utilities to manage chlorine dosing for maximum protection of human health (balancing disinfection needs with control of disinfection by-products).

The use of chlorine as a surrogate for a contamination event has limitations. Extensive sensor networks are needed to protect large networks, and the optimal placement of these sensors is an area of continuing research. Any change in chlorine concentration identified as "anomalous" would be a *nonspecific* indicator of an anomalous condition. There are many benign events that could produce significant changes in chlorine concentrations within real-world distribution systems and lead to false positives. Therefore, field analyses will always be required to identify the cause of the anomalous chlorine data and whether the cause poses a threat to public health. Further, chlorine sensors will not detect contaminants that are not reactive with chlorine, and thus a chlorine sensor system for event detection could give a false sense of security. Chlorine can also react with contaminants, producing by-products that are of concern, and thus booster stations must be used with care.

The research presented here is just a beginning. Significant additional work is needed to evaluate vulnerability in distribution systems, develop anomaly detection programs, manage and interrogate streaming real-time data from chlorine sensors, and optimize the placement and operation of chlorine booster stations.

Acknowledgments This work was supported by NSF Grant Number BES-0329549 and by the Center for Water Quality in Urban Environmental Systems (Water QUEST) at Carnegie Mellon University.

References

ABB AW400 Residual Chlorine Monitoring System. Web page [accessed 2008]. Available at http://www.abb.com/product/seitp330/8c5cdfc61254c6d4c12572f40053747c.aspx.

American Public Health Association (2005) *Standard Methods for the Examination of Water and Wastewater, 21st Edition.* American Public Health Association, Washington, DC.

American Society of Civil Engineers. Report Card for America's Infrastructure (2004a) Web page [accessed 2008]. Available at http://www.asce.org/reportcard/2005/page.cfm?id=24.

American Society of Civil Engineers, American Water Works Association, Water Environment Federation (2004b) *Interim Voluntary Guidelines for Designing an Online Contaminant Monitoring System.* American Society of Civil Engineers.

Bentley WaterCAD. Web page [accessed 2007]. Available at http://www.bentley.com/en-US/Products/WaterCAD/.

Berry, J., Carr, R.D., Hart, W.E., Leung, V.J., Phillips, C.A., and Watson, J.-P. (2006) "On the placement of imperfect sensors in municipal water networks." *Proceedings of the 8th Annual Water Distribution System Analysis Symposium*, Cincinnati, OH.

Boccelli, D.L., Tryby, M.E., Uber, J.G., Rossman, L.A., Zierolf, M.L., and Polycarpou, M.M. (1998) Optimal Scheduling of Booster Disinfection in Water Distribution Systems. *Journal of Water Resources Planning and Management-ASCE*, 124(2), 99–111.

Byer, D. and Carlson, K. (2005) Real Time Detection of Intentional Chemical Contamination in the Distribution System. *Journal of the American Water Works Association*, 97(7), 130–140.

Chemtrol. Chemtrol 265 Free Chlorine Analyzer. Web page [accessed 2008]. Available at http://www.sbcontrol.com/ch265.htm.

Haas, C.N., Jacangelo, J.G., Bishop, M.M., Cameron, C.D., Chowdhury, Z.K., Connell, G.F., Doty, G.A., Finch, G.R., Gates, D.J., Greenberg, A.E., Hoehn, R.C., Huebner, W.B., Jensen, J.N., Lange, A.L., Long, B.W., Moyer, N.P., Nagel, W.H., Noran, P.F., Palin, A.T., Regli, S.E.,

Routt, J.C., Symons, J.M., Thompson, C.K., and Voyles, C.F. (1992) Survey of Water Utility Disinfection Practices. *Journal of the American Water Works Association*, 84(9), 121–128.

Hach, Inc. CL17 Free Chlorine Analyzer Specifications. Web page [accessed 2008]. Available at http://www.hach.com/.

Hall, J., Zaffiro, A.D., Marx, R.B., Kefauver, P.C., Krishnan, E.R., and Herrmann, J.G. (2007) On-Line Water Quality Parameters as Indicators of Distribution System Contamination. *Journal of the American Water Works Association*, 99(1), 66–77.

Hallam, N.B., West, J.R., Forster, C.F., Powell, J.C., and Spencer, I. (2002) The Decay of Chlorine Associated with the Pipe Wall in Water Distribution Systems. *Water Research*, 36, 3479–3488.

Helbling, D.E. and VanBriesen, J.M. (2007) Free Chlorine Demand and Cell Survival of Microbial Suspensions. *Water Research*, 41(19), 4424–4434.

Helbling, D.E. and VanBriesen, J.M. (2008) Continuous Monitoring of Residual Chlorine Concentrations in Response to Controlled Microbial Intrusions in a Laboratory-Scale Distribution System. *Water Research*, 42(12), 3162–3172.

Helbling, D.E. and VanBriesen, J.M. (2009) Modeling Residual Chlorine Response to a Microbial Contamination Event in Drinking Water Distribution Systems. *Journal of Environmental Engineering*, 135(10), 918–927.

Klise, K.A. and McKenna, S.A. (2006) "Multivariate applications for detecting anomalous water quality data." *8th Annual Water Distribution Systems Analysis Symposium*, Cincinnati, OH. August 27–30, 2006.

Kuntze. Kuntze Chlorine Analyzer. Web page [accessed 2008]. Available at http://www.coleparmer.com/Catalog/product_view.asp?sku=9967230.

McKenna, S.A., Klise, K.A., and Wilson, M.P. (2006) "Testing water quality change detection algorithms." *8th Annual Water Distribution Systems Analysis Symposium*, Cincinnati, OH. August 27–30, 2006.

Munavalli, G.R. and Kumar, M.S.M. (2003) Optimal Scheduling of Multiple Chlorine Sources in Water Distribution Systems. *Journal of Water Resources Planning and Management-ASCE*, 129(6), 493–504.

Murphy, B.M. and Kirmeyer, G.J. (2005) Developing a Phased Distribution System, Security Enhancement Program. *Journal of the American Water Works Association*, 97(7), 93–103.

MWH Soft H2OMAP Water. Web page [accessed 2007]. Available at http://www.mwhsoft.com/page/p_product/water/water_overview.htm.

Ostfeld, A., Uber, J., and Salomons, E. (2006) "Battle of the Water Sensor Networks (BWSN): a design challenge for engineers and algorithms." *Proceedings of the 8th Annual Water Distribution System Analysis Symposium*, Cincinnati, OH.

Parks, S.L.I. and VanBriesen, J.M. (2009) Booster Disinfection for Response to Contamination in a Drinking Water Distribution System. *Journal of Water Resources Planning and Management*, 135(6), 502–511.

Powell, J.C., Hallam, N.B., West, J.R., Forster, C.F., and Simms, J. (2000) Factors Which Control Bulk Chlorine Decay Rates. *Water Research*, 34, 117–126.

Prasad, T.D., Walters, G.A., and Savic, D.A. (2004) Booster Disinfection of Water Supply Networks: Multiobjective Approach. *Journal of Water Resources Planning and Management-ASCE*, 130(5), 367–376.

Preis, A. and Ostfeld, A. (2006) "Optimal sensors layout for contamination source identification in water distribution systems." *Proceedings of the 8th Annual Water Distribution System Analysis Symposium*, Cincinnati, OH.

Propato, M. and Uber, J.G. (2004) Vulnerability of Water Distribution Systems to Pathogen Intrusion: How Effective Is a Disinfectant Residual? *Environmental Science & Technology*, 38(13), 3713–3722.

Propato, M., Cheung, P.B., and Piller, O. (2006) "Sensor location design for contaminant source identification in water distribution systems." *Proceedings of the 8th Annual Water Distribution System Analysis Symposium*, Cincinnati, OH.

Shang, F., Uber, J., and Rossman, L. (2007) *EPANET Multi Species Extension Users Manual*, National Risk Management Research Laboratory, US Environmental Protection Agency, Cincinnati, OH.

States, S., Scheuring, M., Kuchta, J., Newberry, J., and Casson, L. (2003) Utility-Based Analytical Methods to Ensure Public Water Supply Security. *Journal of the American Water Works Association*, 95(4), 103–115.

Teledyne ISCO. Orbit Chlorine. Web page [accessed 2008]. Available at http://www.isco.com/products/products3.asp?PL=3024050.

Thompson, S.L., Casman, E., Fischbeck, P., Small, M.J., and VanBriesen, J. M. (2007) "Vulnerability assessment of a drinking water distribution system: implications for public water utilities." *World Environmental and Water Resources Congress 2007*, ASCE, Tampa, FL.

Tryby, M.E., Boccelli, D.L., Uber, J.G., and Rossman, L.A. (2002). Facility Location Model for Booster Disinfection of Water Supply Networks. *Journal of Water Resources Planning and Management-ASCE*, 128(5), 322–333.

United States Environmental Protection Agency. (1999) *Alternative Disinfectants and Oxidants Guidance Manual*. Office of Water, EPA/815/R-99/014, Washington, DC.

United States Environmental Protection Agency. WaterSentinel Online Water Quality Monitoring as an Indicator of Drinking Water Contamination. Web page [accessed, 2005]. Available at http://www.epa.gov/safewater/watersecurity/pubs/watersentinel_wq_monitoring.pdf.

Walski, T.M., Chase, D.V., Savic, D.A., Grayman, W., Beckwith, S., and Koelle, E. (2003) *Advanced Water Distribution Modeling and Management*. Haestad Methods, Inc., Waterbury, CT.

Water Information Management and Modeling Team Integrated Water Quality Security System. Web page [accessed 2007]. Available at http://eh20.saic.com/iwqss/.

Yokogawa. Residual Chlorine Analyzer FC400G. Web page [accessed 2008]. Available at http://www.yokogawa.com/an/chlorine/an-fc400g-001en.htm.

Chapter 12
Biosensors for the Detection of *E. coli* O157:H7 in Source and Finished Drinking Water

Mark D. Burr, Andreas Nocker, and Anne K. Camper

12.1 Introduction

Rose and Grimes (2001) summarized the opinion of a colloquium panel concerning the status of detection and risk assessment in the drinking water industry by stating that "water quality monitoring is mired in the past." They envisioned a future in which, among other things, gene chip biosensors in streams and watersheds could alert downstream water utilities, in real time, that storm runoff from cattle pastures was laden with *E. coli* O157:H7, permitting the utilities to increase disinfection doses and issue boil orders to customers. The scenario sounds like science fiction, but it was advanced by recognized experts in drinking water treatment. The report did not include a timetable for the adoption of this kind of instrumentation nor did it provide many details about how these biosensors would actually function. It did not indicate whether biosensors would need to be able to distinguish viable from non-viable pathogens (Section 12.5). However, it was optimistic that new technologies could and would replace the time-honored practice of culturing fecal indicator organisms as a measure of microbial water quality.

Such an application for biosensors would require them to be real-time sentinels for pathogens, often deployed in non-pristine and remote environments, yet capable of reliable performance for weeks or months with minimal attention. They would likely be installed and maintained by personnel who understood their design and operation only in general terms. The expectation is that most of the time pathogens would be absent from the water stream; thus, pathogen detection would be a rare event. However, in that rare event, the biosensor would be required to distinguish with a high probability the presence of a pathogen versus a false positive. In that respect, these biosensors would be analogous to smoke detectors, and the expectations would be similar. The difference, however, is that biosensors in source or finished drinking water would be subjected to a much harsher environment than

M.D. Burr (✉)
Center for Biofilm Engineering, Montana State University, Bozeman, MT 59717, USA
e-mail: mark_b@erc.montana.edu

R.M. Clark et al. (eds.), *Handbook of Water and Wastewater Systems Protection*, Protecting Critical Infrastructure, DOI 10.1007/978-1-4614-0189-6_12,
© Springer Science+Business Media, LLC 2011

smoke detectors and would have to be incredibly robust. In addition, the optimal distribution or placement of biosensors in environmental waters is not known.

In an earlier review (Nocker et al., 2009), we provided a preliminary evaluation of biosensors for the detection of pathogens in drinking water. Other relevant reviews include those by Teles and Fonseca (2008), Noble and Weisberg (2007), Lazacka et al. (2007), Rasooly and Herold (2006), Mehrvar and Abdi (2004), Deisingh and Thompson (2004), Leonard et al. (2003), and Ivnitski et al. (1999). The literature about pathogen detection using biosensors is vast and growing rapidly. It would be impossible to review all of it; therefore, this chapter focuses on biosensors designed to detect *E. coli* O157:H7. It is a relevant waterborne pathogen with a low infectious dose (as few as 10 organisms; US Food and Drug Administration, 2009), which increases the challenge to develop ultrasensitive biosensors. It is also the organism used for development and testing of a large number of biosensors, so that we can make comparisons among them. Some of the literature reviewed here is derived from the food industry, but it is considered relevant to the water industry as well.

Meeusen et al. (2005) defined biosensors as "analytical instruments possessing a capturing molecule as a reactive surface in close proximity to a transducer, which converts the binding of an analyte to the capturing molecule into a measurable signal." There is great variety in biosensor design and function, but all biosensors have some common features. The specificity of a biosensor is determined by its recognition element. Nearly all biosensors capture whole cells, parts of cells, nucleic acids, or amplification products generated from nucleic acids. Biosensors that use antibodies to capture intact cells or cell fragments are called immunosensors. Biosensors that capture nucleic acid fragments using conventional hybridization are sometimes called DNA or nucleic acid sensors; however, we will refer to them as genosensors (reviewed by Teles and Fonseca, 2008). The biosensor described by Call et al. (2001) was somewhat of a hybrid in that it used antibodies to capture target cells that were then subjected to PCR. This design provided two recognition events for greater specificity as well as amplification of the analyte to increase sensitivity. The fact that both immunosensors and genosensors continue to be developed indicates that neither type is clearly superior to the other.

There have been attempts to develop recognition methods based on other biomolecules (Ngundi et al., 2006); however, to date, almost all biosensors use either antibodies or nucleic acids for capturing the target (analyte). Only one biosensor reviewed here uses a different method of recognition, capture of *E. coli* O157:H7 through its O antigen using a lectin (Lu et al., 2009; Section 12.4). All biosensors process aqueous samples in order to capture and immobilize the analyte at a solid surface. These capture surfaces are very small, usually measured in square millimeters. Most often, the surface is part of, or adjacent to, the transducer. In some cases, capture is onto magnetic beads in suspension. The beads are then manipulated to immobilize them next to the transducer. Certain biosensors are reagentless or label free, which means that binding of the analyte is sufficient to generate a signal. Others, however, require attachment of a label to the captured cell or nucleic acid fragment. The high cost of labeling reagents has encouraged biosensor miniaturization and the development of reagentless biosensors (Lu et al.,

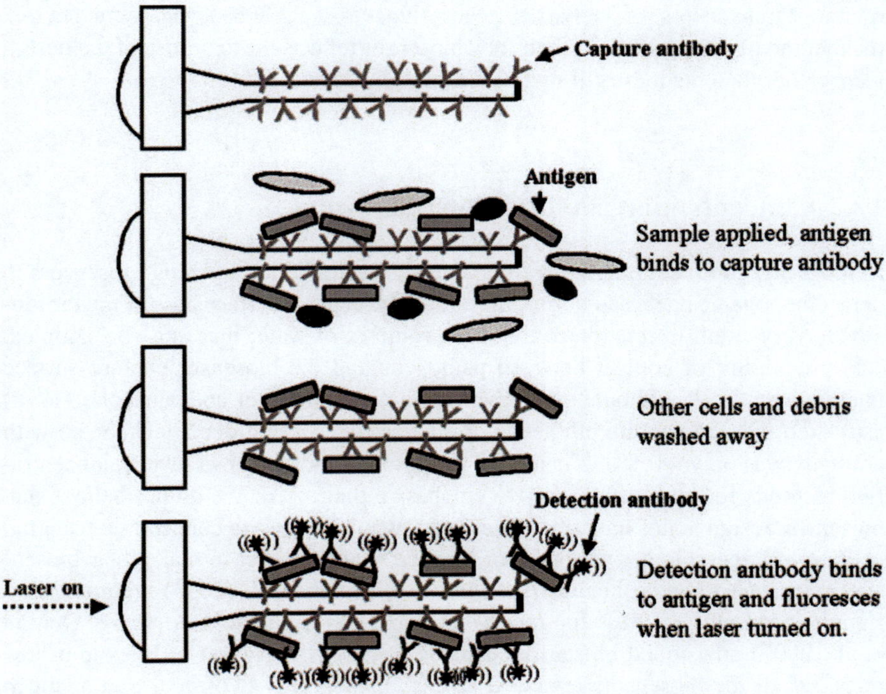

Fig. 12.1 Biosensor sandwich assay. The target antigen is bound by a capture antibody on the fiber optic waveguide. A fluorophore (Cy5)-labeled detection antibody is then attached to form a sandwich assay. The fluorophore is excited by a laser to generate a detectable signal. Reprinted with permission from Lim (2003). Copyright 2003 IEEE

2009). Surface plasmon resonance (SPR) and quartz crystal microbalance (QCM) biosensors (Section 12.4) are label-free biosensors (Su and Li, 2005).

Labels are commonly attached using a "sandwich" format in which the analyte is sandwiched between the capture element and the reporter element (Fig. 12.1). For immunosensors, the labeled antibody is referred to as a secondary antibody. For genosensors, the labeled oligonucleotide is usually called a reporter probe. Sometimes the label is added to the cell or nucleic acid in suspension before capture, but usually the sequence is capture first, then label. In either case, there is no transduction until all elements are in place. Sandwich assays consist of two recognitions of the analyte which increase specificity. There are several labeling options. The label may be a single fluorescent dye molecule; however, to increase sensitivity improved labels have been developed. Among these are liposomes and quantum dots (Section 12.3). The reporter may also be an enzyme which cleaves a substrate to produce a chemiluminescent or electrochemical signal.

Transduction is the process of converting the act of capture to an electronic signal. There are three basic types of transduction, optical, electrochemical, and piezoelectric (mass sensitive). Lazacka et al. (2007) and Teles and Fonseca (2008)

reviewed transducers. In our earlier review (Nocker et al., 2009), we speculated that the method of transduction chosen for a biosensor reflects the expertise of the design team rather than the merits of that particular transducer.

12.2 Concentration Methods for Biosensors

Reidt et al. (2008) summarized a fundamental contradiction in using biosensors to detect pathogenic bacteria in drinking water. The capture surface area of the biosensors is very small in relation to the large volumes of water that must be sampled. The probability of contact between pathogens and the biosensor capture surface is extremely small without some preconcentration. Brewster and Mazenko (1998) estimated that the capture efficiency of an electrode (3 mm diameter) coated with antibody was only ~0.1%. It is not the purpose of this chapter to review concentration methods for biosensors, only to emphasize their need. We do not believe that biosensor design teams have adequately addressed the need to concentrate bacterial cells from large volumes of source or finished water in order to make their biosensors relevant to human health risk. Morales-Morales et al. (2003) recommended sample sizes of 10–100 L for raw water and 1000 L for finished water. Derzon et al. (2009) envisioned biosensors capable of processing water at liters/min flow rates. Of all the biosensors reviewed in this chapter, only two processed as much as 1 L of sample (Campbell and Mutharasan, 2007; Straub et al., 2005). As will be seen, typical biosensor sample volumes are ≤1 mL. For concentrating samples, centrifugation, filtration, and capture on magnetic beads (immunomagnetic separation, IMS) have all been used in conjunction with biosensors.

BEADS (Biodetection Enabling Analyte Delivery System) is a "sample to answer" platform that incorporates concentration and detection (Straub et al., 2005). The system was actually tested using seeded environmental water. One-liter samples of Columbia River water were centrifuged to 1 mL final volume and seeded with *E. coli* O157:H7. This concentrate was pumped through a microchamber in which bacterial cells were captured by antibodies on the surface of magnetic microbeads (5–150 μm diameter). The beads were retained by a magnet while the sample matrix and inhibitors were removed by washing. The beads were then released and flushed by PCR Master Mix into a thermal cycler chamber for amplification of target genes. The PCR products were hybridized to a microarray for detection. Alternatively, total rRNA could be fluorescently labeled and hybridized directly to the microarray. Using the PCR option, the authors could detect 10 *E. coli* O157:H7 cells.

Lim and Zhang (2007) reviewed the use of magnetic microbeads used in conjunction with microfluidics systems. Magnetic microbeads have a large surface-to-volume ratio. They can be manipulated using magnets. Antibodies or nucleic acids as capture molecules can be immobilized on their surfaces. Among the disadvantages of IMS concentration listed by Brewster and Mazenko (1998) were the high concentration of beads necessary (~10^7 beads/mL), the poor recovery of target bacteria (~50%), and the high cost. They estimated that the maximum practical sample volume to process by IMS was only about 2 mL. Zordan et al. (2009) captured *E. coli* O157:H7 on iron oxide magnetic beads that at 1.5 μm diameter were approximately the same size as the bacterial cells they were intended to capture. Optical density measurements showed that 90% of the beads could be recovered from suspension using magnetism. Atomic force microscopy indicated that one to four beads were attached to each *E. coli* cell. The final volume of cells concentrated by IMS was 0.5 mL, but the authors did not indicate the initial volume, i.e., the volume reduction ratio. Tu et al. (2005) used magnetic beads to capture cells from only 200 μL volumes of an extract from seeded ground beef. Su and Li (2004) used IMS to capture bacterial cells, but only from 1 mL samples (using 20 μL of the bead suspension). It does not appear that magnetic beads alone can solve the problem of concentrating pathogens from large volumes of water.

Reidt et al. (2008) showed that low bacterial cell numbers could be reproducibly recovered from seeded micromechanical filters containing an Si_3N_4 membrane with precise and exact 450 nm pore size. In their laboratory tests, 20 mL samples were filtered and eluted (with 5 mL sterile phosphate-buffered saline, PBS). Using plate counts, they estimated almost total recovery of cells from the filter. They did not test their filtration with large volumes of environmental waters nor discuss filtration of samples containing a matrix of humics, debris, and non-target cells. Morales-Morales et al. (2003) could filter up to 10 L of source water through hollow fiber filters, leaving 250 mL of retentate which was then pelleted by centrifugation. Filter performance was not significantly affected by turbidity. Recovery of bacteria was >85% and was consistent over a range of seeding rates. Filters could be reused up to 40 times. We believe that some combination of centrifugation, filtration, and IMS can effectively concentrate samples for biosensors, and that biosensor researchers should specify how samples will be concentrated.

12.3 Recent Technological Advances

This section highlights some recent advances in technologies that will undoubtedly contribute to the improvement of biosensors. These advances have already been incorporated into some biosensor designs (Section 12.4).

12.3.1 Liposomes

A strategy to make biosensors more sensitive has been to increase the brightness of optical signals. Liposomes have been used for this purpose because these nanovesicles are capable of encapsulating 10^5–10^6 dye molecules (Chen et al., 2006; Baeumner et al., 2004). Therefore, replacing a single label with a liposome promises significant amplification of the signal from a biosensor. Liposomes are spherical vesicles with a lipid bilayer (Fig. 12.2). They are self-assembling in the laboratory when lipids and dye molecules are mixed; however, they require

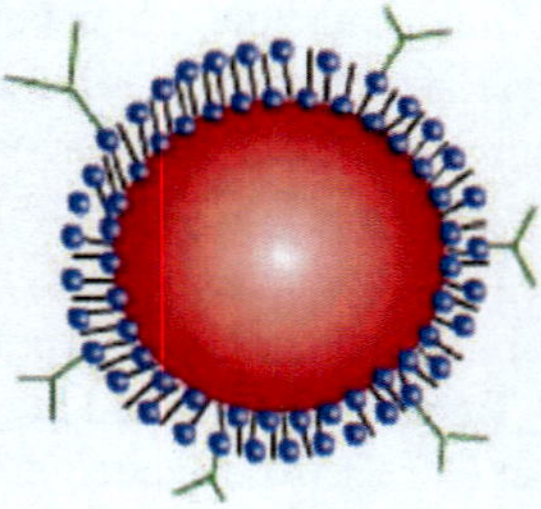

Fig. 12.2 Diagram of a general liposome structure. Lipids form a bilayer entrapping an aqueous core. Biorecognition elements can be tagged to the outside membrane and highly water-soluble marker molecules can be entrapped in the inner volume. Reprinted with permission from Edwards and Baeumner (2006). Copyright 2006 Elsevier

considerable expertise in processing (Baeumner et al., 2003). Their surfaces can be functionalized with antibodies or oligonucleotides to create sandwich reporters. Often the liposome is lysed with detergents as part of the protocol; however, lysis is not always necessary. Lysis may increase brightness by diluting dye molecules that might otherwise self-quench in unlysed liposomes because of their close proximity to each other. Edwards and Baeumner (2006) reviewed studies that reported signal enhancements of 10- to 1000-fold when liposomes replaced conventional fluorescent dyes. However, that signal enhancement is actually less that would have been predicted. The large size of liposomes (300–600 nm) creates steric hindrances that may prevent quantitative binding of liposome-tagged reporters to captured analyte. Liposomes are not limited to containing fluorescent molecules. They may also be used to encapsulate electrochemical or chemiluminescent tags (Edwards and Baeumner, 2006). Liao and Ho (2009) used liposomes to contain redox markers used in an electrochemical transduction.

12.3.2 Quantum Dots

Several problems inherent in conventional fluorescent dyes have been overcome by a relatively new labeling technology, colloidal luminescent semiconductor nanocrystals or quantum dots (QD) (Sapsford et al., 2006) (Fig. 12.3). QD range in diameter from 2 to 10 nm. They are commonly manufactured with a CdSe core (2–8 nm diameter) and a ZnS shell (~2 nm thick). QD have narrow emission spectra that depend entirely upon core diameter and are far removed from their absorption spectra (>100 nm). When a mixed population of QD (differing in core diameter) are excited at the same wavelength, their individual emission spectra can be separated, which makes QD ideal for multiplex assays (detection of multiple analytes simultaneously). Compared to the fluorescent rhodamine label, CdSe/ZnS QD are 20 times brighter, 100 times more photostable, and have emission spectra one-third the width (Yang and Li, 2006; Chen et al., 2006). Hahn et al. (2005) found that QD were

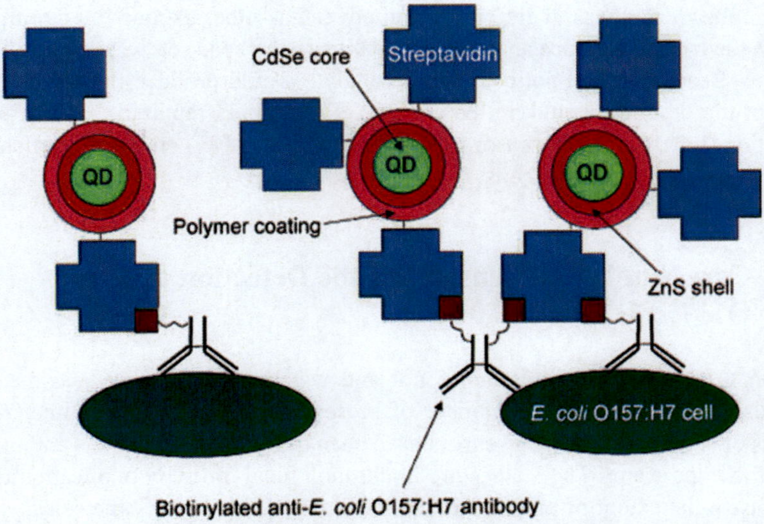

Fig. 12.3 Labeling of *E. coli* O157:H7 cells with biotinylated anti-*E. coli* O157:H7 antibodies and streptavidin-conjugated CdSe/ZnS core/shell QD. Drawing is not to scale. For clarity, only a few antibodies per cell and a few streptavidin molecules per QD are shown. Reprinted with permission from Hahn et al. (2005). Copyright 2005 American Chemical Society

about 100 times brighter than the fluorescent dye FITC. QD–streptavidin conjugates can be purchased and coated with biotinylated antibodies or biotinylated oligonucleotides in the laboratory. Goldman et al. (2006) discussed current limitations of QD for immunosensors. As manufactured, QD are inherently hydrophobic. Intricate chemical processing is required to create a hydrophilic surface that is also compatible with the immobilization of capture antibodies. Creating QD–streptavidin conjugates in order to immobilize biotinylated antibodies is a solution; however, the orientation of the antibody is difficult to control, resulting in heterogeneity and unpredictable affinities. A logical attempt to advance labeling technology would be to encapsulate QD inside liposomes. However, the large size of QD minimizes any expected gain in signal enhancement since a single liposome could contain only ~3 QD compared to $\sim 10^6$ dye molecules (Chen et al., 2006).

12.3.3 Luminex LabMAP^{TM}

Dunbar et al. (2003) described a detection system (Luminex LabMAP^{TM} system) that has a strategy similar to QD-based biosensors. It also employs magnetic microbead sets, each with a specific "spectral address," which was achieved by varying the proportion of two fluorophores contained within the beads. Each dye-encoded bead set was coated with either an antibody or an oligonucleotide. Following capture of the analyte, the beads were directed past a dual laser. One laser excited the reporter to generate an emission that was proportional to the number of microbeads and was, therefore, quantitative. The other excited the fluorophores within the bead to produce an emission specific to the bead set, which identified the analyte. Beads that had not captured an analyte would produce the second signal but not the first and would not be counted. The authors reported that the limit of detection (LOD) as a genosensor was between 10^3 and 10^5 genome equivalents and as an immunosensor $\sim 10^3$ cells.

12.4 Case Studies: Biosensors for the Detection of *E. coli* O157:H7, 1998–2010

E. coli O157:H7 is an important food- and waterborne pathogen and the target organism for testing a large number of biosensors. This section summarizes the results of studies in which biosensors have been designed to detect this pathogen, in both food and water. The studies are in chronological order by publication date to emphasize the extent of progress in the industry or lack of it. Some selected studies are summarized in Table 12.1. The purpose is to give the reader a sense of the diversity of approaches that have been used to detect this organism using biosensors.

Table 12.1 A comparison of selected biosensors for the detection of *E. coli* O157:H7

Type of biosensor	Description/comment	Sample volume	LOD	Reference
Electrochemical immunosensor	Immunocapture in solution, filtration, and enzymatic cleavage of substrate to generate electrons proportional to analyte concentration. Unbound enzyme contributed to excessive background noise	100 μL	5×10^3 cells/mL	Brewster and Mazenko (1998)
Microarray	Immunomagnetic capture, PCR, and hybridization. Alkaline phosphatase linked to captured DNA strand, substrate cleaved to produce fluorescent signal	1 mL chicken carcass rinsate	168 CFU/mL	Call et al. (2001)
Fiber optic waveguide immunosensor	Commercially available instrument. Positive samples tested for viability using cultural methods	1 mL	10^3 CFU/mL	Tims and Lim (2003)

Table 12.1 (continued)

Type of biosensor	Description/comment	Sample volume	LOD	Reference
Microcapillary flow injection immunosensor	Antibody immobilized inside microcapillary. Sequential flow of sample, secondary antibody–liposome conjugate, rinse, and detergent to lyse liposomes, releasing fluorescent dye molecules	150 μL	360 cells/mL	Ho et al. (2004)
Microplate immunosensor	Capture onto magnetic beads, beads deposited into microplate. Chemiluminescent assay for total *E. coli* O157:H7. Viable *E. coli* O157:H7 detected using assay of intracellular NAD(P)H	200 μL from seeded ground beef	10^3 CFU/mL (total counts)	Tu et al. (2005)
Integrated genosensor	Concentration by centrifugation and IMS. PCR with products hybridized to Luminex beads. Luminex system decodes and quantifies beads passing dual laser	1 L river water pelleted to 1 mL, then seeded	10 cells/L	Straub et al. (2005)
QCM genosensor	Oligonucleotide capture probe immobilized on QCM. Single PCR strand captured and linked to nanoparticle to enhance mass, improving LOD	DNA extraction from 1 mL sample	10^2 CFU/mL	Mao et al. (2006)
Piezoelectric immunosensor	Microcantilever inside flow cell. Sample pumped across capture surface at 1.5 mL/min	1 L	1 cell/mL	Campbell and Mutharasan (2007)
QCM-D immunosensor	Immunocapture, measure dissipation of energy of oscillation of microbalance	<10 mL	10^5 CFU/mL	Poitras and Tufenkji (2009)

Brewster and Mazenko (1998) created an electrochemical biosensor that was essentially a modified ELISA protocol. Different concentrations of cells were incubated in suspension with an antibody–alkaline phosphatase conjugate. The cell–antibody–enzyme complex was then filtered onto a membrane which was placed in contact with an electrode. For detection, a substrate was added (para-aminophenyl phosphate) which was cleaved by the enzyme to generate electrons that were converted to a signal proportional to *E. coli* O157:H7 cell concentration. However, because the working electrode was only 3 mm in diameter, a filter of the same diameter was required so that the two could be joined. This posed a serious limitation on practical sample volumes. When the standard assay volume was 100 μL, the LOD was 5×10^3 cells/mL (500 total cells). Attempts to filter greater volumes resulted in increased noise because unbound antibody–enzyme also became entrapped in the membrane.

Abdel-Hamid et al. (1999) created a microfluidics biosensor that consisted of a cartridge containing a nylon membrane in contact with an electrode. A primary antibody was immobilized on the membrane. A continuous stream of iodide in buffer was pumped through the cartridge. Into this stream, an injector sequentially introduced the sample, then secondary antibody conjugated to horseradish peroxidase, and finally a solution containing peroxide. The peroxidase–peroxide–iodide interaction generated electrons at the electrode in proportion to the number of *E. coli* O157:H7 cells captured. The maximum injection volume was 1 mL and the LOD was 100 cells/mL. The assay time was about 30 min.

A microarray was used as an *E. coli* O157:H7 biosensor that incorporated antibody capture, PCR, and hybridization of the PCR products (Call et al., 2001). *E. coli* O157:H7 cells were seeded into water that had been used to rinse chicken carcasses and cells were concentrated using immunomagnetic beads. The beads were added directly to a PCR reaction that targeted virulence genes. PCR products were denatured and one strand was captured by hybridization to the microarray. During PCR, a biotinylated primer had been incorporated into the captured strand. Following hybridization, a streptavidin–alkaline phosphatase conjugate was bound to this captured strand. A substrate (ELF-97®) was added and cleaved by the enzyme to produce a fluorescent signal. Without any enrichment, 168 CFU/mL *E. coli* O157:H7 was detected in the rinsate in 100% of the tests ($n = 5$), but 55 CFU/mL was detected in only two of five trials. This biosensor offered very high specificity because there were three recognition events.

Tims and Lim (2003) used a commercially available Analyte 2000 (Research International) fiber optic waveguide biosensor to detect *E. coli* O157:H7. Cells suspended in 1 mL samples were captured in the waveguide by immobilized antibodies and detected as a fluorescent signal (Fig. 12.1). Positive samples could then be subjected to two additional assays: enrichment and streaking to test for viability and PCR to confirm *E. coli* O157:H7. For PCR, cells were either dissociated from the waveguide or the disposable waveguide was actually sectioned and placed directly into the PCR reaction. The LOD was 10^3 CFU/mL in a 10 h total assay time.

Gehring et al. (2004) tested an enzyme-linked immunomagnetic chemiluminescence (ELIMCL) method to detect *E. coli* O157:H7 recovered from beef that had been seeded at different concentrations. The system consisted of a holder for

1.5 mL microcentrifuge tubes with an external magnet that could be operated to either retain the beads against the walls of the tube or release them into one of the reagent solutions. In a sandwich immunoassay, bacterial cells were first captured by antibodies immobilized on magnetic beads, then exposed to a secondary antibody conjugated to an alkaline phosphatase. The alkaline phosphatase reacted with an introduced substrate to produce a chemiluminescence whose intensity was read on a simple luminometer. Following optimization and reduction of non-specific binding, the authors were able to achieve an LOD in pristine buffer of $\sim 8 \times 10^3$ CFU/mL. Following a 7 h enrichment, an inoculation level of 10 CFU/g beef could be detected. Enrichment improved sensitivity and indicated viable cells.

Ho et al. (2004) described their biosensor as a microcapillary flow injection liposome immunoanalysis (mFILIA) system. Capture antibodies were immobilized onto the interior surface of the microcapillary during an 8 h static incubation. Non-segmented flow of a carrier stream was then initiated. There were sequential injections of 150 μL sample followed by 10 μL of a liposome–secondary antibody conjugate, a rinse to remove unbound conjugate, and a detergent to lyse the liposome and release fluorescent molecules. The fluorescent signal was proportional to the number of captured cells. The unbound liposomes also produced a signal as they passed the detector, but could be differentiated from the true signal because of their elution time. The LOD of this system was 360 cells/mL. The assay time was 45 min, which included reconditioning the biosensor for the next sample.

Liposome nanovesicles were also used in a "universal" dipstick genosensor developed by Baeumner et al. (2004) (Fig. 12.4). It was composed of both generic

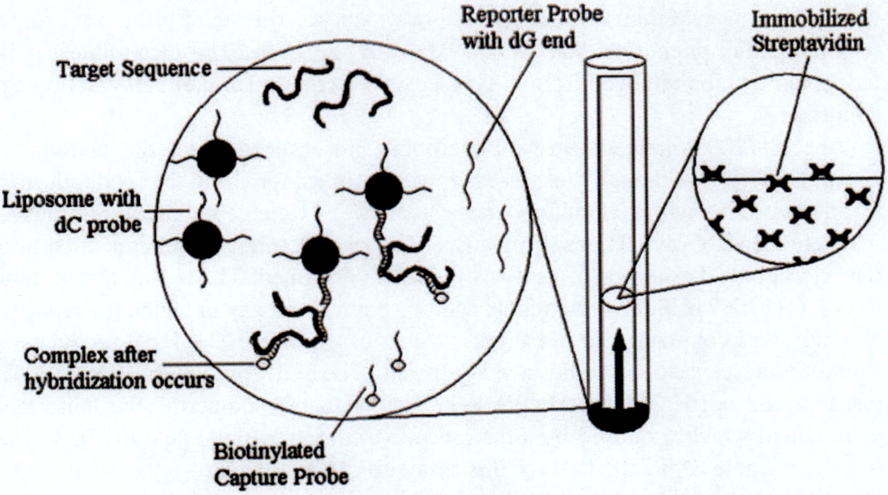

Fig. 12.4 A schematic of the universal genosensor assay. During the assay, both the capture probe and reporter probe hybridize to the single-stranded target. The capture probe is biotinylated so that it can be immobilized onto the dipstick by interaction with streptavidin. The reporter probe is a bridge to a label. It hybridizes in a G-rich generic sequence to a C-rich generic sequence of a labeling oligonucleotide that is conjugated to a liposome filled with dye molecules. Adapted with permission from Baeumner et al. (2004). Copyright 2004 American Chemical Society

(universal) and specific DNA sequence components. It could be easily modified to detect any target by retaining the generic components and switching the specific components. This flexibility was said to minimize optimization for new targets. A single-stranded target sequence, which could be produced by PCR or NASBA (nucleic acid sequence-based amplification) was incubated in solution with three other components: (1) a capture probe (complementary to the target and 5′ biotinylated), (2) a reporter probe (complementary at its 5′ end to the target and consisting of a generic sequence at the 3′ end), and (3) an oligonucleotide complementary to the generic sequence and conjugated to a liposome filled with dye molecules (sulforhodamine B). Following formation of a sandwich that included two specific and one generic hybridization, a membrane strip functionalized with streptavidin was used to bind the sandwich. Reagents in a running buffer lysed the liposomes and released dye molecules. Each nanovesicle entrapped $\sim 10^5$ dye molecules. The detection limit for *E. coli* was 5 fmol of target sequence, equivalent to $\sim 10^8$ copies.

Su and Li (2004) produced a biosensor that combined immunomagnetic separation with quantum dots as a reporter. A 1 mL sample was incubated in solution with 20 μL of superparamagnetic polystyrene beads coated with antibody and 100 μL of antibody–biotin conjugate. This formed a sandwich but without a label. The beads were immobilized by a magnet and washed. The label was added by incubating the beads with a QD–streptavidin conjugate. The LOD was 10^3 CFU/mL. The total assay time was about 2 h.

Zhu et al. (2005) created an optical waveguide immunosensor. Capture antibodies were immobilized on the inside surface of a glass capillary by avidin–biotin interaction. *E. coli* O157:H7 cells seeded into buffer were captured and detected using a Cy5-labeled reporter antibody. The capillary was illuminated at a right angle with a 635 nm wavelength laser and fluorescence was directed by the waveguide to a photomultiplier tube. The LOD was $\sim 10^3$ CFU/mL. The disposable capillary could then be removed from the biosensor for confirmation of cell viability by culturing.

Tu et al. (2005) created a 96-well microplate biosensor to detect total and viable *E. coli* O157:H7. Although this biosensor was designed for use in the food industry, testing was done with cell culture suspensions. Cells were captured onto immunomagnetic beads (7×10^8 beads/mL). Beads were concentrated and deposited into the microplate. Two kinds of assays were then performed. The first detected total *E. coli* O157:H7 using a conventional sandwich immunoassay in which the reporter antibody was conjugated to horseradish peroxidase (HRP). The HRP produced a chemiluminescence upon addition of a substrate. The LOD for total *E. coli* O157:H7 was reported as 10^3 CFU/mL. In the second assay, membrane-permeable substrates were added which produced bioluminescence from intracellular NAD(P)H, which indicated viable cells. The LOD of this assay was 10^4 CFU/mL.

Although not referred to as a biosensor by the authors, the detection system described by Straub et al. (2005) had all the components of one. One liter of Columbia River water was centrifuged to 1 mL and seeded with *E. coli* O157:H7 cells. Cells were captured by IMS in a microchamber. The beads were washed and diverted to a PCR reaction. The reaction products were hybridized to a suspension array, i.e., onto a specific Luminex bead set. The Luminex system then decoded the signal and quantified it. The authors could detect 10 cells entering the system, i.e., 10 cells/L. Seeding directly into 1 L of river water prior to centrifugation was apparently not attempted.

Su and Li (2005) performed a side-by-side comparison of SPR (surface plasmon resonance) and QCM (quartz crystal microbalance) biosensors for detection of *E. coli* O157:H7. SPR sensors are optical transducers that measure a change in refractive index that occurs at or near the optically clear capture surface when analyte binds. QCM sensors are piezoelectric transducers that consist of a capture surface on a microplatform that is oscillated by an external magnet. The resonance frequency at which the microbalance oscillates is a function of the mass bound to it. Both biosensors are label free and reagentless except for the capture antibody. In tests of both instruments, after the antibody was immobilized, a stable baseline was established by pumping a buffer (phosphate-buffered saline, PBS) through the flow cell. Then sample was introduced into the flow cell and incubated statically while the response was measured over time as the cells were captured. The LOD was determined as the lowest concentration of *E. coli* O157:H7 cells for which the signal-to-noise ratio (S/N) was ≥ 3. Sample volumes were 1 mL. The LOD were 10^6 and 10^5 for the QCM and SPR sensors, respectively. An advantage of the QCM is that performance would be less affected by turbid samples.

Radke and Alocilja (2005) tested a reagentless piezoelectric biosensor consisting of a capture antibody immobilized on the surface of a gold-plated electrode. The active capture area was only $\sim$10 mm^2. The electrode was immersed into a 30 mL sample derived from seeded lettuce. Impedance was then measured at frequencies ranging from 100 Hz to 10 MHz. Optimized readings at 1 kHz indicated an LOD of 10^4 cells/mL, which the authors admitted was inadequate to detect *E. coli* O157:H7 at concentrations related to the infectious dose.

Taylor et al. (2005) modified an SPR biosensor assay to increase sensitivity in the detection of *E. coli* O157:H7. The modification was in the preparation of the bacterial cells. They were either untreated, or heat killed then ethanol treated, or detergent lysed. Target cells diluted in PBS/BSA (phosphate-buffered saline/bovine serum albumin) were pumped across the fresh capture surface at 50 μL/min for 20 min (1 mL total volume), followed by a PBS/rinse for 10 min. At this point, the signal was amplified by pumping a secondary antibody through the system. Since SPR detectors respond to anything bound on or near the sensor surface, this "amplification" antibody enhanced the signal. The LOD (using the amplification antibody) was 10^7 CFU/mL for untreated cells, 10^5 CFU/mL for heat-killed ethanol-soaked cells, and 10^4 CFU/mL for detergent lysed cells. The authors believed that detergent lysis created more cell fragments that could be captured and also overcame diffusion limitations since smaller cell fragments diffuse faster.

Subramanian et al. (2006), noting that capture of analytes closer to the surface of an SPR detector enhances the signal, experimented with a self-assembled monolayer to immobilize the capture antibody to the SPR surface. To further improve the LOD, they used a secondary antibody not as a label or reporter but simply to increase the mass of material bound to the surface. The secondary antibody improved the LOD from 10^6 to 10^3 CFU/mL. The authors experimented with renewing the SPR surface by washing with 20 mM NaOH. This process stripped the secondary antibody–*E. coli* cell complex, but left the primary antibody.

Huang and Zhang (2006) created a biosensor that consisted of anti-*E. coli* O157:H7 antibody immobilized on a tin oxide nanowire, with cell binding measured as a change in voltage. Injections of increasing concentrations of the pathogen in increments of 2.4×10^4 CFU/mL produced stepwise voltage changes that quickly stabilized between subsequent injections.

QCM are generally not considered suitable for genosensing because the small mass of nucleic acids, compared to bacterial cells, does not cause a significant response. However, Mao et al. (2006) found that the signal could be enhanced by adding "mass enhancing" Fe_3O_4 nanoparticles to PCR products. A capture oligonucleotide probe was immobilized to the QCM surface. Asymmetric PCR targeting the *eae* gene of *E. coli* O157:H7 was performed using a 10:1 molar ratio of a biotinylated to unlabeled primer. The biotinylated PCR strand was hybridized to the capture probe and a streptavidin–nanoparticle conjugate was coupled to the biotin to form a sandwich. The QCM was monitored in real time for each step including capture probe immobilization. Hybridization by itself caused virtually no noticeable change in resonance frequency. However, when the nanoparticle label was added, the response was considerably enhanced. The authors estimated that the LOD was $\sim 3 \times 10^2$ CFU/mL and was considered a great improvement over gel-based detection. They tested the QCM as a single-use surface, but acknowledged the need to develop methods for renewing the surface between samples.

The piezoelectric immunosensor of Campbell and Mutharasan (2007) was somewhat unique in that the sample volume was 1 L, far greater than for most biosensors. The system consisted of a miniaturized cantilever inside a 7.0 mm diameter plexiglass flow cell (120 μL volume after the sensor had been installed). The flow cell was connected through a manifold to reservoirs for the sample, antibody, and buffer. The system could be operated in continuous flow or batch mode. The antibody was first immobilized in recirculating flow for 2 h. At a continuous sample flow rate of 1.5 mL/min, the LOD was 1 cell/mL in 1 L total sample. A limitation was the excessive noise produced at flow rates $\geq$5 mL/min.

Waswa et al. (2007) modified a commercially available Spreeta™ SPR instrument to create an immunosensor that was able to detect *E. coli* O157:H7 in milk and apple juice at an LOD between 10^2 and 10^3 CFU/mL. They regenerated the sensor surface between samples by flushing the flow cell with 0.12 M NaOH–1% Triton X-100 in distilled water. They acknowledged that unresolved issues included non-specific sensor surface fouling and sample preparation.

The microfluidic immunosensor described by Yacoub-George et al. (2007) used glass capillaries inserted into a chip as both the solid support for *E. coli* O157:H7 capture and the optical cuvette for signal detection. The sandwich reporter antibody was conjugated to horseradish peroxidase (HRP) which acted on a substrate to produce a chemiluminescent signal. The LOD was about 10^4 CFU/mL. Non-specific binding of the HRP-conjugated signal antibody to negative control capillaries contributed to the noise and reduced the signal-to-noise ratio.

The genosensor of Farabullini et al. (2007) operated by sandwich hybridization of specific PCR amplicons between a surface-tethered capture oligonucleotide probe and a biotinylated reporter oligonucleotide probe. The signal probe was then bound by streptavidin conjugated to alkaline phosphatase. Enzymatic oxidation of a substrate resulted in an electrochemical signal. The authors demonstrated the potential to detect *E. coli* O157:H7 and *Salmonella* spp. amplicons in the 1–10 nM range, but PCR amplification was by standard methods and using a conventional PCR thermal cycler. In that respect, their biosensor replaced gel electrophoresis for analysis of amplicons, but with perhaps greater sensitivity and specificity (provided by the additional hybridizations).

Lu et al. (2009) experimented with a magnetoelastic (piezoelectric) biosensor for *E. coli* O157:H7. The sensor was wireless and required no internal power supply. It was caused to vibrate by application of an external magnetic field. The vibration frequency was related to the mass bound to the sensor. The sensor surface was functionalized with mannose. *E. coli* O157:H7 was captured through a bridge between its O antigen, the lectin concanavalin A, and the bound mannose. The reaction volume was 2.5 mL. It was a static system with a 2.5 h incubation. The LOD was about 10^2 cells/mL and the response was linear between 60 and 6.1×10^9 cells/mL. The authors believed that non-specific binding to mannose was negligible.

The photodiode array (PDA) biosensor of Song and Kwon (2009) was a conventional sandwich immunosensor with the exception that the PDA microchip was the platform for immobilization of the capture antibody and also the phototransducer. The secondary antibody was conjugated to alkaline phosphatase. Transduction was based on the absorbance of a blue precipitate produced by the enzyme when a substrate was added. The LOD was reported as 10^4 *E. coli* O157:H7 cells. The authors warned that erroneous readings could result if the operational pH varied beyond a range from 7.0 to 7.5.

Poitras and Tufenkji (2009) tested a QCM-D biosensor for *E. coli* O157:H7. Like earlier QCM biosensors, the detection was based on a mass change upon capture of cells by an immobilized antibody. However, instead of measuring the change in the resonance frequency of QCM oscillation that results when cells were captured, the authors measured a different property, the dissipation factor (D, energy dissipation response). They claimed that D was a better quantitative parameter for measuring the binding of viscoelastic analytes such as bacterial cells. Furthermore, they found that the initial value of the dissipation response (D_{slope}) was more sensitive to *E. coli* O157:H7 concentration than the dissipation shift accumulated over a 60 min exposure of the QCM to *E. coli* O157:H7 cells. This realization would greatly shorten the time for taking measurements. However, the LOD was still $\geq 10^5$ CFU/mL. The authors described the biosensor as a real-time detection device; however, only small volumes (50 μL/min for 2 h) of seeded samples were pumped across the QCM. It could certainly not process raw samples in real time without considerable preconcentration.

Wang et al. (2009) created a genosensor for the detection of *E. coli* O157:H7 in which a capture probe complementary to one strand of the *eae* gene was covalently immobilized in the nanopores of an aluminum anodized oxide membrane by a complex process. Asymmetric PCR of *E. coli* O157:H7 (50:1 molar ratio of primers) produced predominately ssDNA complementary to the capture probe. Following a 2-h hybridization, the ssDNA target was extended to form dsDNA, the purpose being to increase the signal as measured by cyclic volammetry. This research was only reported to establish the proof of concept. A functioning biosensor has not yet been developed.

Eum et al. (2010) attempted to improve the sensitivity of SPR detection of *E. coli* O157:H7 by introducing gold nanorods (GNR) into an otherwise standard sandwich immunoassay. GNR have optical properties that were tested in this study for their potential to enhance the SPR signal. In the protocol, capture antibody was fixed to the SPR surface, excess antibody was washed away with PBS (phosphate-buffered saline), BSA (bovine serum albumin) was added to block any free binding sites, and excess BSA was removed with a second PBS wash. *E. coli* O157:H7 cells (100 μL of 10^5 CFU/mL) were added and incubated for 10 min, followed by a PBS wash. At this point, the captured cells were detected either by a conventional secondary antibody or a GNR–secondary antibody conjugate, followed by a final PBS wash. The authors reported a fourfold improvement in sensitivity by incorporation of GNR.

To summarize, the biosensors described in this section represent a great diversity in design and function. Most of them have not been tested with environmental samples, most of them process very small sample volumes, and most have LOD that are too high for them to be useful in protecting the public health. The biosensors described by Straub et al. (2005) and Campbell and Mutharasan (2007), which can analyze 1 L samples and have LOD $\leq$ 1 cell/mL, are the most promising, in our opinion.

12.5 Biosensors for Viability Testing

Regulators, policy makers, and experts on risk assessment should be the ones to decide if biosensors for the detection of *E. coli* O157:H7 and other organisms must distinguish viable from non-viable cells. One view is that only viable cells indicate a public health risk and that detection of *E. coli* O157:H7 without regard to viability would produce many false positives. The opposing view is that there should be zero tolerance for *E. coli* O157:H7 and that detection of any cells, alive or dead, warrants a response. Zordan et al. (2009) noted that even non-viable *E. coli* O157:H7 cells may retain active endotoxins. This section describes attempts by biosensor design teams to confirm the viability of *E. coli* O157:H7. Some research about viability testing for non-pathogenic *E. coli* as an indicator organism is also included.

Immunosensors are inherently more compatible with viability testing than genosensors, for the simple reason that captured cells may retain their viability. For viability testing that relies on culturing, sample handling and concentration steps must not cause cells to become unculturable. Reidt et al. (2008) demonstrated that a micromechanical filter could be used for quantitative filtration and recovery of cells from the filter as culturable cells. However, the bacterial cells tested were from laboratory broths, which may not have experienced the same stresses as cells recovered from environmental samples. The influence of disinfectants added to finished water was also not considered.

The immunosensor of Zhu et al. (2005) used a disposable glass capillary as the capture surface. Samples that produced a positive response from the biosensor were confirmed as viable by transferring the capillary to a culture medium. The immunosensor of Tu et al. (2005) incorporated two assays for *E. coli* O157:H7. The first produced a chemiluminescence from total cells. The second produced an NAD(P)H-mediated bioluminescence from viable cells. Tims and Lim (2003) detected *E. coli* O157:H7 using a fiber optic immunosensor. Cells from positive samples were dissociated from the capture surface and cultured for viability. Johnson-White et al. (2007) used a glass slide array immunosensor to capture and detect *E. coli* cells and a Live/Dead BacLight Bacterial Viability kit (Invitrogen) to test for viability. Two fluorescent signals were discriminated, one for *E. coli* and the other for viable cells.

For genosensors, a nucleic acid template must be selected that indicates viability. DNA is excluded because it may persist in non-viable or even lysed cells. LaGier et al. (2005) believed that rRNA from the fecal indicator *E. coli* was a suitable template for viability tests. However, Baeumner et al. (2003) believed that only mRNA, because of its short half-life (generally several minutes), should be used to confirm viability. In their assay, non-O157:H7 *E. coli* were exposed to heat shock for 5 min at 41°C in a heating block. Total nucleic acids were immediately extracted and NASBA (nucleic acid sequence-based amplification) was performed using primers targeting mRNA for a heat shock protein (clpB, transcribed only in viable cells in response to heat shock). NASBA is an isothermal amplification method that only uses RNA templates, thus ensuring that the transcript (mRNA) rather than the gene (DNA) was amplified. The amplified mRNA was detected in a sandwich assay by hybridizing to an immobilized capture probe and a liposome-tagged reporter probe. The assay increased sensitivity at two steps, the NASBA amplification of the original mRNA and the signal amplification from the large number of encapsulated fluorescent sulforhodamine B molecules. Specificity was also increased by separate hybridizations of the mRNA to both a capture and reporter probe. The LOD was reported as 5 fmol per sample, which the authors claimed was equivalent to 40 CFU/mL of *E. coli*. Modification of this biosensor to detect *E. coli* O157:H7 would require identification of heat shock gene sequences specific to that pathogen. The authors used seeded samples and did not discuss actual applications in water quality monitoring.

Zhao et al. (2006) used a very similar approach. Their system relied on heat shock to induce transcription of the heat shock protein gene *GroEL* as a means to detect viable *E. coli*. In their tests, heat shock, cell lysis, hybridization of transcripts, RT-PCR, and electrochemical detection were all performed using conventional laboratory equipment; however, they conceptualized a microchip system in which all these steps could be performed in a single chamber with microports for introducing and removing reagents and buffers used in the various steps. In their protocol, sample, reagent, and reaction volumes were $\leq 25\ \mu L$.

Straub et al. (2005) described an automated microsystem that could extract total RNA from 1 L of river water, add a fluorescent label, and hybridize the RNA to a microarray for detection. The authors admitted that rRNA may not be a good indicator of viability because its secondary structure may retard its degradation by RNases even in non-viable cells. They were hopeful that their system could be developed to target mRNA transcripts of genes that indicate recent cell viability.

12.6 A Strategy for Field Testing Biosensors by the Water Industry

Most of the biosensors described in this chapter have not been field tested. Furthermore, most design teams have apparently not imagined how their biosensors would operate in actual practice. We discuss here how biosensors might be operated to monitor source or finished water for the presence of *E. coli* O157:H7 and how these biosensors should be tested for such an application.

12.6.1 Real-Time (Continuous) Sampling

Theoretically, biosensors could be operated either in real time or for analysis of discrete (grab) samples. These two modes of operation are fundamentally different approaches. Real-time analysis is continuous. There is no defined sample size. Raw water is made to flow across the biosensor surface. Pathogen capture is cumulative. The in situ detection of *E. coli* O157:H7 in a watershed as described by Rose and Grimes (2001) sounds like real-time detection, because presumably the physical contact between the target bacterial cells and the biosensor at concentrations above some LOD triggered an immediate alarm, much like a smoke detector. For real-time analysis, the binding of the analyte must be sufficient to elicit a response. There must not be any additional intervention to determine whether an analyte has been captured. Therefore, all types of sandwich and reporter assays, although they may be rapid, are not real time. Among biosensors, QCM and SPR types have the *potential* for real-time applications because they are designed to detect the physical act of analyte binding and do not require labeling, which is a separate step.

Despite their theoretical potential, real-time biosensors capable of monitoring source or finished water in situ do not currently exist, to the best of our knowledge, and are unlikely to be developed because of inherent problems. In our opinion, the biggest challenge to real-time monitoring is the large sample volumes that must be processed by instruments that to date have been designed to accommodate only very low flow rates (Derzon et al., 2009). The small sample volumes that biosensors are capable of processing are related to the small capture surface necessary for transduction (a single spot, essentially) and the kinetics of analyte capture. Furthermore, some instruments that do indeed measure target cell capture in real time also require rinses with buffers (Taylor et al., 2005), so that the flow of sample is not continuous. Real-time instruments are label free, which means that binding of the analyte produces a direct response. However, label-free biosensors do not distinguish specific from non-specific binding, which means that general biofouling creates a false positive. Non-specific binding is manageable in the laboratory when small volumes are pumped through real-time biosensors with fresh capture surfaces. Some real-time biosensors have additional compensation for biofouling, e.g., a parallel reference channel identical to the primary channel except that it does not contain a capture antibody (Taylor et al., 2005).

Because of the potential for biofouling, biosensor surfaces would need to be renewed periodically, whether or not they had captured *E. coli* O157:H7 cells. It is common practice to block non-specific sites on all types of biosensor capture surfaces with BSA (bovine serum albumin) after the capture antibody or nucleic acid probe has been immobilized (Abdel-Hamid et al., 1999; Brewster and Mazenko, 1998). Still, it is difficult to imagine online real-time monitoring of drinking water. Biosensor design teams should be challenged to test the performance of any real-time biosensors over long exposure to raw water and treated water, both with and without seeding. We believe that biofouling would produce signals that over time could not be discriminated from true positives.

12.6.2 Discrete (Grab) Sampling

There are good reasons to limit the use of biosensors for discrete samples rather than deploying them in the environment as real-time monitors. Biosensing is more convenient in the laboratory where a single biosensor could process many discrete samples collected from different locations. In addition, discrete samples can be concentrated. Current culturing protocols use discrete samples, so comparisons could more easily be made between them. For these and other reasons, even biosensors that have the potential for real-time analysis (e.g., QCM and SPR biosensors) are likely to be used to process discrete samples. The very small volumes processed by most biosensors (~50 μL to ~1 mL) probably indicate that they have been designed as downstream detection systems to be used on samples that have already been concentrated. An advantage of using biosensors for grab samples is that the capture surface can be completely renewed between samples. Some capture surfaces have been designed to be disposable (Zhu et al., 2005) or renewed between measurements using a cleaning solution (Rijal et al., 2005).

12.6.3 Pilot-Scale Testing

No biosensor technology merits broad deployment by the water industry without first passing rigorous pilot-scale testing, and we are not aware of any published reports of such research. We believe that biosensors could be tested at a few pilot-scale water treatment plants. In these settings, they could be used to test both source and finished water. The decision about which biosensors should be studied would depend upon how commercially developed the technologies are, the willingness of the designers to subject their instruments to such rigorous testing, and the likelihood of successful operation as determined by a review panel. Side-by-side comparisons of a variety of biosensors could be accomplished. *E. coli* O157:H7 would be a likely test organism because it is a relevant pathogen and because so many biosensors have been designed to detect it. The facilities should have the capability and certification to use and contain this organism.

Ideally, the systems could be seeded in a manner that mimics a natural introduction of *E. coli* O157:H7 into water. In all cases, the seeding should be blind as well as spatially and temporally randomized (Ligler et al., 2007). The operators of the biosensors should have no knowledge of seeding schedules, *E. coli* O157:H7 concentrations, or sites of introduction. We would recommend long periods of time before and between seedings to assess the rate of reporting false positives. In this mock real-world testing, biosensor operators would be challenged to match the performance of parallel conventional cultural methods, which would be the standard. Since biosensors characteristically process only small volumes of sample, the design teams would need to pair their instruments with a method for concentrating bacterial cells and handling the associated matrix and debris, just like in the real world. The volume of raw or finished water to sample would be left to the discretion of the biosensor operators. Larger volumes might need to be sampled for some instruments to improve sensitivity.

Blind testing is probably more important than biosensor design teams realize or acknowledge. Consider how most biosensors are currently tested, with seeded samples applied in small volumes, i.e., slug doses, at times and at concentrations known to the researchers. Buffers are generally used as the sample matrix and do not contain competing non-target bacteria, disinfectants, or other components that would normally be in a matrix unless they have been specifically removed. In most cases, testing is done with fresh capture surfaces. Now imagine a blind test when the operator has no reason to expect pathogen capture but must be able to report with confidence true positives and true negatives. Biosensor baselines are characteristically noisy. Biosensor surfaces would not be expected to remain clean even during periods when target cells are not captured. Would they not accumulate humic substances and non-specific biofilms? Fouling of biosensor surfaces would certainly cause the baseline to drift. Designers and operators would have to decide how to differentiate a real signal from drift. They would also have to decide how much sample to process before renewing the capture surface. We believe that this kind of testing would indicate to regulators and biosensor designers whether biosensor technology merits adoption by the water industry.

12.7 Conclusions and Recommendations

Based on a review of the biosensor literature, the scenario posed by Rose and Grimes (2001), in which biosensors could alert water utilities in real time of dangerous levels of *E. coli* O157:H7 in source water, remains futuristic and overly optimistic, in our judgment. Although the authors did not propose a timeline for the adoption of biosensor technology, it seems that little substantive progress toward field use of biosensors in the water industry has been made in the decade since their report was published. There are probably several reasons for this.

Currently, much of the literature related to biosensors for detection of *E. coli* O157:H7 and other microorganisms is appearing in engineering and electronics journals, rather than environmental science, water research, and microbiology journals. The authors generally do not appear to be environmental scientists. Much of the emphasis seems to be on different methods of transduction. We believe that the success of the researchers, i.e., their ability to obtain funding and publish papers, does not depend upon producing a working biosensor for the water industry.

The biosensor industry is large and growing. However, in our earlier review (Nocker et al., 2009) we determined that the water industry is perceived by biosensor designers as a minor potential market for their products. It is probably not considered very lucrative either, compared to the food industry, for example. The water industry has probably not expressed much real interest in biosensors and biosensor design teams have generally failed to envision how their instruments would be incorporated into the water industry. This lack of vision is puzzling, considering the large number of biosensors that have been designed to detect *E. coli* O157:H7, a pathogen highly relevant to microbiological water quality.

The emphasis on miniaturization and reduction of reagent volumes has resulted in a generation of biosensors that accommodate microliter to low milliliter scale volumes of sample. The sample is usually seeded with different concentrations of *E. coli* O157:H7, usually in buffer and without most of the background matrix, disinfectant, and debris encountered in natural or finished waters. The only scenario in which it makes sense to test biosensors in this way is to assume that the cells have already been concentrated and removed from their background, by IMS, centrifugation, or filtration. Therefore, we believe that designers of biosensors need to give greater consideration to how samples will be processed. The assumption that *E. coli* O157:H7 cells have already been greatly concentrated is also the only way to reconcile the rather high LOD achieved by most biosensors with any usefulness in the water industry. Many authors have claimed that their instruments are "sensitive"; however, they have often been speaking of improvements in sensitivity compared to gels or microarrays for genosensors or ELISA for immunosensors (Rasooly and Herold, 2006).

According to Lim et al. (2005) "biological detectors can only rarely detect microorganisms directly from samples at or below human risk levels because of their lack of sensitivity." Radke and Alocilja (2005) were among the few authors whose research was reviewed in this chapter who admitted that their piezoelectric immunosensor was not sensitive enough to be useful in protecting the public health. Lazacka et al. (2007) believed that biosensors achieved speed at the expense of sensitivity. Compared to the industry standard that no *E. coli* should be detected in 100 mL of finished water, the LOD of most biosensors probably needs to be improved by at least three orders of magnitude.

In summary, biosensors will not be attractive to the water industry until it has been demonstrated in pilot studies that they can be operated over long periods of time with minimal operator expertise, can be integrated into systems that process water volumes on the liter to cubic meter scale, achieve two to three orders of magnitude improvement in detection limits, and produce responses that are unambiguous.

References

Abdel-Hamid I, Ivnitski D, Atansaov P, Wilkins E (1999) Flow-through immunofiltration assay system for rapid detection of *E. coli* O157:H7. Biosensors and Bioelectronics 14:309–316

Baeumner A J, Cohen R J, Miksic V, Min J (2003) RNA biosensor for the rapid detection of viable *Escherichia coli*. Biosensors and Bioelectronics 18:405–413

Baeumner A J, Pretz J, Fang S (2004) A universal nucleic acid sequence biosensor with nanomolar detection limits. Analytical Chemistry 76:888–894

Brewster J D, Mazenko R S (1998) Filtration capture and immunoelectrochemical detection for rapid assay of *Escherichia coli* O157:H7. Journal of Immunological Methods 211:1–8

Call D R, Brockman F J, Chandler D P (2001) Detecting and genotyping *Escherichia coli* O157:H7 using multiplexed PCR and nucleic acid microarrays. International Journal of Food Microbiology 67:71–80

Campbell G A, Mutharasan R (2007) A method of measuring *Escherichia coli* O157:H7 at 1 cell/mL in 1 liter sample using antibody functionalized piezoelectric-excited millimeter-sized cantilever sensor. Environmental Science and Technology 41:1668–1674

Chen C-S, Yao J, Durst R A (2006) Liposome encapsulation of fluorescent nanoparticles: quantum dots and silica nanoparticles. Journal of Nanoparticle Research 8:1033–1038

Deisingh A K, Thompson M (2004) Strategies for the detection of *Escherichia coli* O157:H7 in foods. Journal of Applied Microbiology 96:419–429

Derzon M, Hopkins M, Galambos P, Achyuthan K, Bourdon C, Brener I, James C, McClain J, Peterson D, Rahimian K, Timlin J, Cullor J, Kaminski M, Peck V, Spink E, Yun C, Ludwig G (2009) Timely multi-threat biological, chemical and nuclide detection: a platform, a metric, key results. International Journal of Technology Transfer and Commercialisation 7: 413–435

Dunbar S, Vander Zee C A, Oliver K G, Karem K L, Jacobson J W (2003) Quantitative, multiplexed detection of bacterial pathogens: DNA and protein applications of the Luminex LabMAP™ system. Journal of Microbiological Methods 53:245–252

Edwards K A, Baeumner A J (2006) Liposomes in analyses. Talanta 68:1421–1431

Eum N-S, Yeom S-H, Kwon D-H, Kim H-R, Kang S-W (2010) Enhancement of sensitivity using gold nanorods – antibody conjugator for detection of *E. coli* O157:H7. Sensors and Actuators B 143:784–788

Farabullini F, Lucarelli F, Palchetti I, Marrazza G, Mascini M (2007) Disposable electrochemical genosensor for the simultaneous analysis of different food contaminants. Biosensors and Bioelectronics 22:1544–1549

Gehring A G, Irwin P L, Reed S A, Tu S-I, Andreotti P E, Akhavan-Tafti H, Handley R S (2004) Enyzme-linked immunomagnetic chemiluminescent detection of *Escherichia coli* O157:H7. Journal of Immunological Methods 293:97–106

Goldman E R, Medintz I L, Mattoussi H (2006) Luminescent quantum dots in immunoassays. Analytical and Bioanalytical Chemistry 384:560–563

Hahn M A, Tabb J S, Krauss T D (2005) Detection of single bacterial pathogens with semiconductor quantum dots. Analytical Chemistry 77:4861–4869

Ho J A, Hsu H-W, Huang M-R (2004) Liposome-based microcapillary immunosensor for detection of *Escherichia coli* O157:H7. Analytical Biochemistry 330:342–349

Huang X J, Zhang Y Y (2006) Electrical determination of *E. coli* O157:H7 using tin-oxide nanowire coupled with microfluidic chip. IEEE Sensors Journal 6:1376–1377

Ivnitski D, Abdel-Hamid I, Atanasov P, Wilkins E (1999) Biosensors for detection of pathogenic bacteria. Biosensors and Bioelectronics 14:599–624

Johnson-White B, Lin B C, Ligler F S (2007) Combination of immunosensor detection with viability testing and confirmation using the polymerase chain reaction and culture. Analytical Chemistry 79:140–146

LaGier M, Scholin C, Fell J, Wang J, Goodwin K (2005) An electrochemical RNA hybridization assay for detection of the fecal indicator bacterium *Escherichia coli.* Marine Pollution Bulletin 50:1251–1261

Lazacka O, Del Campo F J, Munoz F X (2007) Pathogen detection: a perspective of traditional methods and biosensors. Biosensors and Bioelectronics 22:1205–1217

Leonard P, Hearty S, Brennan J, Dunne L, Quinn J, Chakraborty T, O'Kennedy R (2003) Advances in biosensors for detection of pathogens in food and water. Enzyme and Microbial Technology 32:3–13

Liao W-C, Ho J A (2009) Attomole DNA electrochemical sensor for the detection of *Escherichia coli* O157. Analytical Chemistry 81:2470–2476

Ligler F S, Sapsford K, Golden J, Schriver-Lake L, Taitt C, Dyer M, Barone S, Myatt C J (2007) The array biosensor: portable, automated systems. Analytical Sciences 23:5–10

Lim C T, Zhang Y (2007) Bead-based microfluidic immunoassays: the next generation. Biosensors and Bioelectronics 22:1197–2104

Lim D V, Simpson J M, Kearns E A, Kramer M F (2005) Current and developing technologies for monitoring agents of bioterrorism and biowarfare. Clinical Microbiology Reviews 18: 583–607

Lu Q, Lin H, Ge S, Luo S, Cai Q, Grimes C A (2009) Wireless, remote-query, and high sensitivity *Escherichia coli* O157:H7 biosensor based on the recognition action of concanavalin A. Analytical Chemistry 81:5846–5850

Mao X, Yang L, Su X-L, Li Y (2006) A nanoparticle amplification based quartz crystal microbalance DNA sensor for detection of *Escherichia coli* O157:H7. Biosensors and Bioelectronics 21:1178–1185

Meeusen C A, Alocilja E C, Osburn W N (2005) Detection of *E. coli* O157:H7 using a miniaturized surface plasmon resonance biosensor. Transactions of the ASAE 48:2409–2416

Mehrvar M, Abdi M (2004) Recent developments, characteristics, and potential applications of electrochemical biosensors. Analytical Science 20:1113–1126

Morales-Morales H A, Vidal G, Olszewski J, Rock C M, Dasgupta D, Oshima K H, Smit G B (2003) Optimization of a reusable hollow-fiber ultrafilter for simultaneous concentration of enteric bacteria, protozoa, and viruses from water. Applied and Environmental Microbiology 67:4098–4102

Ngundi M M, Kulagina N V, Anderson G P, Taitt C R (2006) Non-antibody-based recognition: alternative molecules for detection of pathogens. Expert Reviews in Proteomics 3: 511–524

Noble R T, Weisberg S B (2007) A review of technologies for rapid detection of bacteria in recreational waters. Journal of Water and Health 03.4:381–392

Nocker A, Burr M, Camper A K (2009) Synthesis document on molecular techniques for the drinking water industry. Water Research Foundation, Denver, CO, USA

Poitras C, Tufenkji N (2009) A QCM-D-based biosensor for *E. coli* O157:H7 highlighting the relevance of the dissipation slope as a transduction signal. Biosensors and Bioelectronics 24:2137–2142

Radke S M, Alocilja E C (2005) A high density microelectrode array biosensor for detection of *E. coli* O157:H7. Biosensors and Bioelectronics 20:1662–1667

Rasooly A, Herold K E (2006) Biosensors for the analysis of food- and waterborne pathogens and their toxins. Journal of the AOAC International 89:873–883

Reidt U, Chauhan L, Muller G, Molz R, Lindner P, Wolf H, Friedberger A (2008) Reproducible filtration of bacteria with micromechanical filters. Journal of Rapid Methods and Automation in Microbiology 16:337–350

Rijal K, Leung A, Shankar P M, Mutharasan R (2005) Detection of pathogen *Escherichia coli* O157:H7 at 70 cells/mL using antibody-immobilized biconical tapered fiber sensors. Biosensors and Bioelectronics 21:871–880

Rose J B, Grimes D J (2001) Reevaluation of microbial water quality: powerful tools for detection and risk assessment. American Academy of Microbiology. Washington, DC

Sapsford K E, Ngundi M M, Moore M H, Lassman M E, Shriver-Lake L C, Taitt C R, Ligler F S (2006) Rapid detection of foodborne contaminants using Array Biosensor. Sensors and Actuators B 113:599–607

Song J M, Kwon H T (2009) Photodiode array on-chip biosensor for the detection of *E. coli* O157:H7 pathogenic bacteria. Methods in Molecular Biology: Biosensors and Bioprotection 503:325–335

Straub T M, Dockendorff B P, Quinonez-Diaz M D, Valdez C O, Shutthanandan J I, Tarasevich B J, Grate J W, Bruckner-Lea C J (2005) Automated methods for multiplexed pathogen detection. Journal of Microbiological Methods 62:303–316

Su X-L, Li Y (2004) Quantum dot biolabeling coupled with immunomagnetic separation for detection of *Escherichia coli* O157:H7. Analytical Chemistry 76:4806–4810

Su X-L, Li Y (2005) Surface plasmon resonance and quartz crystal microbalance immunosensors for detection of *Escherichia coli* O157:H7. Transactions of the ASAE 48:405–413

Subramanian A, Irudayaraj J, Ryan T (2006) A mixed self-assembled monolayer-based surface plasmon immunosensor for detection of *E. coli* O157:H7. Biosensors and Bioelectronics 21:998–1006

Taylor A D, Yu Q, Chen S, Homola J, Jiang S (2005) Comparison of *E. coli* O157:H7 preparations methods used for detection with surface plasmon resonance sensor. Sensors and Actuators B 107:202–208

Teles F R R, Fonseca L P (2008) Trends in DNA biosensors. Talanta 77:606–623

Tims T B, Lim D V (2003) Confirmation of viable *E. coli* O157:H7 by enrichment and PCR after rapid biosensor detection. Journal of Microbiological Methods 55:141–147

Tu S-I, Uknalis J, Yamashoji S, Gehring A, Irwin P (2005) Luminescent methods to detect viable and total *Escherichia coli* O157:H7 in ground beef. Journal of Rapid Methods and Automation in Microbiology 13:57–70

US Food and Drug Administration (2009) Bad bug book: foodborne pathogenic microorganisms and natural toxins handbook. Rockville, MD

Wang L, Liu Q, Hu Z, Zhang Y, Wu C, Yang Mo, Wang P (2009) A novel electrochemical biosensor based on dynamic polymerase-extending hybridization for *E. coli* O157:H7 detection. Talanta 78:647–652

Waswa J, Irudayaraj J, DebRoy C (2007) Direct detection of *E. coli* O157:H7 in selected food systems by a surface plasmon resonance biosensor. LWT Food Science and Technology 40:187–192

Yacoub-George E, Hell W, Meixner L, Wenninger F, Bock K, Lindner P, Wolf H, Kloth J, Feller K A (2007) Automated 10-channel capillary chip immunodetector for biological agents detection. Biosensors and Bioelectronics 22:1368–1375

Yang L, Li Y (2006) Simultaneous detection of *Escherichia coli* O157:H7 and *Salmonella* Typhimurium using quantum dots as fluorescent labels. Analyst 131:394–401

Zhao W, Yao S, Hsing I-M (2006) A microsystem compatible strategy for viable *Escherichia coli* detection. Biosensors and Bioelectronics 21:1163–1170

Zhu P, Shelton D R, Karns J S, Sundaram A, Li S, Amstutz P, Tang C-M (2005) Detection of water-borne *E. coli* O157:H7 using the integrating waveguide biosensor. Biosensors and Bioelectronics 21:678–683

Zordan M, Grafton M, Acharya G, Reece L, Cooper C, Aronson A, Park K, Leary J (2009) Detection of pathogenic *E. coli* O157:H7 by a hybrid microfluidic SPR and molecular imaging cytometry device. Cytometry 75A:155–162

Chapter 13
Guidelines, Caveats, and Techniques for the Evaluation of Water Quality Early Warning Systems

Dan Kroll

13.1 Introduction

The need for analytical techniques to determine if our water has been compromised via intentional or accidental contamination is a crucial component in securing our water supplies. The drive toward supplying safe and pure water for the public has been an ongoing battle since the inception of the drinking water industry. After 9/11, the motivating factor of potential terrorist activity was added to the consumers' demand for a reliable and trustworthy source of water. In the past, most of our analytical muscle was brought to bear at the treatment plant. Little consideration was given to source waters or the distribution system as areas needing constant monitoring. That paradigm has changed with the recognition that all areas of the water supply network are vulnerable to intentional contamination, most particularly the distribution system.

In the past several years, the recognition of these vulnerabilities has led to the rapid development of analytical techniques and instrumentation packages designed to address the monitoring needs of the new security-based paradigm. A number of studies conducted since 9/11 have shown that one of the key techniques capable of addressing the need to be vigilant in all sections of the water supply network is bulk parameter monitoring of basic water quality parameters. These studies have shown that the utilization of multi-parameter monitoring has the potential to indicate the presence of a wide variety of harmful agents in water at levels that would be protective of human health (Hall et al., 2007; Kroll, 2006a; USEPA, 2005).

A number of multi-parameter sensor packages, produced by a variety of manufacturers, have been deployed in source water, treatment pants, and distribution systems throughout the world. Being that in most cases these sensor packages generate data from a number of different instruments and in many cases are deployed at a number of sites throughout the system, they tend to generate vast quantities of data.

D. Kroll (✉)
Hach Homeland Security Technologies, Loveland, CO 80538, USA
e-mail: dkroll@hach.com

R.M. Clark et al. (eds.), *Handbook of Water and Wastewater Systems Protection*, Protecting Critical Infrastructure, DOI 10.1007/978-1-4614-0189-6_13,
© Springer Science+Business Media, LLC 2011

The interpretation of the data generated can be extremely complex and beyond the capabilities of most end users. How are we to determine what readings are normal and what readings represent the presence of a possible intrusion into the system? The use of skilled technicians able to correlate and interpret such data streams is impractical from a logistical and cost standpoint. Due to these factors, the need for computer-aided data interpretation becomes axiomatic. Intelligent algorithms are a necessary addition to bulk parameter monitoring if information useful in the decision making process is to be obtained through this type of analysis. A number of such algorithms designed for this use have been under development by private industry, government programs, universities, and national labs and are becoming commercially available. The question then becomes how to evaluate the effectiveness of these potential early warning systems. There are a number of key elements that should be considered when choosing and deploying such systems.

13.2 Detection Class Requirements

When discussing early warning systems, there is a classification system that is commonly in use in the bio-defense field relating to their effectiveness. These classes include the following:

(1) *Detect to treat*: With these systems there is a very high confidence level in their ability to detect anomalies, nearly 100%, that allows treatment to proceed on those exposed. With these systems the response time tends to be slow and procurement and operating costs are high. Detect to treat systems usually rely upon lab-based methods for specific analytes and have not as yet been developed to the extent that their widespread deployment, in an online basis, is possible.

(2) *Detect to protect*: With these systems there is a high confidence level, >99%, that allows for protection by limiting exposure without the need for confirmatory testing. Response time is faster than detect to treat and cost is less than detect to treat systems. There is some deployment of such systems in online configurations, such as online mass spectrometry systems, but their cost and complexity make widespread use impractical at this point in their development.

(3) *Detect to warn*: With these systems the presumptive confidence level <99%. This would allow for protection by limiting exposure while confirmatory tests are run. The response from these systems is fast and the cost to deploy tends to be lower. The amount of confidence afforded by an EWS is usually a function of analysis time and cost, with higher confidence levels increasing both. An Early Warning System (EWS) needs to respond quickly to be of practical use, and it needs to be relatively low in cost to provide wide coverage. The multiple parameter bulk monitoring systems discussed in the remainder of this chapter tend to fall into this category. They have rapid response and wide deployment possibilities, but their results are not 100% definitive of an actual problem but can be indicative of the need to investigate further.

13.3 Dual Use and Return on Investment

The objective of deploying an EWS is to detect contaminants so that people and property can be protected. In reality, little of an EWS's time is spent detecting these rare contamination events. While the ability to detect contamination is critical, if a system is to be capable of offering a likely return on investment (ROI), it would also be useful if the EWS can provide information that would be useful on a day-to-day basis. Justification of the cost can include

(1) Optimization of daily operations within the system monitored including pumping and dosing schedules,
(2) Providing alarms for operational events not related to contamination threats,
(3) Replace or supplement grab sampling for compliance with other regulations with continuous monitoring,
(4) Document system operation anomalies to assist with planning maintenance activities or planning and justifying major system upgrades,
(5) Provide a tool for less experienced workforce personnel to utilize in recognizing system problems before they become critical,
(6) Building consumer confidence by continuously documenting system water quality,
(7) Act as a continuous record of water quality in dealing with customer complaints and disputes.

Capabilities and functions such as these can reduce the cost of operations for the system being monitored and possibly provide information that can be useful to those people who use the monitored system.

Proper sensor selection can be crucial, not only in effectively detecting contamination but also in providing dual use value. With this in mind, as well as the need to have orthogonal sensors that would be capable of responding to the wide variety of contaminants that could be encountered, a suite of sensors are required. The following types of sensors offer value for event detection in the distribution system as well as dual use value. Different sensor packages may be optimal for source water or treatment plant deployment. A variety of sensors may be considered for each scenario but what additional value they bring to the package as far as event detection and dual use should be carefully considered. Trade-offs to reduce cost and maintenance, such as the use of UV-254 in lieu of TOC, are possible, but come at a price in sensitivity and detection range. Due to cost considerations redundancy should be avoided.

13.3.1 Optimized Distribution System Sensor Package

- *pH* for determining acid/base relationships
- *Conductivity* for determining ionic concentrations

- *Turbidity* for determining particulate matter
- *Chlorine Residual Free or Total* for determining disinfectant concentrations and effects upon these levels exerted by threat agents.
- *Total Organic Carbon (TOC)* for determining fluctuations in the organic molecule content of the water.
- *Oxidation Reduction Potential (ORP)* for determining the balance of oxidizing and reducing species. May be of more use in non-disinfected samples as it may be to some extent redundant with chlorine residual.

In the past several years, a large number of analytical packages composed of the types of sensors listed above have been deployed in a number of distribution systems in the USA and throughout the world. These sensors have been instrumental in detecting a number of occurrences that would have gone unnoticed without monitoring. Events that have been detected include road work events, pressure problems, rain events, variable demand problems, ammonia overfeeds, chlorine feed problems including overfeeds and underfeeds, fluoride overfeeds, bad check valves, flushing problems, inaccurate grab sampling programs, uncalibrated hydraulic models, and main breaks among others (Kroll, 2008).

One interesting event that was detected by such a system was an overfeed of caustic that if not detected could have resulted in serious injury. In this deployment scenario, the plant used caustic feed to control water pH. The system experienced a trigger that when investigated was identified as an operational problem that resulted in the feed of excess caustic. The result was that the overfeed affected the pH and the conductivity of the water, causing the alarm (see Fig. 13.1). The reason behind this was that the vendor from which the caustic was being purchased had delivered the wrong concentration of the solution. No one had checked to see if the concentration was correct before feeding in the material. New procedures were put in place to verify incoming raw materials. The system in place at this facility learned the signature for this type of event and can classify a recurrence of this or similar events in the future if there is another failure in the system and it is repeated.

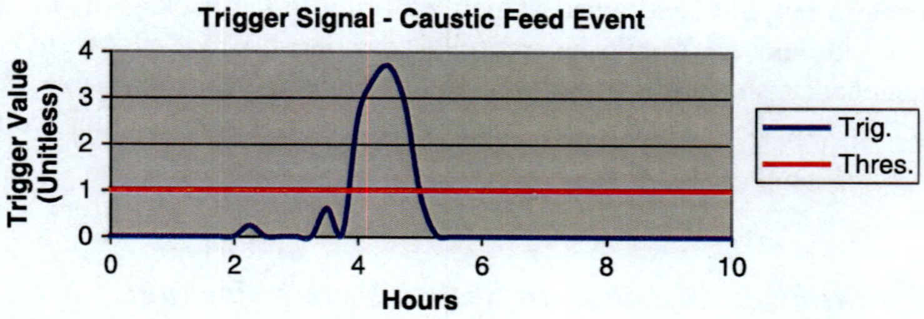

Fig. 13.1 Caustic event

Another good example of dual use occurred when a municipality that had recently installed such a system was faced with a disgruntled customer. The customer, a major industrial user of the city's water supply, was complaining about poor quality water that was adversely affecting their processes, and they were threatening legal action. Fortunately, a multi-parameter security system had been installed near the facility and had recorded data on water quality during the time frame of the dispute. After reviewing the data with the customer, the city helped the customer locate the source of the actual problem, which turned out to be the recirculation loop within the plant, thus saving the city the potential costs of a court battle.

13.4 Operational Attributes

There are a number of operational characteristics that need to be incorporated into the design of such a system if it is to successfully fulfill its role.

13.4.1 Ease of Use

A system that may be absolutely critical in a crisis cannot be difficult to use. User interface operation must be intuitive enough that minimally skilled operators can obtain necessary information without resorting to an operator's manual. During a crisis even simple operations can be difficult so an easy to use system is required. Even highly trained and proficient operators can make mistakes when trying to operate equipment or perform procedures in emergency situations. This failure to accurately complete even well-drilled tasks in stressful situations was amply demonstrated in the American Civil War. Weapons of that time period were predominately cap and ball muzzleloaders. The procedure for loading was to place the powder in the barrel of the gun, then a wad, and finally the ball. After loading a firing cap was attached and the weapon was discharged and the procedure repeated. Many times after a heated battle, when the troops were undergoing weapon inspection, it would be found that the barrels of the guns were completely full of unfired rounds. In the heat of battle, the troops had forgotten to place the firing cap before pulling the trigger and hence kept reloading an already loaded gun. While not as stressful as battle, a water emergency can tax the abilities of even well-trained personnel.

13.4.2 Automated

The system must normally operate without requiring the presence of a human. Human intervention should only be needed for service or maintenance. Due to their placement in remote areas of the distribution system and watershed EWS needs to be able to operate without constant supervision that may be available in treatment plant settings. Appropriate software and command and control structures that allow routine operations, such as instrument checks, to be performed from remote sites rather than having to visit the deployment sites for this type of activity can be a major advantage.

13.4.3 Continuous

The system must run without any long gaps in analysis that enable contamination of significant duration to slip by undiscovered. The maximum time for non-observability should be relatively short – on the order of minutes rather than tens of minutes or hours. This is true not only of data collection but also of data processing. Algorithms for event detection that require long windows of data to make a decision or that require an extended time after the contaminant has reached the sensors to render a determination as to whether or not an event is real are probably not the best choice. Longer response times unduly endanger a larger percent of the population by exposing them to the hazard before remedial actions can be taken. The system also needs to be continuous due to the fact that there is often little or no independent outside warning before an event occurs whether the event is accidental or deliberate.

13.4.4 Reliable

A non-working system is an opportunity for exploitation. A system that has an undue amount of down time, whether as a problem with system design or with failure to perform ongoing maintenance, is a liability. Systems with a proven track record that utilize instruments with a long history of industry use can help to alleviate these concerns. Many manufacturers offer maintenance and service contracts that can help to relieve the local utility of the onus of maintaining the systems.

13.4.5 Cost-Effective

Cost can be one of the crucial constraining factors in design of an early warning system for a utility. Costs can rapidly escalate to unacceptable levels due to the fact that, in most cases, a network design is needed to offer adequate coverage requiring a large number of platforms to be deployed. A good guideline is that amortized cost per day should be comparable to or less than the labor, travel, equipment, and reagents for an existing grab sample program. When determining the cost-effectiveness it is important to take into account the increased data that are being generated and any possible dual use value that can be derived from the deployment of such a system. Nevertheless, costs can still be a major hurdle. Systems that can be deployed in a modular basis can be a good way to spread the expense out over a number of budgetary cycles while still deriving benefits before complete buildout occurs.

13.5 Performance Attributes

Performance attributes are also important in any evaluation of an EWS.

13.5.1 Detection of a Broad Spectrum of Contaminant Classes

There are many contaminants that could cause serious harm if introduced into a drinking water supply. Many of these compounds and substances are readily available to those wishing to do harm. One of the principal problems when designing a monitoring system for water contamination events is the vast number of chemical agents that could be utilized by a terrorist to compromise a water supply system, which tends to preclude monitoring on an individual chemical basis. Chemical warfare agents such as VX, Sarin, Soman; commercially available herbicides, pesticides, and rodenticides; street drugs such as LSD and heroin; heavy metals; radionuclides; cyanide; chemicals commonly utilized in water treatment; and a host of other industrial compounds could be exploited as weapons. There are also a variety of biological agents and biotoxins that could be employed as agents if such an attack were to occur. Which of the agents would be the most likely to be deployed in a terrorist assault is practically impossible to determine. The number of chemical and biological substances that could be used in an attack is immense and when the scenario of mixtures of these substances is explored the combinations become, for all practical purposes, endless.

To be a truly effective monitoring device, an EWS needs to be able to detect any and all of the possible agents that could be encountered as well as their potential mixtures. For instance, a dedicated device capable of detecting ricin is interesting but not very practical. What if an agent other than ricin was utilized? Such a system that was designed to detect specific compounds would be blind to the unexpected or unanticipated. To thwart the diverse number of threats we face, specific sensors would be impractical. This need to detect diverse contaminants requires a realignment of thinking from the traditional development of a sensor for a specific compound or agent. For practical purposes the choice of an array of sensors coupled with algorithms seems to be the only workable solution at this time.

It is important to understand that such multi-parameter instrumentation and software packages are inferential in nature. Being that individual tests are not run to confirm the presence of a specific analyte, an alarm from such a system (even those that have some ability to classify events) does not necessarily mean that an event has occurred. It simply means that the water quality parameters utilized in the system's architecture for determining the occurrence of an event have changed in the same manner as they would in an actual event. These systems are simply a tool for indicating the need for further investigation.

13.5.2 Rapid Detection of Events

Response time is a critical aspect of any system deigned with security in mind. While possibly not as critical for systems deigned for event detection in source water applications, because of the time lag until consumer exposure occurs, a rapid response is crucial when systems are deployed in the distribution system due to the short transport times to the end users. The response time of any detection method that is used on a flowing stream is really a measure of the delivery rate of harmful contaminants. A simple way to view this is given by Eq. (13.1):

$$\text{Lethal doses delivered } (n) = \frac{\text{Stream flow rate } (v/t) \times \text{ response time } (t) \times \text{ concentration } (m)}{\text{Lethal dose } (m/v)} \tag{13.1}$$

Therefore, the longer the response time the more potential for lethal doses to be delivered. A rapid response is to be favored in that it allows more time for effective remediation to be put into place so as to lessen the impact by taking actions such as issuing public warnings and isolating parts of the system. Other approaches to early warning such as the concept of syndromic surveillance appear lacking in ability to rapidly respond when utilized in isolation without the added paradigm of online water quality monitoring . Tracking pharmaceutical sales, counting the number of ambulances arriving at the hospital per hour, or number of body bags per day is an inadequate measure as it is founded on the premise that protection cannot be effective. Thus, a response time in the order of minutes, rather than hours or days, is needed for a chance to protect effectively.

13.5.3 Sensitivity to Concentration of Interest

The goal of an early warning system is not to detect contaminants at extremely low levels that they are regulated by the EPA and other entities. The goal of these regulations is to insure the absence of health effects from long-term or chronic exposure such as the increased risk of cancer over a lifetime of drinking the water. To meet these goals, grab sampling is still an effective means of monitoring. Online EWS monitoring is designed to respond to acute levels of a contaminant that could cause harm immediately or after a short-term exposure. Clearly an EWS must be sensitive to contaminants present in harmful amounts, but quoting a simple minimum detection limit can be misleading. Contaminants change water chemistry and those changes can be analyzed as part of an alarm process that generates a trigger signal. Changes are seen against a background of noise or natural fluctuations in measured parameters. Accordingly, one cannot address sensitivity unless it is relative to a site's noise properties. Being that the duration, magnitude, and pattern of normal water quality deviations will be specific to a site, the concentration of any given agent that can be detected will vary from site to site. Fortunately, military needs have produced a useful tool for simply stating a detector's capability to respond: the receiver operating characteristic (ROC) curve. Such curves allow an operator to select a detection alarm threshold according to local noise characteristics. Note that this allows an operator to increase sensitivity, accepting the higher probability of a false alarm when it is suspected that contamination is more likely. The graphic below shows a representation of ROC space (Fig. 13.2).

An ROC curve is a graphical representation of the trade-off between the false-negative and false-positive rates for every possible cutoff. Equivalently, the ROC curve is the representation of the trade-offs between sensitivity (Sn) and specificity (Sp). By tradition, the plot shows the false-positive rate on the *X* axis and 1 – the false negative rate on the *Y* axis. You could also describe this as a plot with 1–Sp on the *X* axis and Sn on the *Y* axis. A good diagnostic test is one that has small false-positive and false-negative rates across a reasonable range of cutoff values. A bad diagnostic test is one where the only cutoffs that make the false-positive rate low have a high false-negative rate and vice versa.

An evaluator is usually satisfied when the ROC curve climbs rapidly toward the upper left-hand corner of the graph. As shown in Fig. 13.2 (curve through C′). This means that 1– the false-negative rate is high and the false-positive rate is low. We

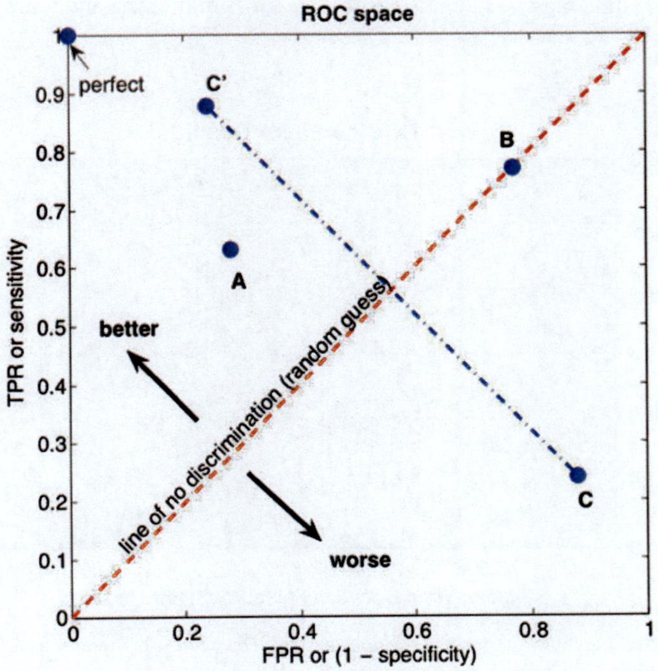

Fig. 13.2 Classic ROC curve space

are less happy when the ROC curve follows a diagonal path from the lower left-hand corner to the upper right-hand corner (curve through C). This means that every improvement in false-positive rate is matched by a corresponding decline in the false-negative rate. You can quantify how quickly the ROC curve rises to the upper left-hand corner by measuring the area under the curve. The larger the area, the better the diagnostic test. If the area is 1.0, you have an ideal test, because it achieves both 100% sensitivity and 100% specificity (perfect). If the area is 0.5, then you have a test that has effectively 50% sensitivity and 50% specificity. This is a test that is no better than flipping a coin (Line B). In practice, a diagnostic test is going to have an area somewhere between these two extremes (curve through A). The closer the area is to 1.0, the better the test is, and the closer the area is to 0.5, the worse the test is. This format works quite well for discrete measurements, such as in test kits but may not be optimized for continuous monitoring technologies such as those used in EWS. With this in mind, an alternate presentation that is more useful when the operational analysis is continuous in time may be used. This alternate presentation makes it easy for the system operator to select the trigger threshold to balance trigger sensitivity against the frequency of triggers caused by system noise.

In the alternate presentation, the false alarm rate function is expressed as mean time between false alarms versus trigger threshold. The hit rate function is translated into the amount of agent, expressed as the percent of LD50 (1 L of water for a 70 kg person), required to give a 100% hit rate, versus trigger threshold (see Fig. 13.3). This format allows the user to select the trigger threshold based on desired trigger sensitivity versus the acceptable time between false alarms due to process noise. This new method may be more useful for operators in running and tuning a systems operation.

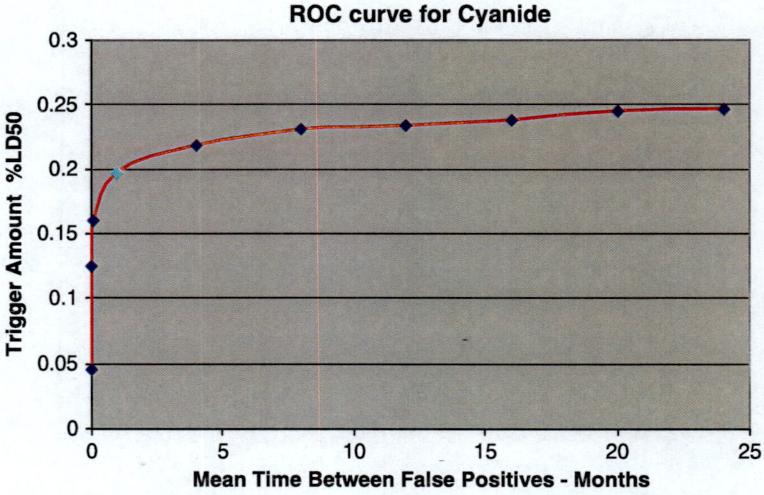

Fig. 13.3 ROC curve format for a continuous system. *Dark blue dots* represent different trigger threshold settings. The *light blue dot* represents the MDL

13.5.4 Specificity

Specificity of contaminant classification can be obtained in two ways. One method would be to analyze for each specific compound or threat agent of interest. As previously discussed, given the immense number of possible contaminants that could be put into a water distribution system by a terrorist, not to mention the equal or even greater number of compounds that could be of dual use interest, looking for specific molecules would require an immense number of sensors. There are so many problems associated with this approach that it quickly becomes untenable. Therefore, a different approach is needed to address the problem of specificity.

The most workable solution is to analyze generally across multiple orthogonal dimensions via mathematical analysis of the data from multiple sensors. The approach of using a manageable set of orthogonal sensors faces only the difficulty of obtaining sufficient information to apply pattern recognition methods that can differentiate between contaminants or classes of contaminants. One advantage of using an orthogonal set of different sensors is that it becomes nearly impossible to find some contaminant that goes by all sensors unnoticed.

While some degree of classification is possible with the multiple parameter inferential-type sensor systems, complete specificity is not possible to the extent that dedicated sensors for specific contaminants would be capable of providing. A positive hit and classification utilizing the inferential method simply means that a water quality change has occurred and that the pattern of that change is consistent with what would be expected if the material being identified were introduced into the water. The actual material may or may not be present. These systems are simply a tool to alert operators to the need to investigate further and the classification is simply a guide to direct that investigation in the most efficient manner possible. Therefore, the use of inferential systems is a trade-off of enhanced range for a decrease in exact specificity.

13.5.5 Reproducibility

An EWS must be reproducible to be trustworthy. When the same condition exists, the same type of alarm should be generated. This can be verified both by playing real and generated data files though the algorithms and also by testing with real materials. There are many non-hazardous substances that could be introduced to the sensors to cause an alarm, but it is also important to test with actual threat agent materials for example: aldicarb, anthrax culture, cyanide, fluoroacetate, nicotine, ricin, sarin, VX, etc. It may not be necessary or even possible to perform these sort of dosing studies due to the extreme hazards they present, but many manufacturers have taken part in EPA studies (ETV, 2005) utilizing these material or will have data available from studies that have been conducted by others.

13.5.6 Low False Negatives, Low False Positives

If a system can be blind to certain classes of chemicals (for example, not visible in the Ultraviolet 254 spectrum) then it represents an opportunity to contaminate the system without an alarm. Systems with multiple types of sensors (assuming that sensor selection has been properly done) are difficult to fool, so that kind of system will normally produce some kind of alarm upon contamination, and false negatives will be very unlikely.

False positives can come from two sources: random system noise and insufficient information during the analysis of event data. Given that systems are not noise free and there will always be some degree of insufficient information, false positives are bound to happen. The ultimate question is the determination of the number per time in a given installation (since noise is site specific) and the acceptable frequency. False positives caused by noise can be reduced by the proper choice of alarm threshold according to ROC curves.

False positives caused by imperfect information will happen in an EWS using an inferential method. Given that a multi-parameter system implies an inferential method that can give false positives, and that an actual contamination event is a very rare occurrence, it should be recognized that there will be more false positives then actual positive events. Thus, any detection must be considered presumptive until follow-up testing verifies or denies the detection.

There has been some confusion in the industry as far as inferential systems are concerned and what constitutes a false alarm. The incidence of alarms when there has not been a significant change in water quality is quite low and is mainly represented by sensor failure. There are a number of alarms that occur due to actual changes in water quality that are not of an emergency nature. These may or may not be classified by systems equipped with classification algorithms as potential problems. This is in fact not a false alarm, as an actual change has occurred, but rather a misidentification. All classifications by inferential systems should be treated as potential classifications only and not as definitive identification. They all require investigations. Systems with a heuristic ability to learn common water quality upsets and classify them as benign is an important factor in decreasing these misidentification-type alarms.

13.5.7 Qualitative

Contaminants can be classified or named if the EWS is specific, but people using the information may not have sufficient training to recognize the nature of the contaminant found from its name. The system should assign a general classification category for simple clarification.

For example, the classification of an alarm as potentially being caused by an organophosphate-type pesticide may be more useful than a proper chemical or trade name. Also if the system is able to automatically link to information about the most likely potential contaminant including toxicity, treatment, analytical methods, and other pertinent data it would help to streamline the process of evaluating the alarm.

13.5.8 Quantitative

Ideally, an EWS would provide enough information to give some quantitative information about any contaminant presumably found in the water system. Such information could be useful in determining the threat level during a given episode and also might be useful if treatment of people or infrastructure becomes necessary.

Systems that are capable of giving a true or extrapolated quantitative number would be the optimum. However, inferential methods make the provision of a definitive quantification difficult. Even so, much useful information can be garnered from the magnitude of the trigger signals provided by such systems. While different in their rate of increase for different contaminants, those with large magnitude can generally be considered to represent a higher concentration than those of smaller magnitude for the same contaminant.

13.6 Coverage Characteristics

Coverage characteristics also need to be considered.

13.6.1 Cost

Cost may not be significantly limiting when selecting an EWS for an icon or high-profile facility, but if the coverage area is large, such as a major metropolitan area, budgets may be constrained, and the degree of coverage becomes a function of the cost per point monitored. Therefore, cost becomes an issue.

Several options are available to limit and offset costs. One of these is selection of diverse instrumentation with differing capabilities for different sites based upon threat scenarios. An extremely sensitive and capable system may be needed for key sites such as the Capitol in Washington, DC, but less capable systems may be adequate for less obvious targets. The concept of a network, as discussed above, with sensitive sensors being backed up by less sensitive packages or even incomplete sensor sets (e.g., just chlorine at some nodes) will help to decrease cost.

Another way of spreading costs out is enhanced that, by their very nature, multi-parameter systems are modular. A monitoring regime can begin with a limited number of sites and small numbers of sensors and can be built upon and expanded as budgets allow. Selection of the types of systems that are amenable to this type of deployment scenario should be a factor.

Finally, a good means of offsetting some of the cost is to select systems with proven dual use capability as was discussed in Section 13.3. It is foreseeable that devices capable of learning and recalling sensor patterns from operational events will become much more than a system that is capable of detecting terrorist activity. They could easily become a critical tool for improving everyday operations.

For example, through many years of experience, the best old hands at treatment plant operations have developed “a sense” for knowing something in the treatment system is amiss. It can be a smell, color, clarity (or lack there of), sound, or just tingling in the nape of the neck. One gains this sense only by extensive experience in a particular facility. These senses do not exist in distribution systems because there has typically been little measurement done upon which to gain these "senses," and therefore, "Bulk Parameter Monitoring in the Distribution System with Interpretive Algorithms" has the potential to become the artificial "sense" able to quickly "learn" the quirks of the distribution system and have those quirks labeled by those with extensive experience so that less experienced employees have the benefit of that knowledge without having to wait 5, 10, or more years. A good phrase to describe this knowledge base would be "institutional intuition" (Kroll, 2006b).

With the aging of the workforce and rapid employee turnover, "institutional intuition" has the chance of quickly dying out. Above and beyond their obvious security benefits, algorithms could be a way to circumvent this loss of knowledge and to build a knowledge base where none has previously existed. This in turn could allow improvements in system operation that may result in cost savings and definitely will result in a higher quality product being delivered to the consumer.

13.6.2 Area of Protection

Note that effective coverage may be a function of the hydraulics of the distribution system within the geographical setting. While "protection of all" may be a laudable goal, reality may constrain the degree of implementation, forcing trade-offs. Coverage characteristics can be a key factor when evaluating systems. The percentage of the population that can be covered is a function of number of units deployed, effectiveness of those units, hydraulics of the water system, and where an assault on the system occurs. Each water distribution system is unique in its configuration and hydraulics, and each community is also unique in its priorities of where and what to protect first. Due to cost constraints and logistical/technical proscriptions on where systems can be deployed, it is virtually impossible to protect every tap. Some water system managers have stated that if they cannot protect everyone then they will just not make the choice at all. This mode of operations is short sighted as any decrease in casualties or damages has to be looked upon as a plus and failure to take action of some sort, no matter how limited, may open up the utilities to a larger liability than deploying existing technology to the best of their ability.

The choice of what type of monitoring system to deploy is not necessarily an either/or decision. The best choice may be a network configuration that deploys different types and cost ranges of sensors in different areas to give the optimum in coverage and capabilities. While not every point will receive complete protection a network approach has the best chance of detecting an event early at its onset and alerting the operators of the system so that they can make the crucial decisions that will be needed to limit the damage being done. After the choice has been made as to what type of monitoring system to deploy, the problem remains of where best to locate the systems in the distribution network. While this is not an easy decision to make it is a necessary one.

The first step is to determine what are the key areas to protect. Thought has to be given to what are likely targets and what are critical assets. These could include such facilities as schools, hospitals, large sporting and entertainment venues, military facilities, government (political) and icon facilities, large office or apartment buildings, and areas of extremely dense population. After decisions have been made as to what areas absolutely must be protected, there remains the question of exactly where to place the sensors for optimum coverage and, when expanding an existing system of monitors to attain more general coverage, where should they be placed to achieve optimal value.

To answer these questions, the EPA's National Homeland Security Research Center (NHSRC), Water Infrastructure Protection Division (WIDP), has initiated a program called the Threat Ensemble Vulnerability Assessment (TEVA) program. TEVA uses the EPANET hydraulic water quality model, as well as the multi-species modules for EPANET, to simulate the fate and transport of contaminants in distribution systems. By considering the uncertainty of potential contamination scenarios, TEVA calculates the statistical distribution of potential health impacts. Consequences of the contamination are estimated by predicting the public health impacts resulting from the ingestion of contaminated water. Using a probabilistic model for ingestion, contaminant-specific dose–response models, and dynamic models for disease progression over time, TEVA can predict the expected health impacts.

TEVA focuses on designing contaminant warning systems to mitigate the effects of the said contamination events. Contaminant warning systems collect information from online sensors to provide an early warning of a contamination event and to reduce public health or economic impacts. The TEVA framework offers several options for optimally locating sensors and allows for the comparison of costs and benefits of various sensor network designs. While software such as TEVA can be helpful in delineating areas of coverage and optimal sensor placement, there are a few other considerations that should be included with an evaluation of an early warning systems deployment.

13.6.3 Communication

Multi-parameter monitoring systems facilitate an unprecedented view of the basic water quality anywhere in the network; however, there are some drawbacks to these monitoring platforms. One deficiency in these systems derives from their widespread geographical deployment. Monitoring of multiple points in a geographical area immediately raises the need for communication at least from the remote points to some centralized facility where the data can be interpreted and actions taken. The EWS can accomplish highly sophisticated interpretation of the local data, but it cannot take actions in a complex situation affecting possibly millions of people. Human interaction and analysis will be required. Thus, an EWS must be structured for secure communication. Both the instrument and the network must have tools in place to affect a high level of security so that information cannot be blocked or false information transmitted on the network.

Monitoring site access for instrument verification, data acquisition, and a wide variety of other housekeeping and emergency response tasks can become daunting when it is understood that a system may be comprised of tens or even hundreds of monitoring nodes. This, combined with the expediency of being able to view the network as a whole so as to be able to correlate and relate patterns of response, makes the need for a hierarchical centralized system of command and control a must. The information from the multiple points must be communicated to an analysis and command center.

Some solutions for this problem are rapidly becoming available. One solution consists of communication and data handling software that gives utilities a bidirectional monitoring and control system that integrates the data flow from water quality monitoring points strategically located throughout a community's water supply network. This gives utility personnel the ability to simultaneously view and minimize response time to critical water quality data anomalies from any Internet browser at any location. The capabilities of the system provide utilities with a virtual command center for water distribution monitoring and control from a location or locations of their choice. This allows for effective response and coordination of multiple sites and functions within the organization.

In the past, utilities could either physically download data from an individual instrument monitor or use a remote service such as VNC (Virtual Network Computing) to view one distribution monitoring site at a time. Every action took a significant amount of time. Only a single computer could access the data. With the new system, multiple utility personnel can simultaneously view all the data together, giving the utility a comprehensive, immediate, and real-time picture. Now, on a single screen, utility personnel can view the current status of all monitoring points in the network, download data, and clear alarms remotely. Users can easily "drill down" into data from an individual monitoring point in the same manner as if they were physically standing at the site. The underlying algorithms at the monitoring sites ability to calculate a fingerprint of a water quality event and its ability to learn, combined with utility personnel's ability to access real-time and historic data, can provide utilities with a deeper understanding of their network's performance. This in turn can allow them to streamline operations, reduce costs and labor, and boost efficiency – all the while further strengthening public health protection.

13.7 Conclusion

A number of criteria and considerations for the selection and deployment of early warning systems have been presented here. These criteria can form the basis for successful selection and deployment of early warning systems for water. As the analytical science behind these systems progresses, they will increase their ability to satisfy all of these factors. As the state of the industry stands today there are systems available that do a good job of addressing all of the criteria, but progress will continue.

References

ETV. 2005. "Multi-parameter Water Monitors for Distribution Systems". A series of EPA Environmental Technology Verification reports performed by Batelle. http://www.battelle.org/PRODUCTSCONTRACTS/etv/verifications.aspx#W12

Hall, John et al. 2007. "On-line Water Quality Parameters as Indicators of distribution System Contamination." *Journal of the American Water Works Association*. Vol. 99, Issue 1, pp. 66–77.

Kroll, Dan. 2006a. "Safeguarding the Distribution System: On-Line Monitoring for Security and Enhancing Operational Performance" *Journal of the New England Water Works Association*. Vol. 120, No. 2, pp. 104–116. June 2006.

Kroll, Dan. 2006b. "Securing Our Water Supply; Protecting a Vulnerable Resource." Pennwell Publishers, Tulsa, OK.

Kroll, Dan. 2008. Let's get real. Real world experiences with real time on-line monitoring for security and quality. Detecting and responding to events. In "Water Contamination Emergencies. Collective Responsibility." Edited by John Gray and Clive Thompson (pp. 68–81). Royal Society of Chemistry, Cambridge, UK.

USEPA. 2005. "Technologies and Techniques for Early Warning Systems to Monitor and Evaluate Drinking Water Quality: A State-of-the-Art Review." US Environmental Protection Agency, Office of Research and Development, National Homeland Security Research Center Research report.

Chapter 14
Protecting Water and Wastewater Systems: Water Distribution Systems Security Modeling

Avi Ostfeld

14.1 Introduction

A water distribution system is an interconnected collection of sources, pipes, and hydraulic control elements (e.g., pumps, valves, regulators, and tanks) delivering consumers prescribed water quantities at desired pressures and qualities. The systems connectivity is often described as a graph with the links representing the pipes and the nodes representing connections between pipes, hydraulic control elements, consumers, and sources. The behavior of a water distribution system is governed by (1) the physical laws that describe the flow relationships in the pipes and the hydraulic control elements, (2) the consumer demands, and (3) the system's layout.

The interest in modeling flow and water quality in water distribution systems stems from three types of circumstances: (1) use of waters from sources with different qualities in a single distribution system serving as a "treatment facility" to mix and convey them, with a blend supplied to its consumers. This situation is common to areas where sources of good quality are limited, and thus there is a need to use alternate water sources, such as saline groundwater, to meet agricultural or industrial water needs; (2) concern in municipal water distribution systems over quality changes such as decay of disinfectants and/or growth of organisms; and (3) deliberate or accidental events in which contaminants enter a drinking water distribution system and are distributed with flow.

Research in modeling water quality in distribution systems started in the context of agricultural usage primarily in arid regions such as the Arava valley in southern Israel where good water quality is limited. In 1990 the US Environmental Protection Agency (USEPA) promulgated rules requiring that water quality standards must be satisfied at the consumer taps rather than at the source treatment plants. This initiated the need for water quality modeling and the development of simulation water quantity and quality modeling tools (e.g., USEPA, 2002) and raised other

A. Ostfeld (✉)
Department of Civil and Environmental Engineering, Technion – Israel Institute of Technology, Haifa 32000, Israel
e-mail: ostfeld@tx.technion.ac.il

R.M. Clark et al. (eds.), *Handbook of Water and Wastewater Systems Protection*, Protecting Critical Infrastructure, DOI 10.1007/978-1-4614-0189-6_14,
© Springer Science+Business Media, LLC 2011

problems and research needs that commenced considerable research in this area to assist utilities meet that goal. Simulation and optimization algorithms for modeling water quality in distribution systems are needed by designers, utilities, and regulating agencies for a number of purposes: (1) planning and design of networks and facilities, (2) real-time operation, (3) monitoring design and operation, (4) simulation of contamination events, and (5) guidelines establishment for planning, design, operation, and monitoring.

Water quality simulation modeling is aimed at studying the changes of water quality substances in time and in space within the distribution system. The major parts of the methods allow the quality model to "ride" on the hydraulic solution. These techniques get as input the time series of the hydraulic simulation and the contaminant concentrations of the sources, which then serve as the starting point for the quality analysis.

The need for optimization exists whenever the solution to a problem is not unique. Common examples for optimization needs in modeling water quality in water distribution systems are design, operation, chlorine control, monitoring, calibration, and since the 9/11 events in the USA water security. The objective of this chapter is to describe issues related to water security within the context of water distribution systems modeling and to highlight future needs and challenges in this area.

14.2 Water Networks Security

Following the events of 9/11 in the USA, the administrator of the USEPA created the Water Protection Task Force (WPTF), which identified water distribution networks as a major area of vulnerability. The need to predict the spread of contaminants in distribution systems and to monitor their concentrations at various system locations, so as to keep the water supplied to the public safe, was declared a major concern.

The threats on a water distribution system can be generally partitioned into three major groups according to the resulting means of their enhanced security: (1) a direct attack on major infrastructure, such as dams, treatment plants, storage reservoirs, and pipelines; (2) a cyber attack disabling the functionality of the water utility SCADA (Supervisory Control and Data Acquisition) system and taking over control of key components which might result in water outages or insufficiently treated water, changing or overriding protocol codes, etc.; and (3) a deliberate chemical or biological contaminant injection at one of the water distribution system's nodes. The threat of a direct attack on major water supply system infrastructure is addressed by improving the system's physical security (e.g., additional alarms, locks, fencing, surveillance cameras, and guarding), which can be assessed by comparing the resulted degree of risk reduction to cost. The American Water Works Association (AWWA) provided comprehensive physical security guidance (AWWA, 2004) aimed at helping water utilities to tailor a physical security policy to their own specific needs.

The threat of a cyber attack can be minimized by employing several basic activities, such as (AWWA, 2004) establishing an optical isolator between communication networks allowing for only one way data traffic; using a router to restrict data transfer to a small number of destinations as regulated by an Access Control List (ACL); using firewalls; installing anti-virus software on all servers and workstations and configuring for daily virus pattern updates; and restricting access to the SCADA control room. However, among the above, the threat of a deliberate chemical or biological contaminant injection is considered the most difficult to address, both because of the uncertainty of the type of the injected contaminant and its consequences and because of the uncertainty of the location and injection time. In principle, a pollutant can be injected at any water distribution system connection using a pump or a mobile pressurized tank. Although backflow preventers provide an obstacle to such actions, they do not exist at all connections and at some might not be functional.

An online contaminant monitoring system (OCMS) was considered by the American Society of Civil Engineers (ASCE) (ASCE, 2004) and by the American Water Works Association (AWWA) (AWWA, 2004) as the major tool to reduce the likelihood of a deliberate contaminant chemical or biological intrusion. An OCMS should be designed to detect random contamination events and to provide information on the location of the contaminants within the system, including an estimation of the injection characteristics (i.e., contaminant type, injection time and duration, concentration, and injected mass flow rate). Once the type of the contaminant and its characteristics are revealed, a containment strategy can be implemented to minimize the pollutant spread throughout the system and to suggest for the system's portions that need to be flushed.

Lee and Deininger (1992) were the first to address the problem of sensor placement by maximizing the coverage of the demands using an integer programming model. Kumar et al. (1997) improved the study of Lee and Deininger (1992) by applying a greedy heuristic-based algorithm. Kessler et al. (1998) suggested a set covering graph theory algorithm for the sensor's layout. Woo et al. (2001) developed a sensor location design model by linking EPANET (USEPA, 2002) with an integer programming scheme. Al-Zahrani and Moied (2001) followed Lee and Deininger's approach using a genetic algorithm scheme (Holland, 1975; Goldberg, 1989). Ostfeld and Salomons (2004) extended Kessler et al. (1998) and Ostfeld et al. (2004) to multiple demand loading and unsteady water quality propagations. Ostfeld and Salomons (2005) extended Ostfeld and Salomons (2004) by introducing uncertainties to the demands and the injected contamination events. Berry et al. (2006) presented a mixed-integer programming (MIP) formulation for sensor placement showing that the MIP formulation is mathematically equivalent to the p-median facility location problem. Propato (2006) introduced a mixed-integer linear programming model to identify sensor location for early warning, with the ability to accommodate different design objectives.

Watson et al. (2004) were the first to introduce a multiobjective formulation to sensor placement by employing a mixed-integer linear programming model over a range of design objectives. Recently the Battle of the Water Sensors (Ostfeld et al., 2006) highlighted the multiobjective nature of sensor placement, with the following multiobjective models: Dorini et al. (2006) developed a constrained multiobjective optimization framework entitled the Noisy Cross-Entropy Sensor Locator (nCESL) algorithm based on the Cross-Entropy methodology proposed by Rubinstein (1999); Eliades and Polycarpou (2006) proposed a multiobjective solution using an "Iterative Deepening of Pareto Solutions" algorithm; Gueli (2006) suggested a predator–prey model applied to multiobjective optimization, based on an evolution process; Huang et al. (2006) proposed a multiobjective genetic algorithm framework coupled with data mining; Ostfeld and Salomons (2006) and Preis and Ostfeld (2006) used the multiobjective Non-dominated Sorted Genetic Algorithm II (NSGA-II) (Deb et al., 2000) scheme; and Wu and Walski (2006) used a multiobjective optimization formulation, which was solved using a genetic algorithm, with the contamination events randomly generated using a Monte Carlo procedure. However, although an OCMS is recognized as the appropriate solution to cope with a deliberate contaminant intrusion, and in spite of the modeling efforts as described above, much of the basic scientific and engineering knowledge needed to construct an effective OCMS is yet partially available: (1) the monitoring/sensor instrumentation tools required to accomplish the detection task, (2) knowledge of the injected contaminants impact on public health, and (3) applications and testing of the modeling algorithms on real water distribution systems.

14.3 Challenges in Water Networks Security

The following challenges are foreseen in the area of water distribution systems security.

14.3.1 Multiobjective Optimization for Sensor Placement

Real-world problems require the simultaneous optimization of multiple, possibly conflicting, objectives which should be optimized simultaneously. In such problems there is no single optimal solution, but rather a set of compromised solutions known as non-dominated or Pareto optimal. This non-dominated set describes the trade-offs among different objectives and may help the designer understand the options available for selecting a solution for implementation. Further research is required for developing and implementing multiobjective optimization methodologies for sensor placement.

14.3.2 Online Contamination Warning Systems Evaluation

To test the performance of any contamination monitoring system (i.e., a given set of sensors), contamination intrusion events need to be selected. Since contamination injections can occur at any node at any time, even for a moderate size of a network the theoretical number of possible injection events is huge, which grows substantially with system size. The challenge is to sample efficiently a predefined rare subset (i.e., a subset of events with a small probability to occur, but with an extreme impact) of the entire set of contamination events, which will provide an "upper bound" (i.e., worst case estimation) for online contamination warning systems evaluation.

14.3.3 Fuzzy Sensor Data Inclusion for Contamination Source Detection

Sensors can be of different types: (1) perfect – transmitting accurate unbiased contamination concentration measurements; (2) fuzzy – delivering fuzzy measured information (e.g., high, medium, and low contaminant concentrations); and (3) "0–1" – indicating a contamination presence/non-presence. The challenge is to develop models taking into consideration partial information of sensors for maximizing the likelihood of detecting sources of contamination intrusion.

14.3.4 Detection Response Modeling

A response to a contamination event involves conflicting objectives such as minimizing the risk that contaminated water will be consumed; minimizing the time needed to return to a regular operation; and maximizing pollutant containment objectives. The challenge is to develop optimization tools to trade-off such goals.

14.3.5 Fuzzy Sensor Data Incorporation for Sensor Layout Design

A common assumption in most sensors' optimal layout models is that sensors are perfect. However, in reality sensors will detect a contaminant only at a specific probability. The challenge is to develop methodologies for incorporating imperfect sensors in the decision process of sensor placement.

14.3.6 Overall Contamination Warning Systems (CWS) Design and Operation

Once a set of sensors is placed it should basically perform three tasks: (1) provide the highest likelihood of achieving a specific design goal (e.g., minimize the expected time to first detection), and once a contaminant is detected (2) provide data for contamination source identification, and (3) be the best layout for implementing a detection response model. A model for trading off in one framework all of these is not available.

14.4 Sensor Layout – Example Application

This section is an example application based on Ostfeld and Salomons' (2004) description of a computer program entitled optiMQ-S for optimal sensor layout.

14.4.1 Main Program Components

Figure 14.1 shows the front and main subscreens of optiMQ-S, built of eight stages: (1) the appropriate EPANET [name.inp] input file is selected; (2) existing monitoring stations, if any, are defined (e.g., at wells, tanks); (3) uneven node probabilities to be injected, if any, are declared (e.g., if a node in the system is highly secured, then its probability to be injected can be set a small number), the default is an even node injection probability; (4) the pollution, demand, and monitoring parameters are defined: NIE = number of injection events – up to a maximum of three; DBIE = duration between injection events – the time duration between consecutive contamination injections (min); LOS = level of service – the maximum contaminated volume exposure to public prior to detection (gallons); MHL = minimum hazard level – the concentration above which the water is considered contaminated (mg/L);

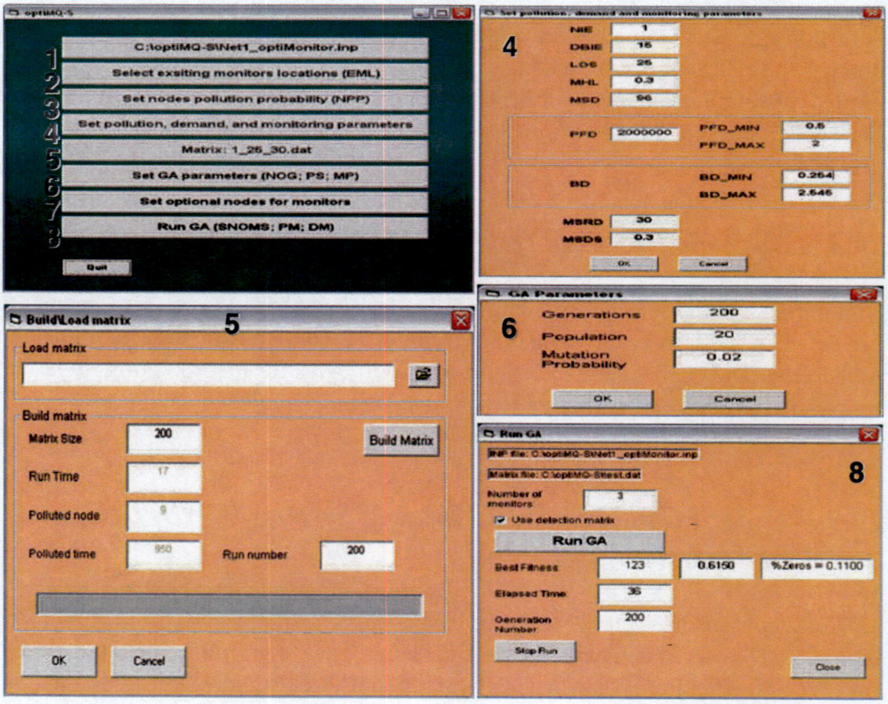

Fig. 14.1 optiMQ-S main menus

MSD = maximum simulation duration – maximum water quality simulation duration (h); PFD = pollutant flow discharge – the pollutant injection discharge rate (mg/min) (e.g., 2,000,000 mg/min = 2 kg/min) for a total of 5 min; PFD_MIN = pollutant flow discharge (PFD) min multiplier (i.e., PFD × PFD_MIN comprises the lower bound of the pollutant flow random injection rate); PFD_MAX = pollutant flow discharge (PFD) max multiplier; BD_MIN = base demand (BD) min multiplier (i.e., BD × BD_MIN comprises the lower bound of the random base demand consumption rate); BD_MAX = base demand (BD) max multiplier [both PFD and BD are randomly selected between their corresponding minimum and maximum bounds, for each pollution event (PE), using a uniform probability distribution]; MSRD = monitoring station response delay – the time duration (delay) required for a monitoring station to detect a contaminant (min); MSDS = monitoring station detection sensitivity – the minimum monitoring station hazard concentration detection ability (mg/L); (5) the RPM is constructed, with the user defining the size of the RPM for pollution events of two or three random injections (for a one injection pollution event, each node in the system is injected at a resolution of 5 min up to a total of 24 h – the demand cycle considered); (6) genetic algorithm (GA) parameters are specified: number of generations; population size; and the mutation probability; (7) nodes at which monitoring stations can be located are defined – this stage allows the exclusion of locations in which monitoring stations cannot be placed; and (8) the numbers of candidate monitoring stations are defined and a GA run is activated. Upon the GA run completion, results are automatically saved in a text file with the entire last GA generation (i.e., the set of all GA last generation solutions), defining the monitoring station's optimal locations and their corresponding detection likelihood (DL) and redundancy (R).

14.4.2 Example Application

The example application is EPANET Example 3 (Fig. 14.2). The system consists of two constant head sources: a lake and a river; three elevated storage tanks, 117 pipes, 92 nodes (consumer and internal nodes), and 2 pumping units. The system is subject to a 24 h demand pattern. The other system data (i.e., detailed consumptions, demand patterns, pipes and pumping units characteristics, tank volumes, and operational rules) are exactly as those used in Example 3 of EPANET (USEPA, 2002). Example 3 is freely available from the USEPA web site when downloading EPANET, and thus the remaining data are not repeated herein.

14.4.3 Nature of a Pollution Event

Figures 14.3 and 14.4 illustrate the nature of a pollution event comprised of a single contamination injection occurring at node 191 at 1:00 AM. The pollutant is injected at a rate of 2 kg/min for 5 min, thus a total of 10 kg of a pure pollutant is injected. The

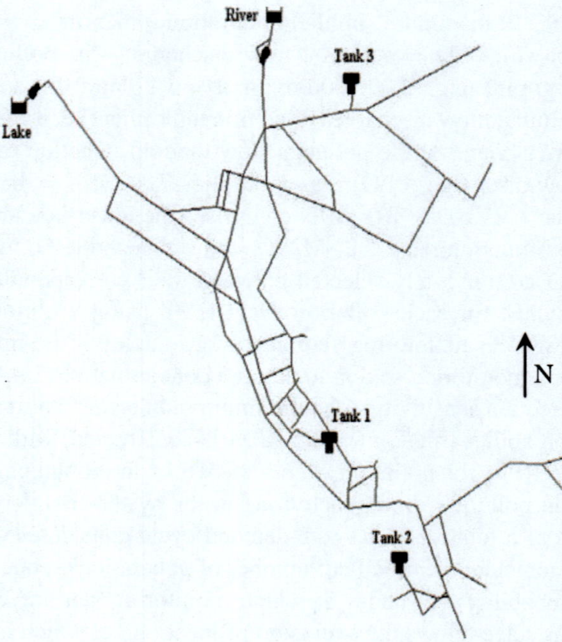

Fig. 14.2 General layout of EPANET Example 3 (USEPA, 2002)

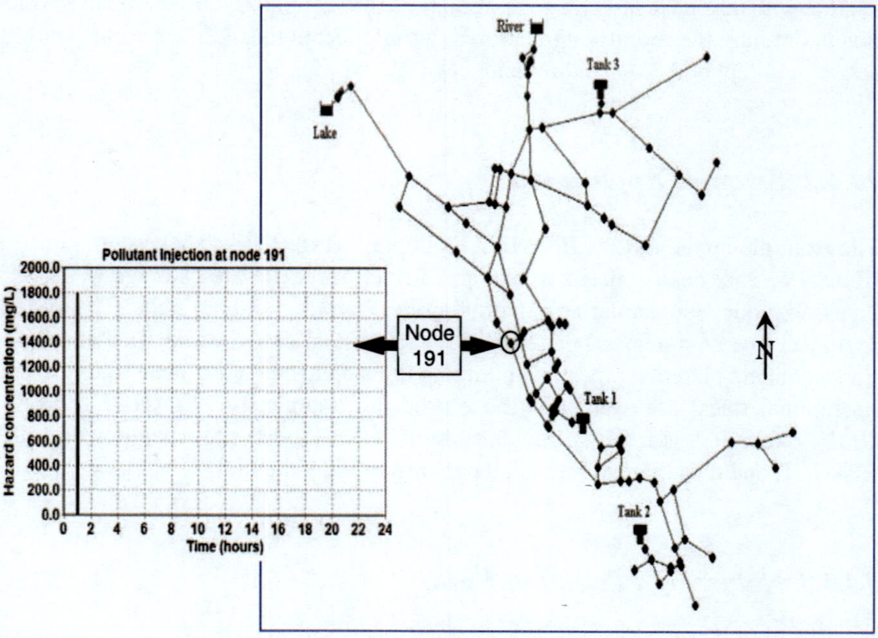

Fig. 14.3 Pollutant injection example event at node 191 at 1 AM: 2 kg/min of a pure pollutant (100% concentration) injection for 5 min (i.e., a total of 10 kg)

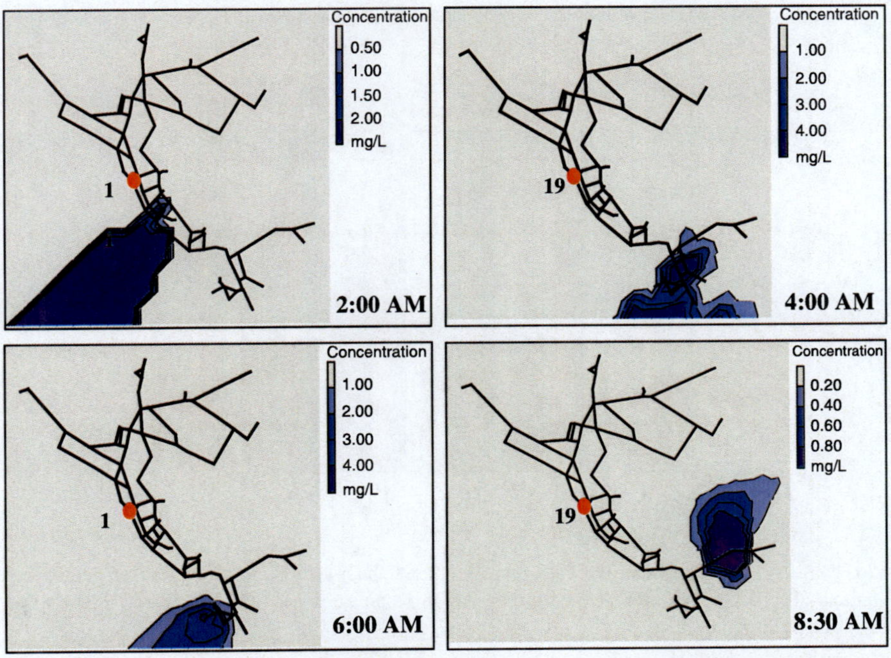

Fig. 14.4 The resulted contamination spreading of the pollutant as of the pollution event at node 191 at 1 AM

result shows a sharp increase of the pollutant concentration at node 191 (Fig. 14.3), and a subsequent pollutant spread downstream, as depicted in Fig. 14.4. The event is completed at around 10:00 AM.

14.4.4 Trade-off Curves

Figure 14.5 shows the trade-off between the detection likelihood (DL) and the total number of monitoring stations (TNOMS) for MSRDs of 0, 30, 60, and 120 (min), subject to EML = existing monitoring locations – at both sources and three tanks (i.e., five existing monitoring stations); NIE = 1; DBIE = not applicable (i.e., NIE = 1); LOS = 100 (gallons); MHL = 0.3 (mg/L); MSD = 96 (h); PFD = 2 (kg/min) for 5 min; PFD_MIN = 1; PFD_MAX = 1; BD_MIN = 1; BD_MAX = 1 (i.e., deterministic consideration of PFD and BD); and MSDS = 0.3 (mg/L). Looking at Fig. 14.5 the following can be stated: (1) there is a substantial reduction (as expected) in the detection likelihood as the monitoring station's response delay increases; (2) as TNOMS increases so does the DL, regardless of the MSRD value; and (3) as MSRD increases the relative improvement to the detection likelihood by adding additional monitoring stations decreases (i.e., the graph becomes

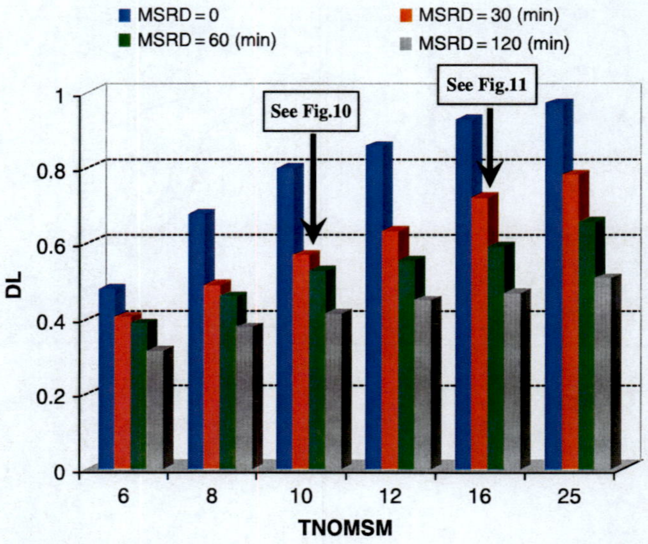

Fig. 14.5 Trade-off between the detection likelihood (DL) and the total number of monitoring stations (TNOMS) for different monitoring station response delay (MSRD) levels, under a one injection pollution event

more and more flat). This is due to the fact that as the MSRD grows the effectiveness of introducing additional monitoring stations decreases, as considerable contaminated water amounts have already been consumed. Consequently, additional monitoring stations will not improve the detection likelihood. Detailed monitoring station locations solutions for two selected runs are further shown in Figs. 14.10 and 14.11.

Figure 14.6 shows the corresponding (to Fig. 14.5) trade-off between the redundancy (R) and TNOMS. It can be seen from Fig. 14.6 that in general (with some exceptions) as MSRD grows the redundancy decreases. The behavior of the redundancy is highly non-linear and much more difficult to predict that that of the detection likelihood.

Figure 14.7 shows the trade-off between the detection likelihood (DL) and the total number of monitoring stations (TNOMS) for different levels of service (LOS): 100, 200, 400, and 500 gallons for a monitoring station response delay (MSRD) of 60 min. The other data are similar to those used for Fig. 14.5. As expected, the detection likelihood decreases as the LOS improves, regardless of TNOMS, and as TNOMS increases the DL relative improvement decreases (i.e., the DL increases exponentially with TNOMS).

Figure 14.8 shows the corresponding (to Fig. 14.7) trade-off between the redundancy (R) and TNOMS. Here as the LOS improves the redundancy decreases. The redundancy behavior is again (see Fig. 14.6) non-linear and much more difficult to anticipate in comparison to the detection likelihood.

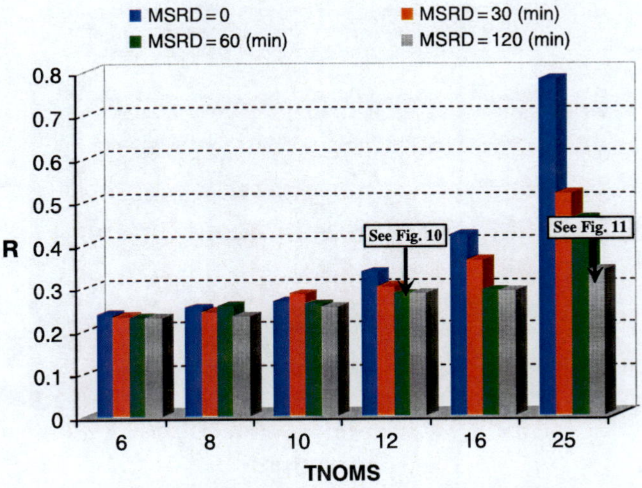

Fig. 14.6 Trade-off between the redundancy (R) and the total number of monitoring stations (TNOMS) for different monitoring station response delay (MSRD) levels, under a one injection pollution event

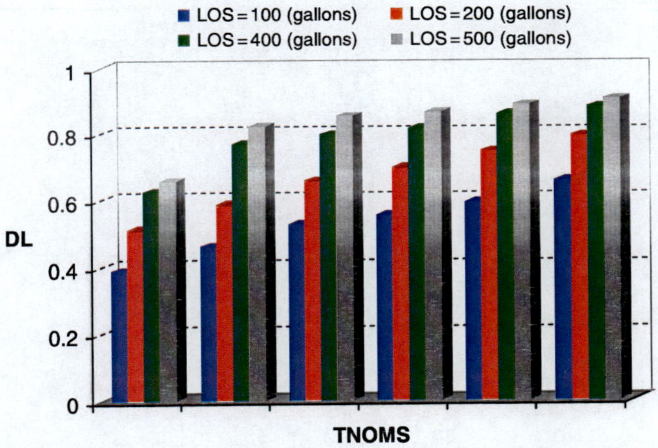

Fig. 14.7 Trade-off between the detection likelihood (DL) and the total number of monitoring stations (TNOMS) for different levels of service (LOS) under an MSRD of 60 min and one injection pollution event

14.4.5 Base Run and Sensitivity Analysis

Figure 14.9 shows the data used for the runs described in Figs. 14.10, 14.11, 14.12, 14.13, 14.14, and 14.15: a base run (BR) (Fig. 14.10), four sensitivity analysis (SA) runs (SA1–SA4) (Figs. 14.11, 14.12, 14.13, and 14.14), and a wide-ranging run using the entire capabilities of optiMQ-S (Fig. 14.15). Figure 14.10 describes the

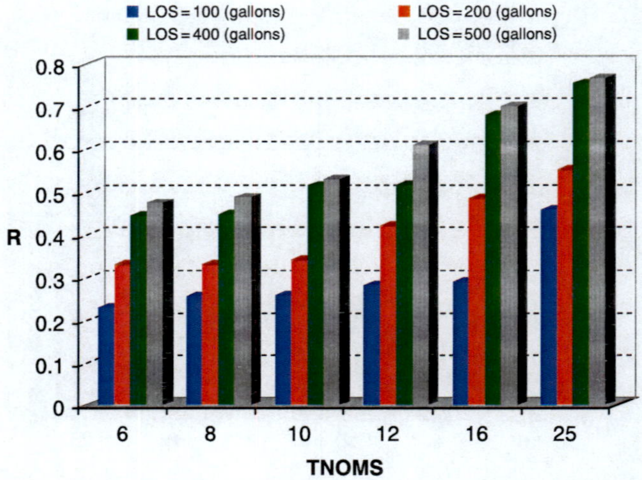

Fig. 14.8 Trade-off between the redundancy (R) and the total number of monitoring stations (TNOMS) for different levels of service (LOS), under an MSRD of 60 min and one injection pollution event

Category	Data	Runs					
		BR	SA1	SA2	SA3	SA4	General
Input	EML	Fig. 11	Fig. 11	Fig. 11	Fig. 11	Fig. 11	Fig. 11
	NPP	Even	Even	Even	Even	Even	Uneven
	NIE	1	1	2	1	1	2
	DBIE	NA	NA	15	NA	NA	15
	LOS	100	100	100	100	100	300
	MHL	0.3	0.3	0.3	0.3	0.3	0.3
	PFD	2	2	2	2	2	2
	PFD_MIN	1	1	1	1	1	0.5
	PFD_MAX	1	1	1	1	1	2.0
	BD_MIN	1	1	1	1	1	0.65
	BD_MAX	1	1	1	1	1	1.85
	MSRD	30	30	30	30	30	90
	MSDS	0.3	0.3	0.3	0.05	0.3	0.2
	SNOMS	5	11	5	5	5	8
Genetic algorithm parameters	NOG	50	50	50	50	50	50
	PS	50	50	50	50	100	50
	MP	0.02	0.02	0.02	0.02	0.02	0.02
Outcome	PMSL	Fig. 11	Fig.12	Fig.13	Fig. 14	Fig. 15	Fig. 16
	DL	0.571	0.723	0.457	0.590	0.572	0.636
	R	0.284	0.363	0.091	0.285	0.253	0.469

Fig. 14.9 Base run (BR) and sensitivity analysis (SA) data. NA = not available; **Runs**: BR = base run; SA1 = sensitivity analysis 1 input; EML = existing monitoring locations (nodes); NPP = nodes pollution probability; NIE = number of injection events; DBIE = duration between injection events (minutes); LOS = level of service (gallons); MHL = minimum hazard level (mg/L); PFD = pollutant flow discharge (kg/min) for a total of 5 min; PFD_MIN = pollutant flow discharge min multiplier; PFD_MAX = pollutant flow discharge max multiplier; BD_MIN = base demand min multiplier; BD_MAX = base demand max multiplier; MSRD = monitoring station response

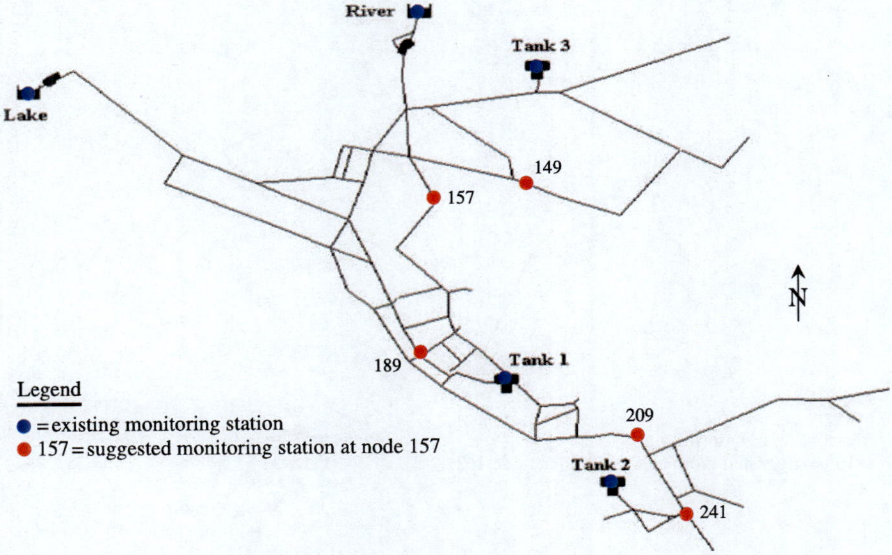

Fig. 14.10 Base run (BR) detailed solution

results of a base run (see also Figs. 14.5, 14.6, and 14.9) for a total of ten monitoring stations: five existing (two at the sources and three at the tanks) and five additionals. The five extra monitoring stations are almost evenly distributed throughout the system. Increasing the total number of monitoring stations to 16 (SA1, Fig. 14.11) (see also Figs. 14.5, 14.6, and 14.9) adds monitoring stations to the north, east, and west parts of the system: nodes 139, 119, 61, 105, 113. In SA2 (Fig. 14.12) the number of injections at each pollution event was increased to two, causing the detection likelihood to drop to 0.457 (0.571 at the BR), the redundancy to 0.091 (0.284 at the BR), and the layout of the monitoring stations to vary. In SA3 (Fig. 14.13) the monitoring station detection sensitivity (MSDS) was increased to 0.05 (mg/L) (0.3 mg/L at the BR), causing the detection likelihood to increase to 0.590 (0.571 at the BR) but the redundancy to remain almost unchanged. The monitoring station's layout is similar to that of the BR (Fig. 14.10). In SA4 (Fig. 14.14) the number of generations was increased to 100 (50 at the BR). As a result of that the monitoring station detection likelihood was almost unchanged, but the redundancy decreased to 0.253 (0.284 at the BR). SA4 demonstrates the flat nature of the objective function response surface: most likely 0.571 (or 0.572) is the global optimal solution; however, there is more

Fig. 14.9 (continued) delay (min); MSDS = monitoring station detection sensitivity (mg/L); SNOMS = suggested number of monitoring stations (additional to EML) genetic algorithm parameters; NOG = number of generations; PS = population size; MP = mutation probability outcome; PMSL = proposed monitoring stations locations (nodes); DL = detection likelihood; R = redundancy

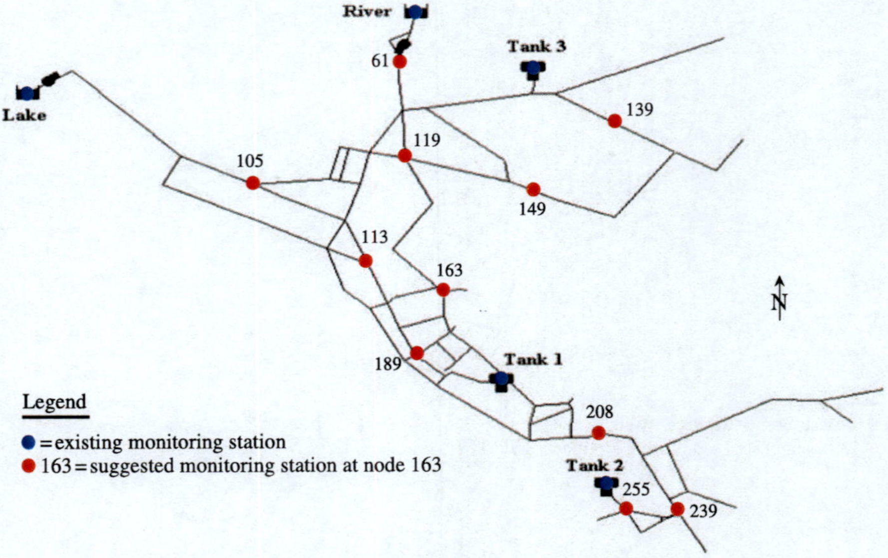

Fig. 14.11 Sensitivity analysis 1 (SA1) detailed solution

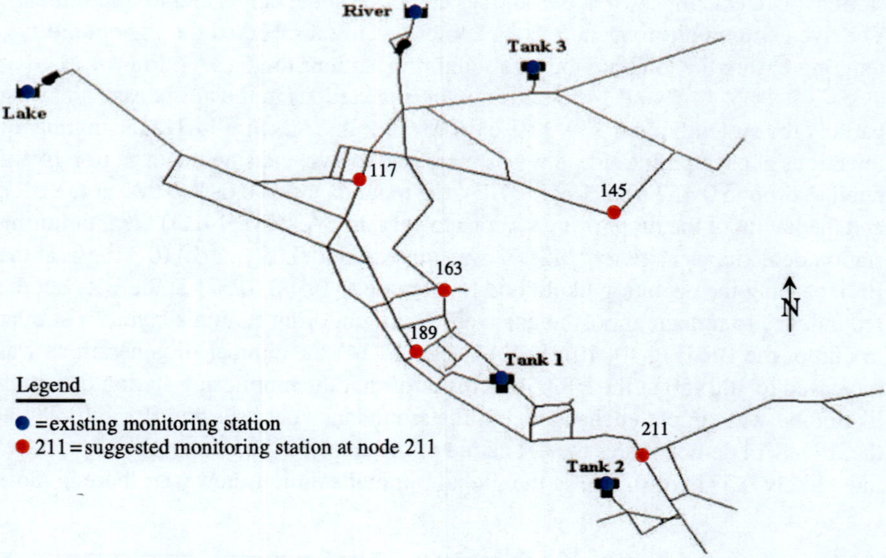

Fig. 14.12 Sensitivity analysis 2 (SA2) detailed solution

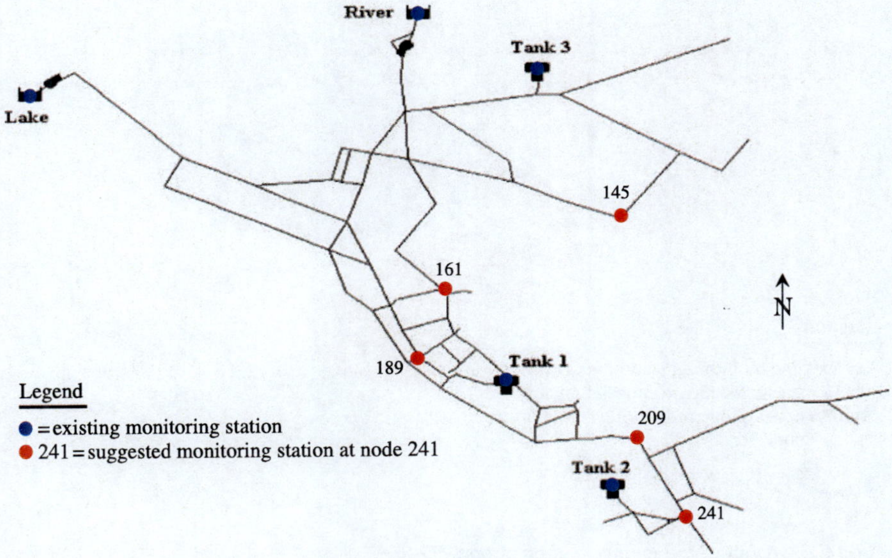

Fig. 14.13 Sensitivity analysis 3 (SA3) detailed solution

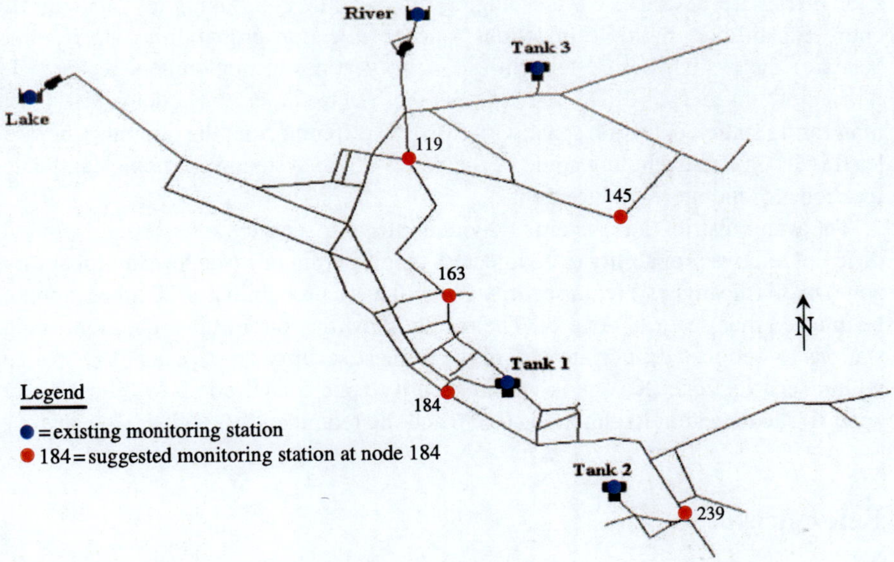

Fig. 14.14 Sensitivity analysis 4 (SA4) detailed solution

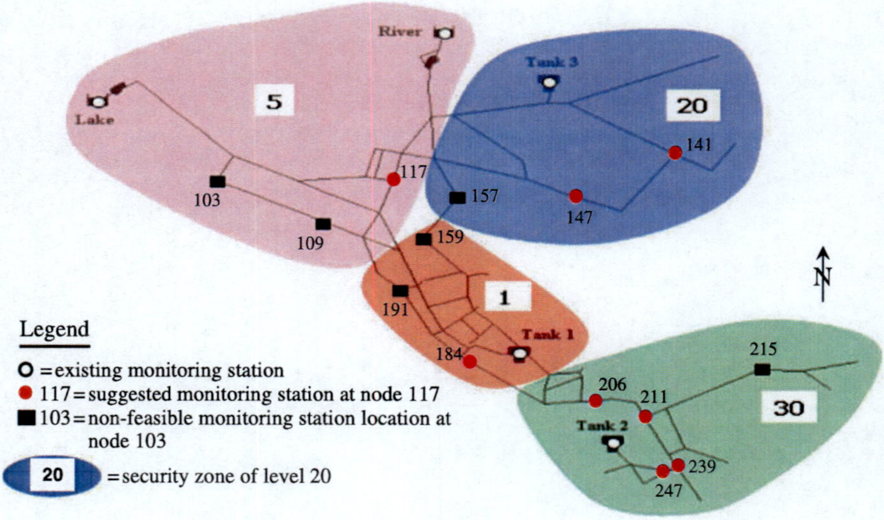

Fig. 14.15 A full wide-ranging solution of optiMQ-S

than one set of monitoring stations that can achieve the same detection likelihood value, with yet different redundancy levels.

Figure 14.15 describes a wide-ranging run of optiMQ-S, taking into account its entire capabilities: multiple injections; uneven injection probabilities at different zones of the system (i.e., establishing "security zones"); randomness of the pollutant flow discharge (PFD); randomness of the consumers base demands (BD); monitoring station detection sensitivity (MSDS) different from the minimum hazard level (MHL); and excluding some of the nodes of the system from being candidate locations for monitoring placements.

The water distribution system is divided into four security zones, each having a different relative probability to be injected, ranging from one (the lowest probability zone) to 30 (the highest) (e.g., security zone 20 has a probability of 20 times more to be injected than security zone 1). The results show that out of the eight monitoring stations to be located, four are within the highest security zone (zone 30), two are within security zone 20, one is within security zone 5, and one is within security zone 1. The detection likelihood is 0.636 and the redundancy 0.469.

References

Al-Zahrani M. and Moied K. (2001). "Locating optimum water quality monitoring stations in water distribution system." In Bridging the Gap: Meeting the World's Water and Environmental Resources Challenges, Proceedings of the ASCE EWRI Annual Conference, Orlando, FL, published on CD.

American Society of Civil Engineers (ASCE) (2004). "Guidelines for designing an online contaminant monitoring system."

American Water Works Association (AWWA) (2004). “Security guidance for water utilities.”

Berry J. W., Hart W. E., Phillips C. A., Uber J. G., and Watson J. P. (2006). “Sensor placement in municipal water networks with temporal integer programming models.” Journal of Water Resources Planning and Management Division, ASCE, Vol. 132, No. 4, pp. 218–224.

Deb K., Agrawal S., Pratap A., and Meyarivan T. (2000). “A fast elitist non-dominated sorting genetic algorithm for multi-objective optimization: NSGA-II.” Proceedings of the Parallel Problem Solving from Nature VI Conference, Paris, France, pp. 849–858.

Dorini G., Jonkergouw P., Kapelan Z., Pierro F. di, Khu S. T., and Savic D. (2006). “An efficient algorithm for sensor placement in water distribution systems.” 8th Annual Water Distribution System Analysis Symposium, Cincinnati, OH, published on CD.

Eliades D. and Polycarpou M. (2006). “Iterative deepening of Pareto solutions in water sensor Networks.” 8th Annual Water Distribution System Analysis Symposium, Cincinnati, OH, published on CD.

Goldberg D. E. (1989). “Genetic algorithms in search, optimization, and machine learning.” Addison-Wesley, New York, NY.

Gueli R. (2006). “Predator – prey model for discrete sensor placement.” 8th Annual Water Distribution System Analysis Symposium, Cincinnati, OH, published on CD.

Holland J. H. (1975). “Adaptation in natural and artificial systems.” The University of Michigan Press, Ann Arbor, MI.

Huang J. J., McBean E. A., and James W. (2006). “Multi-objective optimization for monitoring sensor placement in water distribution systems.” 8th Annual Water Distribution System Analysis Symposium, Cincinnati, OH, published on CD.

Kessler A., Ostfeld A., and Sinai G. (1998). “Detecting accidental contaminations in municipal water networks.” Journal of Water Resources Planning and Management Division, ASCE, Vol. 124, No. 4, pp. 192–198.

Kumar A., Kansal M. L., and Arora G. (1997). “Identification of monitoring stations in water distribution system.” Journal of Environmental Engineering, ASCE, Vol. 123, No. 8, pp. 746–752.

Lee B. and Deininger R. (1992). “Optimal locations of monitoring stations in water distribution system.” Journal of Environmental Engineering, ASCE, Vol. 118, No. 1, pp. 4–16.

Ostfeld A. and Salomons E. (2004). “Optimal layout of early warning detection stations for water distribution systems security.” Journal of Water Resources Planning and Management Division, ASCE, Vol. 130, No. 5, pp. 377–385.

Ostfeld A. and Salomons E. (2005). “Securing water distribution systems using online contamination monitoring.” Journal of Water Resources Planning and Management Division, ASCE, Vol. 131, No. 5, pp. 402–405.

Ostfeld A. and Salomons E. (2006). “Sensor network design proposal for the battle of the water sensor networks (BWSN).” 8th Annual Water Distribution System Analysis Symposium, Cincinnati, OH, published on CD.

Ostfeld A., Kessler A., and Goldberg I. (2004). “A contaminant detection system for early warning in water distribution networks.” Engineering Optimization, Vol. 36, No. 5, pp. 525–538.

Ostfeld A., Uber J., and Salomons E. (2006). “Battle of the Water Sensor Networks (BWSN): a design challenge for engineers and algorithms.” 8th Annual Water Distribution System Analysis Symposium, Cincinnati, OH, published on CD.

Preis A. and Ostfeld A. (2006). “Multiobjective sensor design for water distribution systems security.” 8th Annual Water Distribution System Analysis Symposium, Cincinnati, OH, published on CD.

Propato M. (2006). “Contamination warning in water networks: general mixed-integer linear models for sensor location design.” Journal of Water Resources Planning and Management Division, ASCE, Vol. 132, No. 4, pp. 225–233.

Rubinstein R. Y. (1999). “The simulated entropy method for combinatorial and continuous optimization.” Methodology and Computing in Applied Probability, Vol. 2, pp. 127–190.

USEPA (2002). "EPANET 2.00.10" Available online at http://www.epa.gov/ORD/NRMRL/wswrd/epanet.html (accessed 15 July 2008).

Watson J. P., Greenberg H. J., and Hart W. E. (2004). "A multiple-objective analysis of sensor placement optimization in water networks." In Critical Transitions in Water and Environmental Resources Management, Proceedings of the ASCE EWRI Annual Conference, Salt Lake City, UT, published on CD.

Woo H. M., Yoon J. H., and Choi D. Y. (2001). "Optimal monitoring sites based on water quality and quantity in water distribution systems." In Bridging the Gap: Meeting the World's Water and Environmental Resources Challenges, Proceedings of the ASCE EWRI Annual Conference, Orlando, FL, published on CD.

Wu Z. Y. and Walski T. (2006). "Multi objective optimization of sensor placement in water distribution systems." 8th Annual Water Distribution System Analysis Symposium, Cincinnati, OH, published on CD.

Chapter 15
Protecting Consumers from Contaminated Drinking Water During Natural Disasters

Craig L. Patterson and Jeffrey Q. Adams

15.1 Introduction

Natural disasters can cause damage and destruction to local water supplies affecting millions of people. An authorized team should be in place to manage and prioritize emergency response in devastated areas. Sections 15.2 and 15.3 describe EPA's Disaster Recovery Plan and the steps that the water industry (water utilities, government agencies, non-governmental organizations, academia, and consultants) is taking to tackle potential threats to safe drinking water and drinking water infrastructure.

Emergency response and recovery protocols are designed to reduce the severity of water and wastewater infrastructure damage and ensure safe drinking water after natural disasters. Water system recovery efforts require preliminary preparations, emergency response procedures, and long-term support as described in Section 15.4. Public safety, health, and welfare are top priorities during emergency response activities. Repairs are extremely costly and include the costs of interim operations, cleanup, and other non-capital expenses. Understanding the risks associated with natural disasters in individual communities allows health officials to respond effectively when local disasters become a reality. A case study of EPA emergency response efforts after Hurricane Katrina is provided to bring the impact of major natural disasters on public water systems into focus.

Government agencies including the EPA are supporting the development of small drinking water treatment technologies to bring timely relief to devastated communities. EPA research is focusing on household devices, mobile treatment systems, and disinfection processes as described in Section 15.5 to protect consumers from contamination in drinking water wells, tanks, and distribution systems.

C.L. Patterson (✉)
National Risk Management Research Laboratory, Water Supply and Water Resources Division, USEPA, Cincinnati, OH, USA
e-mail: patterson.craig@epamail.epa.gov

R.M. Clark et al. (eds.), *Handbook of Water and Wastewater Systems Protection*, Protecting Critical Infrastructure, DOI 10.1007/978-1-4614-0189-6_15,
© Springer Science+Business Media, LLC 2011

15.2 EPA Disaster Recovery Plan

EPA has established a Disaster Recovery Plan that includes a voluntary roster of employees willing to participate in disaster response called the Response Support Corps (RSC). RSC employees are trained in all aspects of the Incident Command System (ICS) that facilitates multi-agency, multi-level governmental coordination, comprehensive loss and resource assessment, and a shared vision of recovery plans (Johnson, 2009).

EPA adopted ICS (a nationwide standard for emergency management) because of its interdisciplinary and organizational flexibility. ICS is part of the command and management portion of the National Incident Management System (NIMS) developed by the Federal Emergency Management Agency (FEMA, 2007). NIMS establishes agreements with all public and private entities within a state. The ICS fosters partnerships among local, State, Tribal, and Federal agencies and private industry to handle natural disasters, such as tornadoes, floods, ice storms, or earthquakes.

The five major management functions of ICS organization are as follows:

- Incident Command (has overall responsibility and sets the incident objectives, strategies, and priorities)
- Operations (develops tactical objectives and organization and directs all tactical resources)
- Planning (prepares Incident Action Plans to collect, evaluate, and maintain records on resource status to accomplish incident objectives)
- Logistics (provides support, resources, and other services to meet operational objectives)
- Finance/Administration (provides accounting, procurement, time recording, and cost analyses)

The ICS features related to command structure include chain of command (lines of authority), unity of command (one supervisor per employee), and unified command (multi-agency cooperation) that define reporting relationships and reduce the confusion caused by multiple, conflicting directives. During transfer of command, a briefing form is used to describe the incident situation (map, significant events), incident objectives, summary of current actions, and the status of resources assigned or ordered for the incident or event.

The ICS consists of procedures for controlling personnel, facilities, equipment, and communications. ICS principles are implemented through a wide range of management features including the use of common terminology and a modular organizational structure. The ICS efficiently manages incident resources by

- Maintaining a manageable span of control (supervisory responsibility ranges from three to seven employees)
- Establishing pre-designated incident locations and facilities (command posts, bases, camps, staging areas, mass casualty triage areas, and helicopter landing areas)

- Implementing resource management practices (processes for categorizing, ordering, dispatching, tracking, reimbursing, and recovering resources such as personnel, teams, equipment, supplies, and facilities)
- Ensuring integrated communications (a common plan for interoperable communications)

Personnel and equipment are mobilized or dispatched only when requested by an appropriate authority. An Incident Management Team consists of Command and General Staff members with necessary training and experience to respond to natural disasters of differing magnitudes:

- Type 5: Local Village and Township Level
- Type 4: City, County, or Fire District Level
- Type 3: State or Metropolitan Area Level
- Type 2: National and State Level
- Type 1: National and State Level (Type 1 Incident)

The Incident Command System is designed to manage a wide variety of incidents or events for EPA including multiple types and varieties of natural disasters. Clean and safe drinking water is one of many critical environmental outcomes that result from implementation of the ICS approach within NIMS.

15.3 Water Industry Collaboration

The American Water Works Association (AWWA) National Water/Wastewater Agency Response Network (WARN) provides a conduit for utilities to access mutual aid and assistance during emergencies. WARN helps unite and foster communication between local and state Emergency Management Agencies and the state Primacy Agency. AWWA's WARN web site provides tools, training, and technical assistance for the intrastate deployment of water utility personnel, equipment, materials, and other associated services. Specific guidance is provided on emergency repairs and disinfection of water mains (EPA, 2007).

During state emergencies, mismanagement of utility resources can create delays in water and wastewater infrastructure repair and recovery. WARN is designed to facilitate immediate acquisition of utility resources before the arrival of government aid. WARN establishes an agreement and protocols to access utility personnel with appropriate skills and utility-specific heavy equipment, tools, and supplies (AWWA, 2007). Utility managers that have been impacted by natural disasters recommend being prepared and found it crucial to document and track costs during every aspect of an incident to avoid potential lawsuits and to simplify reimbursement (AWWA, 2009).

EPA uses a funding mechanism called the Emergency Management Assistance Compact (EMAC) for interstate participation in mutual aid and assistance for water and wastewater facilities. EPA recommends that interested parties develop a relationship with state EMAC coordinators to facilitate the deployment of volunteers from public utilities across state lines during a declared state of emergency. EMAC addresses interstate response and liability, workers compensation, and reciprocity issues that arise between utilities from different states (EPA, 2007).

Mock exercises such as the Great Southern California ShakeOut 2008 prepared residents in California for the next major earthquake. The East Valley Water District used United States Geological Survey (USGS) Shaking Intensity Maps to predict severity and damage to the water and wastewater infrastructure. Some of the water issues discussed included thousands of water and wastewater pipeline failures, generator and roadway failures, lack of treatment chemicals with disrupted railways, water shortages from damaged reservoirs, loss of laboratories to test water samples, how to prioritize repair in affected areas, and cross-contamination from wastewater systems. The ShakeOut highlighted the need for better communications between fire and water agencies to provide non-potable water for firefighting, while saving potable water for public consumption (Sturdivan et al., 2008).

15.4 Water System Emergency Response and Recovery

Water system response and recovery efforts are dependent on the magnitude of the natural disaster. Devastating events encompassing large areas require more coordination between local, State, and Federal agencies than small localized events. Public health and safety response efforts are managed by local firefighters, paramedics, and police. In the case of drinking water infrastructure, natural disasters adversely impact the normal operations of electric utilities, water utilities, and city or county health departments. One way for communities to protect the public and minimize loss is to develop emergency operations, response, and recovery plans. This section provides a composite summary of actions taken during Hurricanes Andrew (Murphy, 1994), Hugo (Shimoda, 1994), and Katrina (Patterson et al., 2007) that allowed communities to get life and drinking water in impacted areas back to normal.

15.4.1 Preliminary Preparations

One approach to prepare for potential natural disasters is to develop a strategy using "What if?" scenarios. For instance, utilities can

- Prioritize water systems in areas that are prone to fire, earthquake, flooding, and hurricane damage
- Develop a matrix of potential hazards and determine the vulnerability of water system components
- Estimate damage to vulnerable system components
- Take actions to decrease water system component vulnerability such as capital projects to secure treatment plant processes and distribution system pipelines

Utilities can set up an active training program and conduct regular training exercises. Training sessions should focus on location-specific development of "What if?" scenarios. For instance, using treatment plant plans and distribution system maps to brainstorm response actions or exploring alternative access routes to water system infrastructure to allow workers to maneuver around floods, fires, downed electrical lines, and fallen trees.

Training exercises can also facilitate maintenance and testing of generators, isolation valves, and radio communications. After each training session, utilities should maintain contact information of training program attendees for distribution of emergency response plan updates.

15.4.1.1 Emergency Operations Plan

An emergency operations plan provides a utility or a municipality with a contingency plan for natural disasters based on information gathered during "What if?" training sessions. The plan should summarize the actions that are needed before, during, and after a natural disaster to provide a protocol for response and recovery.

The plan should contain a list of organizations and points of contact to provide technical assistance including local, State, Federal, and contracted personnel. Written agreements can be set up with other water utilities, contractors, and private suppliers including annual provisions for 24 h access and phone numbers of contacts in case of emergencies.

The plan should provide contingency plans to mobilize a command center with established organizational responsibilities. Planning command center operations in advance ensures efficient response and recovery, public health and safety, regulatory involvement, and communication with the water utility personnel. The plan should include options for command center locations in strategic areas with access to power, water, parking, and communications infrastructure. Delivery of emergency resources can be defined with a matrix of appropriate transportation resources that matches the type and location of emergencies. Utilities can develop equipment and supply lists for water treatment and distribution systems and maintain an adequate supply of water tanks, different-sized portable electric generators, and treatment chemicals. Alternative suppliers should be located to handle shortages in fuel for generators and natural gas for treatment plants. Vehicles should be equipped with global positioning system (GPS) units and accurate up-to-date road maps to public water systems.

The plan should forecast the needs of field assessment crews, lab analysts, and accounting personnel. Supply lists for emergency responders can be developed and include health and safety items applicable to local conditions such as drinking water, sun screen, and bug spray. Utilities should identify available mobile laboratories, develop analytical supply lists, and prepare water sampling, monitoring, and analytical protocols. Boil Water advisories can be prepared in advance. Utilities should prepare standardized assessment forms and develop data entry systems with data transfer protocols for assessment information. An efficient and comprehensive emergency response accounting system with procedures and disaster recovery codes is invaluable for tracking FEMA response and recovery costs.

To support a sustainable workforce, utilities should be prepared to house homeless utility personnel, coordinate repair of damaged employee homes, and provide cooked food for utility response and recovery crews at the command post and in the field. Utilities should store food for emergencies and restock food stores annually.

When a natural disaster is imminent, many actions can be taken to strengthen water infrastructure and prevent waterborne disease. For instance, utilities can

- Reinforce exposed facilities such as well houses and pump stations
- Install shutters on windows and clear loose debris from plants
- Sandbag critical water and wastewater infrastructure such as building entrances and pump stations
- Overchlorinate water supplies to protect against waterborne pathogens.
- Top off water storage tanks and close main valves in anticipation of water main and hydrant breaks
- Isolate or shut down exposed pipe at river crossings
- Set electric pressure reducing valves to manual mode

Utilities can also prepare vehicles, equipment, and supplies for approaching natural disasters:

- Fill trucks with gas and furnish equipment to repair flat tires caused by road debris
- Add fuel to and test generator-driven pumps at plants and deliver emergency generators to locations missing generators such as well fields
- Fill fuel tanker trucks with fuel and install appropriate connections to refuel generators
- Equip trucks with service and repair supplies such as T-handle wrenches
- Supplement water supply with contracted water tank trucks

To ensure access to critical information, utilities should secure vital records for use during emergency response. Also, check and secure two-way radios and phone gear for emergency communication. The emergency operations plan is a living document that needs to be updated on a regular basis. Updates ensure that emergency responders are on the same page and that emergency response is as effective as possible.

15.4.1.2 Emergency Response Procedures

Water utility response during emergencies may require the efforts of three groups: a command or liaison group to direct the response, a field investigation group to identify the cause and extent of the damage, and a treatment response group to make repairs. Emergency response and recovery can be broken down into the seven steps as listed in Table 15.1.

Table 15.1 Seven steps for water system emergency response and recovery

Seven steps	Lessons learned
1. Analyze the type and severity of the natural disaster	Conduct system hazard summary and vulnerability analysis Determine status of water system critical components
2. Provide emergency assistance to save lives	Evacuate people in harms way Secure chlorine leaks or severely damaged water towers
3. Reduce the probability of additional injuries and damage	Release water from a damaged reservoir Cut power to downed electric lines Send out Boil Water advisories
4. Prioritize water system demands and make repairs	Provide water to firefighters and medical facilities Set up water distribution staging points
5. Return the water system to normal operation (recovery)	Prioritize and make less urgent repairs Do not lift boil water advisories until full recovery
6. Evaluate the response and preparedness plan	Allow for real-time changes to the emergency response plan
7. Revise the preparedness plan as necessary	Modify emergency operations, response, and recovery plans based on lessons learned

Shimoda (1994)

Effective emergency response communication is critical to the health and safety of the public as well as emergency responders. Emergency responders should notify the media of the location of and phone numbers for the emergency command post and assign management of warnings and advisories to a responsible individual to prevent retrieval of misinformation or inaccurate reports.

Utilities should have contact names and phone numbers for agencies that issue such warnings:

- US Geological Survey (USGS) – Potential earthquakes and volcanic eruptions
- US Army Corps of Engineers – Flood warnings
- US and State Forest Services – Wildfires
- State or Local Health Departments – Waterborne disease outbreaks
- National Weather Service – Hurricanes, tornadoes, and ice storms

Emergency response actions are dependent on the type of natural disaster and the conditions in damaged or destroyed areas. However, there are many common response and recovery actions that apply to water infrastructure damage and destruction.

Utilities should line up work crews and set up schedules to repair uprooted, damaged, and broken water and electric lines. Work crews typically repair hospital water systems first. Next, repair crews bring the least damaged water systems back on line to provide potable water to as many customers as possible. Distribution system water quality parameters are monitored for changes in pressure, flow, pH, tank elevation, and chlorine residual to isolate damaged and leaking sections. Repair crews transport generators to broken water mains to begin pumping water to find and repair leaks. Block by block and door to door surveys of water systems are conducted to repair leaks or shut off service. Repair crews attach spigots to water meters to provide water to residents with damaged household waterlines and set up emergency water stations at strategic locations.

During major natural disasters, assessment and analytical crews are formed to prioritize public water systems by population served, extent of damage, and accessibility and access to alternative water and power sources. These crews also assess and manage risks associated with potential exposure to waterborne pathogens and chemicals. Assessment and analytical crews collect and analyze treatment plant and distribution water samples to interpret water quality data for water system clearance by

- Locating bacteria sample locations with up-to-date copies of maps and information
- Reviewing water sampling and analysis protocols
- Collecting and analyzing chlorine and coliform bacteria samples
- Locating and distributing generators to minimize down-time and pressure loss to water systems
- Immediately notifying water system operators with boil water and water system clearance information
- Providing technical assistance to prioritized water systems following initial assessments

The assessment and analytical crews also prioritize monitoring of source waters based on the extent of contamination. Source water quality data (e.g., coliform, turbidity, inorganics, and organics) are provided to treatment plant operators for acquisition of supplies (e.g., coagulants, polymers, and treatment media). These crews also arrange for delivery of chlorinated potable water tanks and mobile water treatment systems to central locations in heavily damaged and high-risk areas.

Utilities and emergency response and recovery work crews must keep accurate records to document expenditures for reimbursement and historic recordkeeping. The use of infrastructure assessment forms, videos, and photographs is an excellent way to record damage, ensure follow up, and document emergency response actions. To obtain FEMA compensation, participants are required to keep track of the length of pipeline lost and fill out Damage Survey Reports.

15.4.1.3 Long-Term Support

After emergency response and recovery, utilities should update the emergency operations plan based on lessons learned. Utilities can also prepare a schedule for long-term technical assistance and repairs. Emphasis should be placed on assessing water treatment and distribution system integrity and evaluation of options for long-term improvements such as dual systems with water reuse and installation of online and remote monitors. Utilities can also prioritize replacement and repair of large items such as storage tanks, generators, and clarifiers.

15.4.2 Emergency Response Efforts After Hurricane Katrina

In August 2005, Hurricane Katrina quickly engulfed New Orleans, Louisiana, and surrounding communities in one of the largest natural disasters in the history of United States. This section serves as a case study with a chronology of emergency response efforts (Patterson et al., 2007). Hurricane Katrina followed a course east of New Orleans through the parishes of Plaquemines, St. Bernard, and St. Tammany dumping 8–10 in. of rain at a rate of 1 in./h. The outer bands of the storm produced tornadoes as far away as Georgia. In addition to the direct devastation of the hurricane, levee failures resulted in the flooding of New Orleans that was blamed on strong winds, heavy rains, storm surge, and construction flaws. By August 31, at least 80% of New Orleans was flooded with some parts under as much as 20 ft of floodwater (Bohman, 2005). EPA staff, fire fighters, police, and other first responders rescued nearly 800 people from the floodwaters (Zucchino, 2005). Tragically, there were 1080 deaths directly related to Hurricane Katrina in Louisiana (Zucchino, 2006).

15.4.2.1 Damage to the Carrollton Treatment Plant

The main drinking water treatment plant for the City of New Orleans (the Carrollton plant serving over 500,000 residents) sustained water main breaks, fire, and flooding but was brought back on line due to the diligence of plant employees. One worker braved the storm to shutoff one of six water mains to prevent 40–50 million gallons per day of water from leaking out of the distribution system. The high winds and heavy rain drenched, short-circuited, and ignited electrical equipment. A team of workers extinguished an electrical fire after the wind blew out a window in the powerhouse. After the levees were breeched, the Carrolton plant was flooded causing the backup generators to be shutdown. Crews made desperate attempts to prime the boilers both manually and with a fire truck to keep the plant operating. It was not until the plant had enough city power to fire up the boilers that the pumps could draw water from the Mississippi River. By September 11, operators were disinfecting water at the Carrollton Plant with 90 million gallons per day in circulation (Brown, 2005).

15.4.2.2 Water System Emergency Response

A few days after Hurricane Katrina, a multi-disciplinary team of 30 EPA emergency response, research, and water program personnel joined forces with local health and environmental officials to help residents gain access to safe drinking water supplies from hurricane-ravaged parishes surrounding New Orleans. The EPA team met at the EPA Region VI Emergency Response Center in Dallas, Texas, to prepare for assessment of over 400 public water systems in an effort to restore safe drinking water to Louisiana communities. A 2-h briefing included health and safety training to protect relief workers from risks as shown in Tables 15.2 and 15.3.

Vaccinations (hepatitis A/B and tetanus shots) were provided to participants in need at the EPA Region VI medical facility in Dallas. EPA loaded sport utility vehicles, vans, and trucks with emergency response supplies and drove 9 h to Livingston, LA. EPA responders regrouped at the fairgrounds in Livingston, LA, that served as a mobile staging area with available electrical power, phone service, water, and gasoline. The fairgrounds in Livingston (north of Lake Pontchartrain) also provided a centrally located field headquarters within driving distance of damaged areas. Upon

Table 15.2 Health and safety field hazards

Topics	Corrective actions/information
Driving hazards	Defensive driving, seat belts, clearly identify yourself, and watch for Emergency Vehicles. Watch for debris in the road (bring spares)
Boat safety	Use personal flotation devices, ensure proper footing, hip waders for launching boats
Personal protective equipment (PPE)	Hard hats, ear plugs, gloves, outer booties, and watertight boots (steel shank and insole), particulate filter mask, Tyvek or rubber apron, hip waders, and goggles (splash)
Contamination avoidance	Avoid unnecessary contamination. Beware water spray. *Decon* – All personnel and equipment will require decontamination prior to leaving the site
Decon equipment	Brush, bleach and water (10:1), sprayer or sponge, buckets
Electrical hazards	Beware of electrical injuries as power lines are reactivated. Hazardous materials – dislodged tanks, drums, pipes, car batteries, and carbon dioxide "the silent killer"
Explosive atmosphere	Natural gas leaks continue. Numerous oil and gasoline spills
Air monitoring	Hydrogen sulfide, lower explosive limit, carbon monoxide, volatile organic compounds, and radiological compounds
Mosquitoes	Air Force is spraying. Wear insect repellant – prevent West Nile transmission. Symptoms: fever, headache, and body aches, nausea, vomiting, swollen lymph glands or a skin rash. 80% develop no symptoms; <1% develop severe illness
E. coli strains	Gastrointestinal infection, diarrhea, usually little or no fever, and resolves in 5–10 days. The kind of *E. coli* that causes bloody diarrhea and kidney failure is a much more virulent strain (usually comes from eating contaminated meat). To date, that strain has not been reported. Seek medical attention/advice if you are experiencing symptoms
Vibrio vulnificus	Bacteria that targets people with compromised immune systems
Wild/stray animals	Disoriented and dangerous. Snakes in unusual places. Dogs in packs, alligators, rats, fire ants...
Heat stress	POPC liposomes – Heat and humidity is high. Fluids (1:2 energy drinks to water)
Sunblock	Sunburn affects your body's ability to cool itself and causes a loss of bodily fluids

EPA (2005a)

Table 15.3 Health and safety issues

Topics	Corrective actions/information
Work/rest cycles	Work toward 12 h days. Know your limitations. Slip/trip/falls – Field hazards
Accountability	Carry flares. Check-in/call-in when you are expected
Security	Stable but volatile. Follow the direction of the Criminal Investigation Division
Buddy system	Utilize the Buddy System. Report any medical conditions
CISM	Critical Incident Stress Management – Remember to talk with someone
Awareness/HAZCOM	The first step to safety is awareness
Vaccinations	Hepatitis A, hepatitis B, tetanus, and diphtheria toxoid
Hygiene	Practicing basic hygiene is *critical*

EPA (2005a)

Fig. 15.1 EPA's emergency response team

arrival, the group was directed to living quarters at the Livingston Baptist Church gymnasium as shown in Fig. 15.1.

Assessment teams worked from dawn until dusk with temperatures in the nineties, high humidity, and plenty of mosquitoes. Each EPA person was paired with an inspector from a State Rural Water Association (LA, TX, GA, IN, IA, and IL were represented) or with someone from LA Department of Health and Hospitals. The two-person teams visited and assessed public water supplies in the parishes of Ascension, Livingston, St. Bernard, St. Charles, St. Tammany, St. Helena, St. James, St. John, St. Mary, St. Tammany, Tangipahoa, and Washington. Some areas in St. Bernard and St. Tammany had been virtually flattened. Downed power lines and debris were strewn everywhere. Long hours were spent in some areas locating the water treatment systems. In many instances, signs had been blown down or were destroyed. Roads were congested, gasoline supplies were scarce, and cell phone

Fig. 15.2 Lifting a boil advisory during Hurricane Katrina emergency response. Georgia Rural Water Association, New Orleans, LA, September 19, 2005

service was sporadic. Figure 15.2 shows a member of the assessment team displaying a notification that the boil water order imposed after katrina, had been rescinded.

Small drinking water assessment teams were formed to reach impacted public water systems. The objectives of the assessment teams were to assist treatment plant operators and perform initial damage assessments in the parishes east and northeast of Lake Pontchartrain. Water systems were categorized on the basis of the type of the water system (community, non-community, or transient), severity of damage, need, whether they had been visited already by LDHH, and accessibility considering flooding and loss of infrastructure (electrical power and roads). The assessment teams analyzed water samples for free and total chlorine. For those utilities that could still pump water, total coliform bacteria samples were collected as an indicator of the microbiological quality of the water. EPA provided technical expertise setting up and running mobile diagnostic equipment and on-site labs that quickly and accurately determined if local drinking water supplies were safe. After testing was completed, results were sent back to the treatment facilities and the LDHH for decisions regarding public health warnings (boil water and clearance information).

EPA used a drinking water system recovery database to compile information on the status of and damage to individual water systems as listed in Table 15.4. The status codes listed in Table 15.5 were upgraded as recovery efforts progressed.

In a little over a week, the assessment teams had visited and surveyed over 400 water systems in the areas affected by Hurricane Katrina. Major problems included loss of power/loss of pressure, no backup generators (as many as 60 generators were requested), and damage/destruction to treatment plants and water distribution systems. Louisiana Rural Water Association (LARWA) in conjunction with EPA and

Table 15.4 Drinking water system assessment survey

Parish	Date/time of recent visit:	Conventional filtration treatment?
Status codes:	System operational?	PWS source?
Sample collected/date:	System power loss?	Population of PWS:
Assessor 1:	Wells inundated?	Immuno-sensitive populations?
Phone:	Physical damage?	Response: Isolation?
E-mail:	Valves:	Booster disinfection?
Assessor 2:	Pumps:	Increase free chlorine?
Phone:	Electrical equipment:	Re-routing water to customers?
E-mail:	Storage tanks:	Stop of service to customers?
System name:	Pipes:	Additional testing?
PWS ID:	Other:	Type:
System phone:	Backup generator?	Customer complaints?
System address:	Generator in use?	EPA Assistance requested?
Date/Time of Assessment:	Adequate fuel supply?	Date/time for update:
Assessment Text. . .	Disinfection type:	Long-term needs

EPA (2005b)

Table 15.5 Hurricane Katrina system status codes

Code	Definition
OUT	Contact attempted but system unreachable by phone; system assumed down and pressure lost, sanitarians contacting in person; under Boil Advisory (out of contact)
INOP	Contact made with system; system has no power and is currently off-line, assumed pressure lost; under Boil Advisory (inoperable). This category also includes systems that are choosing to stay off-line or rely on backup connections to other systems
GENLP	Currently operating on emergency power/generator and system lost pressure and/or treatment; under Boil Advisory
GENOK	Currently operating on emergency power/generator but system did *not* lose pressure and/or treatment (generator but never lost pressure)
OK	Normal power restored (or never lost) and system never lost pressure and/or treatment (no problem with system)
NEED	Normal power restored; but bacteria samples needed; under Boil Advisory
NEED-RES	Normal power restored. Bacteria samples collected and awaiting results
CLEAR	System online and bacteria samples came back clear (system cleared)

EPA (2005c)

LDHH staged approximately 20 polyethylene water tanks at a central distribution point. When necessary, water was collected from treated sources outside of the damaged areas, chlorinated to 4–5 mg/L of free chlorine, and transported to areas of need. Repair and replacement of large items such as storage tanks, generators, and clarifiers were prioritized to streamline the recovery efforts.

Hurricane Katrina was followed by Hurricane Rita and recovery efforts expanded to include over 1,000 affected drinking water systems. By November 3, 2005, safe drinking water was available to 5 million people. Twenty-six drinking water systems were still under a "boil water" advisory and 59 were inoperable impacting 100,000 people. Most of the remaining inoperable drinking water systems were not coming on line due to complete destruction in many areas (EPA, 2005d). By October 12, 2006, safe drinking water was restored to all areas of New Orleans including the Lower Ninth Ward. However, many neighborhoods went without drinking water for more than a year due to unprecedented damage including a barge that blocked repairs to a major water line (Picayune, 2006).

15.4.2.3 Recovery Cost Estimates

Hurricane Katrina resulted in damage and destruction to local water supplies in Mississippi and Louisiana affecting millions of people. In September 2005, the American Water Works Association (AWWA) estimated that repairs to drinking water infrastructure in the region could cost more than $2.25 billion. Not included in this assessment were costs of interim operations, cleanup, and other non-capital expenses. AWWA based its cost estimate on information from the EPA, state rural water associations, and the Federal Emergency Management Agency (Hedding, 2005).

15.4.2.4 Multi-disciplinary Emergency Response

The early emergency response teams were later joined by a multi-disciplinary group of more than 1000 EPA specialists to work on post-Hurricane Katrina and Rita activities (Zucchino, 2005). EPA evaluated floodwater sediments, well-water safety, vegetative debris burning, air quality, and other ecosystem restoration activities (EPA, 2005e). EPA compiled a list of Katrina contaminants of concern that included *E. coli*, cholera, hepatitis A/B, tetanus, mold, lead, petroleum products, and pesticides (EPA, 2005f). EPA facilitated the removal and management of millions of cubic yards of debris over a 90,000 square mile area (enough debris to fill the New Orleans Superdome eight times). EPA collected and properly disposed of more than 3.2 million unsecured or abandoned containers of potentially hazardous waste, more than 439,000 electronic goods, and over 360,000 large appliances (Zucchino, 2005). To respond to future natural disasters, EPA designated air and water sampling and analysis, risk assessment, monitoring, restoration of infrastructure, options for management of debris and sediments, management approaches for toxic materials, ecological issues, and building reentry as key aspects of recovery efforts (EPA, 2005g).

15.4.2.5 Recommendations Based on Lessons Learned

After Hurricane Katrina, the American Society of Civil Engineers made recommendations on how to effectively manage and reduce the severity of damage in future natural disasters (Howell, 2006). Findings applicable to public water systems included the following:

- Assign the responsibility of emergency management to a single individual
- Implement more effective mechanisms for cooperation between maintenance crews and system designers to upgrade inspections and repairs for improved water system operation

- Upgrade engineering design procedures to place greater emphasis on safety
- Engage independent experts in high-level reviews of critical water and wastewater infrastructure systems

15.5 EPA Research on Portable and Mobile Water Treatment Systems

After natural disasters, the persistence of microbial and chemical contaminants in drinking water distribution systems is a well-documented environmental concern. Loss of electrical power results in loss of pressure in water distribution systems. Without water pressure, contaminated floodwater and wastewater can enter damaged water mains. The potential for cross-contamination increases during repair of water, wastewater, and utility lines in impacted neighborhoods. Contaminated floodwaters can also seep into private and public drinking water wells. There are currently no cost-effective water treatment systems capable of treating every contaminant in drinking water after a natural disaster. Therefore, EPA is studying the individual capabilities of multiple low-cost treatment alternatives and has chosen a multiple barrier approach to protect the consumer.

The US EPA Office of Research and Development has been working with manufacturers of water treatment systems to test, evaluate, and develop commercially available and innovative filtration and disinfection technologies. Studies on the detection and removal of physical, chemical, and microbial contaminants allow emergency responders to fine-tune treatment and analytical techniques and reduce the risk to consumers by providing the best available protection from toxic chemicals and waterborne pathogens in impacted communities and households. Studies on the formation of product-specific byproducts and disinfectant residuals assist emergency responders with the potential impact of residuals after water treatment, storage, and distribution.

EPA is verifying the ability of treatment devices to protect homes, schools, and businesses in devastated areas. Evaluating point-of-entry (whole house) systems and point-of-use (kitchen sink) devices throughout their useful life provides short-term (for emergency response) and long-term (for community recovery) data on treatment capabilities.

15.5.1 Point-of-Use (POU) Devices

POU treatment technologies offer a low-cost option for treatment of drinking water during emergencies. POU devices have become prevalent and provide clean and safe drinking water to individual homes, businesses, and apartment buildings. POU devices are typically easy to install, use, and maintain and can treat a wide variety of physical, chemical, and microbiological contaminants including metals and pesticides. POU technologies can be installed with ease in remote and devastated areas. These devices are designed to provide a final barrier against contaminated water distribution systems and reduce the risk of waterborne disease outbreaks. EPA has conducted short-term studies on the removal capabilities of water filters using *Bacillus subtilis* as a surrogate for *Cryptosporidium oocysts* (Muhammad et al., 2008). A number of other under-the-sink type POU devices employing combinations of membrane and carbon filters have been evaluated under microbial challenges including the bacteria *B. diminuta* and *H. pseudoflava*, and the coliphage viruses fr, MS2, and Phi X 174. A number of organic and inorganic chemical challenges were also conducted (Adams et al., 2008). The POU/POE devices evaluated in these studies showed varying capabilities for the removal of contaminants in water. Some devices showed significant contaminant reductions, but even the best performing technologies had some units from different production lots that showed microbial challenge organisms in their effluents. The POU RO components alone were not absolute microbial and chemical barriers in these studies. EPA is also conducting long-term studies on POU devices such as under-the-sink activated carbon and reverse osmosis systems as shown in Fig. 15.3.

15.5.2 Point-of-Entry (POE) Devices

The American public has become accustomed to water treatment devices in their homes and businesses for removal of drinking water contaminants at the tap. After natural disasters, contaminants from compromised drinking water supplies can enter the body via inhalation and adsorption through the skin during household activities

Fig. 15.3 EPA point-of-use/point-of-entry research area

such as showering or taking a bath. POE treatment technologies provide added protection from contaminants in drinking water during all household and business activities. The US EPA Environmental Technology Verification (ETV) Program and NSF International (NSF) have verified the capabilities of several POE systems employing combinations of RO membranes, UV-ozone simultaneous oxidation process, and carbon adsorption for removal of microbial and organic and inorganic chemical contaminants in drinking water (EPA/ETV, 2004, 2006, 2007).

15.5.3 Pour-Through Devices

EPA is evaluating household microbiological purifiers for use in emergencies. Pour-through devices are challenged throughout their useful life including clean (new), partially clean (50% of life), and dirty (100% of life). EPA is imitating the use of pour-through devices in households and is evaluating their ability to remove arsenic, protozoa, bacteria, and viruses from drinking water supplies. For instance, EPA studied the capability of numerous pour-through devices to remove arsenic from tube wells in India and Bangladesh for the Grainger Challenge (NAE, 2007). EPA is also conducting studies on removal of microbial pathogens using a four-stage pour-through device for household tap water.

15.5.4 Mobile Treatment Systems

The EPA/ETV program has been collaborating on the development of treatment technologies capable of providing thousands of gallons per day of safe drinking water during emergency situations. These mobile water treatment systems are designed to create potable water from water of unknown quality and can be set up to support hospitals, fire stations, police stations, or other critical infrastructure (EPA/ETV, 2007). In light of emergency response needs from natural disasters and contaminated water distribution systems, robust, easy to operate, and effective multiple barrier treatment trains are necessary and essential.

ETV and NSF studied the capabilities of a Mobile Emergency Filtration System (MEFS) to decontaminate water distribution systems. The MEFS treatment train is comprised of dechlorination, a centrifuge for solids removal, media filtration with sand and activated carbon, ultrafiltration, and reverse osmosis treatment technologies (EPA/ETV, 2004).

EPA, NSF, the Department of Defense (DoD), and the Bureau of Reclamation are currently testing the DoD's Expeditionary Unit Water Purifier (EUWP). The EUWP was developed to treat challenging water sources with variable turbidity, chemical contamination, and very high total dissolved solids (TDS) including seawater during emergency situations when other water treatment facilities are incapacitated as shown in Fig. 15.4. The EUWP components include feed pumps, a UF membrane system, a one or two pass RO desalination system with an energy recovery device,

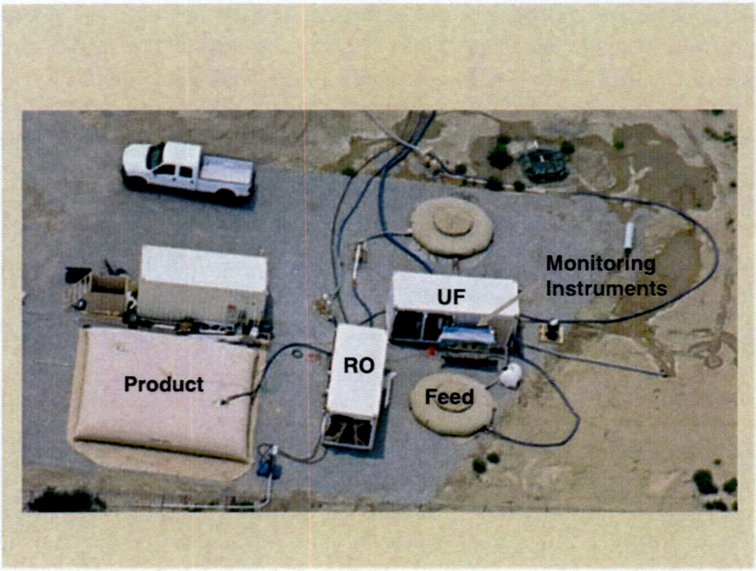

Fig. 15.4 EUWP mobile treatment system field deployment

storage tanks, and product pumps. It has chemical feed systems for optional pretreatment coagulation and posttreatment chlorination. Clean-in-place systems (CIP) are included with the UF and RO skids. Several pilot-scale challenge studies and full-scale field deployment verifications are being conducted to evaluate treatment capabilities for the removal of microbials, particulates, and various organic and inorganic chemicals at low and high concentrations (EPA/NHSRC, 2008; EPA/ETV, 2009a, b).

15.5.5 Disinfectant Research

EPA is currently conducting studies on the disinfectant capabilities of chlorine, chloramine, chlorine dioxide, peracetic acid, UV, ozone, hydrogen peroxide, and simultaneous multiple disinfectants. The research objective is to determine the capabilities of these disinfectants to inactivate protozoa (*Cryptosporidium oocysts, Giardia lamblia*), bacteria (*E. coli, B. subtilis*), and viruses (MS-2) and to investigate the formation of disinfection byproduct residuals. EPA is conducting advanced oxidation process (AOP) research (combinations of UV, ozone, and hydrogen peroxide) to evaluate the degradation of organic contaminants in ground water supplies. EPA is also conducting studies on the capabilities of powdered disinfectants that can be stored in bulk or in packets to be readily available for emergency response teams after natural disasters. Two powders are combined to form chlorine dioxide for inactivation of microbial pathogens.

15.6 Summary

Natural disasters can cause damage and destruction to local water supplies affecting millions of people and can place a tremendous burden on local, state, and national resources. The US Government has taken steps to develop and implement a Disaster Recovery Plan that will facilitate coordination during emergencies. Water industry stakeholders are also collaborating to provide utility access to mutual aid and assistance during emergencies.

Strategies are in place to effectively manage the daunting tasks required to protect consumers from contaminated drinking water supplies and rectify damage and destruction to public water systems after both large and small catastrophes. Prior to an event, utilities and municipalities can use "What if?" scenarios to develop emergency operation, response, and recovery plans that are protective of public safety, health, and welfare and that are designed to reduce the severity of damage and destruction. Cities and small towns are conducting mock exercises to prepare for and minimize liability from location-specific natural disasters.

Government agencies including the EPA are planning ahead to provide temporary supplies of potable water to communities during emergencies. EPA is supporting the development of small drinking water treatment technologies to bring timely relief to devastated communities.

15.7 Notice

The US Environmental Protection Agency, through its Office of Research and Development, funded and managed, or partially funded and collaborated in, the research described herein. It has been subjected to the Agency's peer and administrative review and has been approved for external publication. Any opinions expressed in this chapter are those of the author(s) and do not necessarily reflect the views of the Agency; therefore, no official endorsement should be inferred. Any mention of trade names or commercial products does not constitute endorsement or recommendation for use.

References

Adams, J., Blumenstein, M., and Bartley, B. (2008). *The Reduction of Microbial and Chemical Contaminants with Selected POU/POE Systems*. EWRI World Environmental & Water Resources Congress, Honolulu, Hawaii.

AWWA (2007). Utilities Helping Utilities Workshop Fact Sheet.

AWWA (2009). WARN Leaders Offer Lessons Learned.

Bohman, L. (2005). Climate of 2005 Summary of Hurricane Katrina Overview of Hurricane Katrina, August 24–31, 2005.

Brown, D. (2005). After Storm City's Heart Pumps Again, A21.

EPA, US (2005a). Safety Message Hurricane Katrina, EPA Region 6, September 10, 2005.

EPA, US (2005b). Public Water Supply Assessment Form and Database for Drinking Water Hurricane Katrina Response, EPA Region 6, September 1, 2005.

EPA, US (2005c). Hurricane Katrina System Status Codes, EPA Region 6, September 11, 2005.

EPA, US (2005d). Operational Status of Drinking Water System, EPA Region 6, November 9, 2005.

EPA, US (2005e). EPA Drinking Water Specialists Aid New Orleans Damage Assessment, EPA ORD NRMRL News, November 20, 2005.

EPA, US (2005f). ORD Draft Katrina Needs Hurricane Katrina Recovery Estimates, EPA ORD, September 16, 2005.

EPA, US (2005g). Recommended ORD Post-2005 Hurricane Activities, EPA ORD NRMRL, December 12, 2005.

EPA, US (2007). Interstate Mutual Aid and Assistance: EMAC Tips for the Water Sector.

EPA/ETV, US (2004). Watts Premier M-Series M-15,000 Reverse Osmosis Treatment System.

EPA/ETV, US (2006). Removal of Chemical and Microbial Contaminants in Drinking Water, Watts Premier, Inc., M-2400 Point-of-Entry Reverse Osmosis Drinking Water Treatment System.

EPA/ETV, US (2007). Removal of Synthetic Organic Chemical Contaminants in Drinking Water, RASco, Inc., Advanced Simultaneous Oxidation Process.

EPA/ETV, US (2009a). Removal of Microbial Contaminants in Drinking Water, Koch Membrane Systems, Inc. Targa® 10-48-35-PMC™ Ultrafiltration Membrane as Used in the Village Marine Tec. Expeditionary Unit Water Purifier.

EPA/ETV, US (2009b). Removal of Inorganic, Microbial, and Particulate Contaminants from a Fresh Surface Water, Village Marine Tec. Expeditionary Unit Water Purifier, Generation 1.

EPA/NHSRC, US (2008). Reverse Osmosis Membranes of the Expeditionary Unit Water Purifier.

FEMA, E. (2007). Training Courses: IS-100 – Incident Command System (ICS) and IS-200 – ICS for Single Resources and Initial Action Incidents.

Hedding, K. (2005). Katrina's Damage to Water Systems will top $2.25 Billion. American Water Works Association U.S. Newswire.

Howell, J. (2006). Lessons from Hurricane Katrina. American Society of Civil Engineers Pressroom News.

Johnson, L. (2009). Developing a National Disaster Recovery Framework. Congressional Hazards Caucus Alliance Briefing, Laurie Johnson Consulting.

Muhammad, N., Sinha, R., Krishnan, R., Piao, H., Patterson, C., Cotruvo, J., Cumberland, S., Nero, V., and Delandra, D. (2008). "Evaluating Surrogates for Cryptosporidum Removal in Point of Use Systems." *American Water Works Association* **100**(12): 98–107.

Murphy, M. (1994). "Weathering the Storm: Water Systems Versus Hurricanes." *American Water Works Association* **86**(1): 74–83.

NAE (2007). National Academy of Engineering Announces Winners of $1 Million Challenge to Provide Safe Drinking Water.

Patterson, C., Impellitteri, C., Fox, K., Haught, R., Meckes, M., and Blannon, J. (2007). *Emergency Response for Public Water Supplies after Hurricane Katrina*. EWRI World Water Congress, Tampa Bay, Florida.

Picayune, T. (2006). All of New Orleans Has Safe Drinking Water. Times Picayune.

Shimoda, T.A. (1994). "Emergency Preparedness and Response." *American Water Works Association* 84–92.

Sturdivan, G., Jones, L., and Sellers, S. (2008). The Great Southern California Shake Out. Earthquake Country Alliance, East Valley Water District.

Zucchino, D. (2005). E. Coli Lead Taint New Orleans Flood Waters. Wires Associated Press.

Zucchino, D. (2006). Unfinished Recovery. Los Angeles Times.

Chapter 16
Cyber Security: Protecting Water and Wastewater Infrastructure

Srinivas Panguluri, William Phillips, and Patrick Ellis

16.1 Introduction

The President's Commission on Critical Infrastructure Protection report titled *Critical Foundations* (PCCIP, 1997) was the first report to take a critical look at the vulnerabilities of the nation's critical infrastructure and proposed a strategy for protecting them. This report also identified the water and wastewater supply infrastructure as one of the national critical infrastructures. The report found no evidence of an impending "cyber attack" which could have a debilitating effect on the nation's critical infrastructures. However, the report warned that this finding should not be used as a basis for complacency, as they found widespread cyber capability to exploit infrastructure vulnerabilities. The report warned that the "capability to do harm – particularly through information networks – is real; it is growing at an alarming rate; and we have little defense against it." The report also recognized that in the cyber dimension there are "no boundaries" and the owners and operators of the critical infrastructure such as water and wastewater utilities would need to serve as the "front line" of the security efforts.

Also in 1997, a highly classified internal exercise code named "Eligible Receiver" was initiated by the Department of Defense (DoD) in which a "red team" of hackers from the National Security Agency (NSA) was organized to infiltrate the Pentagon systems. The NSA team was only allowed to use publicly available computer equipment and hacking software. Although many details about Eligible Receiver are still classified, it is known that the red team was able to infiltrate and take control of the Pacific command center computers, as well as power grids, and 911 systems in nine major US cities (PBS, 2004). There are many other well-known hacking incidents that have targeted the military and other critical infrastructure. Shannon and Thomas (2005) discuss two such high-profile incidents named "Moonlight Maze" and "Solar Sunrise":

S. Panguluri (✉)
Shaw Environmental & Infrastructure, Inc., 5050 Section Avenue, Cincinnati, OH 45212, USA
e-mail: srinivas.panguluri@shawgrp.com

R.M. Clark et al. (eds.), *Handbook of Water and Wastewater Systems Protection*, Protecting Critical Infrastructure, DOI 10.1007/978-1-4614-0189-6_16,
© Springer Science+Business Media, LLC 2011

- Moonlight Maze – Beginning in 1998, hackers stole thousands of files containing technical research, contracts, encryption techniques, and unclassified but essential data relating to Pentagon's war planning systems. At the time, this case was the largest intrusion of the DoD network. The subsequent investigations have indicated that the attacks appeared to have originated from the Russian Academy of Sciences.
- Solar Sunrise – Between February 1 and February 24, 1998, a series of DoD unclassified networked computers were attacked. These computers were attacked using a well-known operating system vulnerability. At least 11 attacks followed the same profile on Air Force, Navy, and Marine Corps computers worldwide. The ensuing investigations traced the attacks to two American and one Israeli teenager.

Since the publication of the PCCIP report, many research organizations operating under various mandates have undertaken efforts to understand the complex infrastructure interdependencies especially between water/wastewater infrastructure and the energy infrastructure (electric, oil, and gas) and evaluate mechanisms to protect them. Gillette, et al. (2002) describe four general categories of infrastructure interdependencies (physical, cyber, geographic, and logical) as they apply to the water/wastewater infrastructure. The complex linking of various infrastructures has created new vulnerabilities. In addition, the proliferation of information technology (IT) for organizational efficiency and the increased use of automated monitoring and control systems (e.g., Supervisory Control and Data Acquisition (SCADA) systems) for operational efficiency by the water and wastewater utilities have created additional cyber vulnerabilities that need to be appropriately addressed.

The water supply infrastructure provides water to agriculture, industry (including various manufacturing processes, power generation, and cooling), business, firefighting, and our homes. The wastewater infrastructure is integral to the water supply infrastructure as it treats the wastewater prior to its discharge back to source water streams or lakes to complete the cycle. In fact, one of the most recent and well-documented cyber attack occurred on the wastewater infrastructure at the Maroochy Shire Sewage Treatment Plant in Queensland, Australia. This attack (which occurred between February 9, 2000, and April 23, 2000), resulted in approximately 212,000 gallons of raw sewage to spill out into local parks, rivers, and the grounds of a nearby hotel. The attack was perpetrated by a disgruntled insider, Mr. Vitek Boden, who formerly worked for the Australian firm (Hunter Watertech) that previously installed the radio-controlled SCADA equipment for the plant. During the attack period, Mr. Boden on at least 46 occasions issued unauthorized radio commands to the SCADA system (Abrams and Weiss, 2008).

The recent cyber attack specifically targeting SCADA systems occurred in July 2010 and involved the use of a computer worm named Stuxnet. Various media outlets have suggested that Stuxnet was designed and used to perform a highly complex computer attack mainly targeting Iran's nuclear power infrastructure. In a Senate Homeland Security Committee meeting, Sean McGurk (the head of the Cybersecurity Center at the Department of Homeland Security) described the worm as a "game changer," "incredibly large, and complex code with capabilities never seen before." He went on to describe the worm's specific capabilities as follows (Benson, 2010):

- "This code can automatically enter a system, steal the formula for the product you are manufacturing, alter the ingredients being mixed in your product, and indicate to the operator and your anti-virus software that everything is functioning as expected."

This attack perhaps may be the best example yet to highlight the importance of protecting water and wastewater control systems from malware and other cyber security threats.

Following the attacks of 9/11, on June 12, 2002, the US Environmental Protection Agency (EPA) amended the Safe Drinking Water Act (SDWA) by inserting a new section (Section 1433) that came to be known as the Public Health Security and Bioterrorism Preparedness and Response Act of 2002 (aka Bioterrorism Act, 2002). A key element of this act was to require each community water system (CWS) serving a population greater than 3,300 persons to conduct a vulnerability assessment (VA) of its system to a terrorist attack or other intentional acts intended to substantially disrupt the ability of the system to provide a safe and reliable supply of drinking water. The VAs were the first formal requirement for a utility to identify many vulnerable elements including electronic, computer, or other automated systems that were utilized by the public water system.

Depending upon the identified utility size category, the VAs were due to EPA between March 31, 2003 (those serving populations greater than 100,000), and June 30, 2004 (those serving populations between 3,301and 49,999). Based on the results of the VAs, the utilities were then required to prepare (or revise) an emergency response plan (ERP) within 6 months of completing the VA. However, the subsequent guidance provided by EPA on the development of the VAs or ERPs did not specify any cyber security or IT standard(s) that needed to be met. Therefore, the submitted VAs and ERPs did not uniformly address cyber security requirements. This chapter contains an overview of the sector-specific cyber infrastructure, discusses the current approaches (relevant standards and vendor initiatives) and their key elements, and provides a summary of the recently developed sector-specific cyber security roadmap. Each section is interspersed with examples that document various successes and challenges faced by the water and wastewater sector to meet the requirements of these standards and achieve the goals identified in the sector-specific roadmap.

16.2 Overview of Water and Wastewater Cyber Infrastructure

In order to defend the nation's water and wastewater infrastructure from a cyber attack, it is important to gain an overview and understanding of a typical utility's cyber infrastructure. The following sections define the commonly used cyber infrastructure terminology, connectivity of various infrastructure components, and integration/management of business and SCADA networks.

16.2.1 Cyber Infrastructure Terminology

In this chapter the SCADA acronym is being used to describe utility-wide control systems because SCADA is currently the most common terminology used in the water/wastewater sector. However, a wide range of other acronyms are also used. The term SCADA originally was coined to describe systems that provided central supervisory control and data acquisition for distributed facilities such as water tanks, water pump stations, and wastewater lift stations. Other terms such as Process Control Systems, Distributed Control Systems, Instrumentation and Control Systems are used to describe controls systems for attended facilities such as a treatment plant or other industrial/manufacturing facilities that are typically not geographically spread out.

Likewise, the two major SCADA security standards projects National Institute of Standards and Technology (NIST) Federal Information Management Security Act (FIMSA) Implementation Project and the International Society of Automation (ISA) Specification Project 99 (SP99) use different terms. NIST uses Industrial Control Systems (ICS) in its Special Publication (SP) 800-82. ISA's SP99 uses both Manufacturing and Control Systems and Industrial Automation and Control Systems.

The other commonly used cyber infrastructure terms are defined as follows:

- IT Infrastructure – interconnects and controls the flow of information across computer networks. IT infrastructure includes network routers, switches, gateways, wireless access points, radios, public communications networks, and all types of cable media. IT infrastructure does not include the servers and computers that it interconnects. IT infrastructure can be limited to a single Local Area Network (LAN) serving a single networked application such as SCADA, or it can encompass a Metropolitan Area Network (MAN) spanning a Metropolitan Area and serving an entire municipality or all utility networked applications including SCADA, or it can encompass a Wide Area Network (WAN) spanning a larger geographic area such as those used by large corporations with offices located in multiple cities.
- Business Network – the collection of interconnected devices designed to communicate and share data related to common business activities such as administration of personnel and accounting. Depending on organizational characteristics, this network can reside totally within the water/wastewater utility or be split between the utility and some other governing body such as a County or City IT department.
- SCADA Network – consists of interconnected process control devices and computers that collect, monitor, and adjust process control equipment over a LAN,

MAN, or WAN using wired and/or wireless communications. SCADA network allows operators to monitor and control process units from a central and/or remote location. SCADA networks are often not connected to the business network.

16.2.2 Cyber Infrastructure Connectivity

Historically, business and SCADA networks were separate because the network topologies were vastly different. Even if a utility owner recognized the value of integrating SCADA data into their strategic decision support systems, they could not because of limitation in the network topologies. SCADA systems relied heavily on serial connectivity and very low frequency radio communications that could provide enhanced range and partial line-of-sight connectivity, none of which supported standard IP connectivity desired by business networks. This situation led to the formation of various "islands of automation" that did not share data or resources between them (see Fig. 16.1). This virtual isolation also led to a false sense of security by many SCADA system administrators because they believed that these systems were unconnected to the outside world and thus could not be "hacked" into.

Today, the SCADA and business networks of most medium- to large-scale water/wastewater utilities are connected to provide more integrated operation and to eliminate the islands of automation. The choice of network and communications hardware depends upon the interconnectivity and interoperability requirements of the various software programs that fulfill the utility's business and operations functionality. Typical utility business software systems include financial system, e-mail

Fig. 16.1 SCADA and business networks – "islands of automation" (Panguluri and Scott, 2010)

system, human resource management system, laboratory information management system (LIMS), geographic information system (GIS), and computerized maintenance management systems (CMMS). These systems are not typically located on the same network segment as the SCADA system. The financial, e-mail, and human resources systems are often under the control of a centralized IT group that is not directly part of the utility operation and are normally operated during normal business hours. Other systems such as the LIMS, GIS, and CMMS systems are run by utility staff in cooperation with the central IT group or the vendor that manufactured the software. Lastly, the SCADA system is typically run by utility operations staff in cooperation with the software manufacturer or system integrator. SCADA systems and some utility applications are operated on a 24 h, 7 day per week schedule. Figure 16.2 shows a typical medium-to-large municipal government-owned water and wastewater utility's IT infrastructure with interconnected SCADA and business networks.

Figure 16.2 represents a water and wastewater utility that operates two plants where the business and SCADA networks are connected by routers. The utility's cyber infrastructure includes the following:

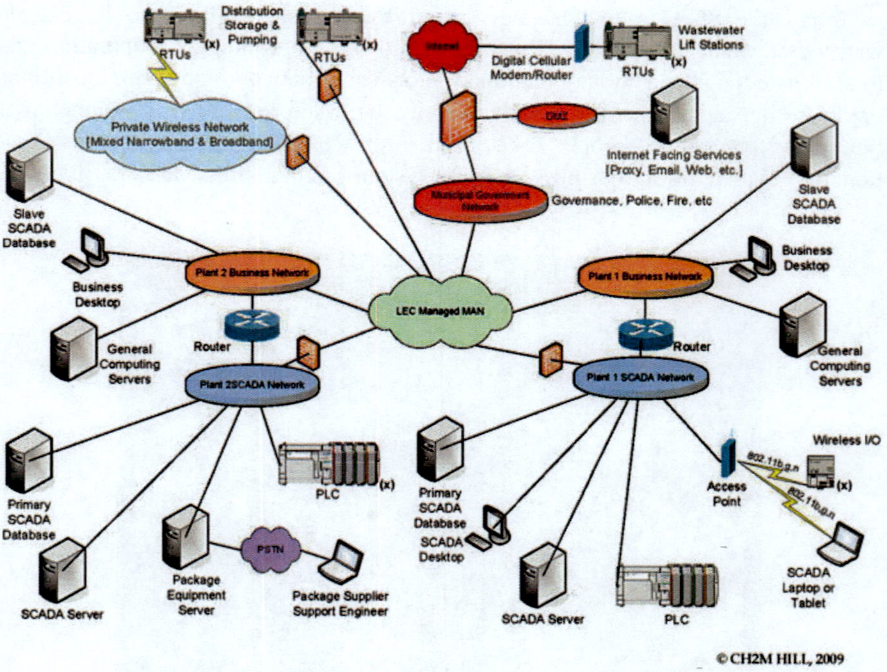

Fig. 16.2 Typical water/wastewater utility's business and SCADA infrastructure and network connectivity (Phillips, 2009a)

- A Local Exchange Carrier (LEC) managed wireline MAN connecting plants to the municipal government network and the utilities' private wireless network. For simplicity, the LEC-provided routers at MAN connections are not shown in the figure.
- A digital cellular-based lift station monitoring network.
- SCADA historical databases mirror servers on the business networks to provide the business network with SCADA status data.

This utility network has a number of clearly visible vulnerabilities. Some, but not all, of these vulnerabilities will be used for discussion later in the chapter (see Section 16.5.2 for some suggested security improvements).

16.2.3 Evolution and Integration of Business and SCADA Networks

According to Moore's law (Moore, 1965), computing power doubles just about every year. In response to these rapid changes, personal computer (PC) hardware operating life cycles are typically 3–5 years before equipment is sent to surplus. In contrast, SCADA technology evolves at a much slower pace, causing most utility system SCADA infrastructures to become obsolete after the 5 year mark when most standard PC equipment gets replaced. For many years, this difference in equipment life cycles was an unforeseen tactical advantage for SCADA systems because most hackers sought their targets through the Internet or via open or unsecured wireless connections. For a short period of time, the serial world of SCADA was enjoying security through obscurity. However, this is quickly changing as newer equipment and software applications began using routable protocols capable of working over TCP/IP connections. Many of today's SCADA systems are also capable of integration with other utility-specific applications such as CMMS and LIMS. This integration requires pushing data to a historian or data warehouse server that is located on the business network.

Increasingly the business case for integrating IT and SCADA systems is becoming hard to ignore. The benefits of SCADA/IT integration include the following:

- Shared Infrastructure – Business and SCADA systems in some cases share MAN or WAN infrastructure to reduce the overall costs for leased or private lines. Proper segmentation of traffic on shared infrastructure can reduce the potential for security breaches.
- Shared Expertise – Common architecture components such as network, database, and security can be managed by trained experts.
- Cheaper Components – SCADA systems can use cheaper transmission control protocol/internet protocol (TCP/IP)-based components.

- Strategic Information Gains – Data for energy management, increased modeling capabilities with connections to LIMS/GIS databases, real-time water quality modeling, forecasting capability, management/regulatory reporting, and providing utility facility status information to Emergency Response Centers.
- Improved Overall Security Integration – Integration of physical security elements such as video monitoring with SCADA allows for 24/7 monitoring by SCADA operators.

These benefits are becoming hard to ignore and now there is a general push in the industry to eliminate the "islands of automation."

16.2.4 Managing Combined Business and SCADA Networks

For some utilities, the business push to integrate business and SCADA networks has led to "turf wars" resulting in greater frustration and a reluctance to realize the true benefits of integrating these two networks. This integration cannot be accomplished without help from the central IT department who, in most cases, do not interface with the SCADA staff on a regular basis. These two groups look at the systems life cycle and availability models in very different ways. Some of the key differences include the following:

- Operational and Reliability Requirements – IT systems handle highly varying volumes of data during regular business hours that typically last between 8 and 12 h, 5 days a week. The SCADA systems generate comparably low volumes of constant data, but operate on a 24×7 basis. These key differences result in the IT systems requiring greater bandwidth and the SCADA systems requiring greater reliability.
- Lifecycle – Network and computing equipment, whether used on a business or SCADA network, typically has a 3–5 year life cycle. This is an acceptable time-frame for business networks as technology advances usually require upgrading and replacement of that equipment. However, on a SCADA network many of the perimeter devices like remote terminal units (RTUs), controllers, input/output devices, and radios typically have an 8+ year life cycle. This makes keeping the network, server, and workstation equipment up-to-date difficult for IT staff since changes could cause an adverse affect on the SCADA system.
- Personnel – IT personnel have greater turnover as their skills are more universally applicable and they have more flexibility in changing jobs. Also, as indicated previously, the software and hardware evolution in IT field is much quicker, requiring greater need for IT personnel to continuously update their skills. Therefore, IT personnel typically possess more certifications that are better understood by business managers and their higher market demand results in higher pay for key IT individuals. Comparatively, the SCADA personnel are generally homegrown staff that learn the IT aspects of SCADA out of necessity. Starting out they may command lower pay as their skills generally tend to have

limited applicability in the immediate outside market. This disparity in pay can lead to some animosity between the IT and SCADA personnel. In some cases, after the SCADA personnel receive specialized training paid for by the utility, they leave for better paying jobs outside the immediate market area or consult with the SCADA integrator/vendor.

Successful management of IT and SCADA infrastructure requires an in-depth understating of these key differences. Recognizing the common structural components (database, server, network), bridging the gap between IT and SCADA personnel, and employing component "experts" to solve problems generally lead to overall system improvement.

16.3 Cyber Infrastructure – Threats, Vulnerabilities, and Attacks

The ever-increasing cyber infrastructure connectivity and the standardization of infrastructure equipment inherently make a utility more vulnerable to cyber attacks. And while the motives of an individual or a group of attacker(s) may vary, the attack tools used and the attack methodology can be very similar. Also, depending upon how well a utility's cyber infrastructure is protected, the outcomes may be very different. The following sections take a closer look at these cyber threats, vulnerabilities, and outcomes. The outcome statistics presented are based on previously documented cyber attacks on critical infrastructure.

16.3.1 Cyber Threats and Vulnerabilities

Cyber threats refer to unauthorized attempt(s) to access infrastructure components (typically SCADA system devices and/or network) using data communication pathways. This unauthorized access can come from a trusted user within the network or from remote locations by unknown persons using the Internet or other remote access mechanisms. Threats to the utility infrastructure can come from numerous sources, including hostile governments, terrorist groups, disgruntled employees, and other malicious intruders. The NIST Special Publication 800-82 (NIST, 2008) specifically lists the well-known sources of cyber threats:

- Hackers – Hackers (aka crackers) break into networks for the thrill of the challenge or for bragging rights in the hacker community. Amateur "crackers" can now download software tools from the Internet that contain attack scripts and protocols that can be easily launched against victim sites. Also, the increase in security measures and IT in general has resulted in the development of more sophisticated attack tools.
- Bot-network operators – They are crackers that take over multiple systems in order to coordinate attacks and to distribute "phishing" schemes, spam, and

malware attacks for profit. The services of these bot-nets are sometimes made available in underground markets (e.g., purchasing a denial-of-service attack, servers to relay spam, or phishing attacks).

- Organized criminal groups (phishers, spammers, and spyware/malware authors) – These groups attack systems for monetary gain. Specifically, they use spam, phishing, and/or spyware/malware to commit identity theft and/or online fraud. Phishing schemes attempt to steal identities, personal information, or trick people to send their money. Spammers distribute unsolicited e-mail with hidden or false information in order to sell products, conduct phishing schemes, distribute spyware/malware, or attack organizations (e.g., denial of service).
- Foreign intelligence services – They use cyber tools as part of their information gathering and espionage activities. Several nations are aggressively working to develop information warfare doctrine, programs, and capabilities, specifically, capabilities that can enable them to seriously damage critical infrastructure.
- Insiders – Historically, a disgruntled insider is a principal source of computer crime. Their knowledge of a target system often allows them to gain unrestricted access to cause damage to the system or to steal system data. The insider threat also includes outsourcing vendors as well as employees who accidentally introduce malware into systems.
- Terrorists – They seek to destroy, incapacitate, or exploit critical infrastructures in order to threaten national security, cause mass casualties, weaken the economy, and/or damage public morale.

Utilities should focus more on "how" their system can be attacked and not "why." To do this, utilities should identify their weakest links that can lead to unauthorized access (e.g., wired or wireless access points, Internet browsing and connectivity, and Universal Serial Bus (USB)-based malware). For utilities where the business and SCADA systems are connected, access can be gained through the business network via the Internet as shown in Fig. 16.3.

Figure 16.3 shows a portion of the network (previously shown in Fig. 16.2) and illustrates a well-publicized hypothetical SCADA attack. The illustrated attack exploits a well-known software vulnerability. As in most real-world publicized cases, a patch for this vulnerability was readily available and simply installing the patch would have prevented the attack from taking place.

The numbered callouts in Fig 16.3 explain each step in the attack. To show relevance to the water sector from this well-known vulnerability, a successful attack depicted in Fig. 16.3 can result in a dangerously high chemical level in the water distribution system.

16.3.2 Cyber Attack Vulnerability Statistics

Between 2004 and 2008, the US Department of Homeland Security (DHS) National Cyber Security Division's Control Systems Security Program performed

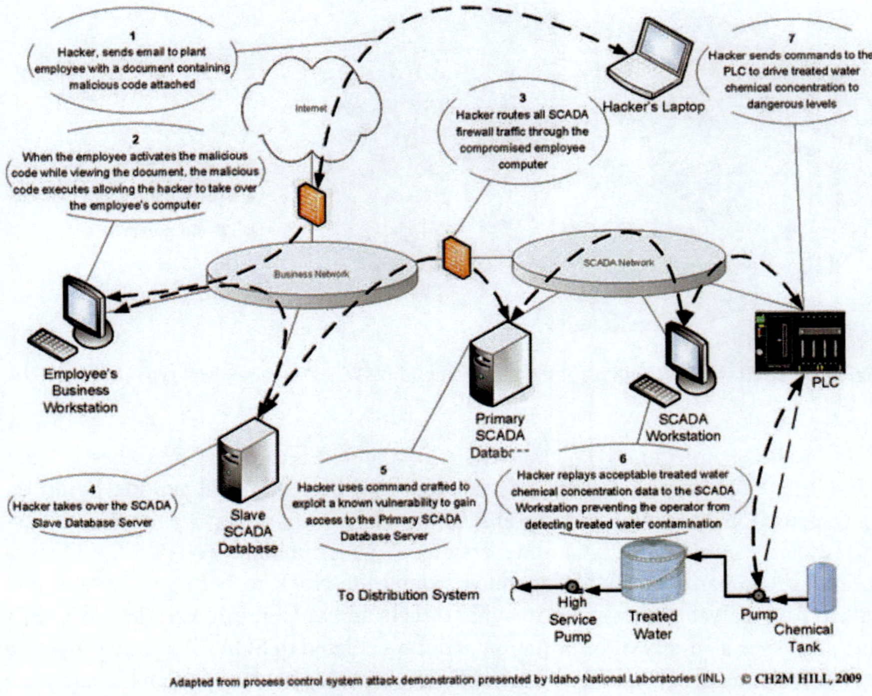

Fig. 16.3 SCADA network attack scenario (Phillips, 2009b)

assessments of 15 SCADA systems and identified 245 vulnerabilities that could put the critical systems at risk from a cyber attack (DHS, 2009). The assessments were designed to test vendor-specific products and services such as custom protocols, field equipment, applications, and services. Onsite system assessments generally assessed how securely external connections, firewall configurations, intrusion detection systems, network architecture, and other components are deployed and installed. Figures 16.4 and 16.5 present a summary of the findings.

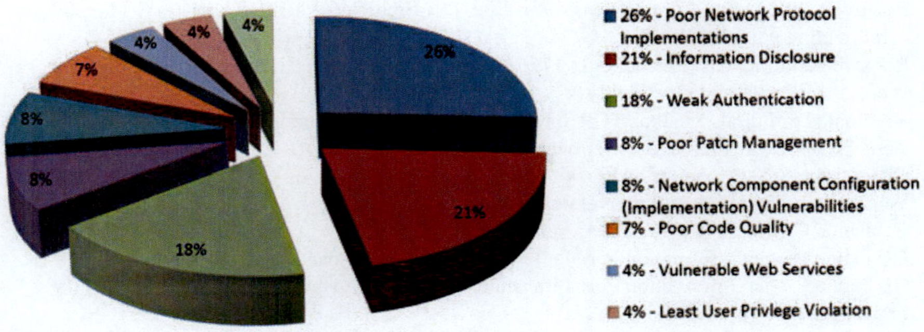

Fig. 16.4 SCADA vulnerabilities assessment findings (DHS, 2009)

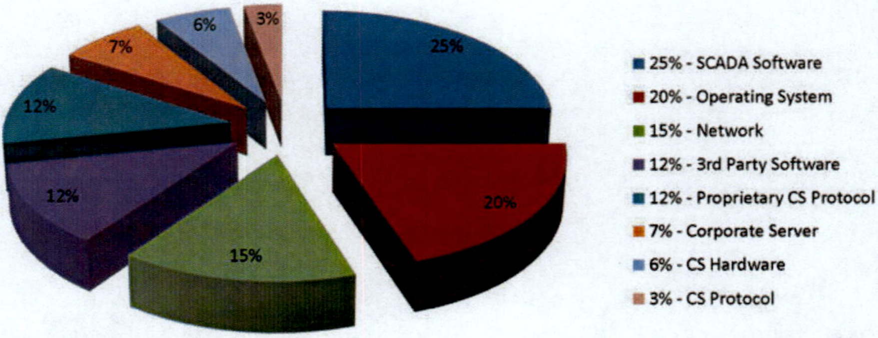

Fig. 16.5 SCADA vulnerabilities assessment findings by SCADA component type (DHS, 2009)

In addition to the DHS, the US Computer Emergency Readiness Team (US-CERT) publishes a variety of technical and non-technical reports based on its information collection efforts and interactions with federal agencies, industry, research community, and state and local governments. The US-CERT is a public/private partnership that operates around-the-clock to help government and industry to analyze and respond to cyber threats and vulnerabilities. The US-CERT also publishes a variety of information and tools related to SCADA systems that are available online at no cost. The vulnerabilities related to control systems security are updated periodically. Table 16.1 presents a summary of control system vulnerability notes extracted from the US-CERT web site (US-CERT, 2009).

In addition to DHS and the US-CERT, there are other organizations such as the SysAdmin, Audit, Network, Security (SANS) Institute, and the Security Incidents

Table 16.1 SCADA vulnerability notes

US-CERT control system vulnerability notes	Date notified
AREVA e-terrahabitat SCADA systems vulnerabilities	Feb 09
GE Fanuc Proficy HMI/SCADA iFIX uses insecure authentication techniques	Feb 09
GoAhead Webserver information disclosure vulnerability	Feb 09
Rockwell Automation ControlLogix 1756-ENBT/A EtherNet/IP Bridge URL redirection vulnerability	Feb 09
Rockwell Automation ControlLogix 1756-ENBT/A EtherNet/IP Bridge cross-site scripting vulnerability	Feb 09
Automated Solutions Modbus TCP Slave ActiveX Control vulnerability	Jan 09
ABB PCU400 vulnerable to buffer overflow	Sep 08
Citect CitectSCADA buffer overflow	Jun 08
Wonderware SuiteLink null pointer dereference	May 08
GE Fanuc CIMPLICITY HMI heap buffer overflow	Jan 08
GE Fanuc Proficy Information Portal allows arbitrary file upload and execution	Jan 08
GE Fanuc Proficy Information Portal transmits authentication credentials in plain text	Jan 08

US-CERT (2009)

Organization™ that maintain and update cyber vulnerability and incident-specific information. SANS develops, maintains, and makes available at no cost the largest collection of research documents about various aspects of information security, and it operates the Internet's early warning system – Internet Storm Center.

16.3.3 Cyber Attacks on Industrial Infrastructure and Outcomes

The Security Incidents Organization™, a non-profit corporation, maintains a Repository of Industrial Security Incidents (RISI). RISI focuses strictly on the industrial automation community. Their database includes incidents that are voluntarily reported by the user community. RISI database also includes the industrial security incident data previously collected under a research project by the British Columbia Institute of Technology (BCIT). The information presented in this section is entirely based on RISI's most recent quarterly report on cyber security incidents and trends affecting industrial control systems (RISI, 2009).

The number of reported security incidents related to industrial automation is increasing worldwide and over 60% of the confirmed or likely incidents in the RISI database have occurred in North America. Figure 16.6 shows the number of reported industrial security incidents (related to industrial automation) occurring in the USA.

It should be noted that the data collection and entry into the RISI database was partially suspended between 2006 and 2008 (which may be reflected on the trend graph presented in Fig. 16.6). Figure 16.7 shows a summary of these incidents in the USA by industry sector.

The water and wastewater sector is number 4 in this list behind power, petroleum, and transportation sectors. Figure 16.8 shows a summary of global security incidents based on the documented point of entry of the attacker.

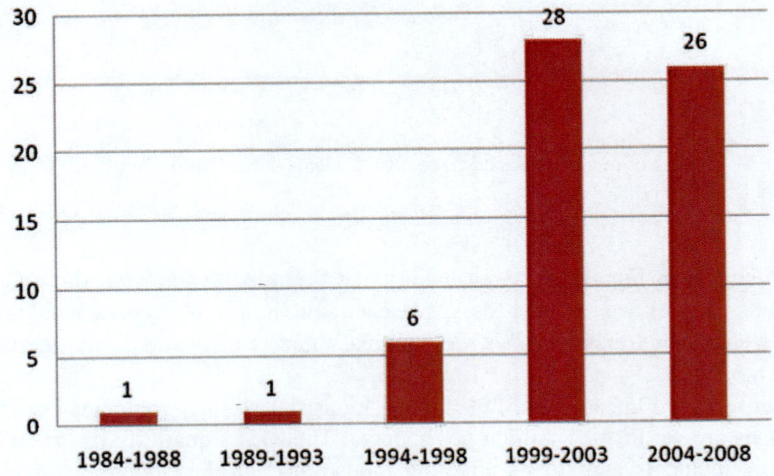

Fig. 16.6 Number of reported industrial security incidents in the USA – summarized in 5-year intervals (RISI, 2009)

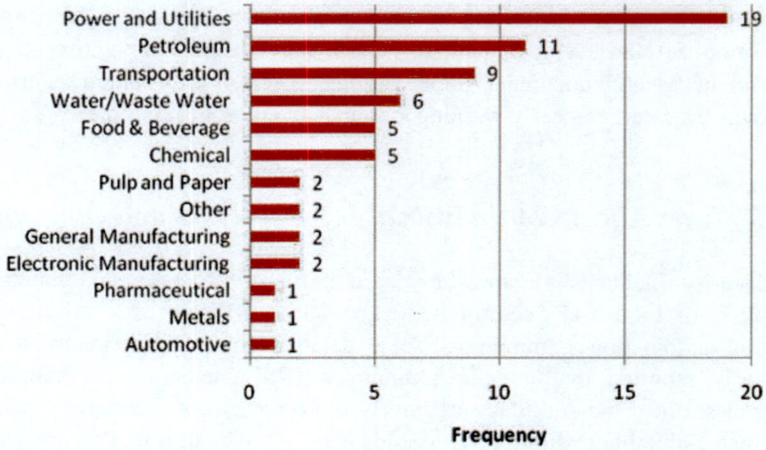

Fig. 16.7 Number of reported industrial security incidents in the USA – summarized by industry types (RISI, 2009)

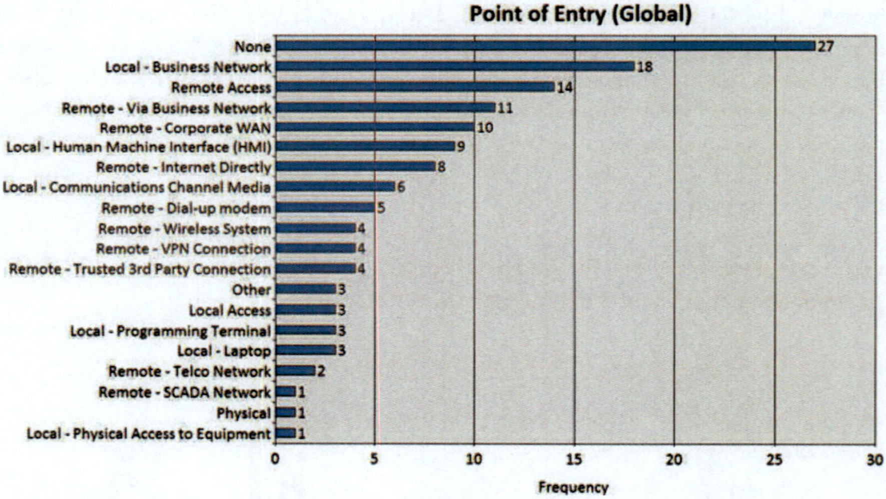

Fig. 16.8 Number of reported industrial security incidents in the world – summarized by points of entry (RISI, 2009)

As shown in Fig. 16.8, a vast majority of these incidents occur through local business network via remote access. The economic impact of security incidents can vary widely. Figure 16.9 shows the financial impacts of incidents summarized in Fig. 16.6.

Figure 16.9 indicates that 29% of the industrial security incidents resulted in damages greater than $1 million per incident. The RISI's quarterly report on cyber security incidents and trends affecting industrial control systems contains additional valuable information such as incident perpetrator, incident detection method,

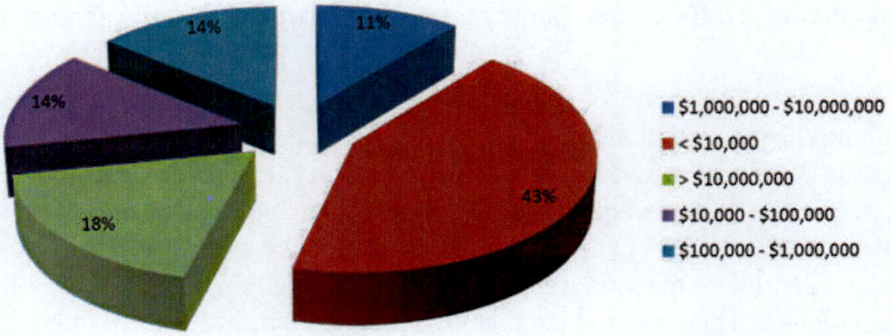

Fig. 16.9 Financial impact of industrial security incidents reported in the USA (RISI, 2009)

general access method, and equipment involved. Each quarterly report includes a detailed summary of selected security incidents. A copy of the full report can be obtained from http://www.securityincidents.org.

16.3.4 Cyber Attack Tools and Scenarios

The availability of sophisticated attack tools has exploded in the past decade and many of these tools are open source and available freely. Over the years the level of technical knowledge required by an attacker to launch a sophisticated attack has decreased significantly. With tools that are freely available on the Internet today, an unsophisticated attacker can launch damaging scripts and malicious code in a fraction of the time it took an attacker with a higher skill set 10 years ago. Fortunately, the same tools that allow hackers to find their way into networks are the tools that utility staff can use to help protect their networks from intruders. Network scanners, password crackers, and wireless sniffers can be used to validate properly secured systems or they can be used to uncover vulnerabilities before an attacker can do the same. Two of the popular and relevant open-source attack tools are discussed here both to serve as examples of what is freely available to potential attackers and what utilities can use to build up their defensive measures:

- Nessus – A vulnerability scanning software that includes tools for configuration auditing, asset profiling, sensitive data discovery, and vulnerability analysis. It also checks for network and configuration vulnerabilities such as open mail relay, missing patches, and use of default password. It can also check for denial-of-service attack against the TCP/IP stack by using mangled packets.
- Kismet – An intrusion detection system for 802.11 wireless networks. It works with any wireless card which supports raw monitoring mode and can sniff 802.11a, 802.11b, and 802.11 g traffic. It works passively without sending any loggable packets to detect the presence of both wireless access points and wireless clients and associates them with each other.

Panguluri et al. (2004) present the following five typical attack scenarios:

1. Scenario #1: An "inside job," a disgruntled staff member or former staff member with privileged information who can be bribed or duped into sabotaging systems and settings or creating access mechanisms for the attacker to gain future access.
2. Scenario #2: Systems equipped with modems or Internet-based remote access points. A "war dialer" or "port scanning tool" can be used to automate an attack and gain access to the system.
3. Scenario #3: This scenario applies to systems equipped with wireless access points with weak encryption technology.
4. Scenario #4: In this scenario a staff member with access to the utility's IT/SCADA systems introduces a virus or other malware by opening an infected e-mail containing a malicious payload. Infected e-mail can come from trusted sources and generally do not require any special action to execute the malware.
5. Scenario #5: A cyber attacker starts flooding the network with useless traffic, thus jamming the links and denying service to users and SCADA devices causing random disruption of service.

With proper training, security, and planning, all of these attack scenarios can be prevented or any resulting damages from successful attacks can be minimized.

16.4 Relevant Standards

In recent years, many organizations have collaborated to develop standards related to cyber security. Overall there are three cross-sector cyber security standards that are relevant to the water and wastewater infrastructure. The most widely recognized worldwide standard is the International Organization for Standardization (ISO)/International Electrotechnical Commission (IEC) Standard 17799. In 2007 it was renumbered as ISO 27002 to better fit within the ISO 27000 Management System. The other two standards are more SCADA system specific, the NIST Special Publication 800-82 and the American National Standards Institute (ANSI)/ISA SP99. In addition to these two standards, for the electric industry, the North American Electric Reliability Corporation (NERC) has developed a Critical Infrastructure Protection (CIP) framework that could also serve as a template for the water and wastewater sector.

16.4.1 ISO/IEC 27002 (Formerly 17799)

The ISO/IEC 27002 was first established as a cyber security standard in 1995; first issued in 2000 (numbered 17799), it was revised in 2005 and renumbered in 2007. The ISO/IEC 27002:2005 is a comprehensive standard that provides general guidelines and principles for initiating, implementing, maintaining, and improving information security management in an organization. The standard outlines objectives, provides general guidance, and contains best practices of control objectives and controls in the following areas of information security management:

- Risk assessment
- Security policy
- Organization of information security
- Asset management
- Human resources security
- Physical and environmental security
- Communications and operations management
- Access control
- Information systems acquisition, development, and maintenance
- Information security incident management
- Business continuity management, and
- Compliance

This standard addresses overall information security management and much of its content is applicable to SCADA systems, but the document does not currently cover all areas of importance from a SCADA perspective.

16.4.2 NIST Publications – Special Publication 800-82

NIST, founded in 1901, is a federal technology agency that develops and promotes measurement standards, and technology. The NIST Special Publication 800 series was established in 1990 to provide a separate identity for IT security publications. In addition, the Federal Information Management Security Act (FIMSA), a US federal law (Public Law 107-347) enacted in 2002, requires federal agencies to develop documents and implement agency-wide programs to provide information security for systems that support the operations and assets of the agency. Standards developed under FISMA establish requirements for federal agencies and also appear to provide excellent guidance that can be applied to the water/wastewater sector.

NIST developed the SP 800-82 guidance to meet its statutory responsibilities under the FISMA and the Homeland Security Presidential Directive 7 (HSPD-7) of 2003. The NIST SP 800-82 complements the NIST SP 800-53's recommendations for security controls for federal IT systems and organizations. The NIST 800-82 is designed to specifically assist in developing and deploying an overall security program for SCADA and distributed control systems (DCS) architecture, supporting devices, such as programmable logic controllers (PLCs), RTU, and intelligent electronic devices (IEDs). The standard document that is freely available includes the following:

- Overview of ICS
- ICS characteristics, threats, and vulnerabilities
- ICS security program development and deployment
- Network architecture
- ICS security controls

Overall, the NIST 800-82 standard emphasizes three key classes of security controls (or countermeasures) defined in earlier standards to mitigate the risk associated with cyber vulnerabilities. These include (1) management controls, (2) operational controls, and (3) technical controls. A brief overview of these control classes is presented below:

- Management Controls – These controls focus on the management of risk and the management of the industrial control system. The main management controls focus around the following areas
  - Risk assessment
  - Developing and implementing a security program
  - System and services acquisition
  - Security assessments

  Management controls must be implemented directly by the water and wastewater utility operating the control system and cannot be included in a control system design project or construction, though significant elements of the work can be outsourced. Still, the management and oversight responsibilities remain with the organization. Also, because these controls are an ongoing responsibility that cannot be funded under capital improvements projects, they must be funded under annual operations budgets and are therefore difficult for utilities from a commitment perspective.
- Operational Controls – These controls are primarily implemented and executed by personnel (utility operators) as opposed to the system (or organization). The main operational controls focus around the following areas:
  - Personnel security
  - Patch and configuration management
  - Checklists
  - Maintenance
  - Network segmentation
  - Incident response and disaster recovery plan
  - Physical and environmental protection
  - Media protection
  - Awareness and training

  Similar to management controls, many of the operational controls must be funded under annual operations budgets for a utility. Some of the focus areas or controls such as media protection, physical and environmental protection, and personnel security can be implemented, but not maintained, under capital projects. For example, providing limited access, environmentally protected computer rooms to house SCADA backup power supplies, network equipment, communications interfaces, servers, and workstations can be included in capital projects and satisfy the aforementioned operational control requirements. Locating workstations in a limited access computer room limits access to workstation media and media ports and requires keyboard/video/mouse/audio extenders to support operator workstations in control rooms.

- Technical Controls – These controls are primarily implemented and executed by the SCADA system through mechanisms contained in the hardware, software, or firmware components of the system. The main technical controls focus around the following areas:

  - User identification, authentication, and authorization
  - Data identification and authentication
  - Device identification, authentication, and authorization
  - Logging and audit
  - Secure communications
  - Access control
  - Intrusion detection and prevention
  - Virus, worm, and malicious code detection

Unlike management and operational control classes, many elements of the technical controls can be implemented, but not maintained, under capital projects. Technical control solutions are the most often associated with security improvements; and it is important to realize that these must be implemented to complement the other control classes to provide the needed "defense in depth" strategy discussed in Section 16.5.

As of September 28, 2010, NIST SP 800-82 standard development process has been completed and the document has undergone the internal review process. An external review is underway and once it is completed the document will be finalized and the final version will be released. [update: The final version of NIST SP 800-82 was released in June 2011]

16.4.3 ISA SP-99

ISA launched Specifications Project (SP) 99 in 2002 to develop control system security standards. The SP99 committee is comprised of over 400 members and 200 companies. The ISA work products are classified into the following four major categories:

1. ISA99 Common (ISA-99.01.xx series)
2. Security Program (ISA-99.02.xx series)
3. Technical Requirements – System Level (ISA-99.03.xx series)
4. Technical Requirements – Component Level (ISA-99.04.xx series)

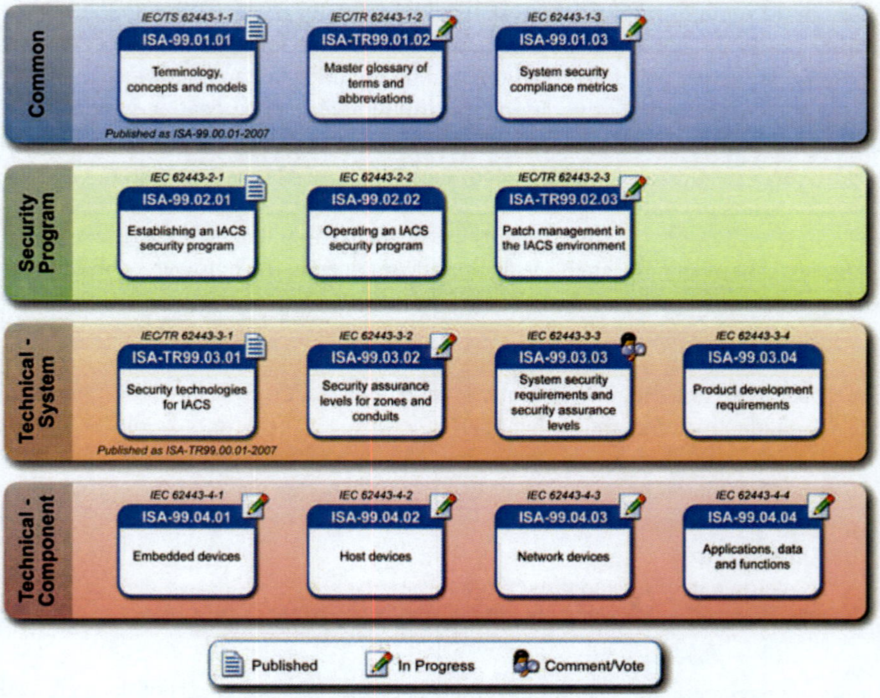

Fig. 16.10 ISA SP99 committee work products (ISA, 2010)

Fig. 16.11 ISASecure designation logo

ISASecure

Figure 16.10 lists the specific SP99 work products under these categories including subjects addressed and status as of October 2009 (ISA, 2010). These standards can be either recommended standards or normative standards depending on the level of acceptance of the standard.

The committee is currently projecting to complete work on the currently planned standards in 2013. ISA is also establishing a Security Compliance Institute and has developed the ISASecure trademark shown in Fig. 16.11 to provide instant recognition of compliant products.

Once products become available with the ISASecure designation, system designers and asset owners will be assured that designated products comply with applicable ISA standards.

16.4.4 NERC CIP

The NERC CIP standards CIP-001 through CIP-009 provide a cyber security framework for the identification and protection of critical cyber assets to support reliable operation of the Bulk Electric System. These standards recognize the differing roles of each entity in the operation of the Bulk Electric System, the criticality and vulnerability of the assets needed to manage Bulk Electric System reliability, and the risks to which they are exposed. Unlike the ISA and NIST standards, the NERC CIP does not include any "how to" content. The security concepts and tools developed under the NIST and ISA standards can be used to achieve NERC compliance. The responsible entities are required to interpret and apply the following standards using reasonable business judgment:

1. CIP-001 – Sabotage Reporting – Requires disturbances or unusual occurrences, suspected or determined to be caused by sabotage, to be reported to the appropriate systems, governmental agencies, and regulatory bodies.
2. CIP-002 – Critical Cyber Asset Identification – Requires the identification (through the application of a risk-based assessment) and documentation of the critical cyber assets associated with the infrastructure that supports the reliable operation of the Bulk Electric System. These critical assets are to be identified.
3. CIP-003 – Security Management Controls – Requires that the responsible entities have minimum security management controls in place to protect critical cyber assets.
4. CIP-004 – Personnel and Training – Requires that personnel having authorized cyber or authorized unescorted physical access to critical cyber assets, including contractors and service vendors, have an appropriate level of personnel risk assessment, training, and security awareness.
5. CIP-005 – Electronic Security Perimeter(s) – Requires the identification and protection of the Electronic Security Perimeter(s) inside which all critical cyber assets reside, as well as all access points on the perimeter.
6. CIP-006 – Physical Security of Critical Cyber Assets – Requires the implementation of a physical security program for the protection of critical cyber assets.
7. CIP-007 – Systems Security Management – Requires responsible entities to define methods, processes, and procedures for securing those systems determined to be critical cyber assets, as well as the non-critical cyber assets within the Electronic Security Perimeter(s).
8. CIP-008 – Incident Reporting and Response Planning – Requires the identification, classification, response, and reporting of cyber security incidents related to critical cyber assets.
9. CIP-009 – Recovery Plans for Critical Cyber Assets – Requires that recovery plan(s) are put in place for critical cyber assets and that these plans follow established business continuity and disaster recovery techniques and practices.

Also, NERC serves as an independent body for the electric industry to ensure compliance with the CIP mandated standards. As stated previously, many of the information system security standards, including standards for SCADA, have been developed and are currently available for use by the water sector. However, there is no independent body (such as NERC) or regulatory agency (such as EPA) for the water and wastewater sector that ensures compliance with a specific (or voluntarily selected) standard. It is completely up to the individual utility to pick and choose which security approaches fit them the best. While this provides great flexibility to the sector, it also leads to the lack of a consistent and acceptable level of security. One approach to providing the needed regulation and enforcement would be to use the NERC CIP model.

16.4.5 International Instrument User's Association (EWE)

EWE comprises of EI, WIB, and EXERA that are instrument users association that collaborate in sponsoring, planning, and organization of instrument evaluation programs. They recently produced the M 2784 – X-10 report (available for download at http://www.wib.nl/download.html) that establishes security requirements for process control domain vendors. This report includes useful requirements that can be used by asset owners in preparing specifications for process control and automation/SCADA systems.

16.5 Implementing Cyber Security – Defense In Depth

By comparison, security and defense concepts have not really changed much since the middle ages. The greatest castle-builders of their time learned the hard way that an enemy could breach even the highest castle walls given enough time and persistence. So, they began building in layers of security that extended beyond the thick walls of the castle itself. These layers were designed to do two basic things, either expose the enemy sooner than they intended or to slow their progress, giving the kingdom more time to muster the troops.

These same concepts are employed in some of the best security plans of our day. By building in layers of protection like e-mail filters, anti-virus, compartmentalization, authentication controls, firewalls, DMZs, intrusion detection, and more, one can slow down an attacker's attempts. In some cases, the attacker's motivation can be reduced by making it seem as if the goal is not worth the effort it would take to capture it. The defensive layers that can be employed include the following:

- Secure network topologies
- Logical network separation
- Effectively employing DMZs
- Limiting physical access
- Restricting privileges

This is an important concept that when properly applied will improve the chances that any failure that does occur as a result of an attack will be more gradual and graceful allowing more time to react.

16.5.1 SANS – Tools That Work

The SANS Institute (SANS, 2010) recommends building the following six layers of defensive walls to protect cyber infrastructure:

1. Wall 1: Proactive Software Assurance – Testing source and binary application code using security scanners with assessments and certifications.
2. Wall 2: Blocking Attacks (Network Based) – Employment of a variety of network-based intrusion prevention and detection systems combined with network behavior analysis, firewalls, enterprise anti-virus, and threat management devices (secure gateways, application firewalls, and managed security services).
3. Wall 3: Blocking Attacks (Host Based) – Implementing a variety of host-based security measures such as endpoint security, network access control, system integrity checking tools, application control, and configuration hardening tools.
4. Wall 4: Eliminating Security Vulnerabilities – Vulnerability management by employing penetration testing and ethical hacking techniques, followed by patch and security configuration management and compliance measures.
5. Wall 5: Safely Supporting Authorized Users – Implementing measures such as identity and access management, mobile data protection and storage and backup encryption, content monitoring/data leak prevention, and virtual private networks (VPNs).
6. Wall 6: Tools to Manage Security and Maximize Effectiveness – Include tools and measures such as log management, event management, media sanitization, mobile device recovery and erasure, security skills development, security awareness training, forensics tools, governance, risk and compliance management tools, and disaster recovery and business continuity planning.

16.5.2 Capital Improvement Projects

One approach being used is adding network security improvement work to capital improvement projects. If the utility has a current vulnerability assessment that includes network improvement recommendations and the network is well documented, the improvements added can be a step-wise approach to implementing the vulnerability assessment recommendations. Without a vulnerability assessment and network documentation, any security improvements implemented are still likely to improve security, but less likely to be the most effective improvements needed. Many of the technical controls identified in the NIST SP 800-82 can be included as capital improvement projects. General technical controls that are very likely to improve network security include the following:

1. De-militarized Zone (DMZ) – A separate small buffer network between a private internal network and an external network (see Fig. 16.12). A DMZ can

eliminate most direct connections between the internal and external networks while providing external access to the internal network.

2. Intrusion Detection Systems/Intrusion Protection Systems (IDS/IPS) – Network-based Intrusion Detection Systems (NIDS) examine network traffic in more detail than routers and packet inspection firewalls. They use a range of detection algorithms to detect and log suspicious network traffic. Network-based Intrusion Protection Systems (NIPS) block suspected malicious traffic. IDS/IPS technology is constantly improving and a single NIPS can protect the whole SCADA system. However, intrusion detection and protection systems still can generate a lot of false positives that increase maintenance support requirements and can restrict network traffic when they become overloaded. NIPS provide improved protection against malicious traffic by

    a. Using better malicious traffic detection methods.
    b. Stopping malicious transactions as they happen.
    c. Protecting against internal attacks.
    d. Providing better forensics.

   Host-based Intrusion Protection Systems (HIPSs) that must be included with the operating system or included as part of a third-party virus protection package are also available. However, like the virus protection application, host-based intrusion protection consumes resources on each workstation and server and can impact ICS responsiveness. Also, host-based systems require updating and maintenance on each host which can be maintenance intensive.

3. Role-Based Access and Single Sign-on – Role-based access control uses computer operating system and application group user policies to limit access to information that required to support the role of the user. Single Sign-on eases user account and group policy management.
4. Wireless Network Access Controls – Uses encryption and authentication mechanisms to discourage intrusion and limit access to wireless networks combined with other network protection techniques to reduce the vulnerability of wireless network. Wireless network access controls are critical for broadband wireless networks with direct network connections.

One example improvement that might be included in a capital improvement project is to improve separation between the utility's SCADA and business networks by adding a third DMZ network, with a firewall connecting the three networks. Figure 16.12 shows this improvement for the example SCADA network previously presented in Fig. 16.2 (Section 16.2.2).

Through this improvement the direct connection requirement between the SCADA and business networks needed to provide database access for both networks has been eliminated (see dashed arched data flow lines). The firewall can also be configured to improve security for RTU digital cellular traffic routed across the Internet.

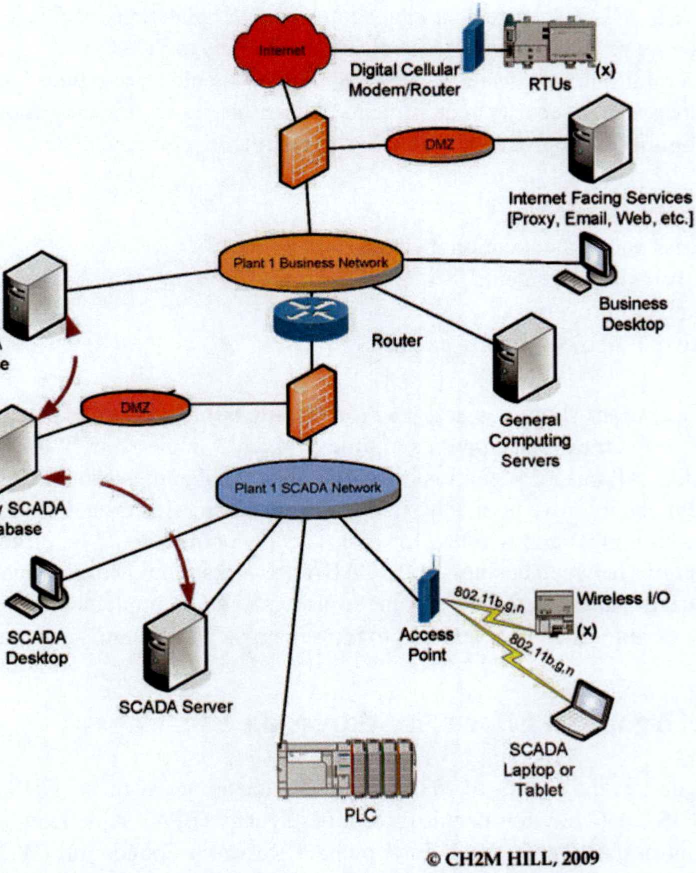

Fig. 16.12 Example DMZ application to improve security (Phillips, 2009c)

16.5.3 Practical Implementation Guide

Although the structure and language across the standards presented differ, the basic guidance provided is relatively consistent; and all are directly applicable to the water sector. Some utilities are not waiting for additional guidance and are applying these and other standards in improving the security of their business and SCADA networks by improving segmentation, adding firewalls, implementing wireless network security, improving access controls, and integrating network security elements into their policies and procedures.

A number of products are also emerging specifically designed for use in SCADA systems. These products are designed to perform reliably in the rugged environments found in SCADA systems. These products could be the leading edge of a wave of development that will make implementing and maintaining network security in SCADA systems easier by providing application-specific solutions that do not require adapting general computing solutions to SCADA.

At a minimum, the water/wastewater utilities should review their VAs and ERPs to address cyber security requirements. Based on the review they should look for opportunities to improve the following infrastructure components:

- Physical security
- Access and authentication methods
- Software improvements
- Privacy improvements
- Network topology improvements

Plant operations should be designed (or redesigned) to run on local control or manual controls for extended periods of time (3–4 days) in the event SCADA network is disrupted. All hazard scenarios including natural calamities should be evaluated to augment and improve upon the existing infrastructure. Emergency operations center (server buildings) and wireless (or radio) access points should be protected. Direct connections between business and SCADA networks should be eliminated or at least minimized and protected. End-point security should be implemented such that it is not possible for the cyber attacker to move data off the system.

16.6 Organized Efforts to Address the Challenges

Since the tragic events of 9/11, many organizations such as DHS, US-CERT, ISA, NIST, US Environmental Protection Agency (EPA), American Water Works Association (AWWA), Water Environment Research Foundation (WERF), SANS and various national laboratories (e.g., Sandia, Idaho, and Pacific) have collaborated on research activities focused on protecting critical infrastructure from cyber attacks. Although EPA is the sector-specific lead for protecting the water and wastewater infrastructure, DHS is the lead federal agency with responsibility for assuring the security, resiliency, and reliability of the nation's IT and communications infrastructure. In addition, many of the SCADA equipment and software manufacturers such as GE, Allen Bradley, Intellution, and Wonderware are undertaking efforts to improve and enhance built-in security in their devices to address the market needs. Furthermore, inspired by the electric sector, the water and wastewater utilities have collaborated to develop a sector-specific roadmap.

A full discussion of all the sector-specific collaborative research efforts is beyond the scope of this chapter. However, some of these key efforts are illustrated in the following sections of this chapter.

16.6.1 Department of Homeland Security

DHS facilitates coordination and information sharing between the federal government, academia, and private sector to reduce cyber attack risks, disseminate threat information, share best practices, and apply appropriate protective actions outlined in the National Infrastructure Protection Plan framework. Some of DHS's key measures designed to prevent future cyber attacks and aid recovery include the following:

- EINSTEIN Program – This program is designed to identify unusual network traffic patterns and trends which signal unauthorized activity. The early identification enables security personnel to quickly respond to potential threats. The program applies to all federal agencies and serves as an early warning system to gain better situational awareness, earlier identification of malicious activity, and a more comprehensive network defense.
- Trusted Internet Connections Initiative – This initiative is to consolidate the number of external connections including Internet points of presence for the federal government infrastructure. This Office of Management and Budget's initiative is designed to efficiently manage and implement security measures to increase protection across the federal Internet domains (.gov).
- National Cybersecurity Center – This center is designed to bring together federal cybersecurity organizations, by virtually connecting and in some cases physically collocating personnel and resources to gain a clearer understanding of the overall cyber security picture of federal networks.
- National Cyber Investigative Joint Task Force Expansion – DHS supported the expansion of this task force to include representation from the US Secret Service and several other federal agencies. This cyber investigation coordination organization overseen by the Federal Bureau of Investigation (FBI) serves as a multi-agency national focal point for coordinating, integrating, and sharing relevant information during cyber threat investigations.
- Control Systems Cyber Security Self-Assessment Tool (CS2SAT) – CS2SAT is a desktop software tool that provides users with a systematic and repeatable approach for assessing the cyber security posture of their industrial control system networks. The CS2SAT was developed under the direction of the DHS Control Systems Security Program (CSSP) by cyber security experts from various national laboratories and with assistance from the NIST. The WERF and Water Research Foundation (WRF) are authorized to distribute the tool only to WERF subscribers and WRF members.
- Cyber Storm – In March 2008, DHS led the largest cyber security exercise (Cyber Storm II), bringing together participants from federal, state, and local governments, private sector, and the international community to examine and strengthen the nation's cyber security preparedness and response capabilities in response to a simulated cyber attack across several critical sectors of the economy. This exercise provided an opportunity to test and evaluate concepts and standard operating procedures developed by various organizations since Cyber Storm I (conducted in February 2006). This exercise series takes place every 2 years to assess preparedness capabilities in response to a cyber incident of national significance. Cyber Storm III is slated for Fall 2010.

Furthermore, the currently confidential National Security Presidential Directive 54 (aka Homeland Security Presidential Directive 23) issued on Jan 8, 2008 (Nakashima, 2008), expanded the intelligence community's role in monitoring Internet traffic to protect against a rising number of attacks on federal agencies' computer systems. The directive called for the formation of the Comprehensive National Cybersecurity Initiative. The program-specific details have been kept confidential by both the previous and current administration.

16.6.2 WaterISAC

The Water Information Sharing and Analysis Center (WaterISAC) was authorized by Congress in 2002 and created and managed by the water sector. Its mission is to keep drinking water and wastewater utility managers informed about potential risks to the nation's water infrastructure from contamination, terrorism, and cyber threats. The mission has been expanded to help utilities respond to and recover from all hazards.

Funded by subscriber fees and matching federal funds from EPA and the Association of Metropolitan Water Agencies, WaterISAC links members through a secure online portal. The subscriber base includes water utilities and state and federal agencies dealing with security, law enforcement, intelligence, the environment, and public health. The WaterISAC is not a government program, but a water industry effort, recommended by the Presidential Decision Directive 63 and Executive Order 13231, whereby all critical infrastructure segments were required to establish information sharing and analysis centers.

The WaterISAC is accessed from the Internet at www.WaterISAC.org. The web site provides a variety of sector-specific information including the following topics that are related to cyber security:

- Alerts on potential terrorist activity.
- Information on water security from federal homeland security, intelligence, law enforcement, public health, and environment agencies.
- Notification of cyber vulnerabilities and technical fixes.
- Research, reports, and other information.
- A secure means for reporting security incidents.
- Vulnerability assessment tools and resources.
- Secure electronic bulletin boards and chat rooms on security topics.
- Summaries of open-source security information.

Many useful SCADA breaches have been reported on the WaterISAC. For example, there was a SCADA vulnerability related to relay control/generator brought to light by Idaho National Laboratory. Subscribers were informed in a timely manner to patch their system and make them immune to such attacks.

Similar to the WaterISAC initiative, the Electricity Sector (ES) also created the ESISAC (www.ESISAC.com) that contains a library of the CIP (see NERC CIP – Section 16.4.4) documents including Security Guidelines, Assessment Methods, CIP Reference Documents, and Archives. The Standards and Guidelines in the Security Guidelines section include information on a number of topics including Control System – Business Network Electronic Connectivity, Cyber – Access Controls, and several other topics related to Cyber Security.

16.6.3 Water Sector Road Map

Following the energy sector's road map, the Water Sector Coordinating Council's Cyber Security Working Group (WSCC-CSWG) sponsored by DHS and AWWA developed a vision and road map. The vision statement is as follows: "In 10 years, industrial control systems for critical applications will be designed, installed, and maintained to operate with no loss of critical function during and after a cyber event."

The Water Sector road map was started in 2007 by two individuals working for public utilities in California. Seth Johnson from Santa Clara Valley Water District and David Edwards from the Metropolitan Water District of Southern California approached the American Water Works Association looking for assistance with creating a security road map for the industry. Until then, water had minimal representation at security venues such as the Process Control Systems Forum (PCSF) and SANS conferences.

A working group was established under the WSCC to develop the road map document. This group, comprised of asset owners from around the country, worked to address the organizational, technical, and administrative challenges of cyber security in the water utility. The mission of the road map was that all SCADA systems would be designed, installed, and maintained to operate with no loss of critical function during and after a cyber event. Their efforts resulted in the first road map document released in March of 2008.

Since that time, the road map has been distributed at various events, including the SANS SCADA Security Conference, the PCSF, and the AWWA Water Security Congress. The short-term goal of the road map is the socialization of its message through workshops held in various cities around the country. The longer term goals include developing SCADA Security Programs, Risk Assessment/Mitigation Strategies, and Partnership and Outreach Programs.

16.7 Unresolved Challenges

In spite of all progress made through various organized efforts, there are many unresolved challenges faced by the water and wastewater utilities. Some of these key challenges are discussed in this section.

16.7.1 Voluntary Standards

None of the relevant standards presented in Section 16.4 are mandated or specifically apply to the water and wastewater sector. Also, unlike the NERC for the electric industry, there is no sector-specific independent body or regulatory agency that ensures compliance with any standard. It is completely up to the individual utility to pick and choose which security approach fits them the best. While this has the potential to provide greater flexibility to the sector, it typically leads to a lack of awareness and vision among the various industry sector participants.

16.7.2 Voluntary Incident Reporting

There is no mandated cyber incident reporting requirement; so all the available information is voluntary, which makes it hard for investigators to get a firm grasp on the exact status of the sector-specific cyber security needs and vulnerabilities. Utility operators may not report minor cyber incursions for thorough investigation, especially if the damage is contained, leading to skewed incident statistics and potentially greater threat at a later date.

16.7.3 Patching Software and Firmware Vulnerabilities

Each utility's SCADA network software and hardware configurations tend to be unique or one-of-a-kind. In addition, water and wastewater utilities do not have redundant hardware/software test beds to test the patches before they are applied on a live system. This makes it impossible to predict potential problems that can be caused during the patching process. Therefore, utilities hesitate and/or wait to apply patches even when they are available.

One example that highlights this issue is the Bellingham control system cyber security incident. On June 10, 1999, a gasoline pipeline ruptured in the City of Bellingham, Washington. Gasoline leaked into two creeks and ignited a fireball that killed three persons and injured eight other persons. The incident caused significant property damage and released approximately ¼ million gallons of gasoline causing substantial environmental damage. Abrams and Weiss (2007) report that the National Transportation Safety Board findings and recommendations cite the unresponsiveness of the SCADA system as one of the causes for pipeline rupture. Specifically, degraded SCADA performance was thought to have resulted from the development work done on the live SCADA system.

Idaho National Laboratory (INL) has built a SCADA test bed for an electric plant where equipment can be tested at component level. INL is in the process of building a test bed for the water/wastewater sector; however, access to this resource is not free. Utilities or SCADA equipment/software vendors have to pay for this service. Some utilities are paying INL for performing security testing in cooperation with the vendors under the pre-condition that any problems identified during the testing process will be patched by the vendor at no cost to the utility. Although SCADA development is increasingly modularized, the utility-specific customizations can result in vulnerabilities that are missed during the implementation phase. Each utility should explore the option for setting up complete virtual systems using third-party software such as virtual machines and SCADA simulators to replicate their specific system configuration.

16.7.4 Financial Constraints

Many of the utilities' annual operating budgets do not accommodate specifically for SCADA security upgrades. Also, as the SCADA life cycle tends to be longer, many utilities need to budget 10–20% annually for upgrades and application support contracts. As the business network machines, the SCADA equipment needs to be replaced on a shorter life cycle schedule. Furthermore, due to financial constraints the utilities tend to not implement "table-top" exercises for the ERP. The training and mock exercises related to cyber security issues tend to be minimal.

16.8 Summary

The information collected by various organizations over the years is clear on one thing – cyber attacks are real and can cause significant damages. Water and wastewater utilities must adopt countermeasures to prevent or minimize the damage in case of such attacks. The greatest challenge for the water and wastewater industry is the large variance among the various infrastructure (or sector)-specific standards and the vendor's approach to meet these standards. The utilities can meet this challenge by voluntarily adopting a comprehensive standard in phases that best meets their security and organizational requirements.

Another approach would be for the sector to follow the NERC CIP model to expedite the development of sector-wide SCADA security regulations. The NIST and ISA standards provide guidance on determining needed improvements and selecting an approach for implementing those improvements; and regulations modeled after NERC CIP could provide the regulation, monitoring, and compliance documentation needed to assure a consistent, acceptable level of security across the water/wastewater sector.

Once developed, the CIP-based standard could evolve over time to address lessons learned, technological changes, emerging standards, and updates to the NIST and ISA standards. Should this approach prove viable, the remaining question for the water sector would be identifying the entity to fill the role filled by NERC for the electric power sector.

While the standards and compliance approaches are being finalized, the utilities should utilize the available resources and prepare/implement programs that are designed to increase cyber security and ensure compliance with the common elements of the aforementioned standards. Utilities should get creative and allocate more funding through annual budgets rather than depend solely upon capital improvement programs to achieve improvements in cyber security.

16.9 Helpful Internet Resources

The information presented in this chapter is intended to guide the water and wastewater utilities to understand the risk of exposure to a cyber attack and provide insights to build better defenses. As technology evolves continually, the best defense against cyber attack is to be prepared and stay current. The following is a listing of web sites where the current information can be obtained and used to minimize and mitigate the impact of a cyber attack.

http://www.cert.org/ – The Carnegie Mellon Computer Emergency Response Team
http://www.us-cert.gov/ – The United States Computer Emergency Response Team
http://www.dhs.gov/files/programs/cybersecurity.shtm – DHS Cybersecurity
http://www.ists.dartmouth.edu/ – The Institute for Security Technology
http://www.sans.org/ – The System Administration, Networking and Security Institute
http://isc.sans.org/ – Internet Storm Center
http://cve.mitre.org/ – Common Vulnerabilities and Exposures
http://www.epa.gov/watersecurity/ – EPA's Water Infrastructure Security
http://www.epa.gov/nhsrc/ – EPA's National Homeland Security Research Center
http://www.waterISAC.org/ – Water Information Sharing and Analysis Center
http://www.isd.mel.nist.gov/projects/processcontrol/ – NIST Process Control Security Requirements Forum (PCSRF)
http://www.isa.org/MSTemplate.cfm?MicrositeID=988&CommitteeID=6821 – ISA-SP99, Manufacturing and Control Systems Security
http://www.sandia.gov/scada/ – The Center for SCADA Security at Sandia National Laboratories
http://www.cisco.com/web/go/ciag/index.html – Cisco's Critical Infrastructure Assurance Group (CIAG)
http://scadahoneynet.sourceforge.net/ – SCADA HoneyNet Project: Building Honeypots for Industrial Networks – Critical Infrastructure Assurance Group (CIAG) Cisco Systems, Inc.
http://www.inl.gov/scada/index.shtml – Idaho National Laboratory – National SCADA Test Bed Program

Acknowledgments The authors would like to acknowledge Mr. John S. Hall of the EPA's National Homeland Security Research Center for his leadership, guidance, and sponsorship of research activities related to the protection of water and wastewater infrastructure. Some of the collection of material information presented in this report was funded under the EPA Contract EP-C-09-041. Any opinions expressed in this chapter are those of the authors and do not reflect the official positions or policies of their employers or EPA. Any use of the information related to products, trade names, or organizations presented in this chapter is at the sole discretion of the reader.

References

Abrams, Marshall and Weiss Joe (2007). Bellingham, Washington, Control System Cyber Security Case Study. Funded by a NIST contract in support of the Industrial Control System Security Project. Available at http://csrc.nist.gov/groups/SMA/fisma/ics/documents/Bellingham_Case_Study_report%2020Sep071.pdf

Abrams, Marshall and Weiss Joe (2008). Malicious Control System Cyber Security Attack Case Study – Maroochy Water Services. The MITRE Corporation, July 23, 2008. Available at http://www.mitre.org/work/tech_papers/tech_papers_08/08_1145/08_1145.pdf

Benson, Pam (2010). "Computer virus Stuxnet a 'game changer,' DHS official tells Senate." CNN National Security Producer, November 18, 2010. Available at http://www.cnn.com/2010/TECH/web/11/17/stuxnet.virus/index.html

Bioterrorism Act (2002). Public Health Security and Bioterrorism Preparedness and Response Act of 2002. Public Law 107-188 (2002).

Department of Homeland Security (DHS) (2008). Fact Sheet: Protecting Our Federal Networks Against Cyber Attacks, April 8, 2008. Available at http://www.dhs.gov/xnews/releases/pr_1207684277498.shtm

DHS (2009). Common Cyber Security Vulnerabilities Observed in DHS Industrial Control Systems Assessments, July 2009. Available at http://www.us-cert.gov/control_systems/pdf/DHS_Common_Vulnerabilities_R1_08-14750_Final_7-1-09.pdf

Federal Information Management Security Act (FIMSA) (2002). FISMA, 44 U.S.C. § 3541, is a United States federal law enacted in 2002 as Title III of the E-Government Act of 2002 (Public Law 107-347, 116 Stat. 2899).

Gillette, J., Peerenboom, J., Whitfield, R., and R. Fisher (2002). Analyzing Water/Wastewater Infrastructure Interdependencies, Presented at 6th Probabilistic Safety Assessment and Management Conference, San Juan, Puerto Rico, June 25, 2002.

Homeland Security Presidential Directive-7 (HSPD7) (2003). Critical Infrastructure Identification, Prioritization, and Protection. Office of the Press Secretary, issued by the White House on December 17, 2003.

ISA (2010). "ISA99 Committee Work Products." Reprinted with permission. Available from ISA at http://isa99.isa.org/ISA99%20Wiki/Work%20Products.aspx

ISO/IEC 27002 (2005). Information technology – Security Techniques – Code of Practice for Information Security Management (Redesignated from ISO/IEC 17799:2005 in 2007). Edited by Dr. Oliver Weissman (Germany) and Dr. Angelika Plate (UK). International Organization for Standardization, Geneva, Switzerland, 2007.

Moore, G.E. (1965). "Cramming more components onto integrated circuits," Electronics Magazine, 19 April 1965, p. 4.

Nakashima, E. (2008). Bush Order Expands Network Monitoring Intelligence Agencies to Track Intrusions, Washington Post, Saturday, January 26, 2008, p. A03.

NIST SP 800-82 Final Public Draft (2008). Guide to Industrial Control Systems (ICS) Security Supervisory, September 2008.

Panguluri, S. and J. Scott (2010). SCADA and Business Networks – "Islands of Automation." Copyright 2010 by Shaw Environmental and Infrastructure, Inc. Reprinted with permission.

Panguluri, S., Phillips Jr., W.R., and R.M. Clark (2004). Cyber Threats and IT/SCADA System Vulnerability, Chapter 5 in Water Security and Safety Handbook, Edited by Larry Mays, McGraw-Hill, New York, NY, 2004.

PBS (2004). Frontline Program titled "Cyber War!" Airdate: April 24, 2003, http://www.pbs.org/wgbh/pages/frontline/shows/cyberwar/

Phillips Jr., W.R. (2009a). Typical Water/Wastewater Utility's Business and SCADA Infrastructure and Network Connectivity. Copyright 2009 by CH2M Hill. Reprinted with permission.

Phillips Jr., W.R. (2009b). SCADA Network Attack Scenario. Copyright 2009 by CH2M Hill. Reprinted with permission.

Phillips Jr., W.R. (2009c). Example DMZ Application to Improve Security. Copyright 2009 by CH2M Hill. Reprinted with permission.

President's Commission on Critical Infrastructure Protection (1997). "Critical Foundations: Protecting America's Infrastructures," The Report of the President's Commission on Critical Infrastructure Protection, October 1997.

Repository for Industrial Security Incidents (RISI) (2009). Quarterly Report on Cyber Security Incidents Affecting Industrial Control Systems – 3rd Quarter Report 2009. Available from RISI at http://www.securityincidents.org

SANS (2010). Internet Security Tools for Defense In-Depth. The SANS™ Institute http://www.sans.org/whatworks/

Shannon, J. and N. Thomas (2005). Human Security and Cyber-Security: Operationalising a Policy Framework. In "Cyber-Crime: The Challenge in Asia." Edited by Roderic G. Broadhurst and Peter N. Grabosky (pp. 327–346). Hong Kong University Press, Aberdeen, Hong Kong, 2005.

US-CERT (2009). Control Systems Vulnerability Notes. Available online at http://www.us-cert.gov/control_systems/

Chapter 17
Real-World Case Studies for Sensor Network Design of Drinking Water Contamination Warning Systems

Regan Murray, Terra Haxton, William E. Hart, and Cynthia A. Phillips

17.1 Introduction

The heightened risk of terrorist attacks on US critical infrastructure has placed the security of the water supply in the same league as the security of US national monuments. There is a long history of threats to water systems and a shorter list of actual incidents at water systems (AwwaRF, 2003; Kunze, 1997; Staudinger et al., 2006). However, public awareness of the threat has increased dramatically since the 9/11 attacks. Although the threat of terrorist attacks might not be a daily worry for water utilities, terrorist threats are of significant concern because of their potentially large public health and economic impacts. Conceivable terrorist threats to drinking water systems include the physical destruction of facilities or equipment, airborne release of hazardous chemicals stored on-site, sabotage of Supervisory Control and Data Acquisition (SCADA) and other computer systems, and the introduction of chemical, biological, or radiological contaminants into the water supply (ASCE, 2004). Contamination hazards might pose a significant threat because they could result in major public health and economic impacts and long-lasting psychological impacts.

In the last several years, water security research efforts have focused on the advancement of methods for mitigating contamination threats to drinking water systems (see, for example, Ostfeld, 2006; AWWA, 2005; Murray, 2004). A promising approach for the mitigation of both accidental and intentional contamination is a Contamination Warning System (CWS): a system to deploy and operate online sensors, other surveillance systems, rapid communication technologies, and data analysis methods to provide an early indication of contamination (US EPA, 2005b). CWSs with multiple approaches to monitoring – like water quality sensors located throughout the distribution system, public health surveillance systems, and customer complaint monitoring programs – are theoretically capable of detecting a wide range of contaminants in water systems. However, CWSs are expensive to

W.E. Hart (✉)
Sandia National Laboratories, Albuquerque, NM 87185, USA
e-mail: wehart@sandia.gov

R.M. Clark et al. (eds.), *Handbook of Water and Wastewater Systems Protection*, Protecting Critical Infrastructure, DOI 10.1007/978-1-4614-0189-6_17,
© Springer Science+Business Media, LLC 2011

purchase, install, and maintain. To make them a viable option, there is a clear need to minimize the investment required by individual drinking water systems.

This chapter focuses on a particular aspect of CWSs: strategic placement of sensors in a water distribution network. In particular, it describes how to use the sensor placement tools in TEVA-SPOT – the Threat Ensemble Vulnerability Assessment Sensor Placement Optimization Tool. This collection of software tools can help utilities design sensor networks (Berry et al., 2008; US EPA, 2009). It was developed by the EPA's Threat Ensemble Vulnerability Assessment (TEVA) Research Team, composed of researchers from EPA, Sandia National Laboratories, the University of Cincinnati, and Argonne National Laboratory.

This chapter presents case studies using TEVA-SPOT and discusses open challenges for application of sensor network design to large-scale real-world drinking water systems. It begins with background information on contamination warning systems and sensor network design, including a brief literature review of current methods. Four sensor network design case studies highlight the differing motivations of drinking water utilities across the USA as well as various methods for integrating optimization with real-world considerations. These cases also illustrate several challenges for real-world applications. The conclusion contains additional open challenges for water quality and distribution experts and for researchers.

17.2 Drinking Water Contamination Warning Systems

Research on methods to mitigate the impacts of contamination incidents has converged over the last several years on the concept of a contamination warning system (CWS) (ASCE, 2004; AWWA, 2005; US EPA, 2005a). The goal of a CWS is to detect contamination incidents early enough to allow for an effective response that minimizes further public health or economic impacts. A CWS is a proactive approach that uses advanced monitoring technologies and enhanced surveillance activities to collect, integrate, analyze, and communicate information to provide a timely warning of potential contamination incidents.

EPA is piloting CWSs through the Office of Water's Water Security (WS) initiative, formerly called WaterSentinel, at a series of drinking water utilities. The WS initiative promotes a comprehensive CWS that is theoretically capable of detecting a wide range of contaminants, covering a large spatial area of the distribution system, and providing early detection to mitigate impacts (US EPA, 2005b). Components of the WS initiative include the following:

- *Online water quality monitoring*. Continuous online monitors for water quality parameters, such as chlorine residual, total organic carbon (TOC), electrical conductivity, pH, temperature, oxidation reduction potential (ORP), and turbidity help to establish expected baselines for these parameters in a given distribution system. Event detection systems, such as CANARY (Hart et al., 2007), can detect anomalous changes from the baseline to provide an indication of potential contamination. The system can also use other monitoring technologies, such as contaminant-specific monitors. The goal is to detect a wide range of possible contaminants.

- *Consumer complaint surveillance.* Water utilities track consumer complaints regarding unusual taste, odor, or appearance of the water, and they record what steps they took to address these water quality problems. The WS Initiative is developing a process to automate the compilation and tracking of information provided by consumers. Such a system, coupled with anomaly detection software, might be able to rapidly identify unusual trends that indicate a potential contamination incident.
- *Public health surveillance.* Syndromic surveillance conducted by the public health sector might serve as a warning of a potential drinking water contamination incident. This surveillance includes information such as unusual trends in over-the-counter sales of medication and reports from emergency medical service logs, 911 call centers, and poison control hotlines. Information from these sources can be integrated into a CWS by developing a reliable and automated link between the public health sector and drinking water utilities.
- *Enhanced security monitoring.* Security breaches, witness accounts, and notifications by perpetrators, news media, or law enforcement can be monitored and documented through enhanced security practices that detect anomalous conditions. This component has the potential to detect a tampering event in progress, potentially preventing the introduction of a harmful contaminant into the drinking water system.
- *Routine sampling and analysis.* Utilities can collect and analyze water samples at a predetermined frequency to establish a baseline for contaminants of concern. This provides a baseline for comparison during the response to detection of a contamination incident. This component requires continual drills or exercises for the laboratory staff and review of procedures so that everyone is ready to respond to an actual incident.

Simply placing a collection of monitors and equipment throughout a water system is not enough to effectively detect contamination incidents. To be effective, a CWS must also manage large volumes of data and provide actionable information to decision makers. Different information streams must be captured, managed, analyzed, and interpreted in time to recognize potential incidents and mitigate the impacts. Each of the above components provides useful information; however, when the data from these components are integrated and used to evaluate a potential contamination incident, the credibility of the incident can be established more quickly and reliably than if any of the information streams were used alone.

Many utilities currently implement monitoring and surveillance activities, but few are operating in such a way as to meet the primary objective of a CWS – timely detection of a contamination incident. For example, although many utilities currently track consumer complaint calls, a CWS requires a robust spatially based system that, when integrated with data from public health surveillance, online water quality monitoring, and enhanced security monitoring, can provide specific, reliable, and timely information for decision makers to establish credibility and respond effectively. Consequence management plans and advanced laboratory capabilities are also required in order to respond to contamination incidents in a timely and appropriate manner. The utility, public health agencies, local officials, law enforcement and emergency responders, and others must coordinate to develop an effective consequence management plan that ensures appropriate response to detection by different components. An advanced and integrated laboratory infrastructure is needed to support baseline monitoring and analysis of samples collected in response to initial detections. Still, the challenge in applying a CWS is to reliably integrate the multiple information streams in order to decide if a contamination incident has occurred. While the primary purpose of a CWS is to detect contamination incidents, dual-use benefits, such as better monitoring of water age and quality in normal circumstances, will likely help to ensure the sustainability of a CWS within a utility.

17.3 Designing Sensor Networks for Contamination Warning Systems

Utilities can consider a number of possible goals for an online sensor system such as minimizing public exposure to contaminants, the spatial extent of contamination, detection time, or costs. These objectives are often at odds with each other, making it difficult to identify a single best sensor network design. Quantifying a sensor placement's performance with respect to these goals allows some comparison of competing placements. The TEVA-SPOT toolkit usually optimizes with respect to a primary objective and can consider one or more secondary objectives. However, there are many practical constraints and costs faced by water utilities that might not be formally modeled. Consequently, designing a CWS is not a matter of performing a single optimization analysis. Instead, the design process is truly a multi-objective problem that requires informed decision making and using optimization tools to identify possible sensor network designs that work well under different assumptions and for different objectives. Water utilities must weigh the costs and benefits of different designs and understand the significant public health and cost trade-offs.

17.4 Literature Review

Researchers have published many papers on sensor placement in drinking water distribution systems in the last several years, including a Battle of the Water Sensor Networks that compared 15 different approaches to this problem (Ostfeld et al., 2008). Sensor placement strategies can be broadly characterized by the technical approach and the type of computational model used. The following categories reflect important differences in proposed sensor placement strategies:

- Expert Opinion: Although expertise with water distribution systems is always needed to design an effective CWS, expert opinion strategies are guided solely by human judgment. For example, Berry et al. (2005) and Trachtman (2006) consider sensor placements developed by experts with significant knowledge of water distribution systems. These experts did not use computational models to carefully analyze network dynamics. Instead, they used their experience to identify locations whose water quality is representative of water throughout the network.
- Ranking Methods: A related approach is to use preference information to rank potential sensor locations (Bahadur et al., 2003; Ghimire and Barkdoll, 2006). In this approach, a user provides preference values for the properties of a "desirable" sensor location, such as proximity to critical facilities. These preferences can then be used to rank the desirability of sensor locations. Further, spatial information can be integrated to ensure good coverage of the network.
- Optimization: Sensor placement can be automated with optimization methods that computationally search for a sensor configuration that minimizes contamination risks. Optimization methods use a computational model to estimate the performance of a sensor configuration. For example, a model might compute the expected impact of an ensemble of contamination incidents, given sensors placed at strategic locations.

Hart and Murray (2010) reviewed over 90 papers related to sensor placement for CWS design. They identified seven steps common to most sensor placement strategies: defining the contamination risk to minimize; describing the characteristics of sensors used in the CWS; selecting the performance objectives; determining the optimization objective; formulating the optimization model; applying an appropriate optimization strategy; and implementing the design. The papers in the sensor placement literature approach each step with varying degrees of complexity and with different optimization and simulation strategies. Hart and Murray divided the 90 papers into 9 groups according to how the authors addressed each step, in particular based on (a) (non)use of contaminant transport simulations to compute risk, (b) sensor failure model, (c) number of design objectives during optimization, and (d) type of optimization objective.

Most of the research literature focuses on new or improved sensor placement optimization methods. However, designing a CWS is not a simple matter of running a computer program; there are many factors to consider when performing sensor placement, including utility response, the relevant design objectives, sensor behavior, practical constraints and costs, and expert knowledge of the water distribution system. In many cases, these factors can be at odds with one another (e.g., competing performance objectives), which makes it difficult to identify a single best sensor network design.

The TEVA Research Team has developed a decision-making process for CWS design that is composed of a modeling process and a decision-making process that employs optimization (Murray et al., 2008b). This modeling process includes creating or utilizing an existing network model for hydraulic and water quality analysis, describing sensor characteristics, defining the contamination threats, selecting performance measures, estimating utility response times following detection of contamination incidents, and identifying a set of potential sensor locations. The decision-making process involves applying an optimization method and evaluating sensor placements. The process is informed by analyzing trade-offs and comparing a series of designs to account for modeling and data uncertainties.

17.5 The TEVA-SPOT Software

The TEVA-SPOT software is a framework for sensor placement optimization that consists of three main software modules which are illustrated in Fig. 17.1. The first software module simulates the set of incidents in a given design basis threat: a set of potential contamination incidents, each characterized by an injection location, time of day, type and concentration of contamination, duration of injection, etc. The second software module calculates the potential consequences of the contamination

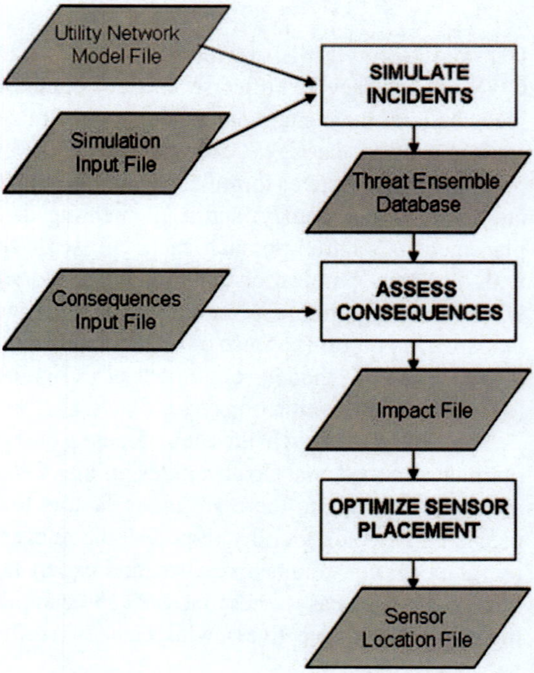

Fig. 17.1 Data flow diagram for the TEVA-SPOT software

incidents contained in the design basis threat. The third software module optimizes the sensor placement.

TEVA-SPOT has been used to design sensor networks for several medium and large US water distribution systems (Morley et al., 2007; Skadsen et al., 2008; US EPA, 2008). The tool has been shown to outperform utility experts in selecting good sensor locations, see for example, Berry et al. (2005) and Ostfeld et al. (2008). Furthermore, the TEVA-SPOT heuristic optimizers have been shown to reliably find near-optimal solutions on real-world applications.

The following features highlight the flexibility and strength of this software:

- *Consequence assessment.* Given a utility network model, and the set of parameter values determined in the modeling process, TEVA-SPOT calculates the consequences of each contamination incident in the design basis threat. The design basis threat is the set of incidents that the sensor network is designed to detect. The consequences are estimated in terms of one or more performance objectives, such as the number of people made ill or the length of pipe contaminated. Typically, TEVA-SPOT considers contamination incidents that occur at every node in the network model. TEVA-SPOT calculates consequences using EPANET for hydraulic and water quality calculations (Rossman, 2000) and models for estimation of exposure and disease progression (Murray et al., 2006).
- *Optimization.* Most utilities have TEVA-SPOT place sensors to minimize the mean consequences for a given objective (averaged over an ensemble of possible contamination incidents). TEVA-SPOT allows users to specify weights, for example, to put more weight on locations with a higher likelihood of contamination; practically, this information is unlikely to be available with any certainty. If the user is most interested in protecting against a few catastrophic contamination incidents, TEVA-SPOT can also minimize the maximum impacts (Watson et al., 2004).
- *Multi-objective analysis.* Researchers and utilities have considered many competing CWS design objectives, e.g., the number of people made ill, the length of pipe contaminated, or the time to detection. TEVA-SPOT can only optimize over one objective at a time, but it does allow the user to specify side constraints that ensure that sensor network designs perform well for more than one objective.
- *Fast, flexible solvers.* To allow for the comparison of designs based on multiple performance objectives and model parameters, TEVA-SPOT needs to be fast and flexible. It has both fast heuristic methods and slower exact solvers that find provable optimal sensor placements. It also has fast methods for computing bounds. This enables users to choose faster methods while at the same time having quantifiable confidence in the quality of the sensor placement selected by the method. For most networks, near-optimal designs can be found in seconds to minutes.
- *Solver scalability.* Users can select from a variety of strategies to ensure that TEVA-SPOT works on large networks with tens of thousands of pipes and junctions: aggregation of problem constraints, aggregation of contamination

incidents, and/or specification of a limited set of feasible junctions for sensor placement (Hart et al., 2008). Furthermore, several of the TEVA-SPOT solvers have been modified to limit the memory required on standard 32-bit workstations. For example, the heuristic solver includes options that explicitly trade off memory and run time.

17.6 Real-World Case Studies

The TEVA-SPOT software was used to help nine partner utilities design sensor networks for CWS based on the modeling process described in Murray et al. (2008b). The steps in this process identify the specific types of sensors to be deployed, the design objectives, and the list of potential locations for the sensor stations. Some utilities already had sensors in place; in such situations, the objective was to identify supplemental locations.

The decision-making process was iterative and involved applying the optimization software to select optimal locations and then verifying the feasibility of those locations with field staff. The software quantified the trade-offs between the locations selected optimally by the software and the "nearby" locations selected by the utility for practical reasons. In addition, the software was used to develop cost–benefit curves for each utility, see Fig. 17.2.

Murray et al. (2009) quantified the benefits to each utility with a simulation study. They considered two realistic terrorist contamination threats: an infectious biological agent introduced over a 24-h period and a toxic chemical agent introduced over a 1-h period. Figure 17.3 illustrates the estimated reduction in economic impacts due to the CWSs deployed or planned for the nine utilities. These economic savings are estimates based on the expected reduction in the number of fatalities given early detection and rapid response as part of the CWS.

A large set of realistic contamination incidents was considered for these utilities; this plot shows the reduction of the mean and 95th percentile of the

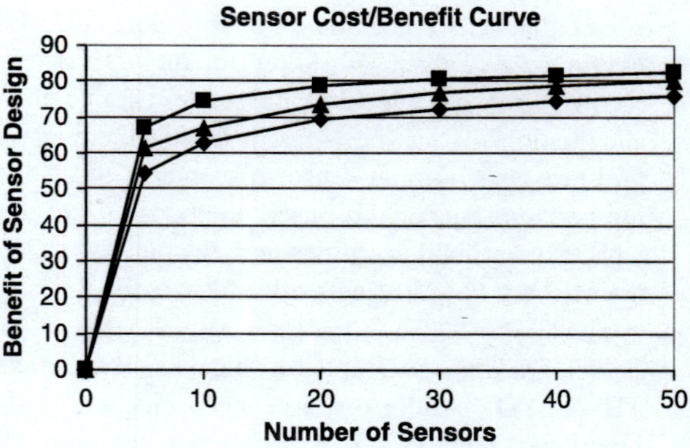

Fig. 17.2 The cost–benefit curves for three utilities show that the benefits of a CWS design (% reduction of impacts) increased as the number of sensors (cost) increased

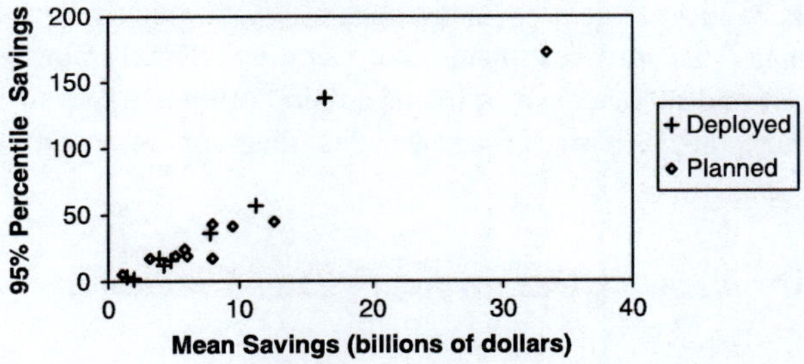

Fig. 17.3 The graph shows economic savings resulting from reduced fatalities (billion $); mean economic savings is plotted versus corresponding savings for the 95th percentile of contamination incidents. Two data points, which represent biological and chemical contamination threat ensembles (not distinguished in the graph), are included for each utility

impact distribution. Fatalities were computed based on contaminant-specific data, after calculating how much contaminant customers at various locations and times throughout the network would consume. Economic impacts as a result of fatalities were computed using a Value of Statistical Life (VSL). VSL is the average value society is willing to pay to prevent a premature death. It does not refer to the value of an identifiable life, but rather the summed value of individual risk reductions across an entire population. In the Groundwater Rule, EPA used a value of $6.3 million in 1999 dollars and $7.4 million in 2003 dollars. To be conservative in this analysis, a VSL value of $6.3 million was used (ATSDR, 2001).

As Fig. 17.3 shows, a CWS can significantly reduce economic consequences of fatalities for biological and chemical incidents. Over the nine utilities, the mean savings are estimated to vary from $1 billion to $33.4 billion with a median of $5.8 billion. However, because an informed terrorist would attempt to maximize the impact of an attack, the mean impacts might not be the best measure. Although the sensor placements were optimized to minimize mean impacts, Fig. 17.3 shows that 95th percentile economic savings were also significant: $1 billion to $171.7 billion with a median of $19 billion.

Additionally, the relationship between the percent reduction of economic impacts resulting from fatalities and the percentage reduction in fatalities was assessed (Murray et al., 2009). These are independent of the values in Fig. 17.3, but are related measures for CWSs. Utilities could have a significant reduction in the number of fatalities, but the percentage reduction is relative to the total number of fatalities without sensors (or, in some cases, with the set of existing sensors). Thus, the percent reduction of economic impacts in these utilities can vary dramatically because of differences in the total number of fatalities in these systems.

Finally, economic impacts incurred by the water utility were estimated. Following a contamination incident, contaminants might be difficult to remove from pipe walls and fittings. In the worst cases, utilities might have to reline or replace contaminated pipes. Therefore, the CWS designs were evaluated to determine the fraction of the distribution network that might need to be replaced. A CWS can reduce the cost of replacement by enabling a utility to quickly contain the spread of a contaminant. This study estimated that using CWSs will reduce the expected decontamination and recovery costs for these nine utilities by up to $340 million. For many utilities, these savings are greater than their annual operating budget.

17.6.1 Sensor Network Design for Greater Cincinnati Water Works

Greater Cincinnati Water Works (GCWW) was selected as the first WS initiative pilot city. EPA's report, "Water Security Initiative Cincinnati Pilot Post-Implementation Status," provides extensive detail on the contamination warning system installed at GCWW (US EPA, 2008). The sensor network design component of the project is summarized here. In this first WS initiative pilot, EPA had an active and direct role in the design and implementation of the CWS. The online monitoring component for GCWW was designed to expand the existing monitoring capabilities. Prior to the WS initiative, GCWW had 40 chlorine analyzers, 3 pH meters, and 2 turbidimeters located at 22 sites in the distribution system. These meters transmitted data over telephone lines to the utility's operations center.

The existing water quality monitoring did not meet all of the objectives of the WS initiative, for example, spatial coverage, timely detection of contamination incidents, or degree of automation necessary for real-time detection. So GCWW installed additional sensor stations at locations identified using the TEVA-SPOT software. The sensor stations measured multiple water quality parameters including free chlorine, TOC, ORP, conductivity, pH, and turbidity. Figure 17.4 shows

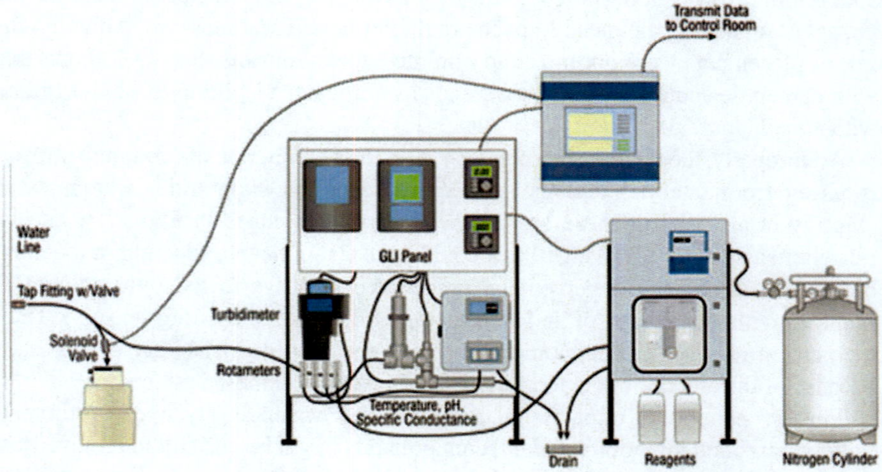

Fig. 17.4 Schematic of one type of sensor station installed at the GCWW WS initiative pilot

a schematic of some of the sensor stations installed at GCWW. The new sensor network was designed to minimize average public health consequences over a large set of possible contamination incidents. The sensor network design process involved three major steps: validating the utility network model, applying the TEVA-SPOT software, and field investigations to finalize sensor locations.

To validate the network model, a tracer study was performed in the field. A tracer (calcium chloride) was injected at four locations in the distribution system. Each injection consisted of at least six 1-h pulses over a 24-h period. Following each injection, conductivity meters were used to measure and record the conductivity signal at approximately 40 locations throughout each study region. The field data were then compared to model predictions in order to assess the accuracy of the model and identify needed improvements.

The TEVA-SPOT software used the validated model to suggest optimal sensor placements for the public health objective. GCWW identified several hundred potential sensor locations that included all utility owned sites (including office buildings), all police and fire stations in the county, as well as certain schools and hospitals. Using geographic information systems (GIS), these facilities were associated with the nearest node in the utility network model. Initially, the design was based on selecting up to 30 sensor stations, although in the end, 17 stations were deployed. The utility located two stations at its treatment plants and the software was used to help identify the best locations for the remaining 15 stations.

The design and implementation was an iterative process. TEVA-SPOT was used to select a list of optimal locations from the list of several hundred potential locations. The hydraulic connectivity of each location was verified on GIS and/or AutoCAD® (Autodesk, Inc.) maps to ensure that the model representation of the facility was correct. Finally a site visit was conducted to locate the exact installation location within the facility, estimate the hydraulic retention time in the pipes from the distribution system main to the monitoring equipment inside the facility, and address any outstanding concerns with that specific location. Sites were also verified to ensure accessibility, physical security, available sample water and drainage, a reliable power supply, and data communications. If at any point a site was deemed unsuitable, it was discarded and the TEVA-SPOT analysis re-run.

If the retention time from the distribution system main to the monitoring equipment inside the facility was greater than 2 h, a water bypass to the sensor station had to be installed (which was not always feasible). A service connection of smaller radius was considered favorable, as was a sensor tap-in location near the point where the service connection met the building (to reduce pipe length). A retention time over 2 h was not recommended, as it has a negative impact on utility response time.

All 17 sensor stations were installed at the locations determined by the above process and have been in operation since 2007. In addition, GCWW is testing the performance of CANARY (Hart and McKenna, 2009), an automated data analysis tool that converts real-time water quality data into alarms that indicate the likelihood of contamination incidents.

17.6.2 Sensor Network Design for New Jersey American Water

In the Burlington-Camden-Haddon system of New Jersey American Water (NJAW), 11 online monitors were installed as part of a collaborative study between EPA, US Geological Survey (USGS), and NJAW. The purpose of the study was to understand the field performance of water quality monitors and the normal variability in background water quality, as well as to identify calibration requirements arising from fouling and sensor drift. This information would later inform the installation of online water quality sensors to support contamination warning systems. In addition, the data gathered over several years at these locations were used to help develop the CANARY event detection software.

Prior to this study, USGS had already worked with NJAW to install two sensor stations to monitor source water as well as two stations in the distribution system. The TEVA-SPOT software was used to select an additional seven monitoring locations in the distribution system. The three main objectives for sensor design were (1) to obtain accurate measurements of the true range and variation in water quality in the NJAW distribution system; (2) to provide protection and early warning of contamination events; and (3) to meet the additional needs and interests of NJAW (operational needs, costs, ease of access, and maintenance).

Because one of the goals of this study was to better understand the variability of water quality within the distribution system, EPANET simulations were performed to predict chlorine residuals throughout the distribution system over a 10-day period, and the nodes were separated into three categories of low, medium, or high chlorine variability. Low variability nodes had a standard deviation of chlorine concentration in the lowest 33%; medium variability nodes fell between 33 and 66%; and high variability nodes were in the upper 33% of nodes.

NJAW and USGS provided a list of seven locations where they wanted to locate sensors in the distribution system. EPA used the TEVA-SPOT software to analyze the expected performance of this "utility design" (UD) and to select three additional designs for comparison. One design was optimized solely for public health protection (PH) and selected locations from all possible nodes in the model. Another design optimized not only for public health protection but also for water quality (WQ) variability. The WQ&PH design required that two nodes have low variability in residuals, three with medium variability, and two with high variability. Finally the third design was a compromise between the UD and the WQ&PH (compromise utility design, CUD) that selected three locations from the list provided by the utility and allowed the software to select the additional four locations from all the nodes in the model.

Figure 17.5 shows the performance of each of the four designs as measured against biological and chemical incidents. The UD was estimated to reduce the mean public health impacts associated with biological incidents by 34% and chemical incidents by 19%. In comparison, the optimal designs for public health protection reduced impacts by 48 and 45%, respectively. Table 17.1 shows for each sensor network design the number of sensors in each category of variability. With the

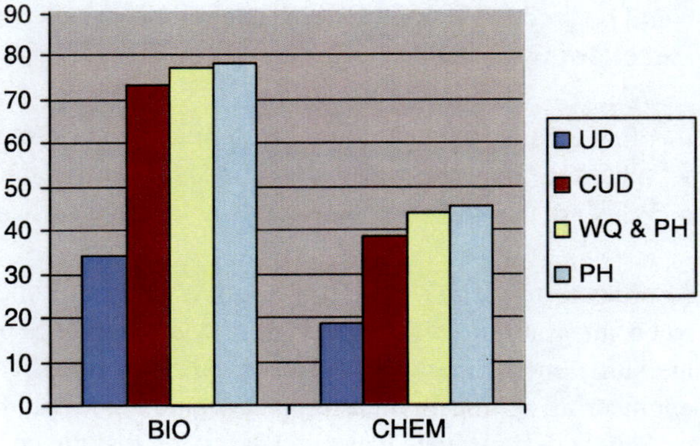

Fig. 17.5 Percent benefit of each of the four sensor designs against biological and chemical contamination incidents (UD=Utility Design, CUD=Compromise Utility Design, WQ&PH=Design for Water Quality and Public Health Objectives, and PH=Optimal Public Health Design)

Table 17.1 Number of sensor locations in each chlorine variability category

Sensor design	Low variability	Medium variability	High variability
UD	2	3	4
PH	3	1	5
WQ&PH	3	3	3
CUD	3	3	3

information provided by this analysis, the utility was able to make a final decision on locating sensors that met not only the objectives of the study to measure water quality but also the needs for public health protection as part of a future contamination warning system.

Following installation of YSI® (YSI, Incorporated) multi-parameter sensors at the selected locations, the USGS maintained the sensor calibrations, manual collection, and quality assurance of the data. Data have been collected for several years and have subsequently been utilized by both EPA and Sandia National Laboratories to develop and refine tools for automated sensor data processing and event detection.

17.6.3 Sensor Network Design for Tucson Water

Tucson Water is an innovative and advanced municipal drinking water system that serves nearly 700,000 customers. Through an EPA Environmental Monitoring for Public Access and Community Tracking (EMPACT) grant, online monitors have been providing near real-time water quality data to the public for several years. At the time of this analysis, Tucson Water was considering the expansion of its online monitoring program to meet new security objectives. The goal of the Tucson sensor network design study was to identify new and/or existing EMPACT locations that could be used as part of a CWS. The preliminary analysis was performed to answer the following questions:

- What are the best distribution system locations to place sensors?
- What are the best EMPACT locations for sensor stations?
- What are the trade-offs between the two different sensor designs?

In order to use TEVA-SPOT, Tucson Water's multiple pressure zone models had to be integrated into a single system-wide model. Separate pressure zone models are sufficient for many utility needs, but water security analyses require a systems engineering approach, focusing on the entire distribution system.

Sensor designs were generated assuming that the goal of the monitoring program was to provide public health protection against a long release of a biological agent or a rapid release of a chemical from any service connection in the distribution system. The sensors were assumed to be water quality sensors capable of detecting changes caused by the two contaminants. The sensor designs are sensitive to response time or the time it takes a utility to effectively respond to a positive detection. Therefore, response time was allowed to vary from 2 to 48 h. A total of 48 different sensor network designs were developed and analyzed.

Figure 17.6 provides sensor trade-off curves for sensor network designs based on the assumption that sensors can be located anywhere in the distribution system (diamond) or only at EMPACT sites (square). These curves demonstrate the trade-off between the number of sensors and the performance of the sensor design, as measured by the percent reduction in mean public health impacts. The results for

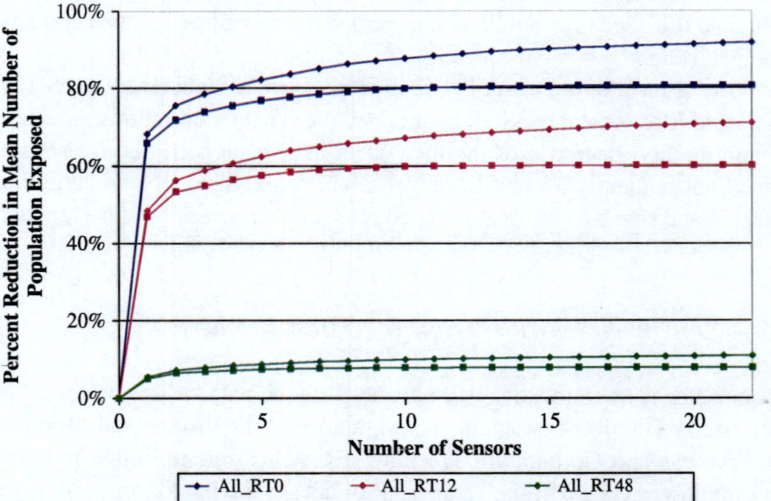

Fig. 17.6 Trade-off curves for the number of sensors versus percent reduction for all locations (*diamond*) and EMPACT locations (*square*) with utility response times of 0, 12, or 48 h

three different response times are also shown in Fig. 17.6, where the blue line is a response time of 0, the pink line is a response time of 12 h, and the green line is a response time of 48 h. These results show that an optimal design can reduce public health impacts caused by a contamination incident from 10 to 92%. The EMPACT design reduces impacts from 6 to 80%. Notably, it is possible to improve the performance of the EMPACT design by selecting one or two additional locations that are not EMPACT locations. Tucson Water is evaluating how to effectively use the designs recommended by TEVA-SPOT to create and implement a sustainable and cost-effective contamination warning system.

17.6.4 Sensor Network Design for the City of Ann Arbor

The City of Ann Arbor undertook a study to design an online monitoring program that could both minimize public health exposures resulting from a contamination incident and detect water quality degradation due to naturally occurring processes, such as nitrification, iron corrosion, and bacterial re-growth. The results of this study can be found in Skadsen et al. (2008) and are summarized briefly here.

The City of Ann Arbor's water system provides treated water to about 130,000 customers, encompasses approximately 49 square miles, and has an average system demand of 15 MGD (million gallons per day) with a range from 7 to 30 MGD. The distribution system is divided into five major pressure districts that have elevated tanks and storage reservoirs to adequately serve the population over a varied topography.

The study involved four steps: (1) analysis of the distribution system, (2) parameter selection and instrument pilot testing, (3) estimation of costs, and (4) proposal of an online monitoring design. The analysis of the distribution system began with an assessment of the accuracy of the existing network model. Following a series of improvements to the model, it was used in both the TEVA-SPOT software and the PipelineNet software (Pickus et al., 2005) to identify good locations for online monitoring. The analysis from both models was overlaid with staff expertise and practical knowledge to determine the final proposed monitoring locations.

The utility identified a list of 27 potential locations that included water utility facilities (pump stations, reservoirs, tanks, and pressure monitoring locations), other city facilities (fire stations and parking structures), and a limited number of private property sites. Each of these field locations was visited to determine its feasibility as a monitoring location. The sites were ranked based on the ownership of the site and availability of a connection to the distribution system, power, communications, and sanitary sewer. Also, access and existing heating, ventilation, and air condition units (HVAC) were included in the assessment. In addition, the availability of historical water quality data was a factor. One location was selected because of the abundance of such data.

The TEVA-SPOT software was used to select the best locations for security monitoring from the list of 27 potential locations. The analysis was performed with a variety of response delay times, 0, 4, 12, 24, and 48 h. Two different contaminants were considered: a fast acting chemical contaminant and a slow acting biological contaminant. The TEVA-SPOT analysis found that a small number of monitors provided significant benefits as measured by the percent reduction in public health impact. The selected locations were spatially distributed throughout the pressure zones ensuring good distribution system coverage.

PipelineNet was used to evaluate potential sensor locations in order to protect sensitive facilities (schools and hospitals) and high-population areas from contaminant attack. This resulted in clustering of sensor locations around the largest of these facilities. Although this might result in increased protection for these sensitive facilities, the remainder of the potential target population was not protected to the extent provided by the TEVA-SPOT sensor designs. The PipelineNet software was also used to determine areas with the highest water quality concern based on the criteria established by Ann Arbor staff, and these were matched against the 27 available monitoring locations. Areas of impacted water quality clustered along the edges of the system and along pressure boundaries, consistent with predictions of high water age areas and previous tracer studies.

The results of the TEVA-SPOT analyses, the PipelineNet results for water quality, and staff knowledge of the system were integrated. Four sites were selected for security monitoring and four locations were selected for water quality monitoring. One of the sites selected for security was the same as a site selected for water quality. This general lack of co-located sites was expected due to the different drivers for security monitoring (protect as much of the population as possible) versus water quality monitoring (find the areas of high water age usually associated with dead end or isolated parts of the distribution system). However, this was considered an important finding by the authors, suggesting that security monitoring locations might not show significant dual benefit in a system where operational concerns are based on water quality effects such as nitrification. The project team was originally interested in the possibility of achieving efficiency in operations and cost savings if the security and quality locations overlapped. However, this was not a requirement for the project.

The project team selected a set of water quality parameters for potential monitoring using a variety of information, including data from US EPA's Test and Evaluation Facility in Cincinnati, Ohio, other research studies, utility surveys, and a workshop (Hall et al., 2007). These studies highly recommended free chlorine and TOC to address water security concerns. Ultraviolet (UV)-254 was selected as a cost-effective alternative for TOC. Other parameters selected for testing included dissolved oxygen, ammonia, chloride, and conductivity.

A variety of criteria were developed to assess instrument performance and acceptability. Of these criteria, accuracy (agreement between lab testing and online instrument results), sensitivity (low-level measurement ability), and variability (presumed normal fluctuation in water quality) proved the most important factors. Other factors, such as calibration ease, frequency, and maintenance are also important but the ability of the units to deliver useful data was deemed the most critical function. Based on pilot testing of instruments, chloride and ammonia were eliminated as parameters for monitoring. This analysis concluded that total chlorine and dissolved oxygen were important parameters for measuring water quality degradation, but total chlorine, UV, conductivity, and dissolved oxygen were recommended for water security.

In Ann Arbor, the costs for monitor acquisition were estimated at $25,000 per installation assuming that each location had four instruments: total chlorine, dissolved oxygen, conductivity, and UV-254 absorbance with the selected manufacturers. Installation costs, including infrastructure and communications, were estimated at an average of $40,000 per location. Operations and maintenance costs were estimated at $7,000 per installation per year. A 10-year life span was assumed for the equipment. Based on these estimated figures, the utility plan included an initial capital investment of approximately $500,000 for eight sensor locations with an annual operating budget, including replacement costs of $106,000. These costs do not include initial design and pilot testing work of approximately $200,000. These figures are important when considering the cost/benefit ratio versus the number of monitors installed. This study resulted in a sensor network design that was proposed to the City of Ann Arbor and is awaiting funding.

17.7 Challenges for Real-World Applications

In the sensor placement field, there are many outstanding research questions (Hart and Murray, 2010). Current application of sensor network design optimization can be challenging and sometimes requires imagination in addition to technical skill. Several common questions that arise when applying sensor network design optimization software are as follows:

- What is the best objective to use for sensor placement?
- How many sensors are needed?
- Should a CWS be designed to protect against high-impact incidents only?
- How can I make sensor placement algorithms work on typical desktop computers even for very large utility network models?

In practice, there are no clear-cut answers to the first three questions because they involve policy concerns in addition to good science.

For demonstration purposes, these questions are considered through the analysis of a simple example network model: EPANET Example 3 network with 92 junctions, 2 reservoirs, 3 tanks, 117 pipes, and 2 pumps. This network is supplied by two surface water sources – a lake provides water for the first part of the day and a river for the remainder of the day. The average water residence time for the network is 14 h, and the maximum is 130 h. Based on the average base demands, the total population served by this network is 78,800. The total length of pipe in the system is 215,712 feet.

In these analyses, sensors are assumed to be "perfect" in the sense that they have a zero detection limit and are always accurate and reliable (no false positives or false negatives and no failures). Utility response to detection of contamination incidents is also assumed to be perfect and instantaneous, meaning that following detection, a "Do Not Use" order is issued and made effective immediately, preventing all further consumption. These assumptions are referred to as "perfect sensors and perfect response." These assumptions are applied in order to provide an upper bound on sensor network performance – the best that is theoretically possible. Even with perfect sensors and perfect response, a sensor network might not detect every event, detect events in a timely manner, prevent all exposure to contaminants, or prevent contamination of pipes. To achieve this perfect performance, in most networks, sensors would need to be placed at almost every junction. This is clearly not feasible in practice, thus the importance of optimally selecting a small number of sensor locations using optimization software.

17.7.1 Selecting the Best Objective

The performance objective is one of the most important parameters to select when designing a CWS; for example, should a sensor network be based on minimizing the population exposed or minimizing the detection time? In practice, the authors have found that sensor network designs based on different objectives can be very different from one another. Thus, it is important to understand differences between the objectives when designing a CWS.

The following six performance objectives are available in TEVA-SPOT: population exposed (PE), extent of contamination (EC), volume consumed (VC), mass consumed (MC), number of failed detections (NFD), and time of detection (TD). In our example, contamination incidents are simulated at each of the 59 nodes with positive user demands. TEVA-SPOT calculates the performance objective for each incident and then finds a single sensor that best minimizes the mean of the performance objective across all of the incidents.

Figure 17.7 displays the sensor locations selected for each of the six objectives in the example network. Note that there is overlap only in two of the six objectives: minimizing NFD or TD results in the same sensor location. The NFD objective aims to detect as many contamination incidents as possible. In this case, Junction-253 (orange circle in Fig. 17.7) was selected and it detects the majority of the contamination incidents since it is at the bottom edge of the network and the majority of the flow paths end there. The flow in this network is from the two sources at the top to the bottom and to the right side of the network; thus, water originating from most injection locations will travel to this node. With this single sensor, 39 of the 59 incidents are detected (20 are not detected), resulting in a 66% reduction in the number of failed detections.

It might seem counterintuitive that the TD objective selected the same sensor location since it is near the end of the flow path, and the time of detection would

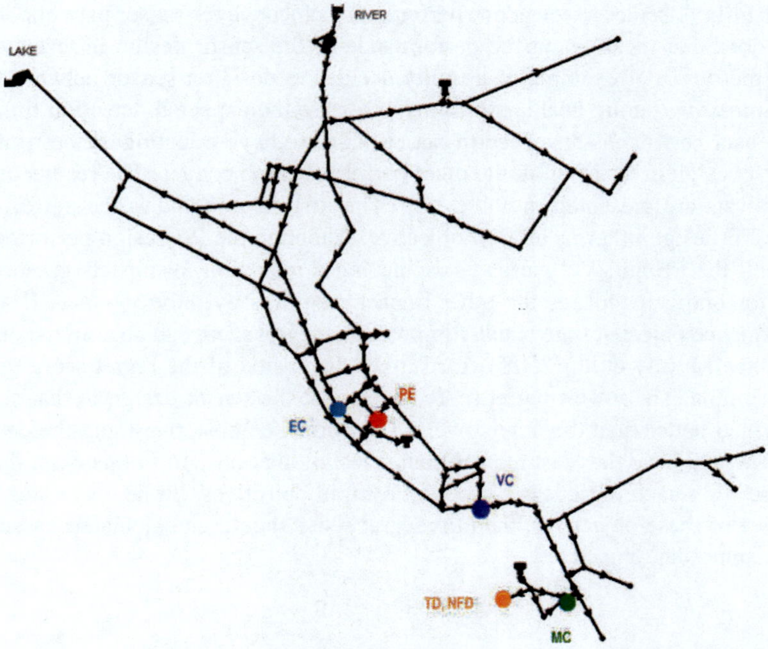

Fig. 17.7 Sensor designs for EPANET Example 3 network based on different performance measures

be quite large. This location was selected because of the way that the TD objective is calculated in TEVA-SPOT. It greatly penalizes sensor network designs for not detecting an incident. If an incident is not detected, the performance metric assigns a detection time equal to the total length of the simulation. Therefore, a shorter simulation time might generate more realistic designs using TD. An advanced option in TEVA-SPOT does not penalize a design for the incidents that are not detected (see Berry et al., 2008). The TD sensor shown in Fig. 17.7 reduces average detection times from 192 h to 69 h for a 64% reduction in detection times.

The sensor location that best minimized the number of people exposed (PE) is Junction-271 (red circle in Fig. 17.7), which is one node upstream of the node with the largest user demand. Thus, the sensor located at Junction-271 would detect all incidents that are along the flow path to the largest demand node. With this single sensor, over all of the 59 contamination incidents, the mean number of people exposed is reduced by 64% from approximately 11,000 people to 4,000 people.

Extent of contamination is another important performance objective for sensor network design, since the economic cost of recovery is linked to the total length of contaminated pipes. The EC sensor location, Junction-189 (blue circle in Fig. 17.7), is in the middle of the network, thereby cutting the largest flow paths in half. A sensor at this node would detect incidents that have the potential to contaminate larger portions of the network. For all 59 contamination incidents, this sensor reduces the mean EC by 47% from approximately 46,800 ft to 25,000 ft.

If a utility decides to select one performance objective for sensor network design, what does that mean about the performance of the sensor design in terms of the other metrics? For example, if a utility decides to design a sensor network based on minimizing public health exposures, what are the expected detection times for that sensor network? This question can be addressed by evaluating the performance of each design in terms of all the other performance objectives. The results of such an analysis are presented in Table 17.2. The first column shows the performance of the PE design in terms of each objective. Although the PE design performs well for both the PE and VC measures (achieving a reduction in impacts greater than 60% for both), it reduces the other impact measures by only 36–44%. If all the performance objectives are equally important, a regret score can be used to compare them; see Murray et al. (2008b) for further discussion of the regret score used in this example. The lowest regret score means that the sensor design in that column performed better than the others over all performance objectives, and the decision maker would have the least regret about selecting that design. In this case, the MC and the VC sensor designs perform best over all objectives. If one only cares about a subset of these objectives, then the regret score should be calculated based only on the important objectives.

17.7.2 Number of Sensors

As the capital costs of sensors can be in the tens of thousands of dollars, and the operation and maintenance costs can be up to 15–30% of capital costs each year, the number of sensors that can be installed as part of a CWS is usually limited by budget constraints. The question of how many sensors a utility needs in order to reliably reduce the risks of contamination incidents has not been answered definitively and requires a difficult policy decision.

Figure 17.8 shows a trade-off curve (the number of sensors versus benefit) based on the PE objective for the EPANET Example 3 network. Without a sensor network, an average of 11,000 people would be exposed to the contaminant out of a total

Table 17.2 Percent reduction achievable for the sensor designs (in each column) based on each performance objective

Performance measure	PE	MC	VC	TD	NFD	EC
Mean PE	64	55	62	44	44	64
Mean MC	36	56	44	56	56	35
Mean VC	91	94	95	92	92	88
Mean TD	38	63	48	64	64	37
Mean NFD	39	64	49	66	66	37
Mean EC	44	22	43	15	15	47
Regret score	0.43	0.27	0.27	0.38	0.38	0.45

Higher percentages reflect better performance

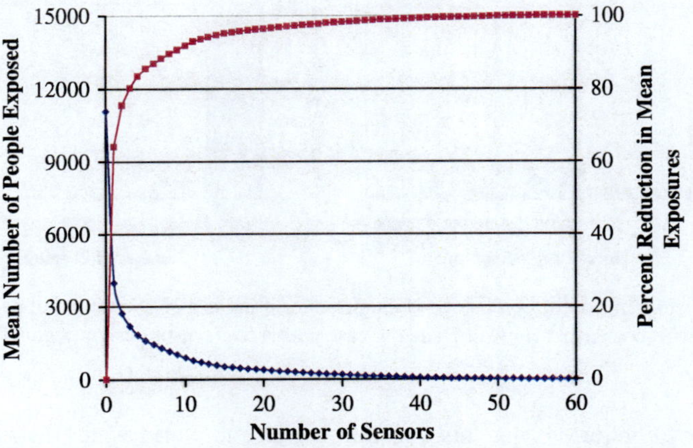

Fig. 17.8 Sensor network design trade-off curve for EPANET Example 3 network based on the number of people exposed (PE) objective. The *blue diamonds* correspond to the left axis which shows the number of people exposed decreases with the number of sensors. The *pink squares* correspond to the right axis which shows the benefit of the sensor network (percent reduction in the number of people exposed) increases with the number of sensors

population of approximately 78,000. A single sensor, optimally located, reduces exposures to 4,000 people on average (for a 64% reduction). Thus, the first sensor prevents an average of (or provides a marginal benefit of) 7,000 exposures. Two optimally located sensors reduce exposures to 2,700 people. The second sensor, then, provides a marginal benefit of 1,300. The third sensor provides a marginal benefit of 500 people. After 10 sensors have been placed, the average number of exposures is reduced to 900 people, but it takes 59 sensors to reduce the average exposure to zero. Note that this would mean placing a sensor at every possible injection location (the 59 nodes with user demands). Each additional sensor yields less and less benefit, reflecting the diminishing marginal returns of larger sensor networks.

Policy makers could draw reasonable yet conflicting conclusions from these results. For instance, a policy maker could say that given budget concerns, placing five sensors provides substantial public health benefit to the system but that no additional costs can be justified (because the marginal benefits decrease dramatically after five sensors). Another policy maker could say that designing a system that would still expose more than 1,600 people (or 1.4% of the population) on average is not acceptable under any circumstances.

If the utility selected a sensor network with five sensors, the number of people exposed is reduced by 85%. Figure 17.9 shows the distribution of PE over the 59 contamination incidents, first in the absence of sensors and then for the five-sensor design. The shape of the two distributions is quite different – the mean and maximum number of people exposed is reduced significantly by five sensors. For this design, 81% of the incidents are detected with the five sensors (i.e., 19% are not), and an average of 18,000 pipe feet are contaminated. Is this level of risk reduction acceptable? Are there additional criteria that should be considered?

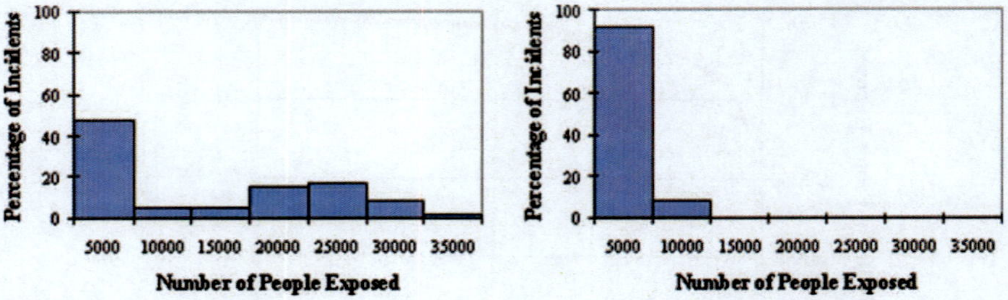

Fig. 17.9 Histograms of the percentage of incidents resulting in a given number of people exposed for the case with no sensors (*left side*) and the case with a five-sensor design (*right side*)

In order to better answer this question, eight real-world water distribution networks that serve from 6,000 to 1.2 million people were examined. The goal was to look for trends among the networks that would help inform the question of how many sensors are needed. Trends are considered for multiple metrics of acceptable risk that impose specific limits on public health objectives (PE), coverage objectives (NFD), and economic objectives (EC). The characteristics of the additional networks can be found in Murray et al. (2008a).

Table 17.3 lists the number of sensors needed to meet specific public health metrics for the eight networks. The first row shows the number of sensors needed to ensure that public health impacts will be less than 10,000 people on average. If a utility is only concerned with limiting exposures to less than 10,000 people, the results show that only one or two sensors might be necessary. However, for lower levels of risk, the number of sensors needed might depend on population.

Figure 17.10 plots the results for PE<1,000. The number of sensors needed to satisfy this metric is plotted against the population of each network (blue diamonds). There does appear to be a trend, although there are two obvious outliers. Upon further examination, it seems that the level of detail in the model might be affecting these results. The right axis is the number of nodes in the model. The two outlier cases represent an extremely detailed model (high number of nodes compared to population – Net 4) and an extremely skeletonized model (low number of nodes compared to population – Net 8).

Table 17.3 Number of sensors needed to achieve public health objective

Metric network	Net 1	Net 2	Net 3	Net 4	Net 5	Net 6	Net 7	Net 8
Population	6.2 K	7.6 K	78 K	142 K	200 K	450 K	840 K	1,200 K
Mean PE < 10,000	0	0	1	0	0	0	1	2
Mean PE < 1,000	1	1	3	19	6	10	38	154
Mean PE < 500	1	2	5	85	21	47	125	[a]
Mean PE < 100	5	5	11	[a]	[a]	[a]	[a]	[a]
Mean PE < 10	24	7	24	[a]	[a]	[a]	[a]	[a]

[a]This metric is beyond the resolution of the utility network model

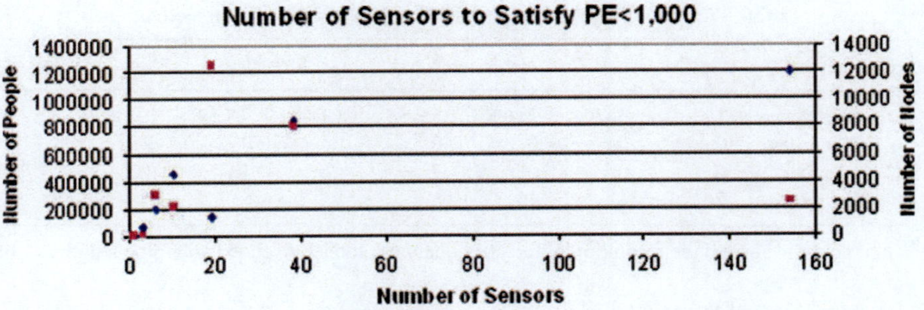

Fig. 17.10 Number of sensors needed to satisfy PE < 1,000 for each of the eight networks versus population (*blue diamonds*) and number of network nodes (*pink squares*)

Similarly, Table 17.4 lists the number of sensors needed to meet several coverage objectives for the networks. The coverage metric measures the percentage of contamination incidents detected by the sensor network (i.e., ID = 1 – NFD). Table 17.4 shows the number of sensors (up to 100) needed to detect from 40 to 90% of incidents. TEVA-SPOT can compute the minimum number of sensors required to meet a specific objective in a single computation. However, for this objective, there was also a strict limit of 100 sensors. Net 4 gives anomalous results for these metrics, requiring significantly more sensors than the other networks. This is the extremely detailed network that produced anomalous results in Fig. 17.10. As the number of nodes increases, so does the number of incidents. Thus, this metric should vary with the level of skeletonization, which is not precisely captured in these results.

Finally, Table 17.5 lists the number of sensors needed to meet economic objectives, measured in terms of the length of pipe contaminated, from 1 to 100 miles.

Typically, water utilities use a combination of budget constraints and sensor network design performance curves in order to determine the appropriate number of sensor stations to install in a distribution network. An analysis of Fig. 17.8 might lead one to determine that five sensors is the most appropriate number for Net 3. However, the results in Tables 17.3, 17.4, and 17.5 demonstrate that a sensor network with five sensors will not be generally sufficient to meet goals for high

Table 17.4 Number of sensors needed to achieve coverage objective (number of incidents detected)

Metric/network	Net 1	Net 2	Net 3	Net 4	Net 5	Net 6	Net 7	Net 8
Incidents	79	9	90	11,000	1,800	2,200	7,000	1,400
Mean ID > 40%	1	1	1	8	2	1	1	1
Mean ID > 50%	2	1	1	30	2	2	3	2
Mean ID > 60%	2	1	1	90	2	4	6	5
Mean ID > 70%	2	1	2	+	4	5	21	11
Mean ID > 80%	3	1	2	+	25	15	75	28
Mean ID > 90%	5	2	6	+	+	+	+	80

+ indicates the sensor placements required more than 100 sensors to meet the objective

Table 17.5 Number of sensors needed to achieve economic objective

Metric network	Net 1	Net 2	Net 3	Net 4	Net 5	Net 6	Net 7	Net 8
Total pipe miles	123 K	64 K	216 K	5.6 M	4.1 M	2.7 M	9.4 M	7.5 M
EC < 100 miles	0	0	0	0	1	0	1	1
EC < 10 miles	0	0	0	7	12	10	25	27
EC < 1 mile	7	4	16	[a]	[a]	[a]	[a]	[a]

[a]This metric is beyond the resolution of the utility network model given existing pipe lengths

coverage, minimal health impacts, and minimal network exposure. Using acceptable risk criteria might persuade the utility to install additional sensors.

The number of sensors needed in a water distribution system is a question of acceptable risk. Acceptable risk must be defined by the water utility, and thus is highly dependent on the detection goals of the community. The risk reduction goals of communities can vary widely from striving to detect only catastrophic incidents to detecting as many incidents as possible (including accidental cross connections). The utility might have broad goals, such as widespread coverage of the distribution system (for example, sensors in every pressure zone), detection of a large number of contaminants, and specific goals, such as preventing events that would be expected to impact more than 100 people. Using a multi-objective analysis might help to improve the performance of sensor designs across several objectives; however, there will always be a trade-off in performance when balancing performance with costs. In order to design and implement an effective contamination warning system, utilities must explicitly consider the performance trade-offs of the system they design.

17.7.3 Sensor Network Design Based on High-Impact Incidents

Frequently, water utilities wonder why research on sensor placement strategies has focused on reducing mean consequences; they ask, "Why not design for high-impact contamination incidents only?" An optimal sensor network design based on minimizing the mean value of a performance measure can still allow many high-impact contamination incidents to occur. Furthermore, most sensor placement optimization is done with the assumption that all incidents are equally likely (uniform event probabilities). This assumption is made because, typically, one does not have information about terrorist intentions; however, this results in an unintended de-emphasis of high-impact incidents.

It is possible to develop sensor networks based on high-impact contamination incidents. Rather than minimizing the mean, the optimization process can attempt to minimize the maximum value, or other robust statistic. A robust statistic is insensitive to small deviations from assumptions (Huber, 2004). For example, the mean statistic is not robust to outliers because a single large value can significantly change the mean.

Although the final determination of the design statistic ultimately rests with policy makers at a utility, the aforementioned factors strongly suggest that, at a minimum, there is a need to understand the differences between and implications of both mean-based and robust sensor designs. To illustrate the relative trade-offs that are possible between mean-case and robust sensor network designs, sensor placement designs that minimize PE with five sensors were examined for EPANET Example 3 network (for a full treatment of this topic on real-world networks, see Watson et al., 2009). Figure 17.11 shows histograms of the simulated number of people exposed during each of the 59 contamination incidents for the mean-case (left) or max-case (right) sensor network design. With five sensors selected to minimize mean impacts, the mean was reduced from 11,000 to 1,600 people and the maximum impact for any incident was reduced from 32,000 to 9,200 people compared to the system with no sensors. The distribution on the left side of Fig. 17.11 exhibits a common feature of sensor network designs that minimize the mean case: the presence of non-trivial numbers of contamination incidents that yield impacts that are much greater than that of the mean. Even with these five sensors in place, there was one contamination incident that exposed more than 9,000 people, and an additional 15 contamination incidents that exposed between 2,000 and 9,000 people.

The right side of Fig. 17.11 shows the distribution of the number of people exposed for a sensor network designed to minimize maximum impacts. With this design, there were not as many high-impact incidents as there were with a sensor network design that minimized the mean number of people exposed. In particular, the highest impact incident exposed 7,600 individuals, in contrast to 9,200 individuals under the optimal mean-case sensor design. However, there were more small-to-moderate impact incidents. The max-case design yielded a mean impact of 2,300 people exposed, representing a 42% increase relative to the mean-case design, which only impacted 1,600 people.

Thus, there is a trade-off involved in switching from the mean- to max-case statistic for optimization – if the mean value is reduced, high-impact incidents can still occur; if the max-case value is reduced, the mean value will increase. In this case, the question for decision makers in water security management is then, Is an 18% reduction in the max-case impact worth the 42% increase in the mean?

It is possible to gain significant reductions in the number and degree of high-consequence events at the expense of moderate increases in the mean impact of a

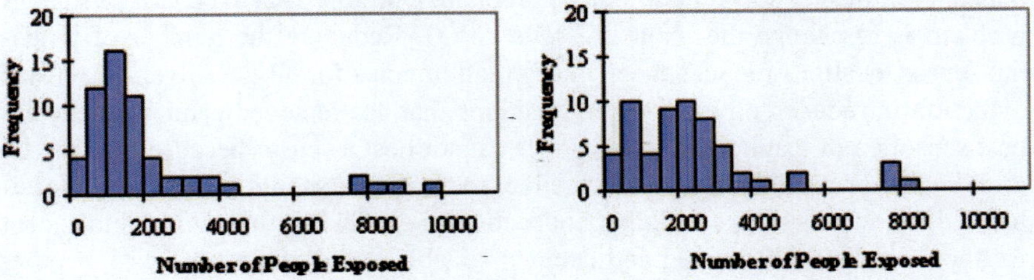

Fig. 17.11 Histograms of the frequency of incidents resulting in a given number of people exposed for the case with a five-sensor network design designed by minimizing the mean case (*left side*) or the max case (*right side*)

contamination event. This can be accomplished through the use of side constraints during the optimization process. For example, if the mean is minimized, the max case can be constrained to be less than some maximum value, so that the resulting sensor network design performs well in minimizing both mean- and max-case consequences. However, in practice these constrained optimization problems may be computationally difficult to solve.

17.7.4 Sensor Placement for Large Networks

Many optimization methods for sensor placement have been developed and tested on small test networks; however, applying them to large real-world networks has proven to be a challenge. One particular challenge is the memory needed to represent large sensor placement problems. TEVA-SPOT implements at least two strategies for handling large networks with limited memory resources:

- Select an optimizer that explicitly manages memory in an efficient manner.
- Reduce the size of the problem by shortening the list of potential sensor locations, the list of contamination incidents simulated, by using skeletonization or aggregation.

TEVA-SPOT has three types of solvers: an integer programming solver (IP), a heuristic solver (GRASP), and a Lagrangian solver (LAG). Although the IP solver will always find an optimal sensor placement, its memory requirements can be very large. The heuristic solver is generally a good first choice for large problems, since it runs quickly and has produced good designs in all experiments to date. If the heuristic fails on a real-world network, but only needs a small amount of additional memory to run, then running the heuristic in sparse mode might be sufficient (see the TEVA-SPOT toolkit User Manual for more information: Berry et al., 2008). The Lagrangian solver uses even less memory than GRASP, but the sensor designs it produces are not as close to optimal as those produced by GRASP.

Reducing the problem size can be done by removing information or restricting options. That means the solution, even if optimal for the reduced problem, will only approximately solve the full-sized problem. One approach to creating a smaller problem is to change the input to TEVA-SPOT. Reducing the number of potential sensor locations reduces the memory requirements for all the solvers. This size reduction introduces no error if the locations that are removed cannot practically host sensors. For example, if some nodes cannot host a sensor because they are on large mains or are otherwise inaccessible, these nodes should be marked infeasible. Utility owners might initially choose to consider all locations infeasible except for those explicitly evaluated and deemed feasible based on cost, access, or other considerations.

Another way to change the input is to reduce the number of contamination incidents in the design basis threat. The selected incidents should represent the original set as much as possible. For example, injection locations should cover all the geographic regions of the network. Currently techniques for automating this process have not been developed.

Users can also change the input by coarsening the network through skeletonization, using, for example, the techniques in Walski et al. (2004) or a commercial skeletonization code. This merges pipes and nodes that are geographically close to create a smaller graph that approximates the hydraulic behavior of the original. However, it will introduce error by dropping some pipes of sufficiently small diameter.

TEVA-SPOT also provides an option, called aggregation, for automatically reducing the size of the mathematical problem that is solved during optimization. Aggregation is only available for the IP and LAG solvers. Aggregation methods group potential sensor locations based on their performance for each incident. This effectively reduces the amount of memory needed to solve the sensor network design problem. When simulations are run with a coarse reporting step, aggregation can save some space without introducing error. The IP solver, for example, will do this automatically. However, if that is not sufficient then users can direct TEVA-SPOT to group nodes with differing, but approximately similar quality. The loss of information means the solver can only approximately solve the full problem. Berry et al. (2007) gave some preliminary results with this method. For a network with 3,358 nodes and 16 times the normal number of contamination incidents, aggregation can reduce the runtime by two orders of magnitude while introducing only 1% error. However, it only reduced the number of impacts by one order of magnitude.

17.8 Discussion

The goal of this chapter is to present case studies and discussions of practical challenges for CWS design that provides some perspective on the opportunities for using optimization tools like TEVA-SPOT to support decision makers who are deploying CWSs. US EPA (2009b) provides a more detailed presentation of these topics, including a description of the decision-making process used in these case studies, a description of the health effects models, and an overview of the mathematical optimization formulations used in TEVA-SPOT.

Hart and Murray (2010) provide a survey of related sensor placement literature, including a critical review of existing research. They conclude by recommending that future research focus on four areas:

- *Solving large-scale problems*: Although research groups have demonstrated the ability to solve large-scale problems, this is not a robust capability that can be readily applied by water security professionals.
- *Improving the quality of input data*: Improved prediction of data – in the form of drinking water demands, population estimates, seasonal operational rules, and an understanding of data uncertainties – is critical to the broad adoption of CWSs.

- *Comparing methods effectively*: Evaluations of new sensor placement algorithms need to include direct comparisons with other optimization strategies in the literature.
- *Decision support*: Decision makers need analytic strategies to assess the impact of data uncertainties, evaluate trade-offs, assess risk, and perform regret analysis.

Although the case studies in this report demonstrate the utility of CWS design tools like TEVA-SPOT, these research areas highlight critical issues that need to be addressed to ensure that water security risks can be effectively mitigated with CWSs.

References

ASCE. (2004). Interim voluntary guidelines for designing an online contaminant monitoring system, American Society of Civil Engineers, Reston, VA.

Agency for Toxic Substances and Disease Registry (ATSDR). (2001). Managing hazardous material incidents (MHMI), Volume 3, U.S. Department of Health and Human Services, Public Health Service, Atlanta, GA. <http://www.atsdr.cdc.gov/mhmi/mmg.html>.

AWWA. (2005). Contamination warning systems for water: an approach for providing actionable information to decision-makers, American Water Works Association, Denver, CO.

AwwaRF. (2003). Actual and threatened security events at water utilities, Project 2810, American Water Works Association Research Foundation, Denver, CO.

Bahadur, R., Samuels, W. B., Grayman, W., Amstutz, D., and Pickus, J. (2003). "PipelineNet: A model for monitoring introduced contaminants in a distribution system." Proc., World Water and Environmental Resources Congress 2003 and Related Symposia, ASCE, Reston, VA.

Berry, J., Hart, W. E., Phillips, C. A., Uber, J. G., and Walski, T. M. (2005). "Water quality sensor placement in water networks with budget constraints." Proc., World Water and Environmental Resources Congress, ASCE, Reston, VA.

Berry, J., Carr, R. D., Hart, W. E., and Phillips, C. A. (2007). "Scalable water sensor placement via aggregation" Proc., Water Distribution System Symposium, ASCE, Reston, VA.

Berry, J. W., Boman, E., Riesen, L. A., Hart, W. E., Phillips, C. A., and Watson, J.-P. (2008). User's manual: TEVA-SPOT toolkit 2.2, EPA-600-R-08-041, U.S. Environmental Protection Agency, Office of Research and Development, National Homeland Security Research Center, Cincinnati, OH.

Ghimire, S. R., and Barkdoll, B. D. (2006). "A heuristic method for water quality sensor location in a municipal water distribution system: mass-released based approach." Proc., 8th Annual Water Distribution Systems Analysis Symposium, ASCE, Reston, VA.

Hall, J., Zaffiro, A. D., Marx, R. B., Kefauver, P. C., Krishnan, E. R., and Herrmann, J. G. (2007). "On-line water quality parameters as indicators of distribution system contamination." Journal of the American Water Works Association, 99(1), 66–77.

Hart, D., McKenna, S. A., Klise, K., Cruz, V., and Wilson, M. (2007). "CANARY: a water quality event detection algorithm development tool." Proc., World Environmental and Water Resources Congress, ASCE, Reston, VA.

Hart, D. B., and McKenna, S. A. (2009). CANARY user's manual, version 4.1, EPA-600-R-08-040A, U.S. Environmental Protection Agency, Office of Research and Development, National Homeland Security Research Center, Cincinnati, OH.

Hart, W. E., and Murray, R. (2010). "Review of sensor placement strategies for contamination warning systems in drinking water distribution systems." Journal of Water Resources Planning and Management, 136(6), 611–619.

Hart, W. E., Berry, J. W., Boman, E. G., Murray, R., Phillips, C. A., Riesen, L. A., and Watson, J.-P. (2008). "The TEVA-SPOT toolkit for drinking water contaminant warning system design." Proc., World Environmental & Water Resources Congress, ASCE, Reston, VA.

Huber, P. J. (2004). Robust statistics, Wiley Series in Probability and Statistics, Wiley, Hoboken, NJ.

Kunze, D. R. (1997). "Assessing utility threats." Security Management, 41(2), 75–77.

Morley, K., Janke, R., Murray, R., and Fox, K. (2007). "Drinking water contamination – warning systems: water utilities driving water security research." Journal of the American Water Works Association, 99(6), 40–46.

Murray, R. (2004). "Water and homeland security: an introduction." Journal of Contemporary Water Research and Education, 129, 1–2.

Murray, R., Uber, J., and Janke, R. (2006). "Model for estimating acute health impacts from consumption of contaminated drinking water." Journal of Water Resources Planning and Management, 132(4), 293–299.

Murray, R., Baranowski, T., Hart, W. E., and Janke, R. (2008a). "Risk reduction and sensor network design." Proc., Water Distribution Systems Analysis 2008, ASCE, Reston, VA.

Murray, R., Janke, R., Hart, W. E., Berry, J. W., Taxon, T., and Uber, J. (2008b). "Sensor network design of contamination warning systems: a decision framework." Journal of the American Water Works Association, 100(11), 97–109.

Murray, R., Hart, W. E., Phillips, C. A., Berry, J., Boman, E. G., Carr, R. D., Riesen, L. A., Watson, J. P., Haxton, T., Herrmann, J. G., Janke, R., Gray, G., Taxon, T., Uber, J. G., and Morley, K. M. (2009). "US Environmental Protection Agency uses operations research to reduce contamination risks in drinking water." Interfaces, 39(1), 57–68.

Ostfeld, A. (2006). "Enhancing water-distribution system security through modeling." Journal of Water Resources Planning and Management, 132(4), 209–210.

Ostfeld, A., Uber, J. G., Salomons, E., Berry, J. W., Hart, W. E., Phillips, C. A., Watson, J. P., Dorini, G., Jonkergouw, P., Kapelan, Z., di Pierro, F., Khu, S. T., Savic, D., Eliades, D., Polycarpou, M., Ghimire, S. R., Barkdoll, B. D., Gueli, R., Huang, J. J., McBean, E. A., James, W., Krause, A., Leskovec, J., Isovitsch, S., Xu, J. H., Guestrin, C., VanBriesen, J., Small, M., Fischbeck, P., Preis, A., Propato, M., Piller, O., Trachtman, G. B., Wu, Z. Y., and Walski, T. (2008). "The battle of the water sensor networks (BWSN): a design challenge for engineers and algorithms." Journal of Water Resources Planning and Management, 134(6), 556–568.

Pickus, J., Bahadur, R., and Samuels, W. B. (2005). "Integrating the ArcGIS water distribution data model into PipelineNet." Proc., ESRI International User Conference, ESRI, Redlands, CA.

Rossman, L. A. (2000). EPANET 2: users manual, EPA-600-R-00-057, U.S. Environmental Protection Agency, Office of Research and Development, National Risk Management Research Laboratory, Cincinnati, OH. <http://www.epa.gov/nrmrl/wswrd/dw/epanet/EN2manual.PDF>.

Skadsen, J., Janke, R., Grayman, W., Samuels, W., TenBroek, M., Steglitz, B., and Bahl, S. (2008). "Distribution system on-line monitoring for detecting contamination and water quality changes." Journal of the American Water Works Association, 100(7), 81–94.

Staudinger, T. J., England, E. C., and Bleckmann, C. (2006). "Comparative analysis of water vulnerability assessment methodologies." Journal of Infrastructure Systems, 12(2), 96–106.

Trachtman, G. (2006). "A 'strawman' common sense approach for water quality sensor site selection." Proc., 8th Annual Water Distribution Systems Analysis Symposium, ASCE, Reston, VA.

US EPA. (2005a). Technologies and techniques for early warning systems to monitor and evaluate drinking water quality: a state-of-the-art review, EPA-600-R-05-156, U.S. Environmental Protection Agency, Office of Research and Development, National Homeland Security Research Center, Cincinnati, OH.

US EPA. (2005b). WaterSentinel system architecture, EPA-817-D-05-003, U.S. Environmental Protection Agency, Office of Water, Office of Ground Water and Drinking Water, Washington, DC.

US EPA. (2008). Water security initiative Cincinnati pilot post-implementation system status: covering the pilot period: December 2005 through December 2007, EPA-817-R-08-004, U.S. Environmental Protection Agency, Office of Water, Office of Ground Water and Drinking Water, Washington, DC. <http://www.epa.gov/safewater/watersecurity/pubs/rpt_post_imp_system_status_wsi.pdf>.

US EPA. (2009a). Tutorial threat ensemble vulnerability analysis – sensor placement optimization tool (TEVA-SPOT) graphical user interface, Version 2.2.0 Beta, EPA-600-R-08-147, U.S. Environmental Protection Agency, Office of Research and Development, National Homeland Security Research Center, Cincinnati, OH.

US EPA. (2009b). Sensor network design for drinking water contamination warning systems: a compendium of research results and case studies using the TEVA-SPOT software. EPA-600-R-09-141, U.S. Environmental Protection Agency, Office of Research and Development, National Homeland Security Research Center, Cincinnati, OH.

Walski, T. M., Daviau, J.-L., and Coran, S. (2004). "Effect of skeletonization on transient analysis results." Proc., The 2004 World Water and Environmental Resources Congress, ASCE, Reston, VA.

Watson, J.-P., Greenberg, H. J., and Hart, W. E. (2004). "A multiple-objective analysis of sensor placement optimization in water networks." Proc., The 2004 World Water and Environmental Resources Congress, ASCE, Reston, VA.

Watson, J.-P., Murray, R., and Hart, W. E. (2009). "Formulation and optimization of robust sensor placement problems for drinking water contamination warning systems." Journal of Infrastructure Systems, 15(4), 330–340.

Chapter 18
Enhanced Monitoring to Protect Distribution System Water Quality

Zia Bukhari and Mark LeChevallier

18.1 Introduction

Terrorism is the unlawful use of force or violence, or threatened use of force or violence, against persons and places for the purpose of intimidating and/or coercing a government, its citizens, or any segment thereof for political or social goals. While acts of terrorism are not new to the United States and span the last 60 years, perhaps the most profound impact was felt by the events of September 11, 2001. Foreign nationals used commercial airlines to orchestrate an attack on the World Trade Center in New York City, which horrified millions of citizens as they witnessed the events unfold through televised footage.

Due to heightened public awareness of terrorist threats in United States, it was considered prudent to implement measures to protect critical infrastructures. Critical infrastructures were defined systems and assets (i.e., banking, electricity, transportation) that are so vital that their incapacitation would have a debilitating impact on the United States. For example, public drinking water distribution systems serve 90% of the US population; however, these systems are readily accessible by the public, making them highly vulnerable to acts of malicious intent. Water, being a fundamental component of life, is necessary for consumption, hygiene, and firefighting purposes. Water treatment processes (i.e., treatment, storage, and distribution) as well as wastewater collection and treatment systems are recognized as critical infrastructures and the US congress included a drinking water component in PL 107-188, the Public Health Security and Bioterrorism Preparedness and Response Act (Bioterrorism Act, 2002). Under the Homeland Security Presidential Directive to protect critical infrastructures (HSPD-7) water utilities in the United States serving more than 3,300 people were required to undergo physical vulnerability assessments and develop Emergency Response Protocols (ERP) that specifically addressed findings of the vulnerability assessments. Based on the findings of the vulnerability assessments, utilities performed physical hardening of their facilities (i.e., fences,

Z. Bukhari (✉)
American Water, Voorhees, NJ 08043, USA
e-mail: zia.bukhari@amwater.com

R.M. Clark et al. (eds.), *Handbook of Water and Wastewater Systems Protection*, Protecting Critical Infrastructure, DOI 10.1007/978-1-4614-0189-6_18,
© Springer Science+Business Media, LLC 2011

keyless entry, alarms, and video surveillance) to deter unauthorized access. This is only the first step to improving security and at best, it can be deemed suitable for delaying access. It does not eliminate distribution system vulnerabilities, especially as distribution systems often extend over hundreds of miles of pipe. The integrity of drinking water quality can be compromised by accidental contamination that may occur as a result of the ageing infrastructure of distribution system pipes and highly variable system pressures. Leaking pipes can cause cross connections or backflow incidents. Where external pressures exceed pressures within the distribution system pipes, potential intrusion of harmful agents into drinking water may occur. Currently, water utilities have no knowledge of the occurrence or frequency of either accidental (cross connections and backflows) or deliberate contamination (including backflow) events until the compromised quality water reaches the customer and is detected via taste/odor issues or following its public health impacts.

In January 2004, President Bush issued Homeland Security Presidential Directive-9 (HSPD-9), which established a national policy to defend the agriculture, water, and food system of the United States against terrorist attacks and other emergencies. HSPD-9 called on the federal agencies responsible for agriculture, food, and water security to "develop robust, comprehensive and fully coordinated surveillance and monitoring systems, . . . that provide early detection and awareness of disease, pest or poisonous agents." For the water industry, the US Environmental Protection Agency (USEPA) was assigned the primary role for guiding water utilities to develop security measures to protect public health from intentional or accidental contamination via the waterborne route. Since then, the USEPA has been developing a comprehensive approach to supplement physical hardening of utilities by examining usefulness of various water quality monitoring tools and emergency response protocols. To aid utilities in selecting appropriate water quality monitoring hardware, the USEPA began conducting reliability assessment of online sensors under the Environmental Technology Verification (ETV) program. As the ETV program was dependant upon voluntary vendor involvement, later this validation format was modified to the Technology Testing and Evaluation Program (TTEP). While this program is an offshoot of the ETV program, it is based on end user needs and has considerable stakeholder input, which is likely to make it more objective a validation framework. The following sections will provide an overview of approaches that utilities may use to detect accidental or intentional changes in water quality within the distribution system.

18.2 Multi-parameter Online Monitors

Given the multitude of agents (i.e., chemical, biological, and radiological) that could potentially contaminate drinking water, an online monitoring program targeting a specific agent or compound would only provide information on that specific compound and would serve little value as an Early Warning System (EWS) for other potentially hazardous agents. To address this, experts in the water industry consider

Table 18.1 Sentinel parameters for distribution system integrity

Parameter	Physical	Hydraulic	Water quality
Routine (primary)			
Pressure	X	X	
Turbidity	X	X (flow reversals)	X
Disinfectant residual		X (water age)	X
Main breaks	X		
Water loss	X		
Color	X (corrosion)		X
Coliforms	X (sanitary, main break)		X (biofilms)
Flow velocity and direction		X (pipes, tanks)	
pH, temperature			X
Secondary			
TOC			X
T&O	X (permeation)	X (water age)	X (biofilms)
Metals	X (corrosion)		X
Nitrite			X (nitrification)
HPC			X (biofilms??)
Tank level/volume		X	

Bold entries **indicate those parameters for which online real-time sensors are available**
From National Research Council (2006)

it appropriate to use various parameters to monitor their responsiveness to changes in distribution system integrity (Table 18.1) that could potentially pose a negative impact on consumers' health. Useful parameters may monitor for physical deterioration (pressure, main breaks, water loss, or corrosion), hydraulic failure (turbidity and complaints of low flow or pressure), and/or water quality aberrations (particulates, tastes, odors, or color).

Water utilities have routinely utilized online analyzers to measure various water quality parameters (i.e., chlorine, turbidity and pH). In a survey of 61 surface water treatment plants (LeChevallier and Moser, 1996) residual chlorine was the most common parameter (87% of the plants) used to monitor treatment plant effluents followed by turbidity (70%), pH (21%), and temperature (7%). While the water industry has routinely used most of these monitors individually, a number of manufacturers (Table 18.2) have developed multi-parameter sensor packages (Fig. 18.1) in panel-type assemblages or into more compact configurations (multi-probe or sondes) for installation at sensitive sites or in discrete locations within the distribution system.

The USEPA has examined reliability of online monitors from different manufacturers under the ETV Program (USEPA, 2004; Hall et al., 2005; Hasan, 2005). The parameters examined either individually or collectively have included free/total chlorine, turbidity, total organic carbon (TOC), oxidation-reduction potential (ORP), specific conductance, ammonia (NH_3), nitrogen (N_2), nitrate (NO_3^-) chloride (Cl^-), dissolved oxygen, pH, temperature, and UV spectrometry. The responsiveness of these parameters (individually or collectively) has been examined with various contaminants (Table 18.3). These studies have shown that useful

Table 18.2 Performance specifications of commercially available sensors

Manufacturer	Dascore	YSI	Hydrolab-Hach	Wallace & Tiernon	Analytical Technology, Inc		Emerson	Hach		ProMinent	In-Situ
Model	Sixcense	6-series	(DS5X, DS5, MS5)	Depolox 3 plus	Series Q45	Model A 15/B-2-1	Model 1055 Comp II Analyzer	WDM Panel or pipe Sonde^{TM}	TOC Process Analyzer	D1C &D2C	Troll 9000
Brief description	Multi-parameter, online, free chlorine or chloramines	Multi-parameter, online	Multi-parameter, online	Free or total chlorine (or chlorine dioxide or ozone)	Multi-parameter, customizable online	Free chlorine or chloramines	Multi-parameter, customizable, online	Multi-parameter, online	Used with WDM	Multi-parameter, online	Multi-parameter, online
Cost (US $)	9,700	15,000	15,000	3,500	<10,000	3,000	15,885	12,000	18,000	7,000	11,200
Free chlorine (mg/L)	0–5	–	–	0–20	0–2, 0–20, or 0–200	0–2, 0–20, or 0–200	0–10	0–4 (DPD)	–	0–0.5, 0–2, 0–10	–
Chloramines (mg/L)	0–20	–	–	–	0–2, 0–20, or 0–200	0–2, 0–20, or 0–200	0–15	0–5 (total)	–	0–0.5, 0–2, 0–10	–
TOC (mg/L)	–	–	–	–	–	–	–	–	<5 to 20,000	–	–
ORP (Volts)	–1.4 to 1.4	–0.999 to +0.999	–0.999 to +0.999	–	–0.999 to +2.0	–	–1.4 to 1.4	–1.5 to +1.5	–	–1.0 to +1.0	–1.4 to 1.4
SC (mS per cm)	0.1–10	0–100	0 to 100	–	0–0.2 and 0–40	–	0–200	0–100	–	0–200	0–200
DO (%)	0–200	0–500	0–200[a]	–	0–40 ppm	–	0–20 ppm	0–20 ppm	–	0.1–10 or 0.1–20 ppm	0–20 ppm
Turbidity (NTU)	–	–	0–3000	–	0.001–4 NTU + other wider ranges	–	0.001–200 NTU	1–100 NTU	–	–	1–2000 NTU
pH	2–12	0–14	0–14	–	0–14	–	0–14	0–14	–	0–12	0–12
Temp (°C)	0 to 50	–5 to 45	–5 to 50	–	–	–		–5 to 50	–	0–100	–5 to 50

[a]LDO: luminescent DO measurements. From Bukhari and LeChevallier (2006)

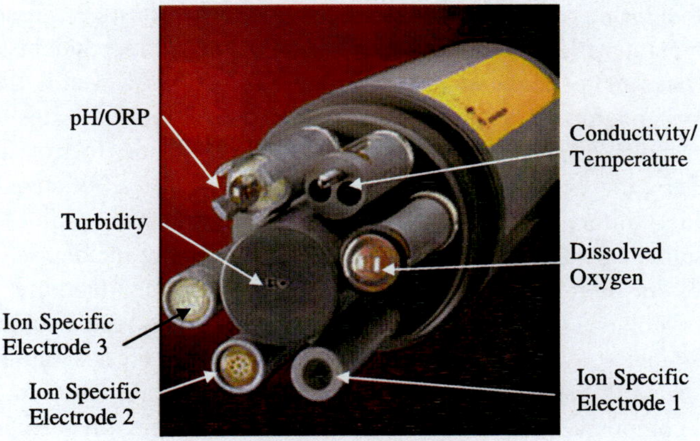

Fig. 18.1 Multi-parameter sonde

Table 18.3 Responsiveness of online parameters to various contaminants

Compound	Free/total chlorine	ORP	TOC	SC	Turbidity	NH_3	N_2	NO_3-	Cl–
Ferricyanide	NC (F+ w/DPD test)	+	++	+	+	F–		F–	F–
Malathion (pesticide)	+	+	+	NC	+				+
Glyphosphate (herbicide)	+	+	+	NC	NC				+
Nicotine (organic)	++		++	NC					
Arsenic trioxide	++	++	NC		+	++	++		
Aldicarb	++		++						
Groundwater	+	+	NC	+	NC				+
Wastewater	+	+	+	+	+				+

Key: ++ = very responsive; + = responsive; F+ = false positive; F– = false negative; NC = no change

Abbreviations: ORP, oxidation/reduction potential; TOC, total organic carbon; SC, specific conductance

From Hall et al. (2005)

parameters for gross contaminant detection include free/total chlorine, ORP, specific conductance, TOC, turbidity, and chloride ions.

Unfortunately, there is scattered information on how frequently utilities use online monitors for the purposes of an Early Warning Systems (EWSs). Information from nine US-based utilities (located in New Jersey, Washington, Florida, Oklahoma, Nevada, California, Utah, Arizona, and Connecticut) indicated varied approaches to online water quality monitoring of distribution systems (Schneider et al., 2010a). The most frequently monitored parameter was pH (9 of 9 facilities) followed by chlorine and conductivity (8 of 9), turbidity (6 of 9), dissolved oxygen and temperature (5 of 9), and TOC (3 of 9). Survey respondents monitored either raw water and/or distribution system water quality; however, it was apparent there was no standardization on parameters for online monitoring, sites for monitor deployment, and on responsiveness to water quality excursions. To provide guidance for the industry, the USEPA implemented the Water Security Initiative (originally Water Sentinel) in three distinct phases. In phase 1, the focus was to identify triggers most suited for identification of adverse water quality incidents: specifically the usefulness of online WQ monitoring (chlorine, TOC, conductivity, etc.); public health surveillance (over-the-counter drug sales, hospital admissions, infectious disease surveillance, 911 calls, poison control center calls, etc.); sampling/analysis (laboratory network-based grab sample analysis of chemical, microbiological, and radiological compounds); enhanced security surveillance (motion detectors, camera door contacts, access hatch detectors, etc.); and consumer complaints (taste, odor, or visual) was assessed. Phase 2 was initiated in June 2006 with piloting at the City of Cincinnati Water Works using all five of the components described in phase 1. Additionally a consequence management plan was implemented to deal with water quality excursions. Initially online monitoring parameters such as conductivity, free chlorine, pH, and turbidity were being monitored; however, turbidity was replaced with online TOC at a later stage of piloting.

The USEPA invested >$30 million to expand piloting from the City of Cincinnati Water Works to four additional larger public water utilities, which are located in California, Texas, Pennsylvania, and New York. The objective of these additional piloting efforts is to better understand the robustness of the EWS, which includes gathering additional operational, maintenance, and calibration information for the water quality monitoring equipment. Following the piloting efforts, phase 3 will focus on guidance and outreach for utilities to adopt the contamination warning systems.

18.3 Maintenance and Calibration

Sensor drift and/or fouling can be major causes for erroneous measurements; however, implementing strict quality assurance programs can help to reduce such errors. It is important to adopt system-specific maintenance protocols to minimize false alarms or erroneous data generation. In one study using at least a quarter of a million data points for each of the selected surrogate water quality parameters (free chlorine, ORP, conductivity, pH, temperature), it was determined that monthly calibrations were the minimum frequency necessary for most parameters to generate reliable water quality data (Bukhari and LeChevallier, 2006). Adopting routine calibration frequencies (i.e., monthly) will ensure optimal performance and timely replacement of individual sensors (or components) where necessary. Employing quality assurance-based criteria has also provided insights into the reliability of various water quality parameters. For example, systems employing membrane-based dissolved oxygen sensors were found to have frequent performance issues that were often associated with compromised membrane integrity. Other parameters (i.e., turbidity) were not adequately sensitive to measuring the low changes in treated water quality. In contrast, sensors for several surrogates (i.e., temperature, oxidation-reduction potential, specific conductance, pH, and free chlorine) indicated reliable performance with >80% of the data meeting the predefined quality control specifications.

From an operational perspective, a suite of sensors targeting selected water quality surrogates (free chlorine, ORP, conductivity, pH, and temperature) would readily allow detection of contamination with inorganic and/or organic compounds and their maintenance cycles would not present a significant operational challenge. While TOC is also an important parameter for detecting the presence of organic compounds, including contaminants of significance to human health, current analyzers are relatively expensive ($18,000–24,000) and maintenance intensive. Perhaps a more cost-effective alternative may be UV spectrometry.

18.4 Sites for Sensor Deployment

From a security perspective, the objective of online monitoring tools is to provide water utility operational staff with real-time information on water quality aberrations, which when used in conjunction with other triggers can serve as early warning tools to protect consumer health. The best case scenario would be to install water quality sensors at each customer's service connection to detect water quality excursions and when an aberration in water quality did occur, the event could automatically trigger a valve to shut down the impacted customer's water supply and alert the utility of this incident. The utility could analyze the water quality excursion data to characterize the contaminant and then feed the contaminant characteristics into verified hydraulic models to simulate its fate and transport through the distribution system. Using this approach would allow the utility to tailor their operational responses to minimize disruptions and specifically target isolation/containment to customers in the vicinity of the incident. Sensor maintenance, calibration, data quality assurance requirements, and sensor costs are some limitations to realizing this ideal scenario. During sensor deployment there are a number of other logistical factors that also need to be considered, which include presence of existent flows, availability of a power source, site ownership (i.e., pump or well stations), and access to sanitary sewer.

Initially, the cost of multi-parameter units exceeded $10,000 with some individual stand-alone units approaching $50,000; however, recently the price has began to stabilize below $10,000. Despite this, it remains cost-prohibitive to have units located at every customer's service connection and strategic deployment of sensors becomes necessary. Possible approaches for sensor deployment include installation at locations providing either proportional coverage of total demand or at specific sources, nodes, and tanks to detect deliberate intentional contamination (Bahadur et al., 2003; Baxter and Lence, 2003; Berry et al., 2004).

Various research efforts (i.e., Sandia National Laboratories and USEPA's Threat Ensemble Vulnerability Assessment [TEVA] program) are also underway to define optimal sensor locations (Lee et al., 1991; Murray et al., 2004; Ostfeld, 2004; Ostfeld and Salomons, 2004; Uber et al., 2004a, b) using hydraulic models. A collaborative study between the USEPA, US Geological Survey (USGS), and American Water (AW) used Monte Carlo simulations to vary parameters, such as the quantity or concentration of contaminant, location of injection, duration (or rate) of injection, and the probability of ingesting an infectious or toxic dose of selected contaminants to generate threat ensembles (collections of many threat scenarios). These threat ensembles were collectively analyzed for estimating health impact statistics, which included mean infections or mean fatalities. The public health impacts of no sensors in the distribution system were compared with benefits of sensors deployed by a utility at convenient locations versus locations determined by using the TEVA designs. During modeling it was assumed that all the nodes ($n = 1{,}834$) were equally vulnerable sites for the introduction of biological and chemical contaminants.

Several contaminant concentrations were evaluated with each sensor design listed in Table 18.4. In these simulations, it was assumed that biological contaminants were injected over a 24 h period whereas chemical contaminants were injected over a 9 h period. The time delay from detection of the contaminant (by the sensor) to the utility implementing a mitigation response (to limit further exposure) was assumed to be zero. The underlying assumption was that the sensor detection and utility-associated response systems were perfect.

In the contamination simulations where no water quality sensors were deployed, the health impacts from a hypothetical biological attack were estimated at 22,287 fatalities (Table 18.4). High variability was noted (see mean and median values in Table 18.4) in the estimated risk calculations for a simulated biological attack, which may have arisen from the numerous assumptions (i.e., dose–response relationships, lethal doses, time to onset of symptoms, and time for effective treatment) that usually form the basis of such hypothetical computer simulations. Additionally, factors such as the time delay in contaminant analysis, identification and verification (i.e., sample collection, transportation, laboratory analysis, and confirmatory analysis), and the delay in implementation of the Consequence Management Plan (CMP) for event mitigation can also influence the numbers of individuals being impacted.

Table 18.4 Public health benefits provided by various sensor designs

Sensor design	Health impacts (fatalities) – biological attack			Health impacts (fatalities) – chemical attack		
	Median	Mean	Max	Median	Mean	Max
No sensors	980	1,544	22,287	158	139	284
Convenient installations	671 (31.5%)	1,015 (34.3%)	5,107 (77.1%)	110 (30.0%)	113 (18.70%)	284 (0%)
TEVA	227 (76.8%)	350 (77.3%)	2,730 (87.6%)	67 (57.6%)	78 (43.0%)	229 (19.0%)

Values in parenthesis are % public health protection relative to the system with no sensors

Deploying water quality sensors at convenient utility-selected locations helped to reduce the fatalities associated with a biological attack from 22,287 to 5,107, yielding approximately 77% protection of public health. In contrast deploying sensors using the TEVA design reduced fatalities associated with a biological attack to 2,730, yielding 87% improvement in public health protection compared to no sensors in the distribution system (Table 18.4).

In simulated chemical attack scenarios, sensor installation at convenient utility-selected locations yielded little (using mean values) or no (using maximum values) public health benefit when compared to complete absence of sensors (Table 18.4). These comparisons may imply that online sensors provide greater benefits against biological attacks than chemical attacks. In reality the interpretation is more complex. It is likely that intentional or accidental contamination of drinking water with significant quantities of a chemical would impart taste, odor, or color, which in itself is likely to serve as a deterrent for most consumers and therefore assist in limiting exposure. Additionally chemical contaminants at levels of significance to human health would likely yield immediate public health impacts in the exposed population. In contrast the public health impacts of a biological attack may not become apparent until organism incubation times (several days to weeks) and/or secondary cases of disease transmission (i.e., from person to person) have taken place. Without the aid of a fully coordinated syndromic surveillance program to complement water quality data generated from online sensors, it would be unreasonable to expect water quality monitoring programs alone to capture secondary cases of biological disease transmission. These data suggest that online sensors can assist in the early warning of drinking water contamination events; however, maximized benefits from online installation of sensors not only need optimized installation sites but also require optimization of the number of sensors in the distribution system. The ideal number of sensors is likely to be system specific and will vary depending upon the distribution system network, the number of service connections, the type of service connections (i.e., primarily residential or commercial), the size of the population being served, and the length of the distribution system pipes. In one system in southern New Jersey, consisting of over 1,300 miles of distribution system pipe, trade-off curves for optimal number of sensors were developed using biological attack simulations and indicated 25 sensors were optimal for approximately 90% public health protection from a biological attack. In contrast simulations of chemical attacks indicated that deployment of 10 sensors yielded approximately 42% protection, with only marginal improvements using 25 sensors (55%) or 40–100 sensors (60%).

18.5 Automated Meter Readers

Ideally, each customer's service connection would be equipped with sensors to monitor water quality excursions and aberrations could actuate valves to shut down supply to impacted customers. Simultaneously the unexpected deviations in water quality could serve as a trigger for notification of utility personnel and/or automated samplers to collect predefined volumes of water for laboratory-based confirmation

Fig. 18.2 Automatic meter reader

of the incident. Considering the present day costs of online monitors, this architecture is unlikely to be feasible from an economic perspective. Another, more viable, alternative may be to develop the above-described conceptual design with the aid of Automated Meter Readers (AMR; Fig. 18.2).

Conventionally residential water meters have been used to measure forward flow into a customer's premises; however, recent advances have enabled development of smart water meters that generate tamper or reverse flow alarms. Typically these advanced meters detect backflow by electronically comparing sequential readings, by rotation of the magnet within the register to detect backflow, or by pulse sensors running "out of sequence" during negative flow. Back pressure and backflow are two major mechanisms by which a substantial volume of water can travel in the reverse direction through a water meter. Thermal expansion can also be a contributor; however, backflow volumes by this mechanism are usually small and often occur over a more prolonged period. The reverse flow capability of AMRs may be useful to detect potential contamination of a system where back pressure is applied intentionally to pump contaminants through the meter and into the drinking water network against the normal pressure (or head) in the distribution system. Backflow occurs when the pressure in the distribution system drops (i.e., due to a main break or power outage) and water passes in the reverse direction through the meter and into the distribution system. If the backflow water is contaminated, it could accidentally compromise the drinking water quality in the distribution system. Although only the intentional back pressure scenario would be indicative of a terrorist activity, use of AMR can also help to potentially mitigate the impact of accidental events.

In a recent study, approximately 1.6% backflow events occurred in 40,000 meter reads (Schneider et al., 2010b). To avoid generation of nuisance alarms from low-level backflow events, it is necessary to establish some acceptable threshold levels. Some manufacturers use reverse flow alarms, which generate alerts only when predefined volumes (i.e., 30 gallons) have passed. Other manufacturers electronically compare readings at specific intervals (i.e., 15 min) and transmit backflow information, which is categorized into one of two levels (i.e., *Level l*: 1–10 gallons in 15 min; *Level 2*: >10 gallons in 15 min). Backflow information collected in this manner needs to be analyzed immediately to provide water security benefits. An Advanced

Metering Infrastructure (AMI) utilizes a fixed network wireless data communication system and facilitates increased monitoring of customer usage (i.e., forward flow), backflow occurrence as well as two-way communication between the AMR transmitting units and the data processing center. This concept allows immediate communication of key data (i.e., accidental or intentional backflow flags) as soon as they are registered. Detection of a cluster of back flowing meters can signal a loss of distribution system pressure due to a main break or a negative pressure transient (Schneider et al., 2010b). However, a single meter flowing backward for a prolonged time could indicate an intentional back pressure event or a deliberate tampering of the meter (e.g., theft of the water). If the meter was associated with an abandoned building, a closed service account, or some other situation where water would not normally be used, this situation would warrant immediate criminal investigation. Moreover, the technology exists (although not widely used) to remotely operate service line valves through a two-way AMI system. With this technology, it would be feasible to isolate or terminate water supply to specific service lines where alerts are generated due to the occurrence of suspicious backflow events.

18.6 Remote Data Communications

To retrieve real-time water quality information from remote locations in the distribution system where water quality sensors or AMR units are deployed, integration of data into an existing Supervisory Control and Data Acquisition (SCADA) network becomes desirable. Majority of the commercially available multi-parameter online monitors are equipped with analog (4–20 mA) or digital data outputs (RS-232, RS 485), thus allowing connectivity via Remote Telemetry Units (RTU), radio transmitters, or cellular/satellite modems. Additionally, reliable transmission of real-time online water quality data can be achieved by using several means: licensed and unlicensed radio frequencies; Ethernet over the local (LAN) and wide area network (WAN); leased lines, frame relay mode; and wireless modems over the public telephone network. In the future, as water utilities migrate completely toward AMR, it will become increasingly popular to shift from drive-by systems to fixed network AMI systems that utilize radio frequencies with solar-powered equipment operation. Such infrastructures may also serve as communication platforms to relay water quality data from remote monitoring locations into the centrally located servers. The advantage of such systems is that they not only serve the normal operating requirements of the water utility, but also provide the additional benefit of providing perimeter-security at each service connection within the distribution system – at essentially little or no additional cost.

18.7 Data Management

When a water quality incident does occur, it is desirable for operators to have immediate notification so as to begin implementing countermeasures to safeguard consumer health. To date, efforts have been focused on establishing the reliability of online monitors, devising procedures for their strategic deployment and data transmission into SCADA. These are incremental steps to improve responsiveness to unanticipated (i.e., intentional or unintentional) changes in drinking water quality, but these alone do not arm the utility with the necessary knowledge to respond appropriately. The situation is exacerbated by the fact that online monitors generate large quantities of data, which render manual data management impossible. Automated data analysis can alert operators to statistically derived water quality anomalies and various manufacturers have developed suites of algorithms for such purposes. Direct processing of water quality data with such software is likely to generate high false alarms, which would potentially erode operator confidence in the real-time monitoring systems. Such a phenomenon can lead to a demise of the Early Warning Systems (EWSs), leaving the distribution system poorly equipped to handle future water quality events.

Because the decisions associated with responding (or failing to respond appropriately) to online monitoring data can have huge social, political, economic, and health implications, processes are necessary to increase the confidence in analytical interpretations of water quality data. Some of the data management approaches that utilities can employ are depicted in Fig. 18.3. Initially, utilities can hook the monitors into SCADA to perform range measurements, data filtering, and simple trending and analysis using manual or SCADA systems. Additional layers could become sequentially more complex and may utilize multivariate analysis or data mining, which when integrated with operational data could help to identify outliers and generate alarms (Fig. 18.3). By comparing real-time data with baseline historical data it would be possible to define and characterize anomalies, allow user-defined and programmable triggers to both the nature and level of contaminant, and

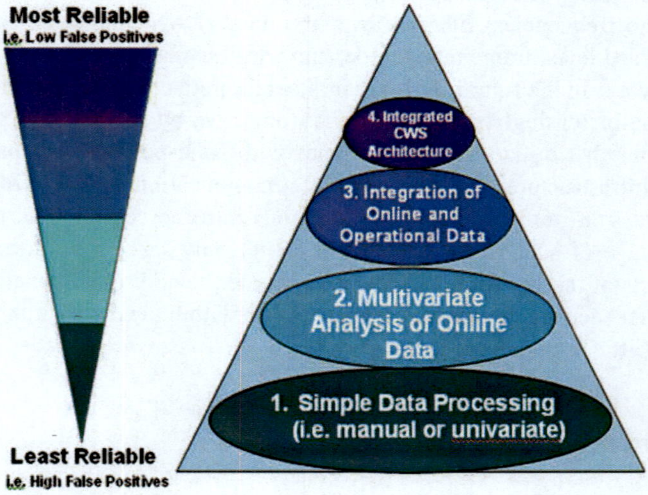

Fig. 18.3 Development of a "tiered" data management system

provide automated notification by means of alarms (on cell phones, pagers, or via e-mail) to facilitate early detection and response.

Several independent companies have developed tools to facilitate data analysis, web server-based data storage, and remote data accessibility. Some sensor manufacturers have also developed software for automated trend analysis; however, these software are intended to aid the manual process of data interpretation and would be placed in step 1 (first tier) in Fig. 18.3. Use of intelligent software that "learns" normal water quality and employs multivariate analyses to identify anomalies in water quality to generate alarms and make recommendations to operators on the appropriate course of action is either under development or in the β-testing phase.

Two software systems that meet these specifications and are currently being tested by the USEPA's Water Security Initiative include the Event Monitor™ Trigger (Hach) and H2O Sentinel™ (Frontier Technologies). Briefly, the Event Monitor™ Trigger (Fig. 18.4) is part of Hach's GuardianBlue® technologies for Homeland Security and approved under the US Department of Homeland Security SAFETY Act, which provides both the utility user and the manufacturer protection from litigations in cases where contamination incidents do occur. According to the manufacturer, data analyses are performed every 60 s using patented algorithms that monitor the rate of change relative to the baseline. This system can also generate a specific water quality "fingerprint," which can then be compared with an updatable agent library that contains fingerprints for a wide variety of threat contaminants (i.e., pesticides, heavy metals, and warfare agents) to enhance the specific identity of contaminants. Additionally this system is also capable of storing a plant library, which can assist in identification of various changes (i.e., cross connections, water aging, biofouling, corrosion by products, and nitrification) that ultimately lead to degradation of water quality and/or aesthetics.

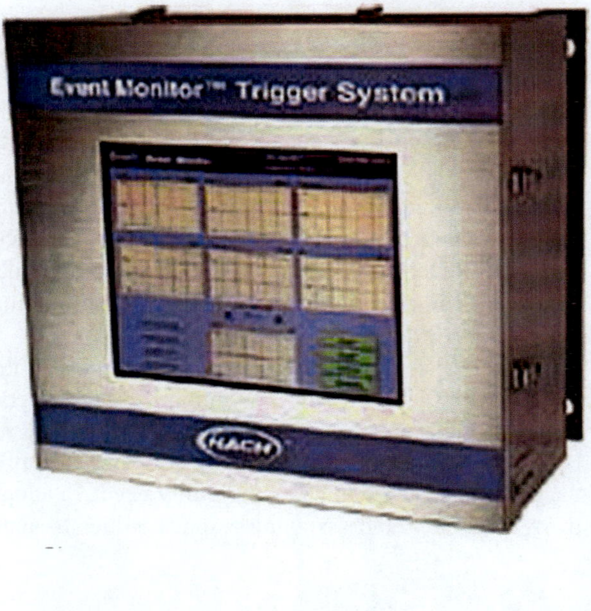

Fig. 18.4 GuardianBlue® Event Monitor™ Trigger System

The H2O Sentinel™ Event Detection System (Frontier Technology Inc) is based on the NormNet™ core technology, which uses statistical/signal processing and Pattern Recognition of Health (PRoH™) combined with Bayesian decision control elements to provide users warning when abnormal conditions are developing. This system employs fully automated training, using a wide variety of (present and past) data collected under normal operating conditions to uniquely characterize sensor performance over a subset of the training data and develops multiple statistical transfer function models (TFMs). Once TFMs are developed they can be stored in a database of models for analysis of real-time operational and water quality information. This is achieved by the system automatically selecting the best TFM from the existing model base using a rapid nearest-neighbor search and then applying this TFM to generate a predicted value for new data points. This function applies a dynamic statistical hypothesis test that "the system variable is normal" to residuals for each of the variables of interest. Based on the outcome of all of the hypothesis tests, the a priori probability that the system is abnormal, and each test's statistical residuals, a dynamic overall probability of abnormality is calculated. At each time step, the tool outputs a floating point number that indicates the probability of abnormality and abnormality severity.

Unlike the Event Monitor™ Trigger, the H2O Sentinel™ is a software-based solution, which has the potential to function as a standalone system or work across existing data network infrastructures, integrate with SCADA, adapt to existing sensor arrays, or interface into existing management systems. While considerable progress has been made on data management tools, to the authors' knowledge, no single software has been exhaustively tested by the water industry in extensive field-scale applications. Field testing will assist in providing data on long-term robustness of these tools and will be important to help utility personnel gain better understanding of the predicative capacity of these statistically sophisticated software.

18.8 Consequence Management Plans

After a potential water quality incident, the water utility goals are to investigate the incident thoroughly, determine its validity, and, where appropriate, carry out the necessary mitigation steps to resume service to the impacted customers as quickly and safely as possible. Following the Bioterrorism Act (2002), water utilities were mandated to perform vulnerability assessments and develop Emergency Response Plans (ERP), which were intended as a compendium of specific Consequence Management Plan (CMP) for a variety of "possible" incidents including distribution system water quality excursions. The purpose of the drinking water distribution system water quality CMP is to minimize the impacts of accidental or intentional water quality excursion by developing, adopting, and regularly exercising a streamlined plan, which is intuitive to all critical water utility personnel. Naturally for the CMP to remain effective, the utility needs to adopt a practice of making this a living document and assigning responsibilities to appropriate personnel to ensure all relevant operational information remains current. Furthermore, the impacts on the drinking water system bring with them significant social, political and logistical pressures and there is the need for the utility to navigate through this smoothly and effectively in an expedient manner to reach the ultimate goal of resuming normal operation as quickly as possible while safeguarding consumer health. To achieve this, the utility will likely require support from external partners (i.e., police, fire, HAZMAT, laboratories, and public health agencies), and establishing/maintaining relationships with such entities is a key to implementing a successful CMP. Using the National Incident Management System (NIMS) structure can help water utilities be proactive, flexible, effective, and efficient and allows sharing of a common structure and language with various organizations (i.e., police, fire, and HAZMAT) so each may work seamlessly with the other to recover from and mitigate the effects of incidents, regardless of cause, size, location, or complexity.

The USEPA framework for a CMP has been modified slightly in Fig. 18.5. The framework is divided into distinct steps where based on one or more initial triggers, the potential threat passes through three distinct threat evaluation stages (possible, credible, and confirmed). While the USEPA has proposed the "possible" stage to allow up to 1 h of consideration, the authors feel the utility needs to err on the side of caution and arrive at a "possible" determination much quicker; ideally within 15 min. This allows navigation to the "credible" determination phase so that the utility can focus on initial site characterization and operational responses to protect consumers. Use of online monitoring equipment or AMR systems would serve as invaluable tools during the "credible" determination of water quality excursions and corroboration with these analytical tools would allow utilities to speed up determination of this threat stage. Where utilities do not have such equipment then site visits will be the only available option. At this point it is important to make a characterization of whether the threat poses a high or low hazard situation. The safety of the utility response personnel needs to be considered first and HAZMAT teams should be used in all situations other than those known to present a low hazard.

The "confirmed stage" requires expanded sampling, enhanced operational responses, and public notifications as appropriate. Where the utility has no internal laboratories, various other resources can be tapped for analytical support. Specifically the utilities should consider advance registration with Environmental Response Laboratory Network (ERLN) or Water Laboratory Alliance by visiting http://www.epa.gov/compendium. Membership can also provide access to National Environmental Methods Index for chemical, biological, and radiological agents at the following site: http://cfpub.epa.gov/safewater/watersecurity/nemi-cbr.cfm.

Depending on the nature of the incident (i.e., catastrophic or terrorism event), the remediation recovery stage may be under the Federal Response Network, which will likely establish an Incident Command System (ICS) with various support agencies. By adopting a NIMS structure, utilities will be prepared in advance for what is required from their response team and what is expected from supporting agencies from the local level to the federal level to respond to the contamination incident. During the "remediation & recovery" stage an additional tool to aid the utilities is the Water Contamination Information Tool (WCIT), which supports

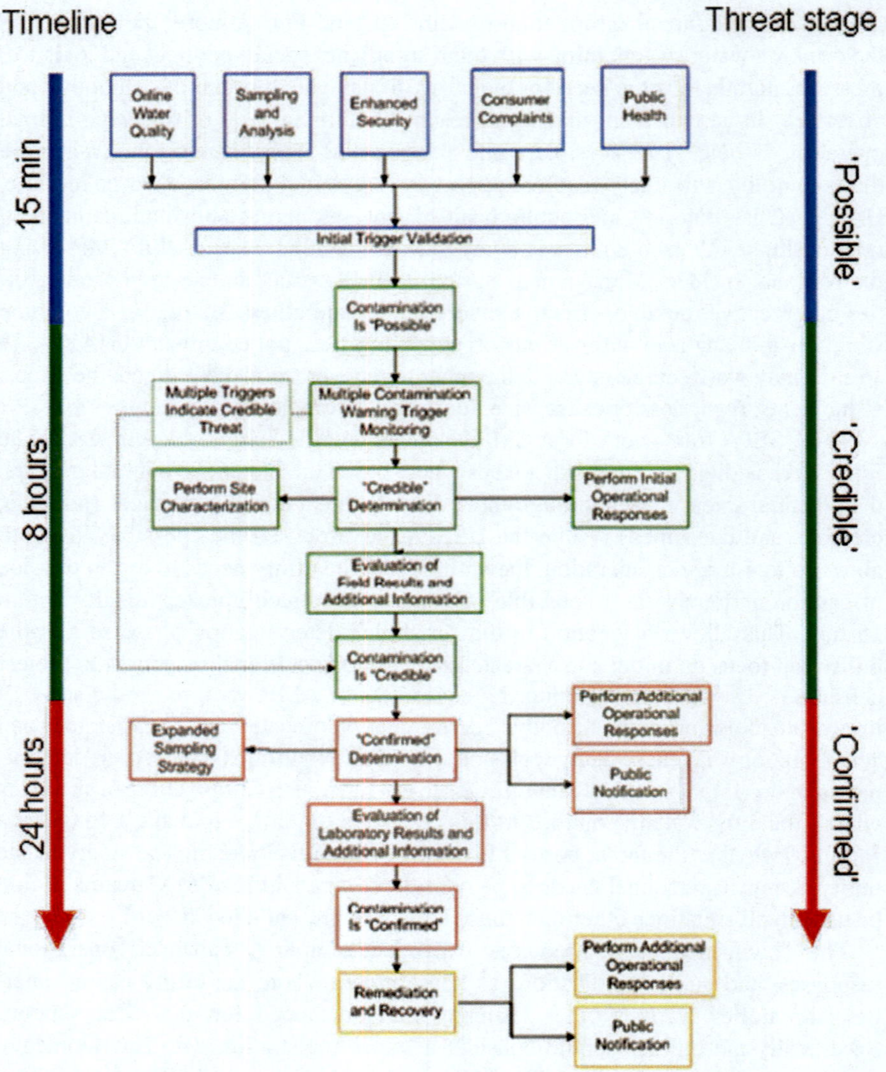

Fig. 18.5 A framework for a Consequence Management Plan

vulnerability assessments, emergency response plans, and site-specific response guidelines. Further details on WCIT are available at http://www.epa.gov/wcit. Additionally where the utility requires multiple laboratories to perform analysis of a single analyte then using the USEPA-developed Standardized Analytical Methods (SAMs) can serve as an additional resource. SAMs are located at http://www.epa.gov/sam/.

While utilities have access to ample information to navigate through evaluation of a water quality threat, it will be imperative for the utility to be thoroughly familiar with each step to respond in an expedient manner. A possible solution is to develop "tear-out" sheets, which can provide guidance to "front-line" utility responders to facilitate navigation from the "possible stage," through the "confirmed" stage to public notification and System Remediation/Recovery. Additionally, and perhaps most importantly, utilities should have a plan in place to keep CMP documents current and regularly exercise their plans to ensure all respondents consider CMP requirements to be second nature.

18.9 Conclusions

Distribution system water quality protection in real time is necessary to ensure accidental or intentional water quality events are detected and appropriately mitigated in a timely manner, before they have significant public health impacts. Based on a concerted effort by the Homeland Security Research Center at the USEPA and independently by various equipment manufacturers, considerable progress has been made in our understanding of monitoring parameters, optimal locations for sensor placement, and optimal number of sensors in a given distribution system. Even though water utilities are probably able to capitalize the expenditure associated with equipment acquisition and deployment, significant resistance exists to adopting these technologies for routine monitoring. Data reliability issues and appropriateness of response steps in the event of water quality alarms have been presented as hurdles to adopting an online monitoring program. The maintenance intensity of the equipment is also deemed an unacceptable burden for the already overworked and understaffed utilities personnel. Conceptualizing the multi-benefits of online monitoring (i.e., distribution system disinfectant optimization, disinfection by product reduction, total coliform or corrosion control) has also gained little traction. Each day that passes after 9/11, the interest in the security of distribution system water quality has been waning. What will it take to get drinking water utilities to adopt these technologies?

It appears that improving the reliability of the monitoring instrumentation or making their performance more robust may not be adequate. Now that the dust has begun to settle, it is becoming clear that no regulations will be forthcoming for real-time distribution system monitoring from a security perspective. Water utility personnel are the guardians of the nation's drinking water and online monitoring, despite their issues, can still be an invaluable tool in their arsenal to ensure their consumer's health is protected. It is important to recognize that anything but real-time water quality data will leave drinking water distribution systems vulnerable. In consideration of these limitations, the conceptual design proposed here is to make use of the AMI as the first line of defense against deliberate backflow events. Where deliberate backflows are suspected, the utility could minimize spread of contaminant by exercising remotely controlled isolation values. Given this scenario, the utility would be aware of the contaminant intrusion site, which when complemented with verified hydraulic models could be used to estimate the contaminant levels at various locations. Both the AMI signal and hydraulic model information would serve to establish a strong possibility of an incident, which when corroborated by strategically located online monitors could establish whether an incident is credible. Once the incident credibility is established, the utility can then confirm results using laboratory-based analytical procedures and, where appropriate, use remediation/recovery measures to ensure the system returns to normal operation as quickly as possible. Without implementing these technologies the utilities, while meeting all regulatory standards, will not be able to guarantee they are safeguarding consumer health.

References

Bahadur, R., Samuels, W., Grayman, W., Amstutz, D., and Pickus, J. 2003. PipelineNet: A Model for Monitoring Introduced Contaminants in a Distribution System. World Water & Environmental Resources Congress, EWRI-ASCE.

Baxter, C.W. and Lence, B.J. 2003. A Framework for Risk Analysis in Potable Water Supply. World Water & Environmental Resources Congress, EWRI-ASCE.

Berry, J., Hart, W., Phillips, C., and Uber, J. 2004. A General Integer-Programming-Based Framework for Sensor Placement in Municipal Water Networks. World Water & Environmental Resources Congress, EWRI-ASCE.

Bioterrorism Act. 2002. Public Health Security and Bioterrorism Preparedness and Response Act of 2002. Public Law 107–188; 107th Congress. http://www.fda.gov/oc/bioterrorism/PL107-188.html.

Bukhari, Z. and LeChevallier, M.W. 2006. Enhanced Monitoring to Protect Distribution System Water Quality. American Water Works Association, Water Quality Technology Conference Proceedings.

Hall, J., Zaffiro, A., Marx, R.B., Kefauver, P., Krishnan, R., Haught, R., and Herrman, J.G. 2005. Parameters for Rapid Contamination Detection in a Water Distribution System. AWWA-Water Security Congress, Oklahoma City.

Hasan, J. 2005. Technologies and Techniques for Early Warning Systems to Monitor and Evaluate Drinking Water Quality: State-of-the-Art Review. http://www.epa.gov/ordnhsrc/news/news120105.htm

Homeland Security Presidential Directive-7 (HSPD-7) 2003. Critical Infrastructure Identification, Prioritization, and Protection. http://www.whitehouse.gov/news/releases/2003/12/20031217-5.html.

LeChevallier, M.W. and Moser, R.H. 1996. Production Facility Management Study: Automation Strategies for American Water System. Company Report, American Water.

Lee, B., Deininger, R., and Clark, R. 1991. Locating Monitoring Stations in Water Distribution Systems. *J. AWWA* 83(7): 60–66.

Murray, R., Janke, R., and Uber, J. 2004. The Threat Ensemble Vulnerability Assessment Program for Drinking Water Distribution System Security. World Water & Environmental Resources Congress, EWRI-ASCE.

National Research Council. 2006. *Drinking Water Distribution Systems: Assessing and Reducing Risks*. The National Academies Press, Washington, DC.

Ostfeld, A. 2004. Optimal Monitoring Stations Allocations for Water Distribution System Security. *In: Water Supply Systems Security*, Edited by Larry Mays (pp. 16.1–16.15), McGraw-Hill, New York, NY.

Ostfeld, A. and Salomons, E. 2004. Optimal Layout of Early Warning Detection Stations for Water Distribution Systems Security. *J. Water Res. Plann. Manage*. 130(5): 377–385.

Schneider, O.D., Hughes, D.M., Bukhari, Z., LeChevallier, M., Schwartz, P., Sylvester, P., and Lee, J.J. 2010a. Determining Vulnerability and Occurrence of Residential Backflow. *J. AWWA* 102(8): 52–63.

Schneider, O.D., Bukhari, Z., Hughes, D.M., LeChevallier, M., Schwartz, P., Sylvester, P., and Lee, J.J. 2010b. "Determining Vulnerability and Occurrence of Residential Backflow". Water Research Foundation Report. ISBN 978-01-60573-114-8.

Uber, J., Janke R., Murray, R., and Meyer P. 2004a. Greedy Heuristic Methods for Locating Water Quality Sensors in Distribution Systems. World Water & Environmental Resources Congress, EWRI-ASCE.

Uber, J., Shang, F., and Rossman, L. 2004b. Extensions to EPANET for Fate and Transport of Multiple Interacting Chemical or Biological Components. World Water & Environmental Resources Congress, EWRI-ASCE.

U.S. Environmental Protection Agency. 2004. Environmental Technology Verification Program: http://www.epa.gov/etv/

Chapter 19
Testing and Evaluation of Water Quality Event Detection Algorithms

Sean A. McKenna, David B. Hart, Regan Murray, and Terra Haxton

19.1 Introduction

Protecting our nation's critical infrastructure from terrorist attacks has become a priority over the last several years. Recent water security research efforts have focused on the advancement of methods for mitigating contamination threats to drinking water systems. A promising approach for the mitigation of both accidental and intentional contamination is a contamination warning system (CWS), a system to deploy and operate online sensors, other surveillance systems, rapid communication technologies, and data analysis methods to provide an early indication of contamination (see ASCE, 2004; AWWA, 2005; US EPA, 2005).

The online monitoring component of a CWS is composed of multiple sensor stations that collect data continuously and transmit it to a central database in a control room, most commonly a Supervisory Control and Data Acquisition (SCADA) database. Various types of sensors, which can be categorized as direct or surrogate, have been considered as part of a CWS. Direct sensors detect specific contaminants whereas surrogate sensors indirectly detect the presence of one or more contaminants through changes in water quality values. For example, pH, chlorine, electrical conductivity, oxygen-reduction potential, and total organic carbon can be considered as surrogate sensors for multiple contaminants. These typical water quality parameters tend to vary significantly in water distribution systems due to normal changes in the operations of tanks, pumps, and valves and daily and seasonal changes in the source and finished water quality, as well as fluctuations in demands. Therefore, an event detection system (EDS) is needed to distinguish between periods of normal and anomalous water quality variability from measures made with surrogate sensors.

Disclaimer: The information in this document has been funded wholly or in part by the U.S. Environmental Protection Agency (EPA). It has been subjected to the Agency's peer and administrative review and has been approved for publication as an EPA document. Mention of trade names or commercial products does not constitute endorsement or recommendation for use.

S.A. McKenna (✉)
National Security Applications Department, Sandia National Laboratories, Albuquerque, NM, USA
e-mail: samcken@sandia.gov

R.M. Clark et al. (eds.), *Handbook of Water and Wastewater Systems Protection*, Protecting Critical Infrastructure, DOI 10.1007/978-1-4614-0189-6_19,
© Springer Science+Business Media, LLC 2011

A critical premise for online monitoring of surrogate parameters is that recognizable variations in water quality signals will occur in the presence of certain contaminants of concern. A number of experiments conducted in laboratory and pipe test loop systems have explored this assumption and concluded that many contaminants cause surrogate parameters to diverge significantly from background levels (Byer and Carlson, 2005; Cook et al., 2005; Hall et al., 2007). In particular, Hall et al. (2007, 2009) tested the response of a number of commercially available water quality sensors in the presence of nine different contaminants introduced to a pipe loop at different concentrations and found that at least one of the surrogate parameters changed in response to the presence of every contaminant.

These laboratory studies indicate rapid and significant changes of water quality parameters (surrogates) in the presence of many contaminants of concern. However, conditions in distribution systems involve much more complex background variations in water quality parameters than found in the laboratory. The purpose of an EDS is to automatically and rapidly distinguish between changes due to the presence of contaminants and changes due to normal background variability.

Typically, an EDS reads in SCADA data (e.g., water quality signals and operations data), performs an analysis in near real time, and then returns the calculated probability of a water quality event occurring at the current time step. A water quality event is defined as a time period over which water with anomalous characteristics is detected. The working definition of "anomalous" can be set by the user by selecting configuration parameters that govern the sensitivity and operation of the EDS. The values of these configuration parameters will vary from one utility to the next and can even vary across monitoring stations within a single utility.

Increasing installation of online water quality sensors in distribution networks and their connection to SCADA systems has significantly increased the amount of water quality data available to system operators and network analysts. As an example, a modest online monitoring system consisting of 10 monitoring stations with five water quality parameters monitored at a 5 min sampling interval will provide 14,400 water quality records per day, or 5.26 million records per year. The possibility of massive amounts of real-time data overwhelming the operators and analysts is real, and automated approaches to making sense of these data are needed. Investment in automated approaches such as EDS will allow a utility to detect and characterize changes in the water quality as well as mine the historical data for recurring patterns and trends. Information derived from these data can then be used to more effectively operate the distribution network.

19.1.1 Background

Event detection from time series data is a research topic in a large number of fields, including tsunami detection, traffic accidents analysis, mechanical component failure, system fault detection, data mining, and network intrusion detection. Based on reviewing developments in these other fields for their relationship and applicability to the EDS problem in water distribution networks, two main categories of event detection can be identified: off-line and online. Off-line, or batch mode, analysis is done on previously collected, or historical, data sets. Online, or real-time, analysis is done in real time on data that are input to the EDS tool as soon as they become available.

Off-line approaches are generally concerned with change point detection (see Raftery, 1994) where change points are defined as the point in time where an abrupt change in the nature of a signal occurs. In off-line analysis, the full data set has already been recorded and data from opposite sides of any proposed change point can be examined to determine if those two data sets are significantly different. In the online scenario, only the data recorded up to the present time are available, and the goal is to identify the change point as close to the time at which it occurs as possible.

Online EDS tools generally consist of a two-stage approach to event detection. The first stage predicts a future water quality value, often based on recently observed water quality values. A wide variety of prediction tools are available, including neural networks, support vector machines, and calibrated water quality models. Our focus here is on traditional time series and multivariate statistical approaches (e.g., Box and Jenkins, 1976; Bras and Rodriguez-Iturbe, 1993). Different statistical models applied to the previously observed data can provide predictions of future water quality values. The process of making the prediction is referred to as state estimation. In the second stage of event detection, the prediction of the expected water quality value is compared to the observed water quality value as it becomes available. The residual between the prediction and the observation is classified to determine if the water quality at that time step is either expected or anomalous. If the residual is relatively small, the predicted and observed water quality values are similar and the water quality is as expected or representative of the background water quality. If the residual is relatively large, the observed water quality value is quite different from what was predicted, and this indicates an anomalous observation. This second stage is called residual classification.

To date, the majority of event detection methods for drinking water distribution networks involve monitoring of surrogate parameters. Current approaches to event detection in drinking water distribution systems are described in Byer and Carlson (2005); Cook et al. (2006); Jarrett et al. (2006); Kroll and King (2006); McKenna et al. (2008); and Yang et al. (2009).

19.1.2 CANARY Software

Here the CANARY software (Hart and McKenna, 2009) is used to examine differences in EDS algorithms and follow-on analyses. The CANARY EDS software has been developed at Sandia National Laboratories in collaboration with EPA's National Homeland Security Research Center (NHSRC). Additional functionality for reducing false alarms has been added to CANARY through engagement with the Singapore Public Utility Board (PUB). CANARY was written using the MATLAB® (MathWorks, 2008) programming language and is distributed as both the MATLAB® source and as an executable program under an open source license. CANARY can be connected to a utility SCADA database directly or through a third-party software connection. All water quality signals contained in the SCADA database can be used as input to CANARY. In addition to water quality data, these signals can also include hydraulic data such as tank levels, flow rates, and valve settings as well as sensor hardware alarms and calibration alarms.

CANARY provides a platform upon which different event detection algorithms can be developed and tested. These algorithms process the water quality data at each time step to identify periods of anomalous water quality and provide the probability of a water quality event existing at that time step. This probability is calculated with respect to the recent water quality values.

CANARY is intended as a research tool to help water utilities and others in the water community better understand normal background fluctuations in water quality and to begin to identify anomalies that are potentially indicative of contamination incidents. To be used as part of a CWS, the utility must integrate CANARY with a well-tested consequence management plan in order to respond effectively and in a timely manner to potential contamination threats.

19.1.3 Chapter Overview

This chapter is divided into several sections. This first section, Section 19.1, provides an introduction and background on event detection systems. The next section, Section 19.2, provides a technical overview of the event detection algorithms and their use within CANARY. Section 19.3 summarizes results of testing and sensitivity analysis of the CANARY algorithms. A discussion of the different parameters analyzed is presented in Section 19.4. The final section, Section 19.5, outlines applications of CANARY and outstanding challenges and research needs. Additional background and details on this work can be found in US EPA (2010).

19.2 Event Detection Algorithms

Two different state estimation algorithms for predicting water quality values have been developed and implemented within CANARY: a linear filter and a multivariate nearest-neighbor algorithm. Detailed descriptions of these algorithms are provided in this section. The residual time series resulting from the application of these algorithms are classified using a threshold comparison that takes into account the relative variability in the background water quality signal. The outcomes of the residual classification over multiple consecutive time steps are combined to provide a probability of an event at each time step using a binomial event discriminator (BED).

19.2.1 Linear Filter

The linear prediction-coefficient filter (LPCF) model uses a linear predictor to estimate the current value of a time series based on a weighted sum of past values. In its most general form, this approach is also known as an autoregressive (AR) model (Bras and Rodriguez-Iturbe, 1993). The more compact form of the AR model is

$$\hat{z}(t+1) = \sum_{i=1}^{P} a_i z(t-i+1) + \delta(t+1) \tag{19.1}$$

where a_i are the estimation coefficients, P is the order of the estimation filter polynomial (number of previous measurements), and $\delta(t+1)$ is the estimation error. The error, or residual, generated by this estimate is

$$\delta(t+1) = z(t+1) - \hat{z}(t+1) \tag{19.2}$$

where a mean-zero Gaussian distribution defines δ. In CANARY, the autocorrelation method of AR modeling is used to estimate values of the parameters, $\boldsymbol{a}$. This formulation is set up as a linear system, $\boldsymbol{Z}\,\boldsymbol{a} \approx \boldsymbol{b}$, where Z is a function of time. Expansion of this equation results in

$$\boldsymbol{Z} = \begin{bmatrix} z(t) & 0 & \cdots & 0 \\ z(t-1) & z(t) & \ddots & \vdots \\ \vdots & z(t-1) & \ddots & 0 \\ z(t-P+1) & \vdots & \ddots & z(t) \\ 0 & z(t-P+1) & \ddots & z(t-1) \\ \vdots & \ddots & \ddots & \vdots \\ 0 & \cdots & 0 & z(t-P+1) \end{bmatrix} \quad \boldsymbol{a} = \begin{bmatrix} \boldsymbol{a}_1 \\ \boldsymbol{a}_2 \\ \vdots \\ \boldsymbol{a}_P \end{bmatrix} \tag{19.3}$$

$$\boldsymbol{b} = \begin{bmatrix} z(t+1) \\ z(t) \\ \vdots \\ z(t-P+2) \end{bmatrix}$$

Here for online operation, all entries in the linear system are updated at every time step and use only the most recent P observations such that $\mathbf{Z}$ has dimension of P. Updating at every time step allows the coefficients, $\mathbf{a}$, to adapt to the changing water quality values contained in the moving window of previous values. To the extent possible within the AR model, non-stationarity and periodicity in the water quality data are captured by calculation of the appropriate coefficients at each time step. Note that this system of equations is solved separately for each water quality variable at each time step.

The solution that minimizes the estimation error through linear least squares is generally solved as $\boldsymbol{a} = \left(\mathbf{Z}^{\mathrm{T}}\,\mathbf{Z}\right)^{-1}\,\mathbf{Z}^{\mathrm{T}}\,\boldsymbol{b}$, where $\mathbf{Z}^{\mathrm{T}}$ is the transpose of $\mathbf{Z}$. The parameter estimation method exploits the fact that there is a direct correspondence between the parameters a and the correlation function of the water quality signals. Consequently, the Yule–Walker equations are used to estimate the parameters by inverting such correspondence. Thus, the correlation coefficients, $\boldsymbol{\rho}$, calculated from the P previous measurements provide a solution for the coefficients in $\mathbf{a}$

$$\begin{bmatrix} 1 & \rho_1 & \rho_2 & \cdots & \rho_{P-1} \\ \rho_1 & 1 & \rho_1 & \cdots & \rho_{P-2} \\ \rho_2 & \rho_1 & 1 & \cdots & \rho_{P-3} \\ \vdots & \vdots & \vdots & \ddots & \vdots \\ \rho_{P-1} & \rho_{P-2} & \rho_{P-3} & \cdots & 1 \end{bmatrix} \begin{bmatrix} a_1 \\ a_2 \\ a_3 \\ \vdots \\ a_P \end{bmatrix} = \begin{bmatrix} \rho_1 \\ \rho_2 \\ \rho_3 \\ \vdots \\ \rho_P \end{bmatrix} \tag{19.4}$$

The subscripts in the above equation indicate the size of the lag spacing, in time steps, for which each correlation coefficient is calculated. Here the correlation coefficients are calculated in the frequency domain using an inverse and forward Fast Fourier Transform (FFT) on the previous P measurements of $\mathbf{Z}$. Since the equation is a Toeplitz matrix, the use of Levinson–Durbin (LD) recursion provides the most efficient solution for $\mathbf{a}$. Once $\mathbf{a}$ has been determined, it is inserted back into Eq. (19.1) and the current value of the signal is estimated.

19.2.2 Multivariate Nearest Neighbor

Another approach to state estimation that uses all water quality signals at each time step simultaneously to define the background state of the water quality is the multivariate nearest-neighbor (MVNN) algorithm (see Klise and McKenna, 2006a, b).

For each time step, all J water quality signals are combined into a vector:

$$\left[Z^{j=1}(t), Z^{j=2}(t), Z^{j=3}(t), \ldots, Z^{j=J}(t)\right] = \mathbf{z}^J(t) \tag{19.5}$$

The vector defines a point in the J-dimensional space at time t. If multivariate clustering is used to define K clusters, or classes, of water quality, the mean coordinate of the kth cluster in the J-dimensional space calculated over the previous P time steps is denoted by $\bar{\mathbf{z}}_k^J(t-P,t)$. Figure 19.1 shows a schematic example of this calculation in $J = 3$ dimensional space. The data in Fig. 19.1a have been classified into five water quality classes and the extent of these classes is shown in Fig. 19.1b. The distance between a new data point, red star in Fig. 19.1b, and the centroid of each existing cluster is calculated as a Euclidean measure.

The MVNN approach does not provide an estimate of the water quality at a future time step, but instead provides a measure of similarity of the sampled water quality with the P previously measured samples contained in the history window. The distance between the new water quality sample, $z^J(t+1)$, and the closest of the previous P water quality samples is measured as the Euclidean distance between

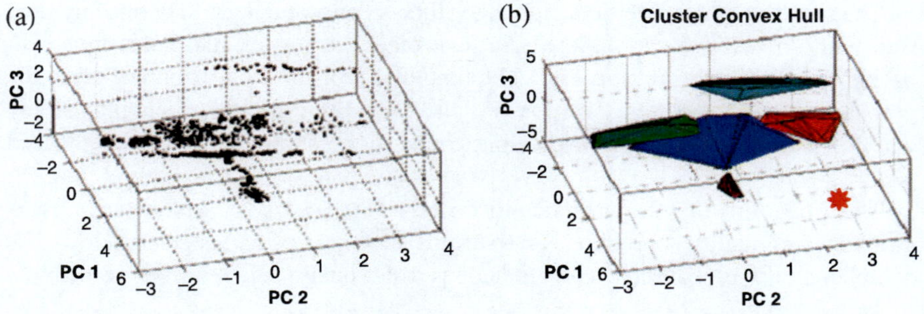

Fig. 19.1 Example of data classification in three-dimensional space. The normalized data in (**a**) are classified into five clusters (**b**), with a new data vector to be compared to the existing cluster centroids shown as a *red star*

the samples within the J-dimensional space. The minimum distance between the points is retained as the distance, Δ, which is compared to the threshold:

$$\Delta = \mathrm{Min}_{i=1\ldots P}\left|\sqrt[J]{\sum_{j=1}^{J} z^j(t+1) - z^j(P-i+1)}\right| \tag{19.6}$$

The Δ value can be calculated between the current water quality value and the mean locations of K previously defined clusters or it can be calculated for every previous sample separately. Work by Klise and McKenna (2006b) demonstrated that as K was allowed to approach P, event detection results improved. Contrary to the linear filter approach described above, the distance calculated with the MVNN is not a function of any individual water quality signal, but is a combined measure of the distance using all signals simultaneously.

19.2.3 Binomial Event Discriminator

Previous application of outlier detection algorithms has focused on classification of the water quality measurement vector (e.g., pH, chlorine, and total organic carbon) at every time step as either an event or background. A result of this approach is large numbers of false alarms tied to significant, but very short-term changes in the water quality, including significant changes most likely due to noise in the SCADA system. The Binomial Event Discriminator (BED) was developed to integrate events over multiple consecutive time steps before declaring the sequence of time steps to be a true event, background water quality, or a change in the baseline of the background water quality. The BED works on the results of any event detection algorithm that produces a binary result (success/failure) for every time step. The BED provides an additional filtering of the data after the LPCF or MVNN algorithms and decreases the impact of any one time step that provides unexpected data.

The result of any outlier detection algorithm is conceptualized to define any time step with an outlier as a "failure" and any other with a residual consistent with background quality as a "success." The binomial probability distribution gives the probability that r "failures" occur in n trials, when the expected probability of any one trial failing is p. The corresponding probability that any one trial will succeed is $q = 1 - p$.

The probability that the water quality observed in the n trials is indicative of background water quality conditions is $P\,(\text{backgrd}) = b\,(r; n, p)$ and is given below. The complementary probability of an anomalous water quality event occurring within n trials is $P\,(\text{event}) = 1.0 - b\,(r; n, p)$.

$$b\,(r; n, p) = \frac{n!}{r!\,(n-r)!} p^r (1-p)^{(n-r)} = \frac{n!}{r!\,(n-r)!} p^r q^{(n-r)} \tag{19.7}$$

In online analysis, the concern is that the number of failures within a specified time period increases toward the positive tail of the binomial distribution of failures. To more efficiently identify such sequences of events, the cumulative distribution function (cdf) of the binomial distribution is used:

$$P\,(r \leq z_c) = \sum_{i=1}^{n(r \leq z_c)} b\,(r; n, p) \tag{19.8}$$

where z_c is the probability threshold value. Using the cdf function ensures that the probability of an event is increasing as the number of failures increases.

The binomial probability distribution describes the outcome of a Bernoulli process which must have the following properties (Walpole and Myers, 1989):

- n repeated trials in the experiment.
- Each trial can only have one of two outcomes: success or failure.
- The probability of failure, p, remains constant from one trial to the next.
- Repeated trials are independent of one another.

Each time step which has water quality data available is considered a trial. In CANARY, a user-defined window length defines the number of repeated trials (the n time steps). The outlier detection algorithms (LCPF or MVNN) are designed to produce a sequence of binary flags (0/1 indicating whether the data from the time step is an outlier or not) as output, which fits the requirement of a Bernoulli process having only success or failure outcomes.

The third requirement for a Bernoulli process is constant probability of failure, p. The use of a threshold that is relative to the current, or recent, variation of the water quality signals maintains a constant failure rate independent of the variation in the water quality. This approach also allows for a much more efficient detection algorithm than can be obtained using a constant threshold such as the "set point" approach often employed (see comparison in McKenna et al., 2006).

The fourth property of the Bernoulli process is equivalent to stating that the values of the estimation error, ε, are uncorrelated in time. Operation of the estimation algorithms under ideal conditions would result in ε being uncorrelated Gaussian noise and, for any z_C, the expected proportion of outliers could be determined from properties of the Gaussian distribution. This proportion corresponds to the probability of any single trial resulting in a failure, p, and could be used directly in the definition of the binomial parameters. However, experience has shown that serial correlation in the errors and other factors do not allow for this interpretation. Experience at multiple testing stations has shown that by keeping $p = 0.50$ and altering both the size of the binomial window and the probability threshold that must be exceeded to declare an event, a wide range of event detection sensitivity can be achieved.

In an operational setting, the BED provides the probability that r outliers could occur in n time steps under background water quality conditions. This probability is calculated at each time step, and if this probability exceeds a user-defined threshold, that time step is considered an event. Within CANARY, a second window length is defined as the number of consecutive time steps beyond the BED window in which every time step must contain an outlier in order to identify a baseline change. The length of the second window is set by the user and not directly tied to the binomial probability distribution.

19.3 EDS Analysis

The performance of the algorithms is particularly sensitive to two parameters: the window size and the threshold. This section provides a step-by-step analysis of both the LPCF and MVNN prediction algorithms applied to data collected at two monitoring stations within an operating US water distribution system. The following water quality event detection issues are investigated:

- Determination of appropriate event detection parameters from background data only (training).
- Simulation of events with different contaminant concentrations for testing the detection capabilities of CANARY's algorithms.
- Application of algorithms with parameters identified in training step for detection of events added to the background water quality data (testing).
- Detailed examination of the events (false alarms and actual events) identified by the CANARY algorithms.
- Evaluation of different parameterizations and the effects on event detection and baseline change identification (sensitivity analysis).

19.3.1 Data Sets

Historical data available from two different monitoring stations within the same distribution network are available for this analysis. Location A has a relatively stable background signal; Location B has similar characteristics to Location A, but with additional periodic variations. For each monitoring station, there are 31 days (22,320 time steps at 2 min intervals) of training data from July 8 through August 7. Each station has four water quality signals: chlorine (Cl), pH, conductivity (CDTY), and total organic carbon (TOC). These training data are shown in Figs. 19.2 and 19.3.

The stability of the background water quality of Location A is noted in the signals shown in Fig. 19.2. The signals vary only gradually throughout the training data period with the exception of a sharp change in pH on July 11 and a sharp change in TOC on July 29. Location B also exhibits relatively stable background water quality (Fig. 19.3). The signals of Location B also exhibit more regular daily periodicity relative to Location A (note the difference in the axes scales).

The training data are used to identify the parameter settings in the event detection algorithms. These algorithms and parameters are then applied to a second set of testing data. Water quality events of varying strengths are added to these testing

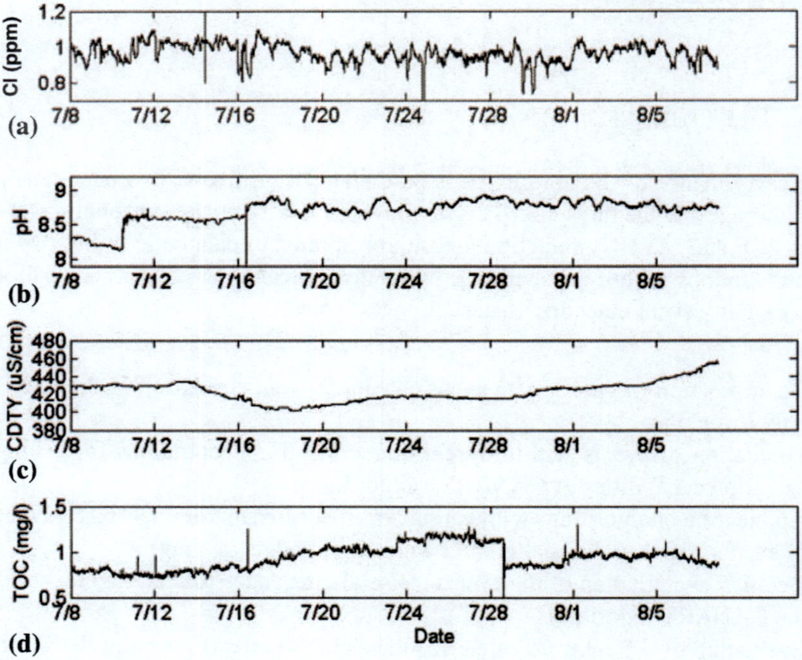

Fig. 19.2 Training data for Location A. The four water quality signals used are **(a)** chlorine (Cl), **(b)** pH, **(c)** conductivity (CDTY), and **(d)** total organic carbon (TOC)

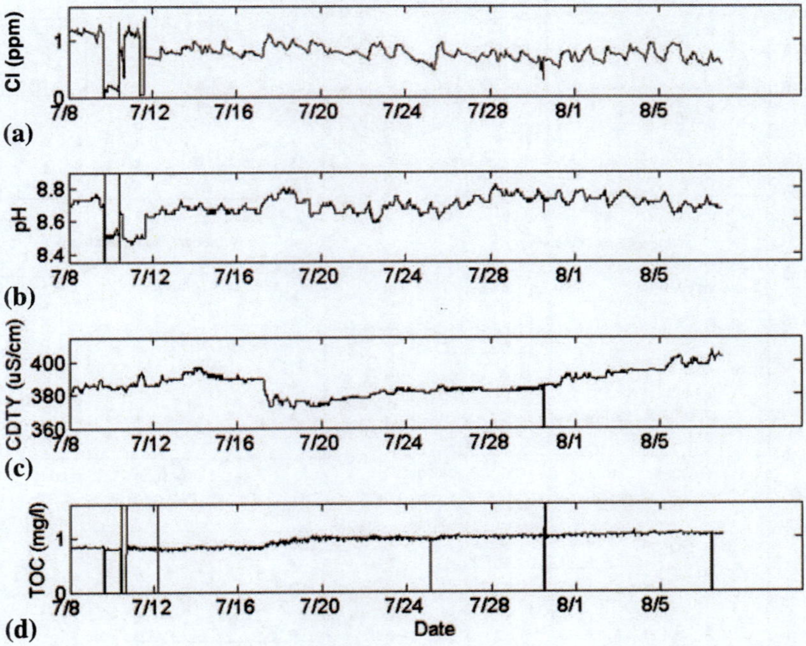

Fig. 19.3 Training data for Location B. The four water quality signals used are (**a**) chlorine (Cl), (**b**) pH, (**c**) conductivity (CDTY), and (**d**) total organic carbon (TOC)

data sets to evaluate the event detection algorithms. It is assumed throughout these steps that the characteristics of the background water quality do not change between the training and testing data sets.

19.3.2 Window Size and Prediction Errors

Determination of the window size defines the number of previous time steps used to predict (LPCF) or compare against (MVNN) the water quality value at the next time step. The values in the window are normalized (mean zero and standard deviation of one) prior to any analysis within CANARY. The best window size is determined by using both algorithms on a training data set to predict each future water quality value. The quality of the predictions is defined by the average absolute value of the residual between the observed and predicted water quality values and the standard deviations of these residuals. These two performance measures are calculated across ten different window sizes ranging from 180 time steps (6 h) to 1800 time steps (2.5 days). Results of these calculations are shown in Fig. 19.4. The parameters controlling the integration of results across time steps using the BED algorithm are held constant across all training runs. These parameters are set such that 14 outliers within 18 consecutive time steps (18 trials) are necessary before an event can be identified.

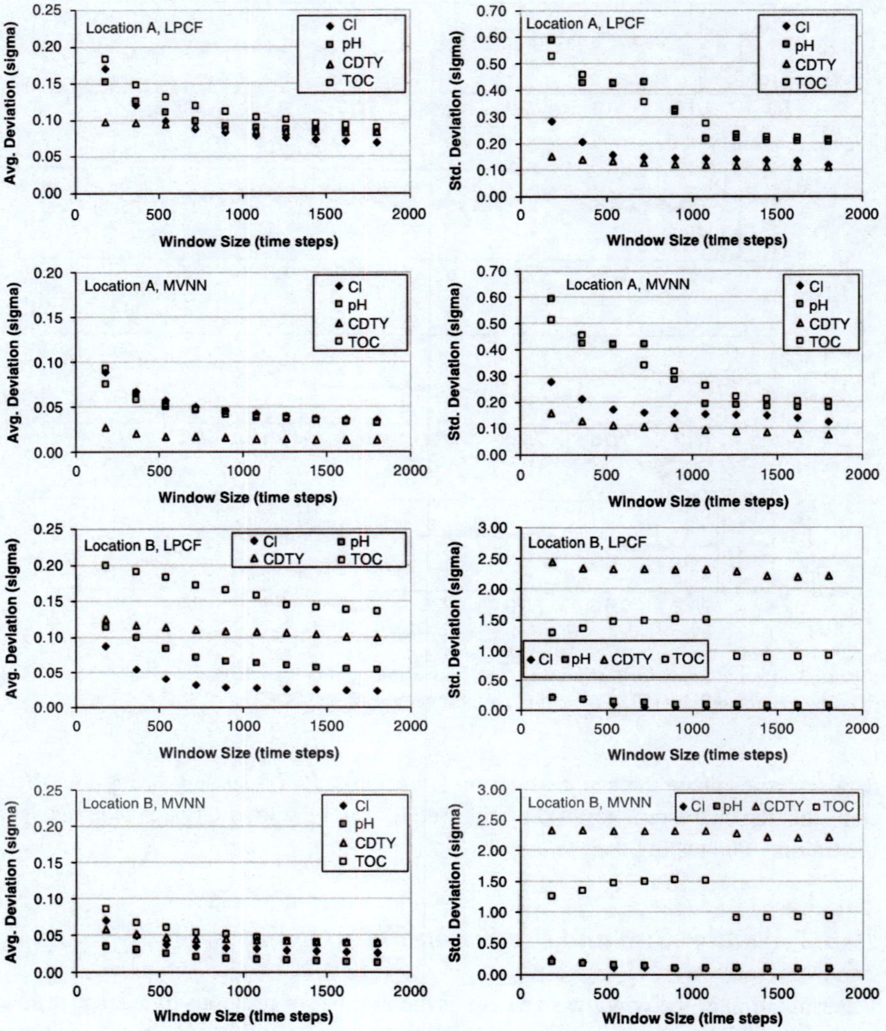

Fig. 19.4 Average deviation (*left column*) and standard deviation (*right column*) of prediction errors as a function of window size for Locations A and B

Figure 19.4 indicates a decrease in performance measure values with increasing window size. Exceptions to this observation are the standard deviation of the TOC and CDTY signals for Location B. These results are attributed to the variation in these signals at early times in the training data sets. The accuracy in the predictions (Fig. 19.4, left column) across the different signals is of the same order of magnitude at both monitoring stations. This result demonstrates how the prediction algorithms in CANARY are able to adapt to different water quality characteristics at different monitoring stations.

In Fig. 19.4, lower values of the average absolute residual and the standard deviation of the residuals indicate increased accuracy and precision, respectively, in the predictions of future water quality values. The MVNN algorithm produces more accurate predictions at both locations. The largest window size (1800 time steps) performs best for both locations, all water quality signals, and both algorithms. Statistical testing showed that the changes in the performance measures from one window size to the next were significant at all window sizes, indicating that even larger window sizes would continue to reduce these performance measures. The drawback of increased window sizes is the longer computational time needed to update the parameters and predict the future water quality at each time step. Experience with other monitoring stations and other water utilities has shown that window sizes between 1 and 2 days are often enough to provide reasonably accurate and useful predictions of future water quality values. Therefore, for the LPCF, a window size of 1440 (2 days) is selected. The results for the MVNN algorithm in Fig. 19.4 generally show the same shape, but have lower values than those for the corresponding LPCF calculations. Based on the similar shapes of the curves, a window size of 1440 time steps is also used for the MVNN algorithm.

19.3.3 Threshold Value and False Alarms

The event detection algorithms require a threshold value to classify residuals as being indicative of either background or outlier water quality. A useful rule of thumb for setting the minimum practical threshold is given by

$$\text{thresh}_{\text{min}} = \bar{\varepsilon} + 2\sigma_{\varepsilon} \tag{19.9}$$

where $\bar{\varepsilon}$ and σ_{ε} are the maximum values of the mean and standard deviation, respectively, of the results across all signals analyzed.

The LPCF algorithm with a window size of 1440 time steps produces a mean deviation of approximately 0.10 and a maximum standard deviation of 0.20–0.25 across the four signals analyzed for Location A (Fig. 19.4). Location B produces higher standard deviation values. Based on these results across the multiple window sizes, the minimum threshold tested here is 0.60. Six threshold values are tested from 0.60 to 1.10 with increments of 0.10. For consistency, the series of threshold values evaluated are held constant for both the MVNN results and the LPCF results.

Both algorithms with a window size of 1440 time steps are used on the training data set for the range of thresholds from 0.60 to 1.10. Threshold values resulting in event declaration on obvious significant changes in water quality while also minimizing events and outliers throughout the rest of the data set were retained (Table 19.1). Even though there are no known water quality events in the training data sets, alarms from CANARY are expected. These alarms are due to significant changes in the background water quality that occur at most monitoring stations. Examples in the training data include the sharp drop in TOC on July 28 (7/28 in

Table 19.1 Event detection parameters used in the analyses

Location	Window	Threshold
Location A, LPCF	1440	0.90
Location A, MVNN	1440	1.10
Location B, LPCF	1440	1.00
Location B, MVNN	1440	1.10

Table 19.2 CANARY results on training data prior to addition of events

	Number of events	Proportion of time steps within events	Average event length (time steps)	Average *P*(event) outside of events
Location A, LPCF	7	0.014	39.7	0.016
Location A, MVNN	7	0.017	45.0	0.016
Location B, LPCF	9	0.014	31.9	0.021
Location B, MVNN	8	0.012	30.4	0.016

Fig. 19.2) or July 29 at Location A and the drop in conductivity at Location B on July 17 (Fig. 19.3).

Training results using the final selected parameters are shown for each station and each algorithm in Table 19.2. Four different measures are used to summarize these results: the total number of events identified by CANARY (i.e., the number of alarms produced); the proportion of all time steps that are identified as events; the average event length; and the average probability of an event *P*(event), for those time steps classified as background (non-event) water quality. The BED parameters used here are the same as in Step 1 and limit the maximum length of an event to 45 time steps.

The results in Table 19.2 show that for Locations A and B, 1.2–1.7% of the time steps are classified as events. For a given monitoring station, the results from the two different algorithms are approximately the same. The average probability of an event outside of the identified events ranges from 0.016 to 0.029. These values are well below the probability threshold of 0.995 and indicate that outside of the events identified, the chances of a false alarm are very low.

19.3.4 Simulation of Water Quality Events

A separate set of testing data is available for each monitoring station from August 8 through September 18 (29,606 time steps, or approximately 41 days). Simulated water quality events are added to these testing data sets to represent changes in water quality that would be observed from the introduction of a small amount of a contaminant into the distribution network. The simulated events change the background water quality by adding a deviation to that background:

$$Z_{\mathrm{E}}(t) = Z_0(t) + E_{\mathrm{ind}}(t) \cdot \delta \cdot E_{\max} \cdot \sigma_Z \tag{19.10}$$

where $Z_{\mathrm{E}}(t)$ is the event-modified water quality value at time t, $Z_0(t)$ is the original background water quality at the same time step, E_{ind} is an event indicator equal to zero at all time steps outside of an event or between zero and one during an event, δ defines a decrease (–1.0) or increase (1.0) in the water quality signal in response to the contamination event, and $E_{\max}$ is a coefficient applied to σ_z, the standard deviation of the water quality sensor data, that determines the maximum amount by which the water quality deviates from the background. An E_{ind} value of 1.0 indicates that the contaminant concentration is at full strength and the maximum change in the water quality sensors is occurring. Values of E_{ind} less than 1.0 indicate time steps within an event at which the contaminant concentration is less than full strength, such as at each end of the event where the effects of dispersion in the pipe have created transitional concentrations of the contaminant between zero and the maximum concentration. The maximum deviation of $Z_{\mathrm{E}}(t)$ from $Z_0(t)$ is plus or minus the quantity $(E_{\max})$ (σ_z).

The initial shape of the simulated contaminant pulse is a square wave. Inclusion of the E_{ind} term in the event simulation allows for the shape of the leading and trailing edges of the contaminant pulse to be modified to represent varying amounts of smoothing that occur due to dispersion and diffusion of the pulse within the pipe network. As an example, Fig. 19.5 shows the values of E_{ind}, fraction of the event strength, as a function of the time step within the contaminant pulse. Both ends of the original square wave of the injected pulse (Fig. 19.5) have been smoothed. The example in Fig. 19.5 has four time steps on each end of the pulse where the concentration is intermediate between the background (0.0) and the maximum strength of the event (1.0). The shape of the transition from background to maximum strength is modeled using a Gaussian cumulative distribution function and the total event length is 34 time steps. Figure 19.6 provides an example of the simulated change in the response of the free chlorine sensor due to the injection of a contaminant.

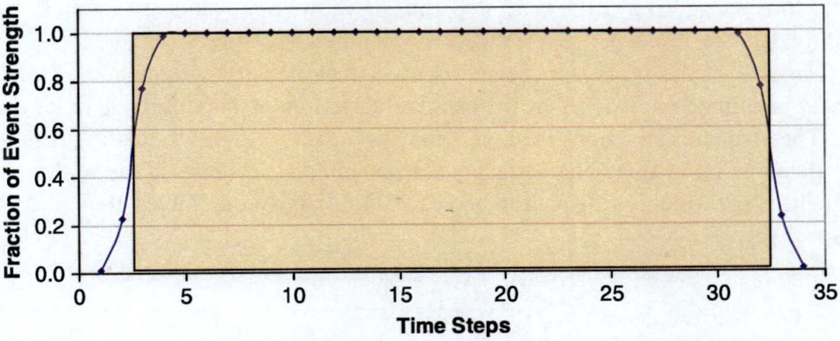

Fig. 19.5 Use of the event indicator (E_{ind}) to define the shape of the event. The *dotted line* represents the shape of the contaminant pulse as witnessed at the monitoring station. The *shaded box* represents the initial square pulse of the simulated contaminant

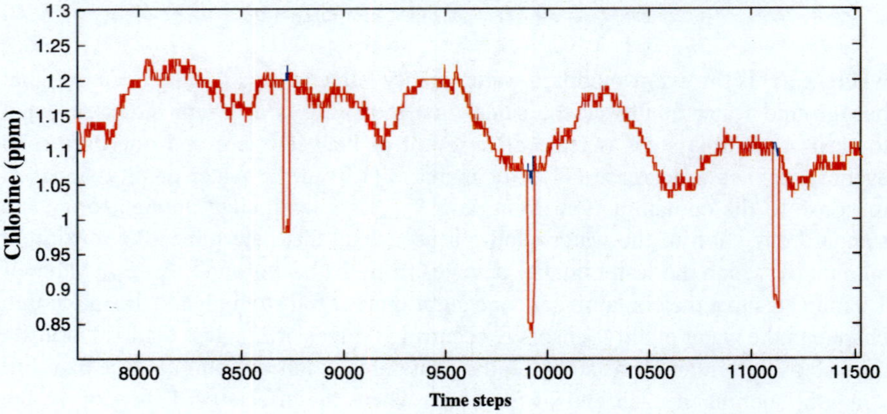

Fig. 19.6 Example of the simulated response of the free chlorine sensor to the introduction of three contamination events. The *blue lines* indicate the original background sensor signal. The time between events is 40 h

The maximum deviation of the sensor response from the background reading in this example is 1.5 times the standard deviation of the signal value ($E_{\max} = 1.5$). This parameterization results in events which decrease the free chlorine concentrations by approximately 0.22 mg/L. The shape of the events is as defined in Fig. 19.5. The spacing between events is 1200 time steps (40 h).

Testing data sets were created by adding simulated events to experimental data. Hall et al. (2007; Table 19.3) showed that many contaminants decreased free chlorine and/or increased total organic carbon. For the majority of the contaminants tested, changes in pH and specific conductance were minimal. For all testing data sets examined here, the shape of the event is shown in Fig. 19.5 with the characteristics that were described above. The effect of an event is to decrease the value measured by the Cl sensor and increase the value measured by the TOC sensor. The first event begins at time step 1501 and the subsequent events begin at intervals of 1200 time steps (40 h) from time step 1501. Twenty-four events are added to each testing data set.

The size of the maximum deviation away from the background water quality signal is defined as $E_{\max}$ times the standard deviation of the observed water quality. The standard deviations of the Cl and TOC data for the two training data sets are given in Table 19.3. The corresponding maximum deviation in the background signal for each monitoring station and each $E_{\max}$ is given in Tables 19.4 and 19.5, respectively.

Table 19.3 Standard deviation of the Cl and TOC signals for both locations

Location	Cl (mg/L)	TOC (mg/L)
Location A	0.1469	0.1635
Location B	0.1818	0.0724

Table 19.4 Maximum signal deviation for each event at Location A

Max event strength (E_{max})	Max Cl deviation (mg/L)	Max TOC deviation (mg/L)
0.50	0.073	0.082
0.75	0.110	0.123
1.00	0.147	0.164
1.25	0.184	0.204
1.50	0.220	0.245
1.75	0.257	0.286
2.00	0.294	0.327
2.25	0.331	0.368
2.50	0.367	0.409
2.75	0.404	0.450
3.00	0.441	0.491

Table 19.5 Maximum signal deviation for each event strength at Location B

Max event strength (E_{max})	Max Cl deviation (mg/L)	Max TOC deviation (mg/L)
0.50	0.091	0.036
0.75	0.136	0.054
1.00	0.182	0.072
1.25	0.227	0.091
1.50	0.273	0.109
1.75	0.318	0.127
2.00	0.364	0.145
2.25	0.409	0.163
2.50	0.455	0.181
2.75	0.500	0.199
3.00	0.545	0.217

For both monitoring stations, the decreases in free Cl range from less than 0.1 mg/L to near 0.5 mg/L. The minimum TOC increases are also less than 0.1 mg/L at Locations A and B. In addition to the calculations done with the simulated event sizes shown in Tables 19.4 and 19.5, the two original data sets (unmodified) are also analyzed with CANARY. Analysis of the unmodified testing data sets corresponds to $E_{max} = 0.0$ and these results provide the baseline event detection results.

The testing data sets from Locations A and B are observed data from an operating utility and as such they have several noteworthy characteristics. Both data sets have a loss of nearly 24 h of data between September 14 and 15. During the periods of data loss, CANARY waits for the data to be available again and then continues to process the new data using data remaining in the window from prior to the data loss. Any significant change in the values of the signals from one side of the data loss to the other will cause an event. This situation causes both algorithms to sound an alarm at the end of the data loss for both stations.

In addition to periods with missing data, CANARY ignores data in which the sensor is off-line. Such periods are identified by CANARY through sensor hardware alarms. In particular, both locations contain sensor hardware alarms indicating that

Table 19.6 CANARY results on testing data set prior to addition of events

Location	Number of events	Proportion of time steps within events	Average event length (time steps)	Average *P*(event) outside of events
Location A, LPCF	5	0.033	37.8	0.009
Location A, MVNN	12	0.040	29.9	0.018
Location B, LPCF	7	0.035	32.9	0.011
Location B, MVNN	16	0.049	37.5	0.025

the TOC sensor is off-line. At Location A, this occurs from September 13 to 14 (about 3.3% of the data). At Location B, the TOC sensor hardware alarm is active for part of September 10 and then from late September 10 through the end of the data set (greater than 18% of the testing data). Throughout periods of TOC sensor hardware alarms at Locations A and B, CANARY will continue processing using the other three water quality signals to detect events. Because the simulated events only alter the Cl and TOC signals, during the periods of TOC sensor hardware alarms, CANARY will only be able to detect events on the basis of the changes in the Cl signal. The impacts of sensor hardware alarms on the CANARY results will be strongest at Location B.

In addition to the loss of data and the TOC sensor hardware alarms, there also appear to be some issues with the pH and CDTY signals at Locations A and B. These signals do not change at all beginning on September 9 or 10 up until the loss of data between September 14 and 15. These signals have no alarms during this period, but this behavior is unusual in water quality monitoring data and may reflect an undiagnosed sensor malfunction or an issue with storing the data values in a SCADA database.

The results of running CANARY on the testing data sets with no events added are summarized in Table 19.6. The performance measures in Table 19.6 are the same used on the training data set as shown in Table 19.2. Across both monitoring stations and algorithms, approximately 3–5% of the time steps are classified as events by CANARY. These results are indicative of the fairly sensitive parameter settings and at least double the proportion of the time steps classified as events in the training data. Such increase might indicate a change in the nature of the water quality signals between the training and testing data.

19.3.5 Event Detection Results

Several different measures are employed to evaluate the performance of the event detection algorithms on the testing data sets. The known times of the simulated events are considered to be the "true" events, whereas the times identified by CANARY are called the "estimated" events. The performance measures are as follows:

- The area under the receiver operating characteristic (ROC) curve;
- The proportion of true events for which there is at least one CANARY detection;
- The proportion of the time steps within the true events that overlap with the estimated events;
- The average delay in the time of detection from the beginning of the true event;
- The average length of the estimated events compared to the same measure for the true events.

The ROC curve has been widely used in evaluating decisions made in medical and engineering applications, including evaluating water quality event detection algorithms (McKenna et al., 2008). The ROC curve defines the trade-off between missed detections (MD) and false-positive decisions in a single curve. The two axes of the ROC curve are defined by the false alarm rate (FAR):

$$\mathrm{FAR} = \frac{\mathrm{FP}}{\mathrm{FP} + \mathrm{TN}} \tag{19.11}$$

and the probability of detection (PD):

$$\mathrm{PD} = 1 - \mathrm{MD} = \frac{\mathrm{TP}}{\mathrm{TP} + \mathrm{FN}} \tag{19.12}$$

where TP and TN are the true positives and true negatives, respectively, as defined by the extent of the simulated event, and FP and FN are the false positives and the false negatives. An FP occurs when CANARY estimates an event when no true event has occurred at the same time. An FN occurs when a true event remains undetected by CANARY. A TP occurs when CANARY estimates an event and a true event occurred at the same time. A TN occurs when CANARY does not estimate an event and there is no true event at that time. Here the decision results: TP, TN, FP, and FN are tabulated at every time step.

The ROC curve provides a single plot that demonstrates the trade-off between FAR and PD across all ranges of the probability of an event, *P*(event). Typically, as the sensitivity of the algorithm is increased, the level of PD increases, but this also results in increased FPs. The area under the ROC curve varies from 0.5, indicating the decision results are only as good as those created by random guesses, to 1.0, which indicates perfect decision making – the case of PD = 1 and FAR = 0. The ROC curve area is used here as a performance measure.

The results of the ROC curve calculations are summarized by the area under the ROC curve for each monitoring station and algorithm (Fig. 19.7a). The areas under the ROC curves increase from approximately 0.5 to 0.7, at an event strength of 0.5, up to, or above, 0.8, at event strengths greater than 1.5. Beyond the event strength of 1.5, the ROC curve areas are nearly constant. Some level of variation exists in the ROC curve areas between the stations and the two different algorithms. At event strengths greater than 1.5, the lowest ROC curve areas (0.8) occur at Location B when using the MVNN algorithm.

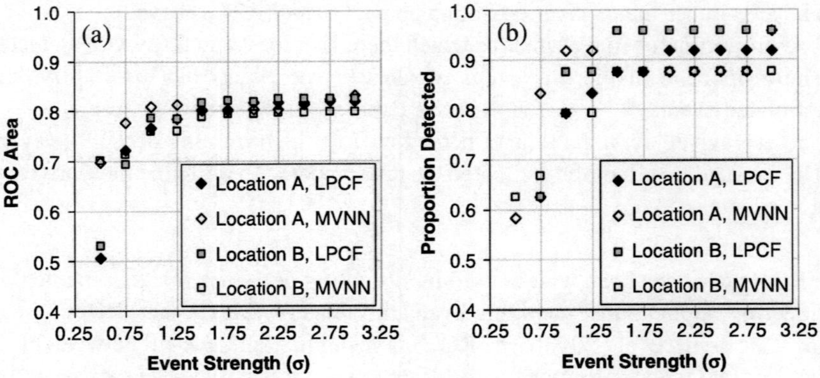

Fig. 19.7 (**a**) Areas under the ROC curves and (**b**) proportions of true events with at least one detection as a function of the event strength in terms of standard deviation (σ)

The ROC curve analysis is conducted by evaluating the decision result at each individual time step. This approach can be misleading as the parameters in CANARY are set to identify water quality events composed of groups of consecutive time steps where the water quality is anomalous. In particular, the settings of the BED algorithm used here require that at least 14 time steps be classified as outliers before an event can be identified. This intentional delay in the event identification works to reduce the number of false-positive alarms, but also creates a large number of time steps where the true event is already occurring prior to CANARY identifying it. These time steps are counted as missed detections (false negatives). This delay in the event identification and the associated – relatively large – number of time steps considered to be false negatives leads to a characteristic shape in the ROC curve.

Examination of individual ROC curves shows a characteristic shape with a steep rise of the curve to a PD value (*Y*-axis) of approximately 0.69 at an FAR of less than 0.05 followed by a change to a more gradual slope to the (1,1) corner. This change in slope is caused by the delay in detection. No matter what threshold is applied to the probability of event values from CANARY, the first 14 time steps, at least, of every true event cannot be detected due to the delay built into the BED algorithm. Therefore, the ROC curve cannot rise any higher along the *Y*-axis. This delay mechanism limits the ability of either algorithm to increase the probability of detection. The impact of changing the BED parameters on the ROC curve areas and the delay in the time to detection is evaluated further in the sensitivity analysis section of this chapter.

In addition to the ROC curve analysis, another evaluation approach is to consider each water quality event as an individual entity and determine the proportion of these events during which CANARY displays an alarm for at least one time step. This approach considers the resolution of the event to be the entire duration of the event and, therefore, is a less precise measure of the event detection capabilities. However, from a practical perspective, the bottom line for event detection is whether or not the events are detected at all, and this evaluation answers that question. For all signal strengths evaluated, the proportion of events that contain at least one time step of an alarm are evaluated and shown in Fig. 19.7b.

When the event strength is 1.5 or larger, the proportion of detected events is greater than 0.85 for both monitoring stations and both algorithms. In general, the LPCF algorithm performs better than the MVNN algorithm by detecting one or two more of the 24 true events, a 0.04 or 0.08 increase in the proportion detected, for each event strength. The best performance occurs at Location B where the LPCF algorithm is able to detect 23 of the 24 events (proportion detected = 0.96) for event strengths of 1.5 and greater.

Results of the other performance measures: proportion of overlap, average delay, and average event length are all consistent with the results discussed above showing that both algorithms are able to identify the majority of events when the strength is 1.5 or larger. To summarize these results, both algorithms indentify an alarm, on average, for about 40% of the time steps associated with each event. This corresponds to alarms for approximately 14 of the 34 time steps in each of the true events. The average delay between the start of the true event and the first alarm is 16–17 time steps depending on the monitoring station and the algorithm. This delay is consistent with the settings of the BED parameters that require a delay of 14 time steps before alarming. Additionally, several more time steps of delay are needed to account for the algorithms not recognizing the first two or three time steps of each event that have transitional concentrations between the background and the full strength. The average event lengths identified are 26–27 time steps compared to the 34 time steps of the true events. This result shows that not only is there a delay in the detections of 16–17 time steps, but that the detections continue beyond the end of the true events by approximately 10 time steps.

19.3.6 Sensitivity Analysis

The BED was proposed as a means of gathering evidence of an anomalous period of water quality across several consecutive time steps (McKenna et al., 2007), but the performance of the BED has still not been rigorously evaluated. The key parameters of the BED algorithm are the number of trials in each binomial probability calculation and the probability threshold compared to the probability of an event, *P*(event). In CANARY, such parameters are defined as bed-window-TS (shown as the "binomial window" in Fig. 19.8) and event-threshold-P, respectively. The bed-window-TS parameter is evaluated to determine its impact on the previously described EDS performance measures and also on the delay in detection of an event. Of particular focus is the question of whether or not changes in the parameterization of the BED can reduce the average delay time between the onset of an event and the detection of that event, while simultaneously increasing, or at least maintaining, the area under the ROC curve.

Calculation of *P*(event) through the binomial model and comparison to the probability threshold of 0.995 specifies that 14 outliers within 18 time steps are necessary in order to declare a water quality event. Here the same results are used to examine

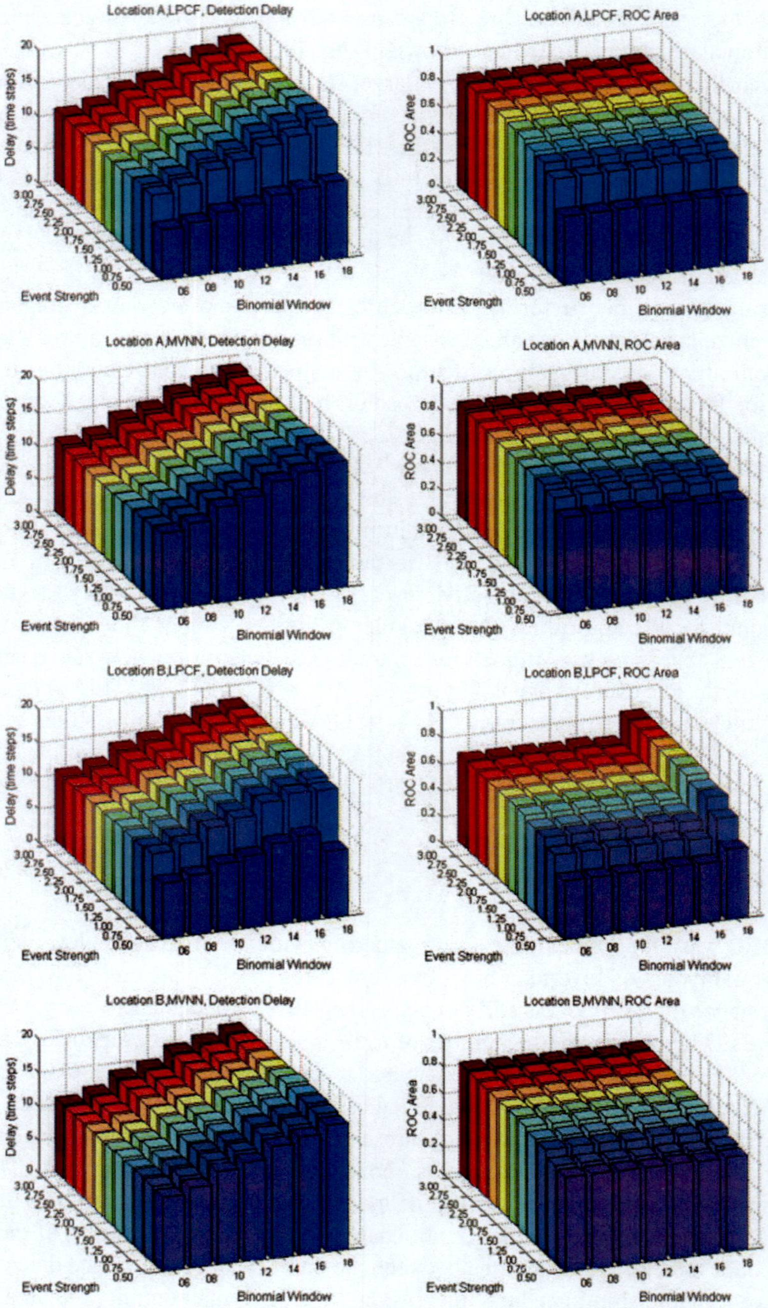

Fig. 19.8 Detection delay (*left column*) and ROC curve areas (*right column*) for both locations and both algorithms

how the performance measures are affected by a decrease in the number of outliers needed for an event declaration. Changes to the number of trials in the binomial experiment are made to decrease the number of outliers necessary for identification of a water quality event. Thus, the value of bed-window-TS is decreased from 18 to 6 in steps of two.

Figure 19.8 (left column) shows the detection delay as a function of the value of bed-window-TS and the event strength. Figure 19.8 (right column) demonstrates the area under the ROC curve as a function of the same two parameters. Figure 19.8 shows both sets of results for both monitoring stations and both algorithms. Several observations are clear from these figures:

- Decreasing the number of trials used in the BED (binomial window) decreases the detection delay in a near-linear manner for all event strengths above 0.5 standard deviations. These results are consistent for both algorithms. This behavior is expected, given that a larger number of trials causes an increase in the delay prior to being able to detect an event.
- The ROC curve area is not strongly dependent on the value of bed-window-TS. For most cases, decreasing the detection delay does not significantly change the area under the ROC curve. A strong exception to this observation occurs at Location B using the LPCF algorithm, since a binomial window value of 18 results in a jump in the ROC curve area relative to smaller values of the binomial window. This jump is due to CANARY identifying 23 of the 24 true events when the binomial window is set to 18 and only identifying 20 of the 24 true events when the value of the binomial window drops to 16 or less.
- The values of the ROC curve areas remain relatively stable as the detection delays decrease. These results indicate that faster times to detection can be achieved for the same decision performance as measured by the ROC curve. The stability of the ROC curve values indicates that a decrease in the delay to detection produces a decrease in the number of false negatives, but it also produces an increase in the number of false positives. This relationship is further explored below.

19.4 Discussion

For both monitoring stations examined here, both algorithms were able to detect more than 90% of the simulated events for event strengths greater than 1.5 standard deviations of the background water quality. Comparison with Tables 19.4 and 19.5 shows that this is roughly a change of 0.25 mg/L in the Cl and TOC signals with the exception of the TOC signal at Location B, where 1.5 standard deviations is approximately 0.11 mg/L. The higher sensitivity for TOC at Location B is due to the more stable signal at that location. Event detections remained at 80% or greater for event strengths between 1.0 and 1.5 standard deviations (a change of 0.15–0.20 mg/L in Cl and TOC). These results are remarkable considering that the daily changes in the background Cl were as much as 0.3 mg/L at Location B. These results demonstrate, on raw data from an operating utility, that both algorithms can be parameterized to provide the necessary sensitivity for event detection while limiting false-positive events to a few percent of the total time steps.

Several issues complicated testing of the event detection capabilities. Some change in the nature of the water quality data appeared to exist between the training and testing data sets. This change roughly doubled the number of false-positive events at both monitoring stations between the training and testing data sets using the same parameters when no events were added to either data set (compare results in Tables 19.2 and 19.6). This change also challenged the underlying assumption that the training and testing data have the same statistical characteristics. Results should improve when the training data are representative of the testing data.

The delay between the onset of an actual event and the declaration of that event by CANARY is controlled by the BED algorithm. The BED algorithm can be considered a post-processor of the outliers determined by the LPCF or MVNN algorithms. The initial probability threshold and bed-window-TS values used in testing resulted in a minimum of 14 outliers before an event could be declared. Given the smoothed leading edge of the contamination events, this requirement generally meant that at least 18 outliers, or 36 min, were needed prior to declaring an event. Changes in the BED parameter, bed-window-TS, reduced the average delay to as little as six time steps (12 min) while keeping the area under the ROC curve the same. This result means that the reduction of false positives created by the decreased delay to detection is offset by an increase in false positives at other points in the data set.

The event detection results show that the differences between the LPCF and MVNN algorithms are minimal. In theory, the MVNN algorithm should require a larger residual threshold to get the same results as the LPCF algorithm, based on the mechanism for calculating the threshold. An example with two signals provides a simple basis for comparison. The normalized distance (residual) between the predicted and observed water quality for each signal is 1.0. The LPCF algorithm will retain the maximum residual for comparison to the threshold. The MVNN algorithm will calculate the Euclidean distance between the current observation and the closest previous observation. The Euclidean distance will be the square root of two or 1.41. For a threshold value between 1 and 1.4, only the LPCF algorithm will identify this time step as an outlier. The residual calculation differences also lead to a broader distribution of residual values from the MVNN algorithm, which is a combination of signals, than from the LPCF algorithm, which only selects a single maximum value at each step. These differences influence the shape of the ROC curves.

Examination of the actual ROC curves calculated with a binomial window of eight time steps across all signal strengths shows a steep rise to a bend in the curve that occurs at a very low false alarm rate, less than 5%. In general, the LPCF algorithm results in lower false alarm rates and a sharper break in slope than the MVNN algorithm. The relative sharpness of the break in the slope is due to the differences in the LPCF and MVNN algorithms as discussed. The wider distribution of residual values created by the MVNN algorithm relative to the LPCF algorithm leads to the smoothed change in slope after the steep rise.

A series of simple calculations using the properties of the simulated true events can provide additional understanding of the values in the ROC curves. For every 1200 time steps, 34 are the true event and 1166 are background. The minimum delay times calculated were near 11 time steps (Fig. 19.8). If the estimated events have a delay of 11 time steps and there are no extra time steps estimated as events at the end of the event – no false positives – then $TP = (34 - 11) = 23$, $FN = 11$, and $PD = 23/(23 + 11) = 0.68$. This value is near that of the break in slope for many of the ROC curves examined. A hypothetical decrease in the delay to 8 time steps increases the PD value to $26/(26 + 8) = 0.76$. An obvious drawback of considering every time step as an independent result when using the ROC curve as an evaluation tool is that the calculation is dependent on the length of the event. If the detection delay remains constant at 8 time steps and the length of the simulated events is simply made twice as long, 68 time steps, the resulting PD value would be $60/(60 + 8) = 0.88$, a 16% improvement over the case of the shorter events.

Examination of the results calculated here shows that false positives occur most commonly by overestimation of the length of the event. If a delay of 11 time steps is needed to identify each event and, at the end of each event, the algorithm continues to estimate an event for 10 time steps beyond the end of the true event, the FAR calculation is $FP = 10$, $TN = (1166 - 10) = 1156$, and $FAR = 10/(10 + 1156) = 0.0086$. Again, this value is close to the break in slope in the ROC curves examined and indicates some delay in the algorithms recognizing background water quality conditions after an event ends.

This sensitivity analysis demonstrates that changes in the EDS parameterization can be completed to decrease detection times at the expense of a higher number of false positives. Operational reasons might exist to bias this trade-off toward faster detection times and higher false positives, for example, during a period of heightened security. Resources necessary to investigate the increased alarms would most likely prohibit operating in this mode for extended periods of time.

19.5 Conclusions

This study documents the testing and evaluation of the EDS component of a CWS using actual water quality data from an operating distribution network. Results show that both algorithms examined are capable of providing the sensitivity necessary to reliably detect events that alter the water quality by at least 1.5 times the standard deviation of the measured water quality signal while limiting false-positive results to 3–4% of the time steps analyzed. Further analysis of the results provided some observations on how to best calculate performance and insight into factors influencing the trade-offs between decreasing the delay to detection (lowering the number of false-negative time steps) and increasing the number of false positives.

Event detection in distribution systems is a classic case of examining a set of noisy signals to detect events that have a low probability of occurrence and may manifest themselves as very subtle deviations from the background signals. In these situations, the required sensitivity of the monitoring algorithm and overlap in the background and event signal signatures will lead to false alarms in the event detection (see Rizak and Hrudey, 2006). Results in the analyses done here show what factors need to be considered in ROC curve calculations done on per time step and on a per event basis. Standardized definitions of "detection" for an event lasting longer than a single time step need to be developed.

To date, EDS analyses have focused on event detection at each monitoring station independently of simultaneous analyses occurring at other monitoring stations within the distribution network. As utilities continue to add monitoring stations within distribution networks, the concept of "distributed detection," where information from multiple monitoring stations is combined in real time to provide an integrated detection capability, will become possible. Recent development and testing of an approach to distributed detection has shown that integration of EDS results across a network can significantly reduce false-positive detections and provide robust estimations of a contaminant source location (Koch and McKenna, 2010).

The data sets examined in this study are relatively stable and do not exhibit significant water quality changes associated with changes in network operations. In cases where the water quality is strongly influenced by changes in hydraulic operations, new approaches are needed to recognize the impact of these changes and integrate operational data streams into the online event detection. Potential approaches to meeting these goals include recognition of recurring patterns in multivariate data streams that are associated with operational changes (e.g., Vugrin et al., 2009) and direct integration of informative combinations of operational signals to temporarily decrease event detection sensitivity during periods of operational change (e.g., Hart et al., 2010).

The CANARY software platform contains all of the algorithms tested in this study as well as additional event detection algorithms and other functionality for analysis of water quality data in both archival and online modes. Additional information on the CANARY software as well as download of the source code is available from https://software.sandia.gov/trac/canary.

Acknowledgment This work was performed under Interagency Agreement DW89921928 with Sandia National Laboratories. Sandia National Laboratories is a multi-program laboratory managed and operated by Sandia Corporation, a wholly owned subsidiary of Lockheed Martin Company for the US Department of Energy's National Nuclear Security Administration under contract DE-AC04-94AL85000.

References

American Society of Civil Engineers (2004) Interim voluntary guidelines for designing an online contaminant monitoring system, American Society of Civil Engineers, Reston, VA.

American Water Works Association (2005) Contamination warning systems for water: an approach for providing actionable information to decision-makers, American Water Works Association, Denver, CO.

Box, GEP and Jenkins, GM (1976) Time series analysis: forecasting and control, Holden-Day series in time series analysis, Holden-Day, San Francisco, CA.

Bras, RL and Rodriguez-Iturbe, I (1993) Random functions and hydrology, Dover, Mineola, NY.

Byer, D and Carlson, KH (2005) Real-time detection of intentional chemical contamination in the distribution system. Journal of the American Water Works Association, 97(7), 130–133.

Cook, J, Roehl, E, Daamen, R, Carlson, K, and Byer, D (2005) Decision support system for water distribution system monitoring for homeland security, Proc., AWWA Water Security Congress, AWWA, Denver, CO.

Cook, JB, Byrne, JF, Daamen, RC, and Roehl, EA (2006) Distribution system monitoring research at Charleston Water System, Proc., 8th Annual Water Distribution Systems Analysis Symposium, ASCE, Reston, VA.

Hall, J, Zaffiro, AD, Marx, RB, Kefauver, PC, Krishnan, ER, and Herrmann, JG (2007) Online water quality parameters as indicators of distribution system contamination, Journal of the American Water Works Association, 99(1), 66–77.

Hall, JS, Szabo, JG, Panguluri, S, and Meiners, G (2009) Distribution system water quality monitoring: sensor technology evaluation methodology and results, a guide for sensor manufacturers and water utilities, EPA/600/R-09/076, U. S. Environmental Protection Agency, Office of Research and Development, National Homeland Security Research Center, Cincinnati, OH.

Hart, DB and McKenna, SA (2009) CANARY user's manual, version 4.1, EPA/600/R-08/040A, U.S. Environmental Protection Agency, Office of Research and Development, National Homeland Security Research Center, Cincinnati, OH.

Hart, DB, McKenna, SA, Murray, R, and Haxton, T (2010) Combining water quality and operational data for improved event detection, Water Distribution Systems Analysis (WDSA) Conference 2010, Tucson, AZ, Sept 12–15.

Jarrett, R, Robinson, G, and O'Halloran, R (2006) Online monitoring of water distribution systems: data processing and anomaly detection. Proc., 8th Annual Water Distribution Systems Analysis Symposium, ASCE, Reston, VA.

Klise, KA and McKenna, SA (2006a) Multivariate applications for detecting anomalous water quality, Proc., 8th Annual Water Distribution Systems Analysis Symposium, ASCE, Reston, VA.

Klise, KA and McKenna, SA (2006b) Water quality change detection: multivariate algorithms, Proc., SPIE (International Society for Optical Engineering), Defense and Security Symposium 2006.

Koch, MW and McKenna, SA (2010) Distributed sensor fusion in water quality event detection, ASCE Journal of Water Resources Planning and Management, 137(1), 10–19.

Kroll, D and King, K (2006) Laboratory and flow loop validation and testing of the operational effectiveness of an online security platform for the water distribution system, Proc., 8th Annual Water Distribution Systems Analysis Symposium, ASCE, Reston, VA.

MathWorks. (2008) MATLAB, 2008b, The Mathworks, Natick, MA. http://www.mathworks.com/products/matlab/.

McKenna, SA, Klise, KA, and Wilson, MP (2006) Testing water quality change detection algorithms, Proc., 8th Annual Water Distribution Systems Analysis Symposium, ASCE, Reston, VA.

McKenna, SA, Hart, D, Klise, K, Cruz, V, and Wilson, M (2007) Event detection from water quality time series, Proc., World Environmental and Water Resources Congress, ASCE, Reston, VA.

McKenna, SA, Wilson, M, and Klise, KA (2008) Detecting changes in water quality data, Journal of the American Water Works Association, 100(1), 74–85.

Raftery, AE (1994) Change point and change curve modeling in stochastic processes and spatial statistics, Journal of Applied Statistical Science, 1(4), 403–424.

Rizak, SN and Hrudey, SE (2006) Misinterpretation of drinking water quality monitoring data with implications for risk management, Environmental Science & Technology, 40(17), 5244–5250.

US EPA (2005) Technologies and techniques for early warning systems to monitor and evaluate drinking water quality: a state-of-the-art review, EPA/600/R-05/156, U.S. Environmental Protection Agency, Office of Research and Development, National Homeland Security Research Center, Cincinnati, OH.

US EPA (2010) Water Quality Event Detection Systems for Drinking Water Contamination Warning Systems: Development, Testing and Applications of CANARY, EPA/600/R-010/036, U.S. Environmental Protection Agency, Office of Research and Development, National Homeland Security Research Center, Cincinnati, OH.

Vugrin, E, McKenna, SA, and Hart, D (2009) Trajectory clustering approach for reducing water quality event false alarms, Proc., of ASCE Annual World Environmental and Water Resources Congress, Kansas City, Missouri, May 17–21.

Walpole, RE and Myers, RH (1989) Probability and statistics for engineers and scientists, Fourth Edition, MacMillan, New York, NY.

Yang, YJ, Haught, RC, and Goodrich, JA (2009) Real-time contaminant detection and classification in a drinking water pipe using conventional water quality sensors: techniques and experimental results, Journal of Environmental Management, 90(8), 2494–2506.

Chapter 20
Water Infrastructure Protection Against Intentional Attacks: The Experience of Two European Research Projects

Cristiana Di Cristo, Angelo Leopardi, and Giovanni de Marinis

20.1 Introduction

Water Supply Infrastructures have the mission to provide water for residential, industrial, and agricultural uses. In particular, drinkable water at residential level must have specific characteristics that assure its safety. The vulnerability assessment of water supply infrastructures is a priority in order to enhance their security against the risks of out of service or contaminations. The supply systems can be damaged by natural hazards or intentional attacks. Regarding an intentional attack, the first question to be answered is, "Are Water Supply Systems vulnerable to this kind of threat?" In terms of accessibility the answer is yes, since in general, there are no stringent security measures or at least not so stringent as for other infrastructures, e.g., power plants. For this reason, in the present work, a procedure for protection system design, developed in two projects, financed in the framework of the European Programme for Critical Infrastructure Protection (E.P.C.I.P.), is described.

In the first project, named D.I.S.W.I.P. (Development of an Integrated System for Water Infrastructure Protection against intentional attacks), the water system of the Sorrento Peninsula, an important tourist site in Italy including Capri island, is analyzed. It serves about 100,000 inhabitants divided into 8 towns. The system has a large number of reservoirs and small tanks, well spread in the serviced area, which can be considered potential points of attacks to the network by means of chemical or biological agents. In the second project (G.L.E.W.I.P. – GuideLines Enhancement for Water Infrastructure Protection against intentional attacks) the Vesuvian water system, which serves about half a million inhabitants in a very urbanized area, is considered.

C. Di Cristo (✉)
Water Engineering Lab (L.I.A.), Department of Mechanics, Structures and Environmental Engineering (Di.M.S.A.T.), University of Cassino, Cassino, Italy
e-mail: dicristo@unicas.it

R.M. Clark et al. (eds.), *Handbook of Water and Wastewater Systems Protection*, Protecting Critical Infrastructure, DOI 10.1007/978-1-4614-0189-6_20,
© Springer Science+Business Media, LLC 2011

The protection system study is approached through a preliminary risk analysis, realized in order to individuate the more adequate measures for its reduction. In particular risk reduction is approached from three points of view:

– prevention against attacks,
– early detection of attack,
– fast and appropriate response, in order to limit damages.

All features are considered and designed using a multi-level approach, in which simple and cheap techniques are first selected.

For the prevention against attacks physical protection methods and monitoring systems are used. Early detection aspect includes the selection of the more appropriate sensor design methodology for the considered system, online monitoring, and data transmission systems. The response to an event is related to the system behavior simulation and a decision support system.

The general procedure is applied to the two considered water supply systems and the protection system was effectively realized and tested for both. In this work, a description of the realized systems is furnished. The primary objective of this chapter is the presentation of projects' philosophy and results. Details on general approaches and procedures can be found in other chapters.

20.2 The Adopted Procedure

20.2.1 The General Layout

The proposed procedure for Protection System Design (PSD) is divided into multiple steps, schematically reported in Fig. 20.1.

An important aspect is to individuate the more probable attack scenarios, considering the characteristics of the system under study. The first step is then represented by Design Base Threats (DBTs). The goal is the identification of possible kinds of attack (ASCE, AWWA, WEF, 2004), individuating who can be intentional in damaging the system and how it can be done. The possible attackers are:

– terrorists,
– criminals,
– saboteurs,
– vandals.

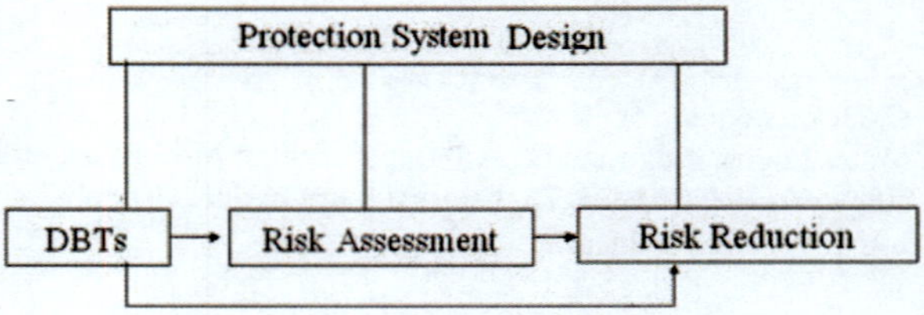

Fig. 20.1 Steps of the proposed procedure

Terror is a state of intense fear. It is often the result of violence committed by one or more groups to intimidate individuals, segment of society, or governments to meet the group's demands. Terrorist groups may have different motivations (political, economical, individual revenge, etc.) and they present new and unfamiliar threats also for the water supply systems. Terrorist organizations are usually well prepared and have also sophisticated means. In some regions or situations, criminal organizations can make profit from a bad working of the water system; in other cases a single person or groups of saboteurs, moved by economical or political reasons, can be interested in damaging it. Finally, vandals can create dangerous situations without particular motivations.

Water facilities are normally protected, with access restricted to authorized personnel only, but sometimes, it may be rather easily breached if the security system is not very effective. Moreover, it is very important to consider that the attack can arrive also by inside saboteurs.

Because a water supply system is often effectively vulnerable to intentional attack, the second question is how such attacks can be realized, i.e., which kind of attack has to be considered. A water supply system must be physically approached in order to threaten it. The threat may take the form of physical attack (e.g., ruptures and bombing), contamination of water (by means of chemical, biological, or radiological agents), and cyber attacks. In fact, as the operations of water supply systems are becoming automated, through computer software for supervisory control and data acquisition (SCADA), they are more open to remote attacks. Among the above-cited threats, contamination of water can be considered as the most dangerous one because of the chance of damaging a large number of people.

Considering one or more possible attacks, the second step is represented by the Risk Assessment, in which an evaluation of the risk is performed. Note that a precise quantitative evaluation of the risk in this context is not possible, since many of the key factors, affecting the calculation, cannot be quantified. Furthermore, some important factors are subjective or highly uncertain. Nevertheless, rough estimates, similar to those made in some vulnerability assessment programs, are possible and useful. The risk definition assumed in the considered procedure is (e.g., Rescher, 1983)

$$R = p \cdot V \tag{20.1}$$

where p is the probability that an adverse event occurs; V is the vulnerability of the system, related to the damage or loss expected if the event does occur, as it is presented in more detail in Section 20.4.2. Probability being dimensionless, the risk units are some of the vulnerability. The probability function ranges between 0 and 1. The reproduction of attack scenarios can be helpful for estimating the vulnerability and then the risk. For this reason the possibility of using software, for simulating the behavior of the system, is a fundamental aspect. In this way, analyzing the different scenarios, the more vulnerable parts of the systems and/or the more critical operating conditions can be individuated.

The third step is devoted to Risk Reduction, analyzing the more adequate measures and evaluating all the alternatives, also considering the economical aspects.

Some considerations about Risk Assessment and Risk Reduction are reported in Sections 20.2.2 and 20.2.3, with reference to a contamination event. Moreover, the presented procedure is described in more detail in Section 20.4, applying it to the water systems considered in the research projects.

20.2.2 Risk Assessment

The risks arising from contamination in water supply systems include

- public health risks that arise from use of contaminated water,
- economic risks to the user community that arise from the use of contaminated water (for example, in industrial processes),
- economic risks to the utility from the loss of saleable water; loss of water for industrial, agricultural, and fire fighting uses; potential legal liabilities; unfavorable audit reports; loss of public confidence; cleanup and repair costs.

The primary concern is the threat to public health from the exposure of people to the contaminant. This exposure can come about through a range of uses including, but not limited to, drinking tap water, such as using tap water in food preparation or for ice, showering, bathing, washing clothes and dishes, using a humidifier, watering lawns, washing cars, and fighting fires.

It is also important to consider that high-value customers from the attacker's perspective, like for example, government installations, schools, and nursing homes, are more suitable targets for an attack. In other words, the target may not be the public at large, but a specific segment of it that holds a high value for the attackers.

Public health risk assessment involves three steps:

- to identify the hazard;
- to evaluate the exposure;
- to determine response to dose.

The hazard is the contaminant, which can be of different types: pathogens, chemicals, and radioactive.

About the exposure evaluation, there are three routes that need to be considered: ingestion, inhalation, and skin contact. Drinking contaminated water from the tap can cause illness and death, if the dose of contaminant ingested is sufficiently high. US EPA posits that a representative value for water ingestion by consumers is about 2 l per day (ATSDR, 2005). This average value depends on the location, demographic factors, the season, etc. Moreover, people who are more physically active probably consume larger quantities. However, when assessing the risk and performing the related calculations, it is reasonable to assume that a nominal person drinks about 2 l of water per day. The water used in cooking or for food preparation can also be indirectly ingested. In fact, the food washed with tap water can retain some contaminant. Moreover, low temperature or brief cooking might degrade or destroy many contaminants, but some others would survive quite well.

Although the ingestion is usually the main exposure path, one should not ignore the inhalation of water vapor or the aerosols from showering, from the use of humidifiers (particularly in winter), the evaporation of water used in washing, and from other sources. The US EPA calculates that the exposure occurs through inhalation under a variety of circumstances. For example, the quantities of volatiles and aerosols through showering once a day is of the same order as that through ingesting 2 l of water per day. Even in the cases where the amount of toxic or pathogenic material inhaled is considerable less than the quantity assumed by ingestion, some contaminants may be more effective in causing illness or death by inhalation than by ingestion. Reasonable and widely agreed values for quantifying the exposure via inhalation for the contaminants of concern in the context of this work are not yet available. Similarly, the exposure by contact may be important too.

Consider a specific attack scenario, for example, the contamination of the water in a tank, it is very important to evaluate the exposure of the population to such attack. This task can be accomplished evaluating the number of people involved, which can be estimated simulating the hydraulic behavior of the Water Supply System. For accomplishing this task a software model able to reproduce the system behavior is a key tool, necessary not only in the Risk Assessment phase. In fact, as is shown in Section 20.4, in the DISWIP and GLEWIP projects the system simulations are used for analyzing different important aspects.

20.2.3 Risk Reduction: Evaluation of Alternatives

A complete contamination prevention of a water system, except in a very few special circumstances, is not possible. However, the increasing of physical security efforts could assist in mitigating the contamination risks by making the entry of pollutants into the system more difficult. Simple measures, as using better locks and/or higher fences, should be first considered for reducing the contamination risk. These measures are referred as *Physical Countermeasures* and they are presented in Section 20.4.3.1. After the physical security enhancement has been considered, a surveillance system, able to detect an attack, can be assessed through access controls and surveillance cameras.

In addition to all these physical surveillance systems, methods for detecting and characterizing a contamination can be assessed, such as a OCMS – Online Contaminant Monitoring System (ASCE, AWWA, WEF, 2004). The fundamental questions to be answered in determining the need for a contaminant detection system are (1) do the potential costs resulting from a contamination event justify the costs for the detection? (2) Does the detection method produce an acceptable reduction in those potential contamination costs?

In addition to online contaminant monitoring, the following detection methods could be considered: the extensions or enhancements of the water quality surveillance program already in use to meet regulation requirements and the observation of medical and pharmaceutical activities in order to individuate anomalous situations. Contamination scenarios, characterized by relatively short but intense events, are likely to evade detection by a traditional grab-sample/laboratory analysis surveillance system. In fact, the samples or measurements cannot be enough to frequently obtain at least one measure, and preferably more than one, during the event. To detect such kinds of events an extraordinary increase in sampling and analyzing activity should be realized. Moreover, the time necessary for laboratory analysis of samples on a routine basis is generally on the order of days. Under emergency conditions this can be shortened, but to adopt such speedy procedures as a routine would invoke substantial increase of the costs, particularly for the required human resources.

A detection scheme based on surveying the medical community and accounting for pharmaceuticals, drugs, and medicaments sold assumes that the public has already been exposed to the contaminant. It is useful as a retrospective examination to characterize the event, which can be used in the treatment of the victims and, therefore, valuable in safeguarding the public health, but it does not prevent the exposure. In summary, the two considered alternative approaches may be helpful, but they do not represent early warning methods; in fact they do not furnish fast and useful information for organizing appropriate responses to limit or prevent the exposure of the public to the contaminant.

An online, near real-time, contaminant detection system can provide early warning. However, such systems involve a number of points of measure, each with several instruments located throughout the priority regions of the system, with a very expensive effort. An OCMS is justified if the likely costs (risks) that could arise from a contamination event exceed the sum of

(a) the costs (risks) of contamination that would remain even with an OCMS;
(b) the costs of establishing and operating the OCMS.

Early contaminant detection in a water distribution system can require a large number of monitoring locations. For this reason an efficient sensor location is important in order to reduce the costs. Determining the more adequate points for locating monitoring instruments (sensor location) for early warning is highly dependent upon the specific characteristics of the water system (Ostfeld et al., 2008). Furthermore, if the purpose of the OCMS is the detection to prevent the exposure, rather than to assist the victims in diagnosis and treatment, monitoring in deep the distribution system would not be cost-effective.

20.3 The Pilot Sites

The two pilot sites are represented by two water systems managed by GORI (Gestione Ottimale Risorse Idriche) S.p.a., which is a mixed public/private water company involved in the management of the overall cycle of the water resources

Fig. 20.2 Geographical location of the pilot sites

(picking up, adduction, distribution to the users, picking up of the flowing back waters, and purification) in a large area located in Campania, a region in Southern Italy (Fig. 20.2), in accordance with the national law n. 36 of January the 5th of 1994 and of the regional law n. 14 of May the 21st of 1997. The principal characteristics of the total area managed by GORI are

- 59 municipalities in the Naples Province
- 17 municipalities in the Salerno Province
- 1,425,429 inhabitants
- 475,853 users
- an extension of 900 km^2
- 4,000 km of water networks
- 2,200 of sewer networks

For the study presented, two pilot sites are selected: the Sorrento Peninsula Water Supply System (SPWSS) and the Vesuvian Water Supply System (VWSS).

20.3.1 The Sorrento Peninsula Water Supply System (SPWSS)

The Sorrento (Sorrentine) Peninsula is located in Campania region in Southern Italy (Fig. 20.2). This peninsula separates the Gulf of Naples, located in the North side, from the Gulf of Salerno in the South side. The peninsula is named from its main town, Sorrento, which is located on the north coast (Gulf of Naples). The south coast is named Amalfi Coast. The island of Capri lies off the western tip of the peninsula in the Tyrrhenian Sea. The peninsula, with the six towns of Vico Equense, Meta, Piano di Sorrento, Sant'Agnello, Sorrento, and Massa Lubrense, and the Island of Capri, with the two towns of Capri and Anacapri, is probably one of the most famous tourist destinations, for both Italians and foreigners. During summer, the area is heavily touristed, also by day tourists from Naples.

In the DISWIP project the main Sorrento Peninsula Water Supply System (SPWSS), which brings water from the sources to the main tanks located near the served towns, is considered. The system has a great number of reservoirs and little tanks, spread in the served area, which are potential points of attacks to the network by means of chemical or biological agents. Moreover, the economical importance of tourism in the area and its fame all over the world suggest the peninsula and the island of Capri as possible attack targets. The water company needs to improve the protection of the system against intentional attacks and to realize a contingency plan in order to manage risk situations.

The Sorrento Peninsula Water Supply System, which is one of the most important systems managed by GORI, serves about 100,000 inhabitants. As previously stated, because of the high tourist interest in the area, population doubles in the summer period. Moreover, there are towns, like Sorrento and Capri, in which the number of tourists in summer is larger than the resident population. This particular feature suggests that it is very important to analyze possible attacks in different flow demand scenarios. However, it may be conjectured that possible terrorist attacks may be in the summer period, in order to maximize the impact.

The principal source of water for the SPWSS is the Gragnano well field. Water is pumped from wells to a large reservoir (Gragnano reservoir – 79 m), with a volume of 16,500 m^3, then from such reservoir it is pumped to other main tanks, from which three principal adduction lines start: the High Line, the Low Line 1, and the Low Line 2. Figure 20.3 shows a map of the SPWSS. Many pumping plants are present in the system, adducting water from the Low Lines to users at high elevations.

20.3.2 *The Vesuvian Water Supply System (VWSS)*

Vesuvio is probably one of the most famous volcanoes in the world. Its surroundings are intensely urbanized, with about half a million of inhabitants living in this area. The Vesuvian Water Supply Systems (VWSS) serves about 400,000 inhabitants in 17 towns. This system is composed of three rings, serving zones characterized by different elevations (Fig. 20.7).

20.4 Protection System Design on the Pilot Sites

20.4.1 *Design Base Threats and Attack Scenarios*

A particular situation about water industry is actually present in Italy, with a large debate about public/private management of the water systems. This situation can

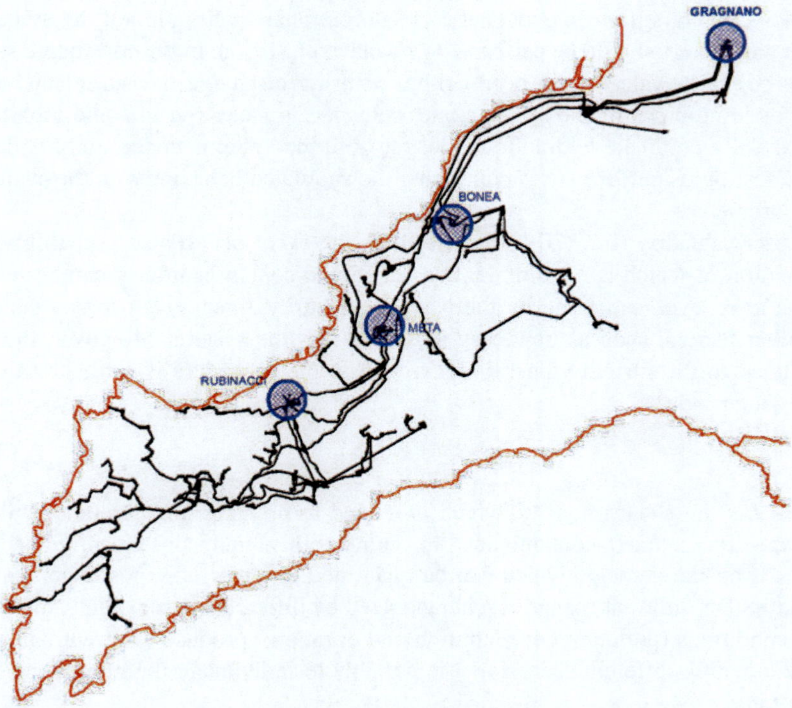

Fig. 20.3 Map of the Sorrento Peninsula Water Supply System

create a motivation for people to attack such systems. So, for each considered system two main attacks can be thought as the most probable:

- saboteur attack, due to revenge motivation against the water company, which is charged "with stealing" water
- criminal attack, due to economical reasons

The design attack scenario assumed for the protection system is the contamination of a tank. This choice is motivated by the elevated number of tanks present in the systems and by the fact that the contaminant intrusion in a water tank is easier. However, since the possible attack scenarios are unpredictable and they can be very different from the assumed one, the protection system must be done in order to be very flexible.

20.4.2 Risk Assessment

In the Risk Assessment phase, it is necessary for reproducing the system behavior in different conditions. For this reason it is important to have a good, user-friendly, and cheap code for hydraulic and water quality modeling. In the projects the well-known EPANET (Rossman, 2000) code is adopted. In order to have results coherent with the working conditions in a real network, a preliminary calibration of the hydraulic input parameters should be performed (Walski et al., 1987). In the considered study, pipe roughness values have been calibrated using discharge measurements available in some pipes of the systems and water levels measured in some monitored tanks. So, a complete hydraulic model for both the systems under study is developed, testing its performances comparing the simulated behavior with the available measurements.

Risk evaluation (Eq. (20.1)) is related to the event occurrence probability, the estimation of which is not simple. For intentional contamination, it can be related to the alert level, estimated by the National Security Agency, but it also depends on other factors, such as how easy it is to access the system. Moreover, the risk is related to the system vulnerability, which in the procedure is estimated by the following product:

$$V = I D \tag{20.2}$$

where I is the event intensity, which is related to pollutant toxicity, and D is the damage. For a fixed contaminant, the vulnerability, and consequently the risk, depends on the damage, which can be quantified in terms of exposed population. In particular, different scenarios, characterized by different source points and injection conditions (pollutant concentration and duration), produce different damages. So, simulating different scenarios it is possible to individuate the more dangerous situations.

The exposed population can be estimated using different parameters; in the procedure three of them have been selected. The first one is mass of contaminant which exits from each node (Clark et al., 1996), computed as

$$\mathrm{Mc}_j = \sum_{k=1}^{T} C_{jk}\, Q_{jk}\, \Delta t \tag{20.3}$$

Mc_j being the mass which exits from node j, C_{jk} and Q_{jk} the concentration and the demand at node j at time step k, respectively, Δt the time interval considered in the simulation, and T the total time steps. A node can be considered "contaminated" if the mass is larger than a fixed level value, which depends on the kind of substance (Nilsson et al., 2005). The second parameter is $C_{\max}$, the maximum concentration reached in a node during the event. Then, a third parameter is proposed, named *contamination time*, T_c, and defined as the period during which the concentration in the node is larger than a fixed dangerous value. Using these parameters it is possible not only to compare different scenarios but also to find the more vulnerable zones of a system.

For the SPWSS, six scenarios with different injection points have been simulated, highlighting that the worse condition is represented by a contamination in the Gragnano tank (Fig. 20.3) in the summer period. As example, the results obtained with an injection of 33.3 mg/s (with the mass point booster option) of a conservative solute for 17 min are reported. In particular, the parameters $C_{\max}$ and T_c suggest three areas with a higher vulnerability: the ones served by the tanks of Bonea, Meta,

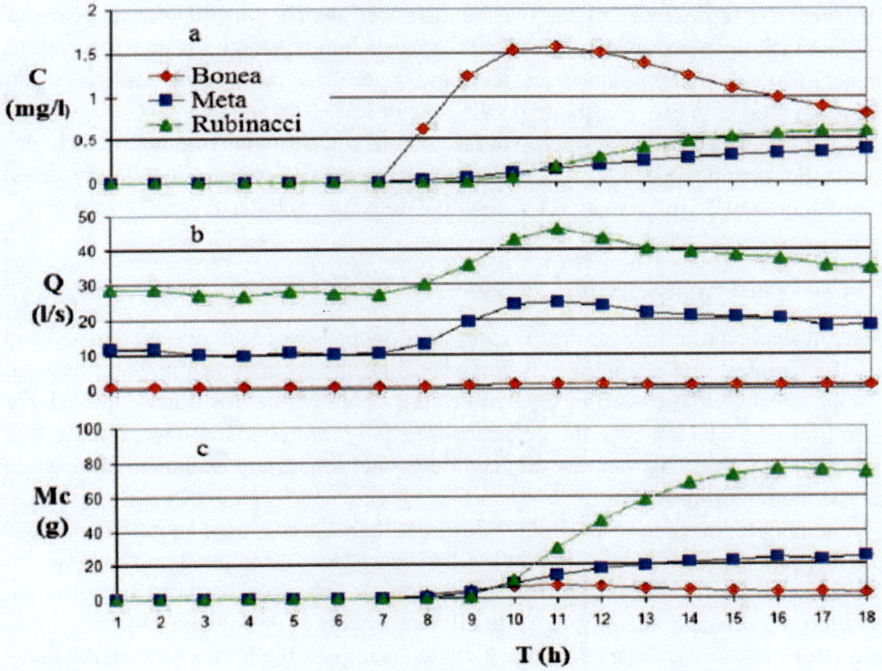

Fig. 20.4 Time history of the concentration, water demand, and solute mass in the three nodes Bonea, Meta, and Rubinacci

and Rubinacci (Fig. 20.3). These parameters, based on the concentration estimate, do not furnish an adequate evaluation of the exposed population. In fact, high concentration values where the demands are small do not represent a large damage in terms of population. In Fig. 20.4 the values of the concentration, water demand, and solute mass in the three considered node during the first 18 h after the beginning of the event are reported. It is evident that the parameter mass (Eq. (20.3)) selects the area served from the tank Rubinacci as the more vulnerable, furnishing an estimate of the damage.

The results obtained from the Risk Assessment phase are used in the Risk Reduction step, in order to adopt the ad hoc physical countermeasures and to realize a surveillance system in the selected sites.

20.4.3 Risk Reduction

The risk reduction is based on four elements (ASCE-AWWA, 2006):

(a) deterrence
(b) detection
(c) delay
(d) response

Deterrence is represented by the overall measures, which can convince an offender not to attack the water system. Detection consists in using methods which permit to detect an attack and/or characterize it. Delay is performed using systems, which can retard an attack, in order to facilitate the Response.

All countermeasures considered in the present work use one or more of such elements. For instance a locked door and a high fence are deterrence and delay systems, simultaneously. Similarly, a CCTV is both a detection and deterrence system.

20.4.3.1 The Physical Countermeasures

The physical countermeasures are represented by passive protection methods. They are the simplest and cheapest way for increasing the security of the water system. The physical countermeasures have to be considered before any other approach for their limited cost. However, their efficacy is usually limited to low-level threats, like the ones performed by vandals. Locked doors and fences are examples of possible physical countermeasures.

The adopted definition is slightly different than the one used in ASCE-AWWA (2006), where also the physical surveillance system is included in the physical countermeasures. However, herein a conceptual difference between the physical countermeasures and the physical surveillance system is introduced, since the latter permits a detection in case of an attack, which is impossible for the former. All facilities in the pilot sites considered in the projects are equipped with adequate physical countermeasures.

20.4.3.2 The Dynamic Response Concept

The architecture of the protection system is designed in order to permit a dynamic response to critical situations. In other words the system has to permit

(1) a fast individuation of the contamination (early warning);
(2) the selection of the appropriate responses, in order to limit damages.

The first task is accomplished designing a surveillance and early warning system, composed of two levels:

- The physical surveillance system
- The Online Contaminant Monitoring System (OCMS)

While the physical surveillance system is realized by controlling accesses to the system, the OCMS design is more complex. Moreover a key aspect of a surveillance system is the data transmission system, as illustrated in Section 20.4.4.

About the second task, the adequate responses have to be selected through the analysis of the possible evolution of a contamination, in order to support the operators in taking decisions for reducing the risk for the population (Decision Support

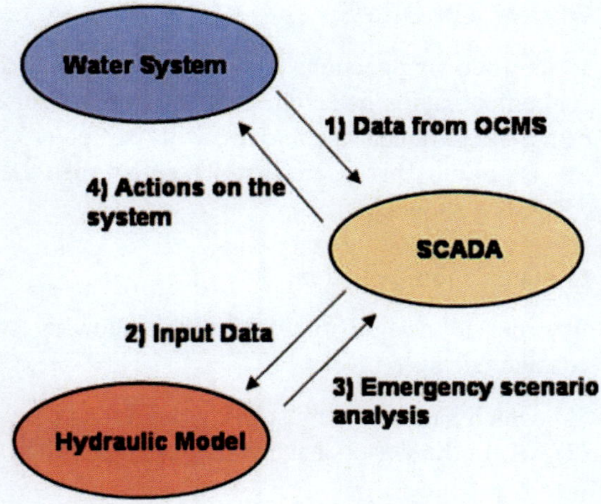

Fig. 20.5 Schematic representation of the Dynamic Response Concept

System, DSS). For both the considered systems, this task is accomplished connecting the hydraulic simulation model with the SCADA system. For the SPWSS the developed hydraulic model is integrated with the existing SCADA system, which provides input data for it. In this way an operator is able to simulate the fate of a contamination, when a warning is provided by the surveillance and early warning system. Using such knowledge and the response protocols an operator is able to choose the best response strategy to the contamination event. The *dynamic response concept* is schematically illustrated in Fig. 20.5: the surveillance system provides data to the SCADA; the SCADA furnishes input data to the hydraulic model; the SCADA receives the results of the analyses of emergency scenarios by the model; then the actions for reducing the risk are implemented on the system.

20.4.3.3 The Physical Surveillance System

The physical surveillance system consists of a set of equipments, which permits a fast detection of an attack. The common feature of a physical surveillance system is the capability to produce an early warning about the presence of an attacker. The simplest surveillance system can be realized with the use of alarmed doors, so that the signal of a door opening is furnished in real time. The control system of the critical accesses is a further improvement. The opening of a door is controlled by a badge, that people of the water company must sweep. This system is very useful against insider saboteurs, since authorized people only can have access to specific facilities.

The above-mentioned systems do not permit a complete control of the activities in the facility. This task can be accomplished only using surveillance cameras. In order to reduce the cost of using these equipments, *motion detection* technologies can be used for the interiors of the facility.

20.4.3.4 The OCMS

In the following sections, some general aspects about the designing of an OCMS for contamination detection, related to the sensor design and location, the selection of the measurement techniques, with particular attention to new solutions, such as the use of biosensors, are first presented; then the OCMS executive project for the SPWSS is shown.

Contamination Detection

In order to design an OCMS the following two major questions have to be considered:

(1) which parameters must be monitored and how (sensors selection);
(2) where the sensors have to be placed on the system.

About the second point, a wide literature review on sensor location problem can be found in Ostfeld et al. (2008). About the first question, it is well known that different instruments are required for detecting different contaminants. The *ideal* instrument would detect any contaminants, identify it, and measure its concentration reliably and accurately. It would be fully functional in all different field operations; it would require minimal response time, maintenance, and housekeeping; it would produce a digital data stream and have the capability of some onboard processing, in order to minimize the data transmission and the analysis requirements. Of course, such an ideal instrument does not exist, but many researchers and companies are currently developing instruments for detecting specific kinds or classes of contaminants in water. However, among the instruments now commercially available, the majority of them have not been tested yet in realistic field circumstances.

Moreover, many instruments do not individuate contaminants directly, but they measure some water parameters, which are affected by them. So, their presence is detected through the measurement of some surrogate parameters. Even if there are many well-tested instruments, which measure a wide variety of water properties with good reliability and accuracy, the relationships between the measured parameters and a specific contaminant, with a specific concentration, are not well understood yet in real conditions. Nevertheless various potential surrogate parameters, which might provide indications about the presence of different kinds of contaminants, are reported in Table 20.1. A contaminant may affect some measurable properties of the water and thus it signals its presence through changes in the surrogate parameter values. Different kinds of changes of the water properties may be produced by a contaminant:

- changes of the chemical properties, due to a reaction between the contaminant and the water constituents;
- changes of optical properties, through absorption, emission, or scattering of light at various wavelengths;
- changes of biological properties;
- changes of mechanical and acoustic properties.

Table 20.1 Potential surrogate parameters for different kinds of contaminants

Kind of contaminants	Chemical	Microbiological	Toxin	Radiological
Surrogate parameter				
pH	X			
Toxicity indicators	X	X	X	X
Turbidity	X	X		
Total organic carbon (TOC)	X	X	X	
Residual chlorine	X	X		
Conductivity	X			
Nitrate, nitrite	X	X		
Dissolved oxygen	X			
Multi-angle light scattering	X			
Phosphate	X	X		
Oxidation reduction potential	X	X		
Biological oxygen demand (BOD)		X		
Biomonitors	X	X	X	
Ammonia	X	X		
Fluorometry		X		
UV254/280	X			
Alpha				X
Beta				X
Gamma				X

For example, some chemical contaminants hydrolyze in water resulting in by-products, which may or may not be less toxic than the original. In the hydrolytic process, the pH of the solution generally changes, because the ion concentration changes. This process may also affect the conductivity of the water as well as the concentrations of other water constituents.

Other contaminants react with chlorine or other constituents of water, producing new chemicals that may or may not be more toxic than the original. These reactions may change pH, electrical properties, oxidation reduction potential, chlorine residual, dissolved oxygen, etc. The effects of a reaction depend upon many factors. The problem is that in a real system not all the important factors and the related effects are known enough for an accurate and unique prediction of the changes expected in surrogate parameters. Furthermore, there are important potential contaminants that neither hydrolyze nor react in any significant degree with water. For these contaminants, only very minimal changes if any in the values of the surrogate parameters may be found.

Another important aspect, related to the contamination detection problem, is represented by the techniques that can be used for measuring the selected surrogated parameters. In the following, some examples relative to the more routinely used parameters are reported.

- *Residual Chlorine*: Chlorine or chlorine compounds are frequently present in water from the disinfection process. Many chemical and biological contaminants react with them, producing changes in the level of residual chlorine or chlorine compounds and increasing chlorine by-products. So, residual chlorine is one of the most sensitive and useful surrogate parameter, since a significant drop of its value could indicate the presence of a contaminant. Unfortunately, without a fairly accurate knowledge of the chemistry of the interaction between the contaminant and the water in the specific system under study, it is not possible to infer much information about the contaminant from a change in chlorine measurement. Therefore, at this stage the only practical approach, which can be used in this case, is performing further analyses in order to understand the causes of the change. Many commercially available instruments measure residual chlorine.

- *Turbidity*: Particulates substances, like for example pathogenic organisms or microencapsulated contaminants, may be detectable using measurements of the water's optical properties, such as turbidity. The instruments for turbidity measurements estimate the average amount of a collimated light scattering over a defined angular range and it is measured in Nephelometric Turbidity Units (NTU). A sensitivity of 0.01 NTUs is usually achievable. Both the particles size and the suspended solid concentration, as well as the level of dissolved solids, can affect the reading. When measuring suspended solids, the instruments are usually able to detect particle concentrations as low as parts per million (ppm).
- *Total Organic Carbon (TOC)*: The systems employed for evaluating TOC adopt a well-defined and commonly used methodology that measures the carbon content of dissolved and particulate organic matter present in water. Measuring the changes in TOC concentrations is an effective "surrogate" for detecting contamination from organic compounds, such as petrochemicals, solvents, and pesticides. Thus, while TOC analysis does not give specific information about the identity of the contaminant, it is still a valuable indicator of a contamination event. The response time of a TOC analyzer may vary, depending on the manufacturer, but it usually takes from 5 to 15 min to get a stable, accurate reading. Online TOC analyzers are designed to operate in remote locations, without a continuous surveillance by an operator. However, to operate reliably, the instruments require regular calibration, inspection, and maintenance by a technically skilled personnel.
- *pH*: The pH measurement in water is one of the oldest and most useful measurements. Along with changes in conductivity measurements, changes in measured pH give an indication of the changes in the ionic constitution of the water. Even if many benign additives can cause changes in the pH values, this information can still give useful indications of a contamination event, if used in combination with the measurements of other parameters.
- *Oxidation Reduction Potential (ORP)*: The oxidation reduction potential is related to the concentration of oxidizers or reducers in a solution. It provides an indication of the solution's ability to oxidize (accept electrons) or reduce (donate electrons) another material. Since oxidizers and reducers are relatively unstable in a solution, the ones present in a system have generally been intentionally added for a specific purpose. In particular, the addition of an oxidizer raises the ORP value, while the addition of a reducer lowers the ORP value. The ORP measurements can be realized using an electrode similar to that for the

pH measurements. The electrode has to stay in solution and it requires routine cleaning and calibration. It has to be replaced on a regular basis, every 1–2 years, depending on the operating conditions.

- *Conductivity*: The conductivity in water is affected by the presence of inorganic dissolved solids such as chloride, nitrate, sulfate, and phosphate anions or sodium, magnesium, calcium, iron, and aluminum cations. Some organic compounds do not conduct electricity very well and have a low conductivity in water. Some organic molecules hydrolyze or dissociate in water, with some of them producing ions that lead to an increase of the conductivity. The conductivity is also affected by the temperature: a higher temperature generally produces a higher conductivity. The conductivity measurements are generally very reliable and conceptually very simple: the electric current between two electrodes, across which there is a known voltage, is measured.

An alternative method for a rapid individuation of a toxic substance in water is represented by the use of biosensors. This technology does not identify or determine the concentration of the contaminant, but serves as a screening tool to quickly determine whether the water is potentially toxic. The rapid toxicity technology uses bacteria (for example, *Vibrio fischeri*), enzymes (for example, luciferase), small crustaceans (for example, *Daphnia magna*), fishes, or specific chemicals. In the absence of toxic contaminants, these substances, either directly or in combination with reagents, produce a background level of light or use dissolved oxygen at a steady rate. So, a change of the color or of the intensity of light produced or a decrease in the dissolved oxygen rate can indicate the presence of a toxic contaminant. Some of these methods are actually commercialized, but further developments are needed to find species that respond in a calibrated way to a broad range of contaminants – chemical and pathogenic – and that require little maintenance and housekeeping.

Among the different organisms, which can be used for monitoring the toxicity, bacteria-based biosensors are actually more suitable for drinking water. In fact, they usually respond to toxic substances more quickly (e.g., minutes), while higher level organisms, such as fishes, may take several days to produce a measurable effect. Bacteria-based biosensors have recently been incorporated into portable instruments, making field testing practical and with a rapid response. Since the residual disinfectant, eventually present in the water, can affect the response of some organisms, it should be removed before the water sample is passed to the toximeter.

The OCMS Design for the Pilot Systems

The procedure used for the OCMS design of the SPWSS takes into account technical, economical, and operational aspects. First of all, monitoring stations are placed close to the reservoirs, in order to economize and simplify their installations and to protect these sites, which are more vulnerable to intentional attacks.

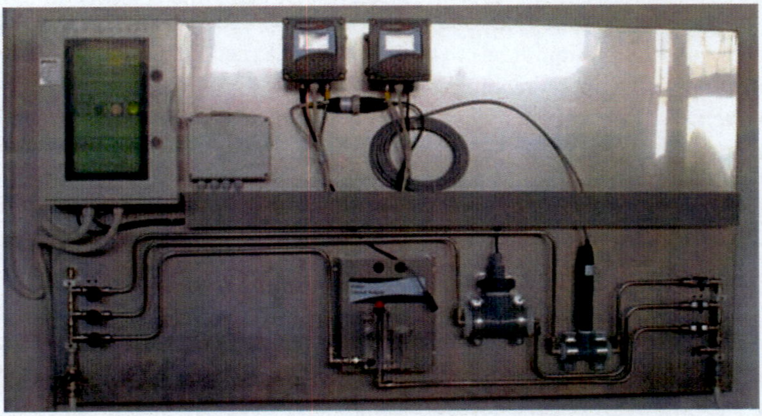

Fig. 20.6 Picture of a standard installation of the monitoring system

The most "critical" reservoirs, which have to be monitored, are selected considering the results of the Risk Assessment phase and through the "*maximum coverage criterion*" proposed by Lee and Deininger (1992).

On the SPWSS the monitoring system is realized by installing seven monitoring platforms. This number is fixed from economical considerations. Each platform is able to transfer all data to the SCADA system and is equipped with the instruments necessary for monitoring the following parameters: chlorine residual, conductivity, pH, and temperature. Figure 20.6 shows a picture of the adopted standard installation.

Because of the high cost of online installation of biosensors (€30,000–50,000 per monitoring point), the application in such configuration is not used for the pilot sites, in which a high number of monitoring points are required. Anyway, some biosensors, in field-portable configuration, are acquired to verify a contamination, eventually detected by the OCMS.

20.4.4 The Data Transmission System

A very important aspect of the surveillance system is represented by the data transmission. The data transmission system has to be

- fast, in order to assure an early warning of an attack event;
- reliable, in order to avoid missing of data;
- safe (i.e., not easy to attack). In fact it is clear that the out of service of the data transmission system corresponds to an out of service of the overall surveillance system.

Different technologies were evaluated for selecting the data transmission system to adopt on the pilot sites. A first restriction is represented by the complex morphology of both pilot sites. Moreover, the data network should be immune to the so-called single point failure, i.e., it should largely rely on redundancy, in order to be robust to some hardware failures and/or intentional attacks.

The impressive progress of the information and communication technologies, which took place in the last decades, has tremendously broadened the set of available solutions for the realization of a communication network. Here, attention is focused on the strategies more suitable for the considered scenarios and applications.

A first classification is between the wired networks (i.e., networks using wires, such as twisted copper pairs and optical fibers) and the wireless networks (i.e., networks using antennas and radio propagation to communicate). Moreover, a communication network may be either private (i.e., the user of the network is also the owner of the hardware equipments) or public (i.e., the network is owned and managed by a telecommunication company acting as a service provider, while the owners of the network are customers paying money for the network usage). Among the available strategies, the following have been selected:

1. *Optical network:* These are wired networks, wherein the cables are optical fibers.
2. *ADSL-based networks:* This is another wired solution, based on the use of the existing fixed telephone network cables to convey data.
3. *IEEE 802.11 (Wi-Fi):* This is a wireless solution. The IEEE 802.11 is a family of wireless communication standards for the deployment of wireless local area networks.
4. *High performance radio LAN (HIPERLAN):* the Hiperlan standard is currently available in two versions, hiperlan/1 and hiperlan/2. This standard implements a Dynamic Frequency allocation System (DFS), which permits an automatic choice of the less interfered channel with a considerable improvement of the system performances. Another advanced feature of this standard is the implementation of the Transmit Power Control (TPC). In this way, the Hiperlan devices have the capability to adapt their transmit power to the current channel conditions, in order to reduce the radiated power when the interference is negligible. This strategy permits a considerable reduction of the radiated power, thus contributing to an overall reduction of the power "on air," with, eventually, less interferences and better performances.
5. *WiMax:* Another standard which is gaining more and more attention is the IEEE802.16, also known as WiMax.
6. *The cellular network:* Another possible strategy is to resort to data connection services offered by cellular networks. Indeed, it is well known that the second generation of the cellular systems may support data rates up to some tens of kilobits per second, while the UMTS system can support up to fractions of one megabit per second and more. On the other hand, however, this solution appears to be very expensive, since it depends on an external communication structure, which the Water Company should pay for.

With regard to the considered pilot systems, in agreement with the Water Company technical staff, a wireless solution is more advantageous; indeed, the wireless solution is much more flexible than a wired one and, moreover, it makes possible to build a *proprietary* communication network with a reasonable expense.

The preliminary phase of the communication network design is represented by a detailed analysis of the territory cartography, for a better understanding of the land morphology and of the distribution of the GORI sites on the area. Then, the most critical links for the realization of the wireless communication network are individuated. As result of this preliminary study, a combination of hiperlan and Wi-Fi technologies is selected as the best solution, with the use of the hiperlan for the long-distance links and the use of the cheaper Wi-Fi suited for the less critical links. In particular, the selected hiperlan2 solution is known to exhibit higher performances with respect to the Wi-Fi; moreover, hiperlan2 standard is equipped with the so-called seamless roaming.

20.4.5 The Response

The choice of the response to adopt in case of a contamination event is a difficult task. Many response procedures cannot be acted by just the Water Company. In particular, when an attacker is individuated in a facility, a fast advice must start for the Police and other security organizations. However, some response procedures have to be pursued by the water companies.

If a water contamination occurs, with the suspect of a possible danger for the consumers, the interruption of the service is required. In this case a reduction of the number of the "false-positive" signals is a very crucial point, in fact a service interruption without a real contamination event can reduce the belief of consumers respect to the Water Company. In this task biosensors (see section "Contamination Detection") can be very helpful.

If the monitoring system advices that a contamination event is in act, the source location of the contamination is important in order to reduce its effects. For example, the application to the Vesuvian Water Supply System of the source location procedure proposed by Di Cristo and Leopardi (2008) is presented. Since the procedure is widely described in the original paper, only its application to the VWSS is shown here. In particular, the part of the VWSS, showed in Fig. 20.7 and composed by 101 pipes, 106 junctions and 14 tanks, is considered. It is assumed that the intrusion is possible in every node (i.e. 130 possible intrusion points) and synthetic data are generated considering five sensors, which furnish contaminant concentration every 6 h (Fig. 20.8). Such data are generated simulating a constant intrusion of a passive solute in node 826, close to wells field of De Siervo. In particular, a mass of 2.50 mg/s is inserted using the option "mass point booster" of EPANET. Even if the better sensor location for the procedure is the one that comes from application of the maximum coverage criterion (Lee and Deininger, 1992), a random sensor location is used here. This is because in real world applications, it is common that available sensors (installed for different uses than the source location) are used. In some cases none sensor is presented on the network, but we have only some water quality measures, taken from samplings on the system. The procedure individuate a group of

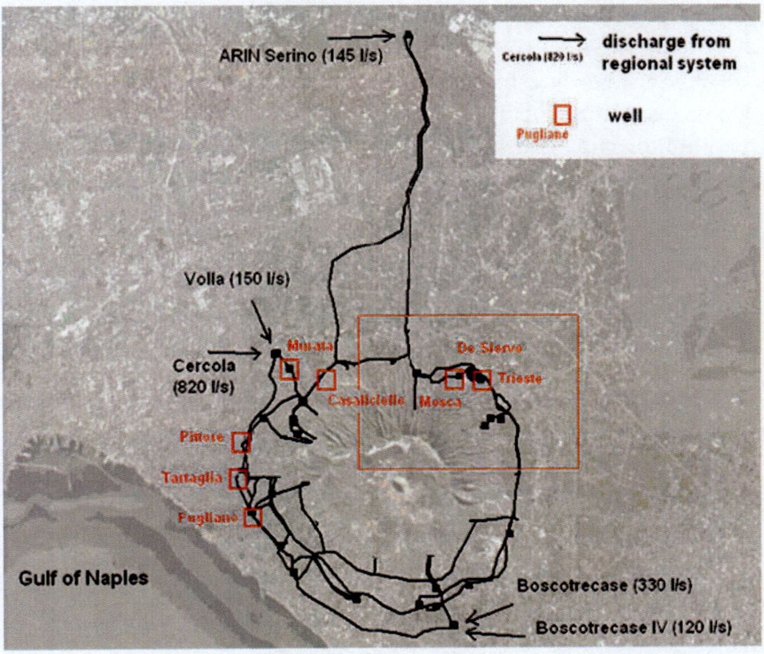

Fig. 20.7 Scheme of the Vesuvian Water Supply System and location of the zone used for the source location test

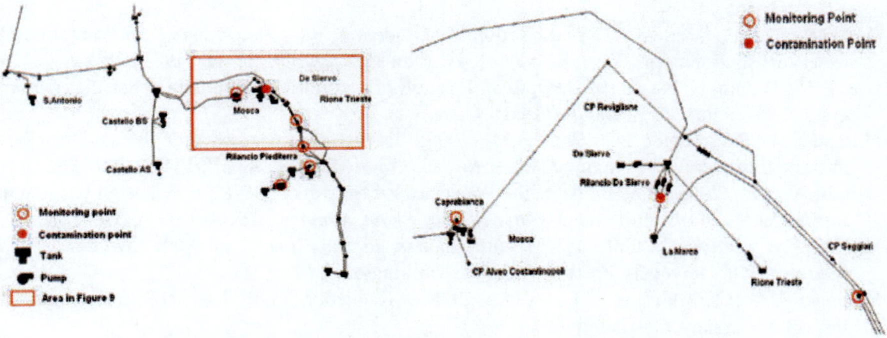

Fig. 20.8 Detail of the zone used for the source location test, with the indication of the sensors and of the intrusion point

62 candidates as source point, composed by less than half of the initial number of points. Though the fitness function evaluation, the source point is correctly located in the node 826.

This application demonstrates that this simple-using procedure (Di Cristo and Leopardi, 2008) can be easily applied to real world situation. Then, the source location procedures can be very useful for a Water Company, in order to study the countermeasures for flushing the network after a contamination event.

20.5 Conclusion

In this chapter a procedure for protection systems design, developed during two research projects, realized in the framework of the European Programme for Critical Infrastructure Protection (E.P.C.I.P.), are described, through the application to two real water systems, characterized by different size and behavior. For each of them, the protection system is effectively realized and tested. These projects are a good example of allocation of public funds for water systems security enhancement. Moreover, the partnership between a University and a private water company represents a "bridge" between theory and practice in water systems management.

Acknowledgments The projects DISWIP (grant n. JLS/2006/EPCIP/018) and GLEWIP (grant n. JLS/2007/EPCIP/022), described in this chapter, were funded by the Directorate General Freedom, Security and Justice of the European Commission in the framework of the European Programme for Critical Infrastructure Protection (EPCIP).

References

Agency for Toxic Substances & Disease Registry – ATSDR (2005), *Public Health Assessment Guidance Manual* (Update)

ASCE, AWWA, WEF (2004), *Interim Voluntary Guidelines for Designing an Online Contaminant Monitoring System*, ASCE, Reston, VA.

ASCE-AWWA (2006), *Guidelines for Physical Security of Water Utilities*, ASCE, Reston, VA.

Clark, R.M., Geldreich, E.E., Fox, K.R., Rice, E.W., Johnson, C.W., Goodrich, J.A., Barnick, J.A., Abdesaken, F. (1996), Tracking a Salmonella servor typhimurium outbreak in Gideon, Missouri: Role of contaminant propagation modelling, *J. Water Supply Resour. Tech.*, 45(4), 171–183.

Di Cristo, C., Leopardi, A. (2008), Pollution source identification of accidental contamination in water distribution networks, *J. Water Resour., Plan Mgmt, ASCE*, 134(2), 197–202.

Lee, B.H., Deininger, R.A. (1992), Optimal location of monitoring stations in water distribution system, *J. Environ. Eng., ASCE*, 118(1), 4–16.

Nilsson K.A., Buchberger S.G., Clark R.M. (2005), Simulating exposures to deliberate intrusions into water distribution systems, *J. Water Resour. Plann Manage., ASCE*, 131(3), 228–236.

Ostfeld A. et al. (2008), The Battle of the Water Sensor Networks (BWSN): A Design Challenge for Engineers and Algorithms. *J. Water Resour. Plann. Manage., ASCE*, 134(6), 556–568.

Rescher, N. (1983), *A Philosophical Introduction to the Theory of Risk Evaluation and Measurement,* University Press of America, Washington, DC.

Rossman, L. A. (2000), *Epanet2 Users Manual*, Risk Reduction Engrg. Lab., U.S. Environmental Protection Agency, Cincinnati, OH.

Walski, T. M. et al. (1987), Battle of the network models: Epilogue, *J. Water Resour. Plann. Manage. Div., ASCE*, 113(2), 191–203.

Chapter 21
Utility of Supercomputers in Trace-Back Algorithms for City-Sized Distribution Systems

Hailiang Shen and Edward McBean

21.1 Introduction

Contaminant intrusion into water distribution systems (WDS) is drawing increasing attention, to protect against both accidental events and deliberate injection. These accidental events include examples of water main breaks, cross-connections, backflows, and pressure transients (Methods et al., 2003). Backflow issues may be caused by back pressure or back-siphon and are potentially a significant threat to WDS. Cross-connection refers to actual or potential connections between potable and non-potable water supply and represents another form of contaminant intrusion into a WDS (US EPA, 2003). As examples of both back pressure and cross-connection, in 2005, in Stratford Ontario, a carwash cleaning agent entered the WDS, resulting in 30,000 people being affected, with 19 seeking medical attention; in 1997, in Guelph Ontario, a petroleum chemical back-siphoned into the WDS, leaving 50,000 affected and some without water for up to 1 week. Even worse, after Sept 11, 2001, deliberate injection of chemical or biological agents at treatment plant intakes, tanks, pump stations, and consumer nodes is considered as one of the most severe dangers to the public (Ostfeld and Salomons, 2005).

A contaminant warning system (CWS) is a possible strategy to deal with the intrusion issue (AWWA, 2004). The CWS relies on optimally placed sensors. Identification of the existence of contaminant(s) within WDS triggers contaminant source identification (CSI) and emergency response procedures.

The CSI procedure requires both speed and accuracy. Contaminants move rapidly with water and may spread about 3 km away from the intrusion source within only 1 h, suggesting it is infeasible for an algorithm to run days to identify CSI results, e.g., the possible intrusion nodes (PINs), where the possible intrusion event occurred. Complicating matters is that various uncertainties exist in modeling for CSI, such as nodal demand and unknown contaminant type. These uncertainties will impact the results of CSI in regard to identifying PINs, missing the true intrusion

H. Shen (✉)
School of Engineering, University of Guelph, Guelph, ON N1G 2W1, Canada
e-mail: shenh@uoguelph.ca

R.M. Clark et al. (eds.), *Handbook of Water and Wastewater Systems Protection*, Protecting Critical Infrastructure, DOI 10.1007/978-1-4614-0189-6_21,
© Springer Science+Business Media, LLC 2011

node, thus requiring the CSI procedure to reflect these uncertainties which greatly expand the complexities of analyses.

An efficient way to address the two requirements of CSI is to utilize a data mining procedure. The procedure compiles an off-line database containing the simulation results of an array of possible events and then mines the database for CSI as soon as a sensor alarms. The merit of a data mining approach is that it can be accomplished very efficiently. Herein, the abilities of data mining are extended to incorporate the impact of uncertainty. In addition, a geographic information system (GIS) toolkit, integrating the online data mining procedure, is developed to facilitate quick emergency response.

21.2 Literature Review

Currently, various types of methodologies exist for CSI. The first type used, for example, by Shang et al. (2002), proposed a methodology to trace back a contaminant particle in discrete time, given a sensor's first detection time and concentration; however, this procedure cannot determine the contaminant release history although it is suitable as a pre-step to reduce the search space for an optimization procedure.

A second methodology, a simulation-optimization method based on reduced gradient method (e.g., Guan et al., 2006) and genetic algorithm (GA), involves considerable runtime due to the necessity to simulate large numbers of injection events using EPANET. To accelerate the GA optimization procedure, parallel GA has been proposed (e.g., Sreepathi et al., 2007), which allows simulation of intrusion events with EPANET in parallel; the parallel GA procedure has the following limitations: (i) to utilize this procedure online, a water utility may need to maintain parallel computing facilities or hardware routinely, since the time of an intrusion event is never known a priori, and hence the facilities may be required at any time. Applying cloud computing or supercomputers maintained by other organizations may be other options when job waiting times can be guaranteed to be short; (ii) there is no guarantee for the GA to converge to the global optimum, i.e., the true intrusion node may not be identified; and (iii) there may be the need for simulating duplicate intrusion events, resulting in need for extensive computational power.

Use of a neural network is another alternative (e.g., see Kim et al., 2008), which applies sensor response and intrusion events data as the input and output of the neural network; this method has only been tested in a pilot network and the scale-up to a large network may require considerable off-line neural network training time and online computation time from the trained neural network model.

Perelman and Ostfeld (2010) proposed a Bayesian network in CSI. The three steps are (i) group all network nodes into clusters, within each cluster no link flow can reverse; each cluster is connected by link(s), (ii) quantify the local conditional probability by simulating injection events, and (iii) given a cluster where a contaminant is observed or not observed, find the probability of the source cluster (i.e., source nodes). However, under nodal demand uncertainties, the flow direction in each link may vary, leading to different grouping of cluster nodes and the subsequent source nodes identified.

Wong et al. (2010) applied manual sampling to gradually reduce the number of possible sources as more manual samples are available. For real-time application, the contaminant type(s) is rarely known. Accordingly, instead of measuring the suspicious contaminant(s) directly, it is necessary to measure some indicators of contamination, such as chlorine, turbidity, and others. However, in the case of manual sampling, there are insufficient background measurements of selected indicators to identify whether these indications really indicate the existence of contamination in the WDS or are natural variations; this results in high false-positive/negative rates in determining the existence of contamination which, in turn, impacts the ability to reduce the number of possible sources. This procedure works well when the contaminant can be easily identified by aesthetic parameters, such as turbidity, color, and smell which are easily detectable by manual sampling.

Another procedure is data mining (e.g., Huang and McBean, 2009; Shen et al., 2009a, b), involving "mining the database" with structural query language (SQL). This approach is described more fully in Section 21.3, as this procedure is able to be applied to real (city-sized) WDS.

21.3 Methodology

This section discusses the existing data mining procedure and extends its ability to incorporate uncertainty analysis. The extension is made feasible with the capacity of parallel computing in Shared Hierarchical Academic Research Computing Network (SHARCNET). The PIN identification requires the analyses to be completed online as soon as a sensor alarms; thus displaying the PINs on a WDS map would greatly facilitate emergency response. On further coupling with Global Positioning System (GPS), GIS can be applied for real application. Herein, a GIS toolkit based on ArcObjects/VBA/MySQL is developed to integrate the online data mining procedure. Integrating GPS into the GIS toolkit is ongoing.

21.3.1 Existing Data Mining Procedure

The overall data mining procedure employed consists of three steps. First, a database is populated, which contains the array of possible intrusion events (i.e., the combination of injection nodes, injection times, durations, and mass rates) and their corresponding simulated sensor detection information in a scenario. The assembly of this information is very time-consuming due to the large number of simulations of possible intrusion events, and this is particularly relevant for a city-sized WDS. Most importantly, this overall database is completed off-line, before a real intrusion event, and is only "mined" during an event when the CSI is alarming. Hence, the initial effort of compiling the data does not count in the data mining online run time.

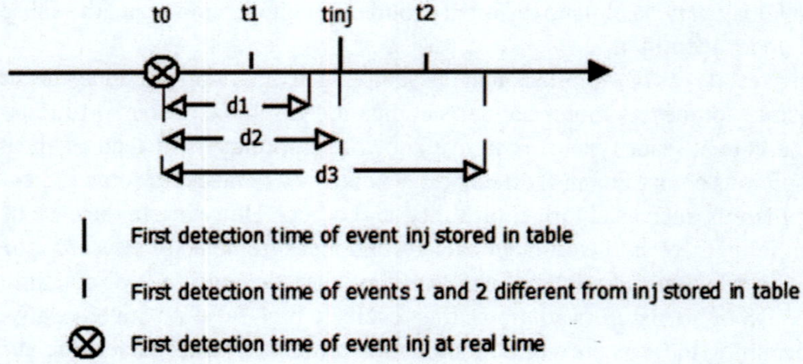

Fig. 21.1 Analysis of Event inj

Second, in the case when sensor "*S*" alarms, PINs are selected by querying against the pre-populated database table with an SQL: "select injection events which result in first detection time in database at sensor '*S*' between time *t–m* and *t*+*m*," where *t* is the observed first detection time at sensor "*S*" and "*m*" is an offset value from *t*. Figure 21.1 demonstrates the rationale to determine the offset value for a single event inj. If the offset value is set to d_1, a false event 1 is selected while the true event inj is missed; by setting the offset value to d_2 (equal to $(t_j - t_0)$), the true event inj as well as event 1 is identified by SQL; given an offset value d_3, in addition to events 1 and inj, another false event 2 is identified. Thus, among the three distances, d_2 is the best, since it can identify the true intrusion event inj by comparing with d_1 and select the lesser number of false events by comparing with d_3. These offset values exist for every online scenario and every event. The "*m*" value is characterized in a statistical manner.

The third step is to quantify the probability of each PIN as the true intrusion node and to check the existence of priority nodes which are upstream of important facilities, where priority nodes are identified herein as schools, hospitals, and governmental offices, as examples. Priority nodes are identified as requiring higher emergency response priority due to the larger consequence in case impacted by contaminants. It is noted, however, that discussion of the third step on how to select priority nodes, a subset of the PINs identified right after the current sensor alarm, as well as the methodology to quantify the priority degree is not covered in this chapter; details can be found in Shen et al. (2009a, b).

21.3.2 Extension to Uncertainty Analysis

To understand the impact of uncertainty on the possible events or PINs, it is necessary to "learn" the behavior of the contaminant under various uncertainties, i.e., to simulate the array of events under various scenarios. However, the simulation of even a single scenario for a large WDS is very time-consuming, suggesting it is

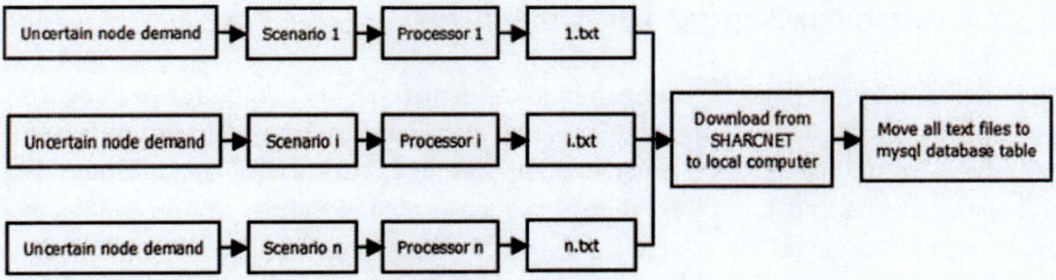

Fig. 21.2 Parallel computing flow chart of parallel computing

impossible to simulate multiple scenarios in serial computing. Parallel computing is introduced to simulate scenarios in parallel, thus reducing the scenarios simulation time linearly by the number of processors applied.

SHARCNET is one of the facilities for parallel computing. SHARCNET consists of over 13,000 cores or processors and is employed herein to implement parallel simulation of various scenarios. The parallel database construction process is illustrated in Fig. 21.2.

For purposes of discussion here, only nodal demand uncertainty is addressed; the methodology of incorporating other sources of uncertainties is the same. In EPANET, within a hydraulic time step, nodal demand is quantified by the multiplication of its base demand by the pattern factor in a time step and is reasonably characterized by the normal distribution (Babayan et al., 2005). Herein, to generate random demands obeying the normal distribution for each node, within each time step, the mean value is set to the pattern factor and the standard deviation is set to 10% of the mean value. It is noted that the probability of generating negative random pattern factors is 7.6E-24 (see in Appendix), a very small probability. Hence, in case a negative number is generated, it is set to zero, which will not impact the normality of the generated random numbers. Simulation of the array of events in scenario "*i*" is completed in the processor *i*. The events and corresponding sensor's first detection times as well as concentrations are stored in the text file i.txt. All generated text files are downloaded from SHARCNET to a local personal computer and then moved to MySQL database tables for the uncertainty analyses.

21.3.3 Number of Scenarios Required

It is impossible to simulate an infinite number of scenarios. A cut-off number has to be identified to be served as a benchmark to quantify the false-negative rate and to compute the "*m*" value for each sensor. The array of events is the same in simulating various scenarios and thus, by increasing the number of scenarios, the number of detected events within the array would increase. There may be a number of scenarios of simulation *N*, after which, few new events would be detected by increasing the number of scenario simulations. The number *N* is referred as the point of diminishing marginal return.

21.3.4 Impact of Storing More Scenarios

Two cases, namely I and II, are examined to identify an "*m*" value for each sensor. In Case I, only the simulation results of the events in the first scenario are stored, and the resulting database table is named as "table_1." In Case II, in addition to the 1st scenario, the 2nd as well as the subsequent scenarios are stored in a database table, named "table_2."

These two cases will also be applied to characterize the impacts of storing more scenarios in the database table. The impacts include (i) reduction in the false-negative rate of each sensor, (ii) the variation of "*m*" value, and (iii) the numbers of false possible intrusion events and PINs. By applying database table "table_2" instead of "table_1" for CSI, three situations 1, 2, and 3 may happen to the "*m*" value of a specific sensor "*S*": reduced, increased, or unchanged, respectively. To illustrate the three possible situations at specific sensor "*S*" clearly, a total number of two events (events 1 and 2) and three scenarios (1st, 2nd, and 3rd) are examined.

In situation 1, only event 1 is detected; while in situations 2 and 3, both events 1 and 2 are detected. The time t_i and u_i represent the first detection time of the events 1 and 2 at sensor "S," respectively. In the following sections, a statement "t_1 happens in real time" has two meanings: (i) event 1 is the true event, as indicated by *t* (instead of *u*) of t_1, and (ii) the 1st scenario happens, as suggested by the subscript 1 of t_1.The offset value analyses are illustrated in Table 21.1 and Fig. 21.3.

21.3.4.1 Case I

As listed in Table 21.1, for example in situation 1, t_3 has an offset value $|t_3 - t_1|$. If "t_3 happens in real time," we can execute SQL against "table_1": "select injection events that can result within first detection time at sensor *S* between time $t_3 - |t_3 - t_1|$ and $t_3 + |t_3 - t_1|$" to identify the true event 1; likewise, the offset values are computed for t_2 and t_3.The offset values in situations 2 and 3 are calculated in the

Table 21.1 Offset values calculation in cases I and II

Cases	First detection time of the three scenarios	Offset values at situation 1	Offset values at situation 2	Offset values at situation 3
Case I (store the 1st scenario)	t_1	0	0	0
	t_2	$\|t_2 - t_1\|$	0	0
	t_3	$\|t_3 - t_1\|$		$\|t_3 - t_1\|$
Case II (store the 1st and 2nd scenarios)	t_1	0	0	0
	t_2	0	0	0
	t_3	$\|t_3 - t_2\|$		$\|t_3 - t_1\|$
	u_2		0	0
	u_3		$\|u_3 - u_2\|$	$\|u_3 - u_2\|$

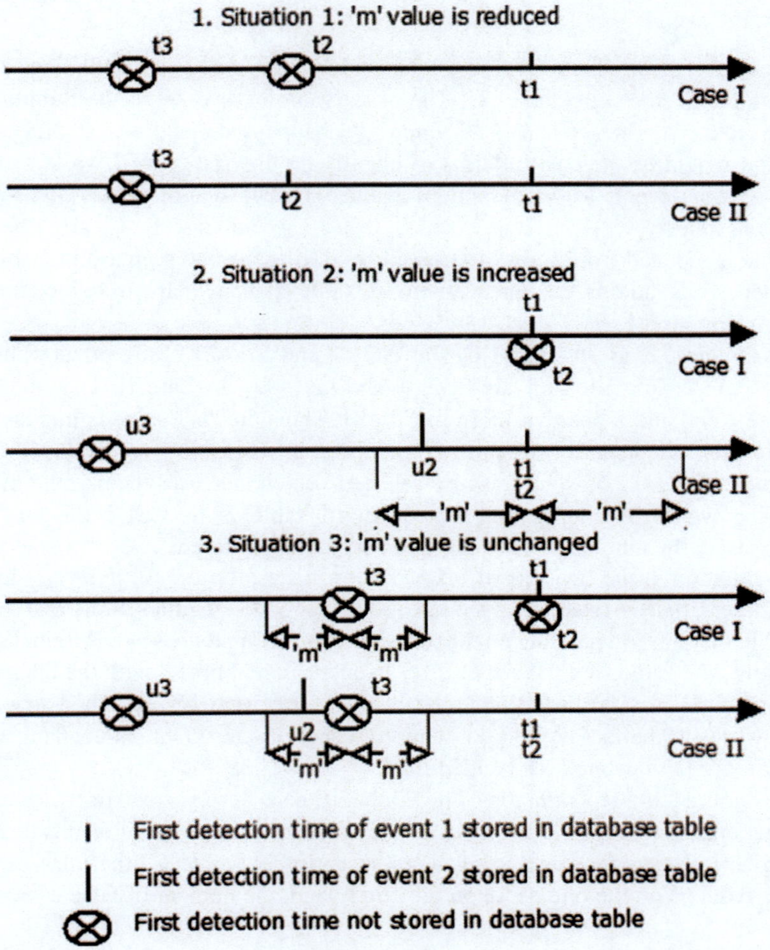

Fig. 21.3 Offset values analysis in Cases I and II

same manner.. It is noted from Fig. 21.3 that in situation 2 for the 3rd scenario, event 1 is not detected by sensor "*S*"; thus, its offset value in Table 21.1 does not exist in Case I. In situation 1, the "*m*" value would be the 95% quantile of the offset values 0, $|t_2 - t_1|$, $|t_3 - t_1|$. To explain the "*m*" value, 95% of events have offset values less than the "*m*" value; in other words, we have 95% probability of identifying the true intrusion event and accordingly, the true intrusion node in the PINs selected from "table_1," if the true intrusion node is really stored in "table_1."

Nevertheless, the database table, "table_1" does not store the event simulation results of the 2nd scenario. The impacts of storing the 2nd scenarios in the database table "table_2" are discussed in Section 21.3.4.2.

21.3.4.2 Case II

In situation 1, from Table 21.1, the offset values of t_1 and t_2 are zero since t_1 and t_2 are already stored in "table_2." The offset value of t_3 is set to the minimum of $|t_3-t_1|$ to $|t_3-t_2|$, which is $|t_3-t_2|$, since if "t_3 happens at real time," an offset value $|t_3-t_2|$ would be the best in terms of identifying the true event 1 while avoiding false possible events to the maximum extent. The two sets of offset values are 0, $|t_2-t_1|$, and $|t_3-t_1|$ in Case I, and 0, 0, $|t_3-t_2|$ in Case II. Clearly, the 95% quantile is reduced from Case I to Case II by storing the 2nd scenario in "table_2." In other words, adding the 2nd scenario in "table_2" can "help" to reduce the "*m*" value in situation 1.

In situation 2, as suggested by Table 21.1 and Fig. 21.2, the offset values in Cases I are 0, 0, while they are 0, 0, 0, and $|u_3-u_2|$ in Case II. The "*m*" value in Case I is 0 and a positive value in Case II. Thus, the "*m*" value is increased by storing the 2nd scenario. Suppose "t_1 happens in real time," as displayed in the subplot 2 of Fig. 21.3, event 2 is also selected, which increases the number of false intrusion events. By adding the 2nd scenario in "table_2" for CSI, if the "*m*" value is increased, the number of false intrusion events is increased.

In situation 3, the two sets of offset values in Case I are 0, 0, $|t_3-t_1|$, and the values are 0, 0, $|t_3-t_1|$, 0, $|u_3-u_2|$ in Case II. The 95% quantiles of the two sets are possibly the same, depending on the two values $|t_3-t_1|$ and $|u_3-u_2|$. Herein, the two resulting "*m*" values are the same. If "t_3 happens in real time," with the unchanged "*m*" value in Case II, in addition to the event 1, false event 2 is identified as well. In other words, if the fact is that adding more scenario(s) results in an unchanged "*m*" value, more false events will be identified.

It is noticed in both situations 2 and 3 that if the event 2 instead of the event 1 is the true intrusion event, "table_1" in Case I misses the true event or results in a false negative. It means that by storing more scenarios in "table_2" the false-negative rate is reduced on the one hand; on the other hand, the number of false events may increase.

In summary, by storing more scenarios in "table_2," the "*m*" value of each sensor may be decreased (in situation 1), increased (in situation 2), or unchanged (in situation 3); with increased or unchanged "*m*" value, the number of false intrusion events will increase. In the case studies below, no observations show the "*m*" value is increased, which suggests that the possibility of existence of increased "*m*" value is very low. It does not mean situation 2 is not happening, but its "increased" effect on the "*m*" value is declined by the "reduced" effect of situation 1 and "unchanged" effect of situation 3.

21.3.5 False-Negative Rate

Two types of false-negative rates A and B exist. False negative rate A refers to true events which are not identified due to their not being stored in the database for CSI; false-negative rate B characterizes the fact that true events are missed due to the statistical characterization of the "*m*" value as the 95% quantile of the offset values.

Scenarios are stored one by one from the 1st scenario into a database table; thus after the ith scenario is stored, there is a total number of i scenarios in the database. To define the false-negative rate A of a specific sensor, after the storage of the ith scenario, the total number of detected unique events by a specified sensor among the N (the point of diminishing marginal return) scenarios is calculated and denoted as K. The database now includes k unique events detected by the sensor. The ratio of $(K - k)/K$ represents the false-negative rate A of the sensor.

The false-negative rate B is only related to the quantile specified as the "m" value and is irrelevant to the false-negative rate A. In this chapter, the 95% quantile is set as the "m" value, which means for 95% probability, a true event can be identified, and with 5% probability, the true event is missed.

False-negative rate is defined in Eq. (21.1):

$$\mathrm{FNR} = 1 - (1 - \mathrm{FNR_A})(1 - \mathrm{FNR_B}) \tag{21.1}$$

where FNR = false-negative rate, $\mathrm{FNR_A}$ = false-negative rate A, and $\mathrm{FNR_B}$ = false-negative rate B.

For example, after storing the 25th scenario into "table_2," 20% of events of the N scenarios are missed at sensor "S," and 95% quantile is set for the "m" value of sensor "S." Then, current "table_2" contains 80% events, or have 80% chance to identify a true event if it is really stored in "table_2"; if the true event is stored in "table_2," we have 95% chance to identify it as discussed in previous sections; hence, we have 80% × 95% (i.e., 76%) chance to identify a true event with "table_2." Therefore, the false-negative rate would be 100 – 76%, or 24%.

21.3.6 GIS Toolkit Development

For rapid emergency response in the case of a sensor alarm, it is required to integrate the developed online data mining portions into geographic information system (GIS) to visualize the PINs and priority nodes on a map and to list probabilities of PINs and priority degree of priority nodes. A toolkit is developed in response. Figure 21.4 shows the developed graphical user interface (GUI) based on ArcObjects/VBA/MySQL. Note that both the false-negative rate quantification and the "m" value computation are completed off-line and thus do not require integration into the GIS toolkit for online application. This chapter only demonstrates the abilities of the GIS toolkit in PIN identification and displaying the locations of PINs, although the functionalities to display priority nodes and to list probabilities and priority degrees have been completed.

21.4 Case Studies

The City of Guelph WDS is utilized to demonstrate the procedure. The WDS consists of 3,420 nodes and 4,272 links. A total of 300 scenarios are simulated. The simulation of a single scenario takes approximately 3 days. In serial computing, the

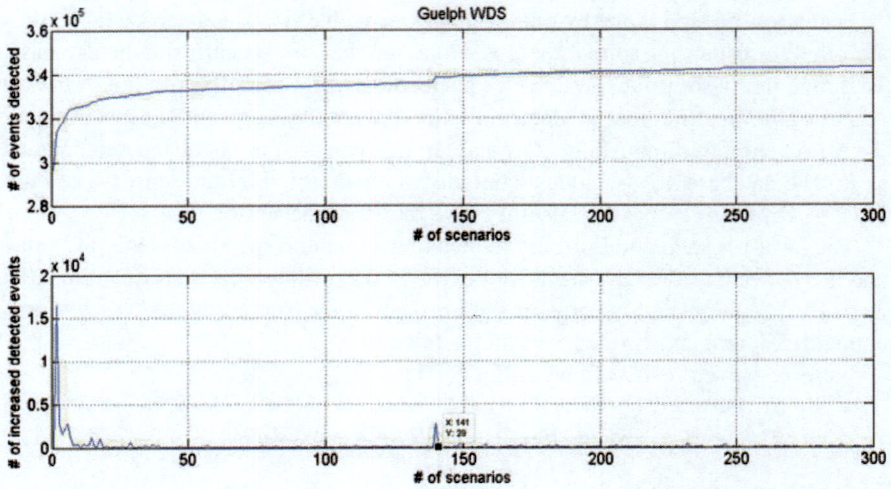

Fig. 21.4 Guelph WDS return curve

runtime for these 300 scenario simulations would be 900 days, which is infeasible for the subsequent false-negative rate quantification and "*m*" value calculation. Given the power of parallel computing of SHARCNET, the run time of the scenarios is only 3 days.

The return for storing more scenarios in "table_2" for the Guelph WDS is presented in Fig. 21.5. With increased numbers of scenarios, more new events are detected by the sensor network, which implies fewer events are missed by the sensor network, i.e., a lower false-negative rate of each sensor is obtained. It is interesting to note that there is a point of diminishing marginal return, 141, in Fig. 21.5.

Statistical analyses for the Guelph WDS in Cases I and II are summarized in Table 21.2. For example, for the sensor node index 1,899, the false-negative rate is 11.3% if "table_1" is applied for CSI, and this number is decreased to 6.4% if a total number of 25 scenarios are stored in "table_2" and utilized for CSI; for Case I, on setting the "*m*" value to 990 min, there is 95% confidence that the identified PINs contain the true intrusion node. The reason for an "*m*" value as large as 990 min is that nodal demand uncertainty greatly changes the flow paths of the contaminant. In addition, on incorporating other uncertainties, such as unknown contaminant type, the value may become even larger. Three points are observed through Table 21.2 by increasing the numbers of scenarios in "table_2": (i) the false-negative rate is always reduced, (ii) the "*m*" value of each sensor is reduced or kept unchanged, and (iii) when 80 scenarios are stored in "table_2" for CSI, the "*m*" values of all sensors are reduced to zeroes. Two factors may contribute to zero values: (i) the water quality step in EPANET is set to 30 min, thus the resolution of the first detection times is 30 min, or "grouped" to one-half an hour, e.g., 8:30 AM, 9:30 PM, etc. and (ii) by storing more scenarios in "table_2," the chances of a real-time scenario matching the one in "table_2" increases, and in turn, the first detection times at sensors have

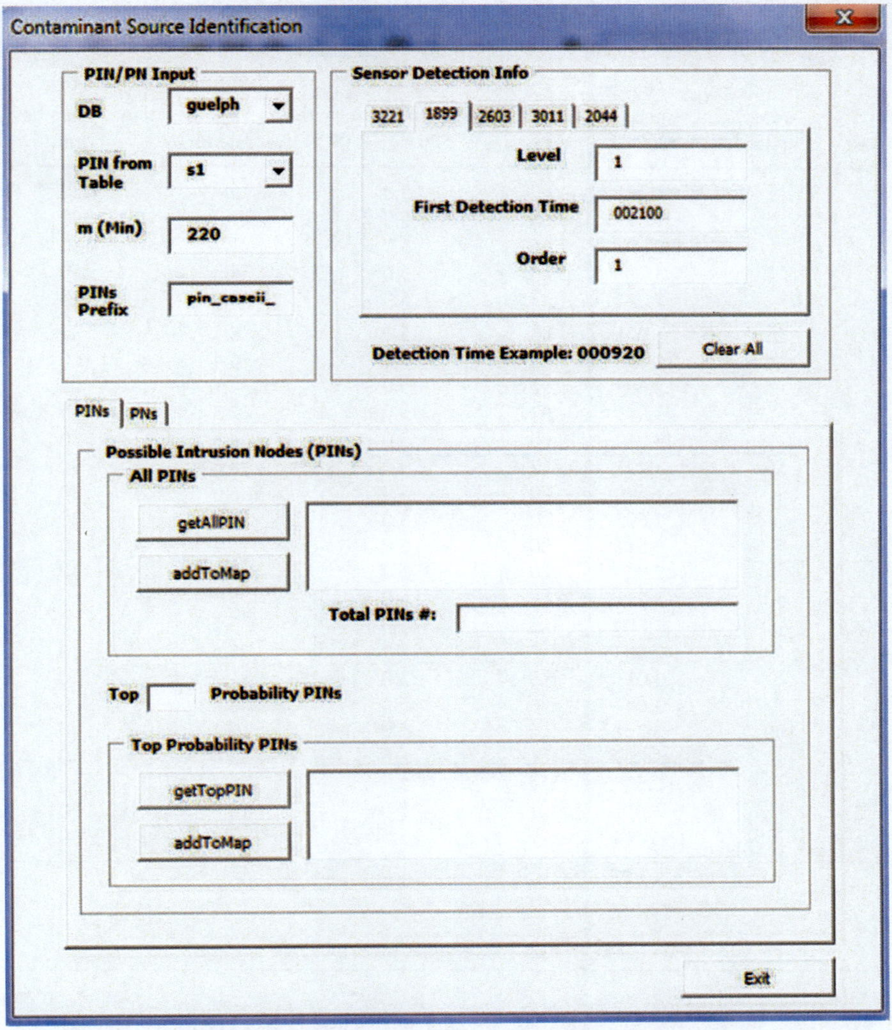

Fig. 21.5 Developed GIS toolkit GUI

increasing chances to be the same as one in "table_2." Thus, the offset values get more chances of becoming zero; the 95% quantile, i.e., the "m" value, may go to zero, which is what happens when 80 scenarios are stored in "table_2."

To test the online data mining procedure, an event happening at node index 42 and 8:00 AM is simulated to obtain its online sensor alarm info; the event is detected first by sensor node index 3,011 at 9:00 PM. Listed in Table 21.3, the online run time of the data mining procedure is less than 2 min and thus acceptable for online application. Storing the 25th through the 50th scenarios reduces the "m" value from 30 to 0 min, and the number of false PINs is reduced from 124 to 123. Three points

Table 21.2 Statistical analysis for Guelph WDS

Cases	Number of scenarios in database	Sensors node indices	False-negative rate A (%)	False-negative rate B (%)	False-negative rate (%)	*m* value – 95% quantile (min)
I	1	3,221	12.0	5.0	16.4	480
		1,899	6.6	5.0	11.3	990
		2,603	17.6	5.0	21.7	780
		3,011	20.3	5.0	24.3	450
		2,044	15.7	5.0	19.9	510
II	25	3,221	3.5	5.0	8.3	30
		1,899	1.5	5.0	6.4	60
		2,603	2.2	5.0	7.1	60
		3,011	7.7	5.0	12.3	30
		2,044	5.0	5.0	9.8	30
	50	3,221	3.2	5.0	8.0	30
		1,899	1.0	5.0	6.0	30
		2,603	1.0	5.0	6.0	30
		3,011	7.0	5.0	11.7	0
		2,044	4.0	5.0	8.8	30
	80	3,221	2.8	5.0	7.7	0
		1,899	0.5	5.0	5.5	0
		2,603	0.6	5.0	5.6	0
		3,011	6.6	5.0	11.3	0
		2,044	3.6	5.0	8.4	0
	100	3,221	2.6	5.0	7.5	0
		1,899	0.4	5.0	5.4	0
		2,603	0.4	5.0	5.4	0
		3,011	6.1	5.0	10.8	0
		2,044	3.1	5.0	7.9	0

Table 21.3 PIN identification analyses

# of scenarios in its "table_1" or "table_2"	"m" value (min)	# of possible intrusion events	# of PINs	Online runtime (seconds)
1	450	1,440	190	42
25	30	363	124	37
50	0	293	123	41
80	0	343	129	53
100	0	350	130	77

may be concluded from Table 21.3: (i) if the "*m*" value is unchanged, the number of PINs or the number of false PINs will increase, (ii) if the "*m*" value is reduced, the number of false PINs is usually reduced, and (iii) no data show the "*m*" value is increased for any sensor, which suggests low likelihood of increased "*m*" value.

To select the best number of scenarios in terms of obtaining the least number of false events for CSI, it is necessary to address the variation of "*m*" value for each sensor. If 50 scenarios are applied for CSI, and sensor node index 3,011 detects contaminants in real time, the least number of false events may be obtained; however, for the other sensors, they can all get the least number of false events with 80 scenarios, with which number their "*m*" values are decreased to zero; accordingly, the determination of the best scenario number becomes the number of false events trade-off between different sensors.

Figure 21.6 presents the PINs in Cases I and II. For the legend, e.g., "pin_casei_com_1" is the PINs in Case I after the first sensor alarm and "pin_caseii_com25_1" shows the PINs in Case II after the first sensor alarm, where a total of 25 scenarios are stored in "table_2" for CSI. The true intrusion node index 42 is identified.

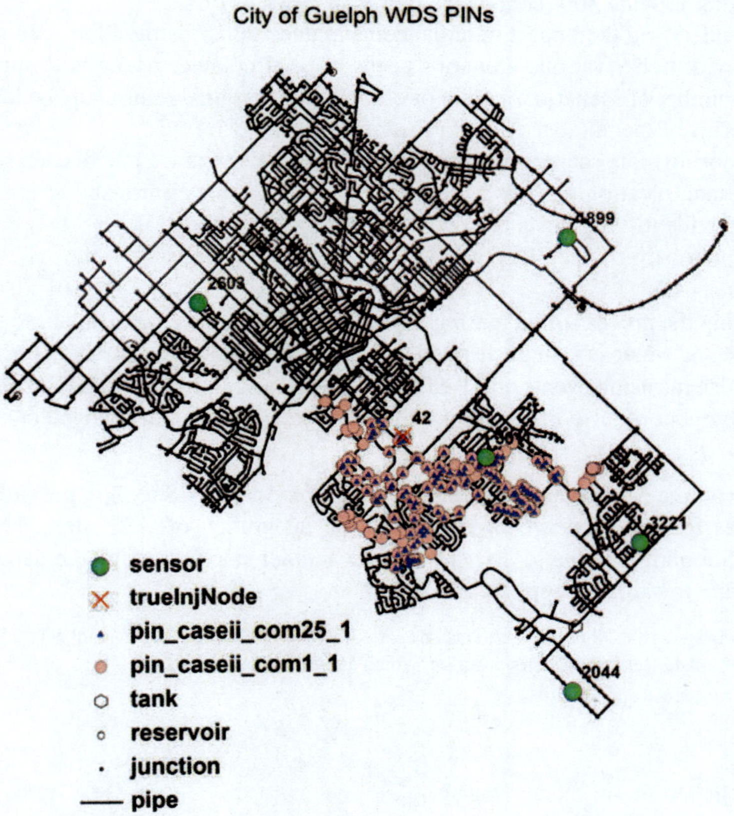

Fig. 21.6 PINs in Cases I and II

21.5 Conclusions

The parallel computing ability of a supercomputer (SHARCNET) enables the simulation of a number of scenarios under nodal demand uncertainty simultaneously, in a reasonably short time, and hence provides the possibility to incorporate the characterization of nodal demand uncertainty impact in a data mining procedure for CSI, i.e., analyses of the false-negative and false-positive issues. Parallel computing also provides a way to incorporate other uncertainty sources in false-negative/positive analyses. Without access to parallel computing, the ability to resolve the false-positive/false-negative issues is infeasible.

The following points are concluded for false-negative/positive analyses for a WDS:

- Nodal demand uncertainty greatly impacts the complexity of understanding contaminant behavior in WDS, as shown from the large "*m*" value of each sensor. It suggests that CSI procedures only relying on a single scenario (or model run) will not provide high-confidence results in terms of PINs.
- To understand the impact of nodal demand uncertainty on the PINs, contaminant behavior under various scenarios needs to be simulated. Along with increasing the number of scenarios, a point of diminishing marginal return may be identified in terms of the number of scenario simulations.
- By storing more scenarios in the database, the false-negative rate of each sensor in PIN identification will always decrease, meaning a lower probability is obtained for not identifying the correct point of intrusion.
- The number of false PINs or false intrusion events is related to the "*m*" value of each sensor.
- During the process of increasing the number of scenarios in the database for CSI, if the "*m*" value is kept unchanged (at zero or other positive numbers), the number of false intrusion events or false PINs will increase; if the "*m*" value is decreased, the number of false PINs will usually not increase or kept unchanged or reduced.

The online data mining procedure is integrated into a GIS toolkit, providing possibilities for rapid emergency response. The integration of GPS, street layer into the GIS toolkit is in the to-do list, which is another step in moving the data mining procedure toward real application.

Acknowledgments This research was supported by the NSERC strategic grant STPGP 336126 and the Canada Research Chairs program, which are greatly appreciated.

Appendix

To get the probability of generating negative random numbers, the following notations are applied:

x= random pattern factor, obeying normal distribution,
$\overline{x}$= original pattern factor,
s= standard deviation,
y=normalized random pattern factor, obeying standard normal distribution.

Pattern factor makes no sense in negative values; every original pattern factor $\overline{x}$ is non-negative.

When $\overline{x} = 0$, $s = 0.1$, $\overline{x} = 0$, thus, the generated random number would be always zero.

When $\overline{x} > 0$,

$$y = \frac{x - \overline{x}}{s} \tag{21.2}$$

Since s equals to $0.1\,\overline{x}$, Eq. (21.2) is converted to

$$y = \frac{x - \overline{x}}{0.1\overline{x}} \tag{21.3}$$

Thus, the probability of generating random negative pattern factor is

$$\begin{aligned} P\,(x < 0) &= P\,(\overline{x}\,(0.1\,y + 1) < 0) \\ &= P\,(y < -10) \\ &= 7.6\mathrm{E} - 24 \end{aligned} \tag{21.4}$$

References

American Water Works Association (AWWA). (2004). "Security guidance for water utilities." http://www.awwa.org/science/wise Accessed Oct 14 2009. AWWA

Babayan, A., Kapelan, Z., Savic, D., and Walters, G. (2005). "Least-cost design of water distribution networks under demand uncertainty." Journal of Water Resources Planning and Management, 131(5), 375–382

Guan, J., Aral, M.M., Maslia, M.L., Grayman, W.M. (2006). "Identification of contaminant source in water distribution systems using simulation-optimization method: case study." Journal of Water Resources Planning and Management, 132(4), 252–262.

Huang, J., and McBean, E. (2009). "Data mining to identify contaminant event locations in water distribution systems." Journal of Water Resources Planning and Management, 135(6), 466–474

Kim, M., Choi, C.Y., and Gerba, C.P. (2008). "Source tracking of microbial intrusion in water systems using artificial neural networks." Water Research, 42, 1308–1314

Methods, H., Walski, T.M., Chase, D.V., Savic, D.A., Grayman, W., Beckwith, S., and Koelle, E. (2003). Advanced water distribution monitoring and management. 1st ed. Haestad Methods, Waterbury, CT

Ostfeld, A., and Salomons, E. (2005). "Securing water distribution systems using online contamination monitoring." Journal of Water Resources Planning and Management, 131(5), 402–405

Perelman, L., and Ostfeld, A. (2010). "Bayesian networks for estimating contaminant source and propagation in water distribution system using cluster structure." Water Distribution System Analysis 2010, Tucson, AZ, September 12–15

Shang, F., Uber, J.G., and Polycarpou, M.M. (2002). "Particle back tracking algorithm for water distribution system analysis." ASCE Journal of Environment Engineering, 128(5), 441–450

SHARCNET: www.sharcnet.ca. Accessed on Aug 25 2010

Shen, H., McBean, E., and Ghazali, M. (2009a). "Multi-stage response to contaminant ingress into water distribution systems and probability quantification." Canadian Journal of Civil Engineering, 36(11), 1764–1772

Shen, H., McBean, E., and Ghazali, M. (2009b). "Contaminant source identification for priority nodes in water distribution systems." Dynamic Modeling of Urban Water Systems, monograph 18, CHI, Guelph

Sreepathi, S., Mahinthakumar, K., Zechman, E., Ranjithan, R., Brill, D., Ma, X., and Laszewski, G.V. (2007). "Cyberinfrastructure for contamination source characterization in water distribution system." Computational Science ICCS 2007, Part I, LNCS 4487, 1058–1065

US EPA. (2003). "Cross connection control manual."

Wong, A., Young, J., Hart, W.E., McKenna, S.A., and Laird, C.D. (2010). "Optimal determination of grad sample locations and source inversion in large-scale water distribution systems." Water Distribution System Analysis 2010, Tucson, AZ, September 12–15

Chapter 22
Water/Wastewater Infrastructure Security: A Multilayered Security Approach

Laurie J. Van Leuven

22.1 Introduction

Once an owner/operator recognizes their system's threats and vulnerabilities, they need to develop a plan for how they will mitigate those vulnerabilities and lower the risk to the system and increase their resiliency. Elements that will need to be considered when developing a security program include budget, staffing resources, regulatory requirements, coordination with capital system improvements, criminal activities, acts of vandalism, and intuition. This chapter will help utilities assess their current level of security, identify areas for improvement, and outline the components of a multilayered security program.

22.2 Where to Begin?

In the early days of water systems security (ca. 2002), the conventional wisdom was to focus efforts on identifying and assessing vulnerabilities and then develop a response plan for how the system would address an attack on its vulnerable assets. Now expectations are rising. It is no longer enough to just know which asset is vulnerable and document how various entities would respond. New expectations for critical infrastructure protection have catapulted the water industry into the realm of risk-based performance measures for an entire system and targeted measures for each critical and vulnerable asset. This leaves owner/operators with a long laundry list of security improvements and not enough dollars or resources to address them all.

The prioritization exercise is crucial to developing a game plan. A comprehensive vulnerability assessment will usually identify options for consideration and recommendations for risk reduction measures. However, it is quite likely that the

L.J. Van Leuven (✉)
Seattle Public Utilities/U.S. Department of Homeland Security (DHS), FEMA, Washington, DC, USA
e-mail: laurie.vanleuven@gmail.com; Laurie.VanLeuven@fema.gov

R.M. Clark et al. (eds.), *Handbook of Water and Wastewater Systems Protection*, Protecting Critical Infrastructure, DOI 10.1007/978-1-4614-0189-6_22,
© Springer Science+Business Media, LLC 2011

list of recommendations will be overwhelming to a utility with budget-conscious and skeptical officials and limited resources. Utilities should understand that these challenges are not uncommon. It is rarely feasible for any water system to employ all recommended security countermeasures. So, the next challenge is selecting a starting point. The good news is many water systems already have some baseline security measures in place.

22.2.1 Evaluation of Existing Countermeasures

It is important to evaluate existing security countermeasures and get a clear understanding of how effective current security is at reducing risk. For example, a wastewater treatment facility that was constructed in 1999 most likely has a fence line around the perimeter of the facility and locking mechanisms on the access points. Here are some questions that might help assess the ability for an existing measure to thwart a successful attack:

- How tall is the fence?
- What type of material was used?
- Is it climb resistant?
- Does it have anti-climb toppers such as outriggers of barbed wire?
- How close together are the posts and how deep is the cement?
- What about the doors?
- What type of access control is in place?
- Does it operate with simple mechanical keys and if so how are the keys controlled?
- Is it locked at all times or are there operational periods when doors remain unlocked or propped open?
- Is there a keypad code or a proximity card reader?
- Are there alarm points on the doors and if so are they monitored and by whom?
- Who responds to alarms and how long does response take?

This type of detailed assessment of each security countermeasure will help interpret the effectiveness of existing security. If existing security countermeasures are weak, simple upgrades and policy or procedural changes may be a cost-effective strategy. Another good place to begin is expanding other protective programs that may rely less on technology or physical security and may employ a more personnel/human resources approach to the task.

22.2.2 Protective Programs

There are several programs that may already be in existence or programs that could be started very easily with limited resources that can contribute to lower risks and improved resiliency. Countering the sheer volume of potential adversaries and threats to a water system that could enact a plot 24 h a day, 7 days a week, requires the assistance of individuals with knowledge about the consequence, a security mindset, and a mechanism to report anomalies.

22.2.2.1 Suspicious Behavior Reporting

Since a premeditated large-scale attack on a water asset will involve pre-attack planning and surveillance, a utility that can detect and report suspicious activities to law enforcement has a better likelihood of thwarting the attack. Ensuring that critical assets have the benefit of active eyes and ears from all employees, contractors, customers, and passersby is a valuable strategy. These programs can be very cost-effective, since they do not require large dollar investments of equipment or technology. Instead they can be achieved through training, outreach, and information sharing. Some utilities refer to these programs or campaigns as "Water Watchers" or "If You See Something, Say Something." These programs are most effective when they take a grass roots approach to achieve two-way dialogue. It may involve creation of a brochure, topics at community meetings or block watch sessions, and it should always provide a consistent mechanism for 24/7 reporting (not just dumping the caller into a voicemail box). The simple objective of these programs is to encourage people to immediately report suspicious activity on or around water or wastewater assets. This will in turn facilitate quicker assessment and response.

22.2.2.2 Source Water Protection Programs

Sources of drinking water include lakes, streams, rivers, springs, and ground water. The watersheds where these drinking water sources are located are often in remote areas; however, they also may reside in suburban locations that can be heavily populated. Regardless of where these valuable resources are situated, source water protection is an important step in the delivery of safe drinking water. Protecting drinking water sources requires the combined efforts of many partners such as public water systems, communities, resource managers, and the public. A comprehensive source water protection plan includes source water assessments, citizen involvement, protection planning, and use planning and stewardship.[1] Many communities across the USA have active neighborhood groups who seek out opportunities to engage in civic and environmental causes. Collaborating with neighborhood groups can be a good way to increase public awareness and build a cadre of active citizens who will report suspicious behavior and activities.

---

[1] Environmental Protection Agency, Source Water Protection, http://cfpub.epa.gov/safewater/sourcewater/, Accessed May 10, 2010.

22.2.2.3 Water Quality Monitoring

Water quality monitoring is another program that benefits the overall security of a drinking water system. Many different local, state, and federal agencies are involved in monitoring the health and quality of water. In addition to water system employees or contractors, other participants who might actively engage in efforts to ensure healthy water include researchers, students, and volunteers. Initiatives adopted by various entities to examine water quality may aid in detecting anomalies.

A water quality monitoring program can be structured in a way to increase the likelihood of detecting a contamination event. The quantity and locations of monitoring sites should take into account the threats and vulnerabilities of the entire system. The more frequent the monitoring activities are, the better the odds of detecting contamination early enough to save lives and prevent further illness among the population. It is important to note that the type of agents and potential contaminants that are regularly screened through standardized water quality monitoring often do not include the agents that would be most likely be used by a terrorist to poison a water system. There are limited numbers of labs that are prepared or equipped to detect the type of harmful biological, chemical, and radiological agents that would be used in an intentional attack. The remainder of this chapter will focus on implementation strategies of an active and effective security and emergency management program.

22.2.2.4 Security and Emergency Management Programs

Many utilities initiated and/or formalized their Security and Emergency Management programs immediately after the events of 9/11 to protect their critical infrastructure and their ability to provide essential functions. A formal security program should focus its goals on safeguarding employees and the community; protecting infrastructure from theft, vandalism, and terrorism; and ensuring that interruption in services to customers will be restored quickly. The field of physical security design has become a rapidly changing discipline, with many technological advances designed to reduce risks and increase detection capabilities. These capabilities provide a varying level of effectiveness to deter criminal behavior and enable a utility to initiate the appropriate level of law enforcement and security response.

Prior to initiating a capital investment program, utilities must prioritize what measures they will take via some type of risk methodology approach that justifies their decision-making process. Different utilities will base their decisions on different security drivers.

22.3 Security Drivers

There are many internal and external factors that may drive a water or wastewater utility to invest in security. An example of an internal driver is the business need to limit financial losses by improving security at a facility that has been hit repeatedly by criminals stealing equipment, copper, or other electronic components or wires. An example of an external driver is a regulatory agency that mandates certain actions such as conducting vulnerability assessments and the development of an emergency action plan. The way utilities prioritize security improvements depends on the culture and the decision-making process of the organization.

22.3.1 Risk-Based Justifications

The asset management approach within the water and wastewater sectors is a growing trend. With many competing demands for funding to address aging infrastructure, and customer expectations on the rise, the asset management concept is gaining ground as a mechanism to analyze risk and weigh the costs and benefits of significant investments. Asset management can be described as a total life cycle investment approach to managing infrastructure, from cradle to grave. This approach uses and develops knowledge management tools and results in robust decision support systems. These decision support systems incorporate several types of analysis, such as rigorous cost–benefit and economic analysis, analysis of work orders to evaluate performance in the field, and risk of failure analyses. Utilities that practice an asset management approach typically gravitate toward risk-based decisions.

22.3.1.1 Risk Assessment Tools

The Department of Homeland Security (DHS) has partnered with the Environmental Protection Agency (EPA) and other federal agencies and water sector industry associations to develop various different types of risk assessment tools and guidance documents. Here is a sampling of risk assessment tools available to water utilities.[2]

- Vulnerability Self-Assessment Tool (VSAT™)
- Water Health and Economic Analysis Tool (WHEAT)
- Water Contaminant Information Tool (WCIT)

Water utilities will need to review and evaluate each risk assessment tool/methodology based on the features, ease of use, depth of process, and outputs provided. The best assessment tool is one that is sustainable with existing in-house expertise and produces reliable and consistent results that are understood and accepted by organizational decision-makers.

In addition to the tasking of water utilities to assess the risks of their systems and assets, local state and federal agencies with homeland security missions such as law

[2] More information about these and other risk assessment tools and resources is available at http://water.epa.gov/infrastructure/watersecurity/techtools/.

enforcement, fire service, and emergency management divisions are also expected to maintain databases of critical infrastructure for their jurisdictions. The standard database used for this purpose is called the Automated Critical Assets Management System (ACAMS).[3] Ideally, the risk assessment tools used by individual utilities should be compatible with ACAMS to prevent duplication of database and analysis efforts.

While the model of a risk-based decision system might sound like the best approach, there is a lot more to it when security investments are on the table. Oftentimes, the rationale for which facilities or assets a utility needs to secure first takes on a less complicated approach, aka, listening to that gut feeling. As demonstrated in Fig. 22.1, security drivers fall under three different types of decision justifications.

22.3.2 *Must Do*

Security drivers that fall into this category represent mandates from federal or state agencies. Even though water and wastewater utilities are most commonly represented in the Water Sector (one of the Nation's 18 critical infrastructure sectors), several of their assets and facilities actually fall into dual national infrastructure

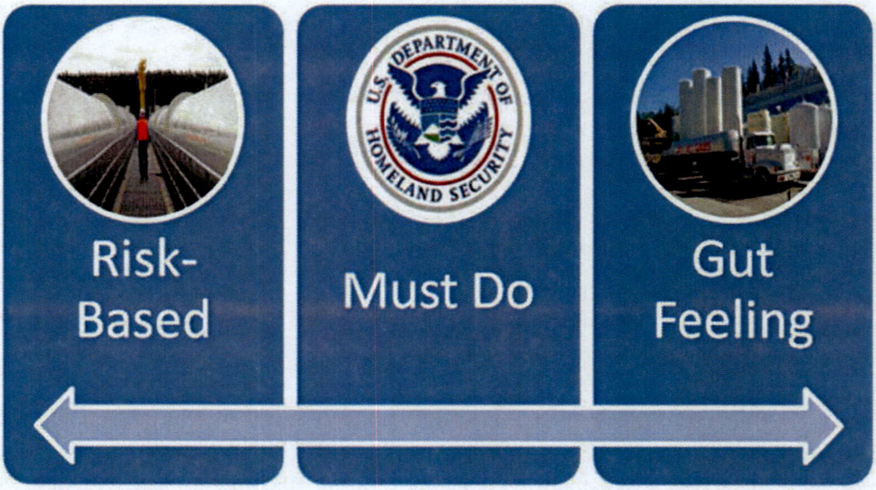

Fig. 22.1 Justifications for security investments

[3] ACAMS accounts are granted by DHS to state and local emergency responders, emergency managers, homeland security officials, and other personnel with official infrastructure protection responsibilities following authorized ACAMS and PCII training. Water and wastewater utilities should contact their local law enforcement agency and inquire about use of ACAMS data for their critical assets.

sectors. While the transmission, distribution, storage, and collection components including pumping stations all fall within the water sector, other parts of the system might be categorized under other sectors.

22.3.2.1 Regulatory Agencies and Mandates

To better understand all of the different regulatory agencies with an interest in how system assets are secured, we will examine the regulations involved with operating a dam. A drinking water utility that owns and operates a dam to manage its source water needs to coordinate with several different agencies in the realm of security. If the dam also generates hydroelectric power, there are security requirements from the energy sector, such as the Federal Energy Regulatory Commission (FERC). Depending on the quantity and the risk tier level of the energy-generating dam, there may be requirements by the North American Energy Regulatory Commission (NERC). The energy sector itself (also one of the 18 infrastructure sectors) is regulated by the Department of Energy. Regardless of energy generation at a dam, all dams fall into the Dam sector (another one of the 18 infrastructure sectors) that is regulated by the Department of Homeland Security (DHS). To further complicate the regulatory spectrum, if any water or wastewater facilities use or store above threshold quantities of any substances of concern such as gaseous chlorine, that individual facility lands in the chemical sector. The chemical sector is also regulated by DHS.

In some states, federal regulatory agencies have deferred the role of administrator to state agencies such as state departments of health. Figure 22.2 provides a snapshot of the spectrum of security system regulators that one drinking water utility might need to manage.

Fig. 22.2 Examples of water security regulators

22.3.2.2 Chemical and Water Security Legislation

Security regulations for critical infrastructure have been getting progressively more stringent, focusing not only on preparedness and response but now with expectations also on prevention, protection, and deterrence. The drinking water and wastewater treatment sectors use dangerous chemicals that terrorists could exploit to cause dire consequences including death, injury, serious adverse effects to human health, the environment, critical infrastructure, public health, homeland security, national security, and the local and national economy.[4] Due to the high risk involved with the manufacturing, transportation, storage, and use of such chemicals or substances of concern, the Department of Homeland Security embarked on regulations to reduce the risk and consequences of an intentional act to target chemical facilities.

In 2007, the Department of Homeland Security (DHS), Office of Infrastructure Protection, rolled out the interim final regulations on the Chemical Facility Anti-Terrorism Standards (CFATS), 6 CFR Part 27, requiring high-risk chemical facilities including water and wastewater treatment facilities, to comply with the Homeland Security Appropriations Act of 2007.[5] This Act was revised in 2009 and sets forth regulations and compliance mandates of security practices at chemical facilities.

Drinking water facilities were initially granted an exemption on the regulations due to conflicts within the Safe Water Drinking Act, which requires that surface water sources be disinfected via chlorination. However, proposed legislation[6] could enact new chemical and water security regulations and amend the Homeland Security Act of 2002 to resolve previous regulatory conflicts. This regulation could significantly expand requirements for vulnerability assessments and protective measures to reduce the risk of intentional acts of sabotage to water and wastewater facilities. Utilities should keep apprised of proposed legislation in this area, as the topic continues to surface and regulatory conflicts have not yet been resolved via legislation and/or mandates.

22.3.3 Gut Feeling

While some security countermeasures required by regulators are nonnegotiable, the identification of additional security investments and the methodology for prioritization can be subjective. In other words, owner/operators who have in-depth knowledge and experience with individual assets might have a sixth sense about risks to the system and how much damage an evildoer could levy through relatively

---

[4] H.R. 2868 – 111th Congress, Chemical and Water Security Act of 2009, http://energycommerce.house.gov/Press_111/20091001/hr2868_billtext.pdf.

[5] H.R. 5441 – 109th Congress: Department of Homeland Security Appropriations Act, 2007, http://www.govtrack.us/congress/billtext.xpd?bill=h109-5441

[6] H.R. 2868 passed the House in November 2009. The Senate referred the bill to the Committee on Homeland Security and Government Affairs. As of November 2010, there have been no further actions taken or anticipated on this bill.

simple tactics. These insights and intuitions should not be discounted. A vulnerability assessment might return a recommendation that gives high priority ranking to large post-treated drinking water storage facilities such as a reservoir or clear wells, based on many hatches that facilitate direct access to finished drinking water. However, talking with the water system operator about the assessment might reveal that there are vaults and other hatches under the radar that provide direct access to the drinking water in remote areas that are not part of an aboveground facility. These vaults and hatches may not have been captured in a formal VA, but they represent a significant vulnerability that could be easier to exploit by a nefarious actor with ill intent.

If a water system operator tells you, "If you really want to know how to contaminate the drinking water. . ." you should pay careful attention to what he or she tells you. A good strategy to reveal this type of gut feeling is to engage with those who know the operational systems best in an exercise of "Black Hat, White Hat."

22.3.3.1 Black Hat, White Hat Exercise

This exercise requires two simple steps and documentation. Begin by instructing the operator to, "Place a black hat on your head and imagine that you are an evildoer. What would you do to attack this asset?" This takes a security assessment out of the realm of statistics and analysis and instead captures that intuition how an actual person might carry out an intentional attack. This captures the scenario that might keep an operator awake at night or make the hairs on the back of their neck stand up.

Make sure to follow up with the second half of the exercise. Be sure to ask the operator, "Now place a white hat on your head and describe how you would protect against that scenario." The type of information gained from this exercise can be very enlightening and the operators might offer a simple approach to protecting the asset. Security planners should capture this information and incorporate it into their decision-making process.

Clarifying which assets need security improvements and placing them in a rank-ordered list will help utilities develop a programmatic approach for how to sequence investments over several years if necessary. Determining the appropriate security measure and level of investment required to reduce the risks is the next step in the process.

22.4 Security Countermeasures

Deciding on which type of security countermeasures to incorporate is no easy task. There are guidance documents available that provide suggestions for developing and maintaining an active and effective security program. The EPA published a document that highlighted 10 common features of effective security programs of water and wastewater utilities.[7] The document was developed in collaboration with the national drinking water advisory council working group in 2005, and today it still serves as a great resource for owners/operators seeking ways to protect their critical infrastructure.

There are a wide variety of security countermeasures that contribute to a multilayered security program. This next section will focus on physical security countermeasures. More comprehensive approaches to security will be addressed at the end of this chapter.

22.4.1 Physical Security Considerations

Utilities that embark upon a physical security program intended to delay or detect malevolent parties whose actions may otherwise defeat the mission of the utility will find that there is no one-size-fits-all approach. Since utility assets are extremely unique, securing those assets takes an individual approach that compensates for the physical structures, the geography and topography of the site, and the ongoing operations and maintenance activities. Technology is advancing rapidly in the area of physical security systems for critical infrastructure, which is a good thing. However, selecting a technology that will become an enterprise system such as a CCTV system requires a significant investment and careful planning. The type of physical security countermeasures can be grouped according to the capabilities they are designed to achieve. These groups include countermeasures intended to (1) deter, (2) delay, (3) detect, (4) assess, and (5) respond.

22.4.1.1 Deter

Investments in equipment or strategies that result in an obvious presence of security measures in an area can discourage an adversary from attacking a facility. Discouraging an individual with ill intent might make them seek opportunities elsewhere or they might abandon their attempt due to negative consequence of being caught. Deterrence can include no trespassing signage, on-site staff that conduct frequent site inspections, and Crime Prevention Through Environmental Design (CPTED) principles such as lighting, open spaces, and activity generators. Deterrents may also include a formal security force comprised of in-house resources or contract security officers that perform roaming patrols or fixed posts at access points to highly critical assets. These security countermeasures help to reduce the occurrence of general low- to moderate-level crimes. They are not generally considered part of an electronic integrated physical protection system with a predictable level of effectiveness.

---

[7] Environmental Protection Agency, Office of Drinking Water, October 2008, http://www.epa.gov/safewater/watersecurity/pubs/brochure_watersecurity_featuresofanactiveandeffective.pdf

22.4.1.2 Delay

Delay tactics in physical security are primarily attempts to slow down a perpetrator, by creating a sufficient time lag between detection of an instruction and the point at which the attack becomes successful, ideally with the intent of preventing the attack. The ability to delay a threat until law enforcement arrives is rarely feasible with water and wastewater facilities, since many sites are in suburban or more rural areas with limited number of police responders. Delay capabilities therefore should be structured to provide adequate delay time to detect, assess, notify law enforcement and utility decision-makers, and initiate mitigation strategies. Delay tactics typically include perimeter barriers such as fences, gates, and vehicle barriers and building barriers with protective measures around windows and doors and stringent locking devices that limit access to those with legitimate business need.

22.4.1.3 Detect

Detection is the point at which a potential attack is discovered. Detection can occur in real time through visual assessment from an individual, such as when an employee, contractor, or citizens witnesses someone breaching a security system. Detection can also be the moment when someone notices that there has been an intrusion at a facility after the fact such as seeing the evidence of cut fence line, a broken door lock, or a ladder propped up against a building. These forms of detection are valuable indicators that something bad has or is about to happen to the facility or asset, but they do little to prevent something from happening and cannot be counted on to detect in real time 24/7. Many forms of automated security system features can provide ongoing real-time detection capabilities, such as access control systems with electronic sensors and alarm points (perimeter gate contacts, door and hatch contacts, and motion or heat detection devices) are more reliable for round-the-clock detection coverage.

The second component of detection is monitoring. Building a network of sophisticated access control devices and sensors is meaningless if there is no one there to monitor the alarms. The monitoring capability of a physical security system may be an in-house monitoring and response center that is staffed 24/7 or it could be that alarm points are routed to a third-party alarm monitoring company. Both options provide necessary coverage to generate a law enforcement response; however, the level and quality of assessment to generate the appropriate level of response is greater with an in-house monitoring staff.

22.4.1.4 Assess

It is not enough to have systems in place that deter and detect; there needs to be an assessment capability that has the tools to understand what caused the intrusion or security violation. The assessment capabilities rely on individuals who can distinguish between false alarms and real security events, weigh the potential consequences, and notify and mobilize the appropriate level of response.

Assessment tools are achieved through closed-circuit television systems synced with alarm contacts on access points. It is highly unlikely that a utility will have the resources to employ staff for ongoing live video monitoring of each CCTV camera. Therefore, it is important to have an integrated security system that includes access control contacts, motion detection, and other alarm points that signal an intrusion. Recording and retaining image storage of CCTV footage is an important factor. Digital video images will be critical in analyzing a security event and will become invaluable when trying to understand how and when an incident occurred. Utilities should make sure they have adequate video retention for evidentiary purposes and to facilitate investigations after a security incident. The digital video recorder (DVR) will also enable the video to be accessed for alarm call up by a surveillance monitoring operator when a detection point enters into an alarm condition. CCTV video can also be utilized for assessment and verification at an intercom when an individual is requesting access to a critical site.

22.4.1.5 Achieving Assessment with In-House Monitoring

In-house alarm monitoring requires investments in a robust access control system that includes software and hardware expenditures. These integrated systems include equipment in the field with connectivity to a centralized server and multiple workstations with access to the system platform and graphic user interface. Integration of alarm events is accomplished through alarm point detection and routing of video images through a selection of predetermined camera positions. Some cameras may be programmed to record only when triggered by motion-based activity. Those video images may then be called up through alarm event monitors where an operator will assess and make a determination of whether a response is warranted.

If the utility has in-house resources tasked with monitoring security alarms through a monitoring center, the operators responsible must have extensive security training. These operators also should be prepared to receive suspicious behavior and incident reporting from employees, contractors, and customers, while monitoring the integrated security system.

22.4.1.6 Respond

A utility monitoring operator with access to an integrated security system needs to be prepared to initiate the appropriate level of response depending on their assessment. Whenever there is confirmation of suspicious individuals or incidents, an operator should immediately refer the incident to law enforcement. Unsecured facilities are indicated when an intrusion alarm will not reset. If this occurs some type of physical response by utility staff is required to secure the asset. Other alarm conditions may indicate a maintenance issue that will need to be addressed during regular business hours or assessment may indicate that employees or contractors may be operating the system inappropriately.

For example, if an employee has a mechanical key that will open a facility door, they may skip the step of using an electronic key card that would log the event and shunt the security alarm from being triggered. Hence, the employee could generate an alarm condition that the system would characterize as a door being forced open. The monitoring operator will need to be able to distinguish real security events from these non-threatening false alarms that may only require minor follow-up reminders with individuals. False alarms can be a significant downside to an integrated security system. Depending on the assessment, the types of security responses required can be grouped into four areas: (1) suspicious circumstances, (2) unsecured facilities, (3) escort provisions, and (4) system maintenance.

Suspicious Circumstances

Conditions for response activities that fit this area include but are not limited to detection of a crime in progress, confirmation of intrusion into a secured area, discovery of theft or vandalism, intentional sabotage of water system assets, any verbal or expressed threat, discovery of weapons or contaminants on site, eliciting of sensitive information, and photography of unusual subject matter (i.e., a chemical truck, valves, pipelines, etc.). Any of these circumstances needs to be reported to a law enforcement agency and a local Fusion Center if appropriate. It is also wise for a utility to have a central point of contact responsible for security to interface directly with law enforcement agencies during a police response and investigation.

Unsecured Facilities

If it is suspected that a critical asset may be unsecured, it is imperative for a utility to mobilize a response ASAP. Unsecured facilities that have direct access to drinking water or any type of chemicals on site are particularly concerning. Regulatory agencies who oversee the water sector may require a facility be taken off-line pending water quality samples if a hatch is found unsecured. Assessment capabilities might be able to alleviate concern by confirming that a door or hatch was unintentionally left open by an employee, thereby eliminating the need for more drastic response. However, the event should be logged and documented and the access point(s) will need to be re-secured.

Escort Provisions

There may be occasions when a utility receives an irregular access request. This may be anything from a group of school children requesting a tour of their local water or wastewater facility, to a cell phone vendor who needs to access their equipment located at a water storage facility, to a citizen whose cell phone or sunglasses accidently landed in a secured area. Unless a utility follows a strict policy of background screening on every single visitor, which would be cost and time prohibitive, all of these examples would require escorts by a utility employee. Oftentimes these requests and provisions to mobilize an escort can be coordinated through a 24/7 monitoring center.

System Maintenance

Integrated security systems involve a lot of separate devices that have to work together. Various circumstances may contribute to security system outages or instability that will need to be repaired. Power outages, power surges, and various testing of other electronic equipment could create a connectivity problem with security equipment or servers. Effective systems will generate an alarm when the system is malfunctioning. Monitoring operators should not ignore these indicators, but rather distinguish them as a different type of needed response, such as by generating a service ticket for a utility employee or a vendor to repair the system.

Any utility that embarks upon an electronic access control system will need to take deliberate steps to ensure successful implementation by

- Having strong security policies in place requiring the appropriate use of the security system
- Training authorized users on how to effectively operate the security system
- Providing advanced security and threat training to monitoring and response staff
- Establishing an ongoing maintenance program to keep alarm points in good working condition
- Repairing known problems with the security system expediently
- Evaluating response protocols

Many utilities have already adopted physical security measures, installed access control and CCTV surveillance systems, and improved their security response capabilities through contracted guard services or internal resources. However, the protection of vulnerable assets involves a multilayered approach beyond physical security.

22.5 Multilayered Security Approach

As important as physical security is, it would be impossible to build a fortress around all utility assets to prevent every possible attack. A more appropriate strategy to water system security is to enlist a combination of measures that will protect systems through various activities. Ensuring that a multilayered approach is both effective and sustainable requires dedicated in-house resource(s) to manage a proactive security program and high-level support from the organization's top-level leadership. The remainder of this chapter identifies activities that contribute to a comprehensive security approach and are categorized into four basic capability areas: prevention, protection, preparedness, and response.

22.5.1 Prevention

Prevention of a security incident or a plot to attack a system can occur at different phases, times, and locations. Strengthening the mechanisms that reduce crimes of opportunity and disrupting or preventing a premeditated attack is ideally the most successful element of a security program. Most prevention activities require partnerships with employees and contractors, other entities, and the general public. An effective multilayered security approach incorporates a combination of interagency, organizational, and public involvement and awareness and includes investments in the development of policies and procedures and physical, chemical, operational and design controls to increase overall program performance. The following programmatic elements can help prevent security incidents.

22.5.1.1 Intelligence Activities

Water and wastewater utilities should regularly engage with their local intelligence community. Establishing relationships with local law enforcement, FBI field offices, state fusion centers, and local emergency management offices is a crucial step in building partnerships that will help facilitate quick information sharing when warnings and indicators of a possible attack or security incident are present. Water and wastewater agencies should also become familiar with WaterISAC (Information Sharing Analytic Center) to access and subscribe to emergency management and security industry news and threat warnings pertinent to the water sector.[8] In addition to developing relationships with Intelligence agencies, water utilities need to receive, review, and analyze intelligence report and products to determine if their threat environment has changed and determine if additional security measures and actions need to take place. An effective first step is to make contact with local law enforcement agencies and a FBI representative assigned to a utility's jurisdiction and request a briefing on information sharing networks and counterterrorism activities for critical infrastructure. Joining already existing committees and networks will reveal additional opportunities for collaboration.

22.5.1.2 Access Control

The practice of securing entry points into water and wastewater facilities is the foundation of prevention measures. Utilities need to limit access to their facilities only to those with an authorized business need. Access control systems include issuance of credentials to individuals necessary to enter specified secured facilities and computer networks with sensitive information. The credentialing system, locks

[8] WaterISAC is a community of water sector professionals who share a common purpose: to protect public health and the environment. https://portal.waterisac.org/web/

and barriers, and alarm points all need to be actively managed to prevent unauthorized access, damage, and interference to premises, equipment, systems, materials, and information.

22.5.1.3 Screening People

Verifying the identity and business need of persons who request access to critical water facilities is a crucial prevention step. Utilities need to conduct background investigations to determine suitability for employees and contractors (both initially and ongoing). Utilities can mitigate the insider threat by assuring only trustworthy and reliable personnel are employed by the organization and granted access to critical systems needed for their duties and responsibilities. Policies and procedures that support background investigations prior to granting unescorted access help prevent or deter potential threats to assets and security, including classified information. Individuals who request access to or a tour of a critical asset (such as a dam or treatment facility) should be tracked or registered (with a valid driver's license or ID) and escorted by a utility employee at all times while on the premises. Remember, individuals who do not have a business need to enter critical facilities create unnecessary risks for the utility.

22.5.1.4 Investigations and Law Enforcement

All levels of theft and criminal activities contribute to a utility's overall risk spectrum and can help identify trends and weaknesses in existing security measures. Incident reporting, documentation of circumstances, and tracking the resolution of each criminal, civil or administrative case, provide valuable insights to utilities that demonstrate areas that may need additional security measures, new policies or procedures, or a stronger emphasis on police collaboration (i.e., some law enforcement agencies may have a metal, copper, or wire theft task force). Whenever the incident is criminal, local law enforcement should be involved, but utilities should ensure that they also track the incident and regularly follow up with authorities on case progress. It is advisable to work closely and aid efforts to prosecute offenders, including instances of theft by employees. Swiftly dealing with administrative and criminal acts may help prevent additional criminal activities that could escalate beyond current levels.

22.5.1.5 Security Planning and Risk Mitigation

Prevention also includes building a comprehensive security program that evaluates risk, identifies and prioritizes risk reduction measures, establishes standards for physical security countermeasures, and provides project management for implementation, including funding requests, project approval, project management oversight, cost control, and effectiveness.

22.5.2 *Protection*

Due to the sheer number and location of critical components in any drinking water or wastewater system, it is impossible to prevent all security incidents from occurring. Protection includes the efforts taken by a water utility to place a calculated security emphasis on specific components of its systems. The following security program elements can help protect specific infrastructure, assets, and systems.

22.5.2.1 Asset Classification

The concept of asset classification is simple. Most utilities have a clear indication of how to group or classify their assets by the function they serve, such as pump stations, treatment facilities, and storage facilities. Less common is a more formal process for documenting detailed information about each asset, which would help management and employees better characterize the on-site features and risks associated with each asset, that should influence its priority ranking in case of a system failure. For example, finished drinking water storage tanks should be classified not only by primary function (storage) but also by the chemicals stored on site, the presence of telecommunication antennas on the tank, the capacity of storage, the proximity to emergency response agencies, and the criticality or redundancy of service within a pressure zone or the entire system.

When tackling the task of asset classification, the employees who regularly tend to the operational needs of the asset will be aware of the quantities and types of chemicals used on site and the frequency and particulars of chemical deliveries, while other people within the organization may not have a clear understanding of those details. Until that information is collected for all assets, the task of prioritizing risk across all parts of the utility's systems will be incomplete. It is vital to have operational and maintenance staff serve on a security assessment team to identify, review, and prioritize assets and to help develop appropriate protection standards and plans.

22.5.2.2 Assess Vulnerabilities

As mentioned in Chapter 2, applying a risk assessment methodology will help the utility prioritize assets, identify internal and external threats and vulnerabilities, prioritize risks, and prioritize countermeasures to mitigate threats, risks, and potential losses.

22.5.2.3 Security Countermeasures

Based on vulnerability assessments and asset classifications, the security assessment team should determine and recommend the most cost-effective physical security countermeasures. This team should work collaboratively with the security program/planning team to request funding for and provide project management oversight for implementation of physical security countermeasures to protect assets and related supporting infrastructure from threats.

22.5.2.4 Patrols and Monitoring

Another important element of a security program is providing random facility security patrols and 24/7 monitoring of access control systems and burglar alarms. The emphasis of real-time monitoring helps prevent, detect, deter, and mitigate threats and unwanted or criminal acts. Patrols and premise checks should be a regular part of an ongoing security program and should include incident reporting and tracking of anomalies. Security officers may also be assigned 24/7 to fixed posts to ensure perimeter security around a highly critical and vulnerable asset. The ability to increase patrols and physical security monitoring is vital during times of elevated threat conditions based upon local intelligence or the Department of Homeland Security's National Terrorism Advisory System.

22.5.3 Preparedness

We often hear about the necessity of emergency preparedness activities related to natural disasters, but it is also important to incorporate preparedness into a comprehensive security program. Preparedness activities help the utility consider in advance which steps it will need to take if a security breach or incident occurs. Planning, testing, and training are the key elements of security preparedness.

22.5.3.1 Pre-gathering Information

Utilities should gather emergency contact information for every partnering agency or organization that might have a role in responding to a security incident. This includes security and other utility staff and contractors, fire service and hazardous material agencies, local law enforcement, local and state emergency management agencies, neighboring utilities, water sector regulatory agencies, the FBI, and the organization's Department of Homeland Security – Protective Security Advisor. Individual asset or site security fact sheets with critical information about the asset along with photographs, maps, and detailed information about chemicals on site are invaluable for emergency responders. Having this information readily available will help the organization convey time-sensitive information and help aid response efforts.

22.5.3.2 Contingency Planning

All emergency planning efforts should be well coordinated to ensure the capability to respond to emergencies or threats that would require integrated security actions and the continued performance of essential security functions when normal operations are disrupted by natural disaster or other emergencies. Times of civil unrest such as protests and rioting present additional security concerns for publicly owned utilities. An organization's Continuity of Operations Plan (COOP) should incorporate contingencies for various security scenarios. Plans should address provisions for how security will be maintained if the power is out, if facilities are shut down, and the plan for increasing security if there are heightened security indicators, alerts, and warnings.

22.5.3.3 Testing, Training, and Exercises

A testing, training, and exercise (TT&E) program is a key element of a comprehensive security and emergency management program. Utilities should develop an ongoing training plan for its employees and contractors and frequently test the organization's capabilities through exercises designed to identify areas from improvements. Specific training may include National Incident Management System (NIMS), Incident Command System (ICS), position-specific skills training, threat-specific incident training, and other emergency management topics. Exercises can be designed as seminars, orientations, drills, tabletop, functional, or full-scale exercises. All exercises should be followed up by after action reports (AARs) which serve as the basis for future training and exercises and should be used to make improvements to existing plans, policies, and procedures.

The main objectives of a TT&E program are to

- Assess and validate plans, policies, and procedures
- Ensure that utility staff are familiar with standard operating procedures
- Ensure that utility staff are sufficiently trained to carry out essential functions
- Test and validate equipment to ensure both internal and external interoperability
- Discover planning weaknesses and resource gaps
- Clarify roles and responsibilities and improve coordination
- Practice using alert systems and communication protocols
- Improve readiness for a real incident

22.5.3.4 Threat-Level Planning

There are several mechanisms available to detect a change in a water utility's threat level. Internal statistics on security incidents along with bulletins from the intelligence community, law enforcement agencies, and industry associations are essential to understanding when the threat environment is on the rise. Utilities should be aware of the National Terrorism Advisory System, which replaced the color-coded threat-level Homeland Security Advisory System in April, 2011.[9] Utilities need to

[9] US Department of Homeland Security, National Terrorism Advisory System, http://www.dhs.gov/files/programs/ntas.shtm

plan for how security measures might be increased during elevated threat levels. Utilities should document what type of additional or emphasis security measures they will put into place when the threat level rises. Examples of emphasis measures might include increasing security patrols, suspending visitor access, locking down all facilities, and adding fixed guards to entry points.

22.5.3.5 Mutual Aid Agreements

The time it takes to develop a mutual aid agreement with neighboring utilities and local government agencies is well worth the effort. Utilities should not assume that mutual aid will be available without advanced agreements in place. Mutual aid agreements help facilitate the sharing of staff, equipment, expertise, and other resources to ensure adequate response to incidents and resource tracking for cost reimbursements.

22.5.4 Response

The capability for a water or wastewater utility to respond to and recover from a disaster, be it a natural disaster or a human-caused event, will directly impact the community that relies on the essential functions of water. Readiness and response plans and capabilities are critical components of a comprehensive security program. The following programmatic elements can help a utility become more resilient and lessen the time it takes to reconstitute services to the public.

22.5.4.1 Response Actions

Upon the discovery of a security incident, whether it is a security breach, tampering or sabotage of infrastructure, theft of equipment or supplies, suspicious behavior report, or an indication of a cyber security intrusion, immediate response actions are necessary. Each incident needs to be assessed and appropriately investigated. Security staff should be available 24/7 to respond to a security incident and prepared to mobilize additional security resources to implement on-site security when necessary to secure critical infrastructure that may have been compromised. Confirmation of suspicious activity and crimes in progress should always be reported immediately to local law enforcement agencies.

22.5.4.2 Incident Management

If the incident has resulted in damaged infrastructure or compromised the utility's ability to deliver services, the utility should mobilize an incident response/management team to stabilize the situation, assess the damage, and prioritize objectives and further actions. Utility employees will need to serve in the role of an emergency responder and in support of other first responders (fire, law, and emergency medical). Depending on the size and scale of the incident, the utility responders may need to coordinate their response activities with a larger incident management team led by life safety agencies or the local FBI field office.

22.5.4.3 National Incident Management System (NIMS)

If a water or wastewater utility is publicly owned (local, municipal, county, special purpose district) they will need to ensure that they are compliant in Federal mandates for emergency management training. Homeland Security Presidential Directive #5[10] requires that state and local governments be in compliance with the National Incident Management System (NIMS)[11] as a condition of federal preparedness assistance and for the local jurisdiction to qualify for homeland security grant funding.

For response, utility responders need to (1) coordinate with other first responders and emergency management agencies; (2) participate in incident management system organizations; and (3) support command and control operations of the incident consistent with the National Incident Management System (NIMS).

22.6 Summary

Security experts agree that the human factor poses the greatest single source of risk for any asset. A solid security program begins with hiring the right people and building partnerships that leverage the shared missions of other agencies with a homeland security mission. Security programs that employ a variety of strategies and tactics to address the full spectrum of threats or events, both natural and manmade, will improve their ability to prevent and withstand threats and recover from incidents to ensure prompt resumption of essential services to the community.

---

[10] US Department of Homeland Security, Homeland Security Presidential Directives, www.dhs.gov/xabout/laws/editorial_0607.shtm

[11] Federal Emergency Management Agency, National Incident Management System Facts, www.fema.gov/emergency/nims/index.shtm

Chapter 23
Vulnerability of Water and Wastewater Infrastructure and Its Protection from Acts of Terrorism: A Business Perspective

Dave Birkett, Jim Truscott, Helena Mala-Jetmarova, and Andrew Barton

23.1 Introduction

Potable water and its supply is essential for contemporary human existence, with wastewater services recognized as being crucial for the maintenance of public health, particularly with ever-increasing population densities. Therefore, threats to water infrastructure whether from natural causes referred to as all hazard origin or external intervention such as terrorist related sources pose significant risks to society and human health.

As water is essential for the smooth functioning of society and operation of vibrant commercial, industrial, and agricultural economies, water and wastewater systems and infrastructure may be considered, according to Gleick (2006), as potential terrorist targets, more specifically within industrialized nations.

For the purpose of this chapter, water infrastructure is described as specific components of water and wastewater systems. These components can be defined as follows:

1. Water source (reservoirs, tanks, bores, and underground aquifers).
2. Water treatment plants (WTPs), where water is chemically treated, filtered, and disinfected.
3. Water distribution systems including pump stations and storage facilities.
4. Wastewater collection systems including pump stations and retention facilities.
5. Wastewater treatment plants, where water is chemically, biologically, and physically treated.

Any of these water system functions (Fig. 23.1) may be subject to external natural incidents, for example, floods or earth quakes, infrastructure failure through inadequate engineering design, unmanaged deterioration over time, or human interference.

D. Birkett (✉)
Truscott Crisis Leaders, Wembley Downs, WA, Australia
e-mail: dbirkett@crisisleaders.com

R.M. Clark et al. (eds.), *Handbook of Water and Wastewater Systems Protection*, Protecting Critical Infrastructure, DOI 10.1007/978-1-4614-0189-6_23,
© Springer Science+Business Media, LLC 2011

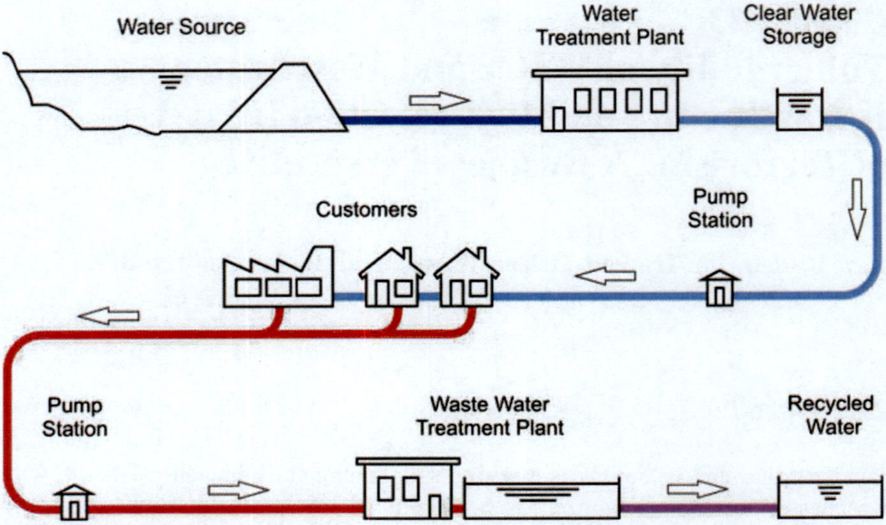

Fig. 23.1 Typical water distribution system and wastewater collection system schematic

The cultural and societal concept of public concern related to human interference with or compromise of potable water is invariably eidetically visualized as associated with bulk water storages or WTPs which may be the case for naturally induced failures. These are, however, partially protected by virtue of the dilution effect of the large quantities of held water and the added benefit of filtration and treatment at downstream plants, which provides some assurance to mitigate the effects of poisons and toxins. The issues related to physical bomb attacks on reservoirs and WTPs may also be mitigated by raised levels of physical security and more frequent testing of various risk and security plans.

According to Isenberg (2002), in terms of external vulnerabilities, the real terrorism risk exposure resides in water distribution systems, which is in stark contrast to the concept of large reservoirs and WTPs.

Although the authors are well aware of methodologies and risk exposures related to physical attacks across the spectrum of water and wastewater infrastructure, it is not intended in this chapter to provide the focus or attention to underline this identification.

23.2 Historical Background

As a recognized point of vulnerability within various societies and clusters of human habitation, water and wastewater systems have been subjected to levels of threat or damage since ancient periods of time. From the 6th century BC, there have been numerous documented potable water risk incidents. A historical sample is listed in Table 23.1 displaying a comparison of incidents related to all hazard and terrorist

Table 23.1 Selection of historical water threats and incidents 6 BC to 2002 AD

Year	Location	Incident	Description	Terrorism/other
2450–2400 BC	City of Umma, Middle East	Water diverted from adjacent state of Umma as tactical strategy by Urlama, King of Lagash	Denial of water service as political strategy	Form of terrorism to gain power over other state in lieu of warfare
600 BC	Cirrha City	Rye ergot inserted into local water supply. Cirrhaeans then became violently ill	Solon of Athens besieged Cirrha for a wrong interpretation to the Temple of Apollo. This facilitated the capture of Cirrha City	Biowarfare or bioterrorism
1748 AD	New York, USA	Angry mob burnt down a ferry house on the Brooklyn Shore of the New York East River	Revenge action for unfair allocation of water rights from the New York East River	Form of terrorism
1907–1913	Owens Valley of California, USA	Repeated dynamiting of water aqueduct system due to local concern over water being diverted to meet the needs of the growing city of Los Angeles	Farmers within the agricultural area of Owens Valley took extreme action to preserve the limited water reserves for agriculture in lieu of the city diversion	Form of terrorism taking extreme action to secure limited water supplies from being diverted from agricultural use to larger urban population
1961	Kiev, Ukraine	Dam failure subsequent to heavy rain	Babi Yar loam pulp dump dam failure led to 1,500–2,000 deaths of villagers	An unplanned event in the dam operation leading to imprisonment of senior officials
1963	Vajont Dam, North East Italy	Dam failure from overtopping	Landslide into dam caused overtopping, leading to flooding of several villages and deaths of 2,000 people	An unanticipated event due to existing geological instability on edges of dam storage
1979	Morvi Dam, Gujarat, India	Dam failure, subsequent to heavy rain and massive flooding	Machchu-2 dam failure resulting in the deaths of 1,500–15,000 villagers	Weather-induced event exceeding the design parameters of the dam
1993	Milwaukee, USA	*Cryptosporidium* water-born parasite	*Cryptosporidium* parasite commences a life cycle within the digestive tract of domestic animals such as cattle, which in this case washed into the water storages. This parasite can survive filtration and water disinfection. It caused severe illness and resulted in over 100 deaths	An unintended consequence of stock agriculture within a catchment area of water storage

Gleick (2006)

origins with details of consequences in each case. In consideration of multiple deaths, and asset deterioration or failure, it is apparent that the number of incidents originating from natural causes exceeds those under the category of terrorism.

As a variant view, terrorist acts may appear more visible due to media amplification and public interest. Interestingly, Gleick (2006) provides some evidence that the all hazard origin incidents have resulted in worse consequences. A relevant documented example is the outbreak of the *Cryptosporidium* water-born parasite in Milwaukee in 1993, which resulted in the deaths of over 100 people and documented sickness in another 400,000. Gleick (1993) also suggests that due to the scarcity of water, increasing density of the mega cities phenomena, and some sharing of aquifers, rivers and lakes, there will be increased conflict associated with water in the future.

Increased global water conflict tends to produce a potential future environment of elevated risk and threat levels. From the business perspective, it is considered that there is minimal variation in relation to the preventative strategies to reduce the risk exposure between an all hazard incident and a terrorist attack event.

23.3 Identification of Terrorism on Water and Wastewater Infrastructure

Globally, and in consideration of the heightened awareness of terrorism since the major terrorist incident in New York (2001), water and wastewater service providers have focused on physical security and research regarding the protection of water as an essential service and valuable commodity. Water is likely to be a resource at the center of human conflict for the indefinite future (CIA, 2009). Threats to water infrastructure can be divided into three main groups: (i) threats without action, (ii) simulated attacks, using colorants, for example, and (iii) real attacks (Druisiani as cited in Hayward, 2002).

The most commonly raised questions within water agencies are *What defines an act of terrorism which could potentially threaten water and wastewater? Who conducts the acts of terrorism? What is the reason for terrorist activities? What are the threat characteristics?*

A definition of terrorism from The Australian Commonwealth Criminal Code Act 1995 is expressed as: *. . . an action or threat of action that causes serious physical harm or death to a person, or endangers a person's life or involves serious risk to public health or safety, serious damage to property or serious interference with electronic systems* (Australia Federal Government, 2004). As outlined in the definition above, the terrorist threat to water appears to conform with the broad category related to threats against public health. However, there are more significant potential economic risks to the public when the significance of the linkages with associated interdependent, industrial, and commercial structures are considered. The example would be electricity generation which is heavily dependent on water for power station cooling systems. Additionally, a large percentage of industry across industrialized nations is operationally linked to water as an essential node of continued operation (Fisher, 2010).

According to Denileon (2001), the delivery of water is dependent on three crucial attributes:

1. There must be adequate quantities of water compatible with the various levels of demand in the system which may be required at any time across the water distribution and delivery system;
2. The water must be delivered within an agreed range of pressure to the various water delivery points across the time of delivery; and
3. The water must be safe to use and consume and delivered to an agreed code or standard.

Denileon (2001) states that the perceived terrorist threat is considered to potentially interfere with any of those delivery parameters and agreed standards. Additionally, the interference of wastewater is significant across large population centers and potentially also poses a serious health risk to large clusters of people in towns and cities, from interruption or acts of terrorism.

23.4 Vulnerabilities of Water and Wastewater Systems

Denileon (2001) accurately defines water threats as affecting the three previously stated crucial delivery criteria of quantity of water, pressure of water, and safety of water to use. He also defines the threats further as follows:

- *Physical destruction of the water system components*, or the disruption of the water supply (considered by observers to be the most likely scenario).
- *Bioterrorism/chemical contamination to water storages*, considered as the most likely from media amplification and perceptions from various films and fictional stories. However, in fact, this is considered technically difficult due to the dilution effect in large water storages and the possibility of observed covert behavior being raised to the attention of water managers.
- *Cyber Attack* The increased use of remote control systems to operate infrastructure has raised the risk and consequences of any potential future external terrorist attack. This type of attack can be coordinated from any external location, remote from the control system or control room. There is a recognized and enhanced physical risk to the control room and most water operators have increased physical security of this system node as an increased dependency on Supervisory Control and Data Acquisition (SCADA) systems occurs over time.

Additionally, in relation to wastewater, Denilion (2001) suggests that a safe and secure waste disposal system could be included and be considered in relation to the health threat potentially posed by any interruption to wastewater disposal for large population centers.

23.5 Responses to Water and Wastewater Terrorism

Subsequent to the 2001 terrorist attack on the Twin Towers in New York, some countries have substantially expanded budget allocations related to counterterrorism and associated intelligence gathering as a form of threat mitigation. In the United States, specific federal acts were implemented which addressed direct threats to the water supply network (Mays, 2007).

In the State of Victoria, Australia, in 2003, the State Government introduced legislation under an Act of Parliament (Terrorism (Community Protection) Act 2003, Part 6) to ensure that all 20 water agencies in Victoria, combined with other designated essential services, have a terrorist risk management plan. This plan, which is reviewed and tested on an annual basis, includes a scripted desktop counterterrorist exercise simulation. These exercises test realistic scenarios across the full spectrum of likely and rare events including potential interdependent essential service failures and anomalies.

23.5.1 Conventional Commercial Business Behavior

Conventional commercial business behavior changes and transforms over time, reflecting the social and potential criminal/terrorist influences that arise to threaten or influence the commercial outcome.

Progressive and innovative water agencies, as commercial entities, understand this issue, and that learning faster than the terrorists provides the only sustainable competitive advantage. Just as chief information officers are evolving toward chief knowledge officers, it is highly likely that corporate affairs managers, managers of the immediate past, will evolve toward chief intelligence officers and become managers of the future, to reflect progressive commercial business behavior.

To place water agencies in a leading position in business they should ensure that they are able to function in a "non-business as usual" environment and to capitalize on negative events. Benchmarking and learning from mistakes and successes of others tends to place the progressive water agency in an optimal position relative to perceived external threats. All executives emerge with cohesion when they participate in extreme leadership sessions establishing that they can align their communications with the operational response.

23.5.2 Future Mitigations to Threat Perceptions

In consideration of forward looking commercial business behavior, preparedness for a terrorist event within the area of water and wastewater systems can be approached in a conventional commercial business all hazard process (Fig. 23.2). This business practice is illustrated within the context of evaluations of crisis preparedness against international practices and production of crisis control procedures. This commercial

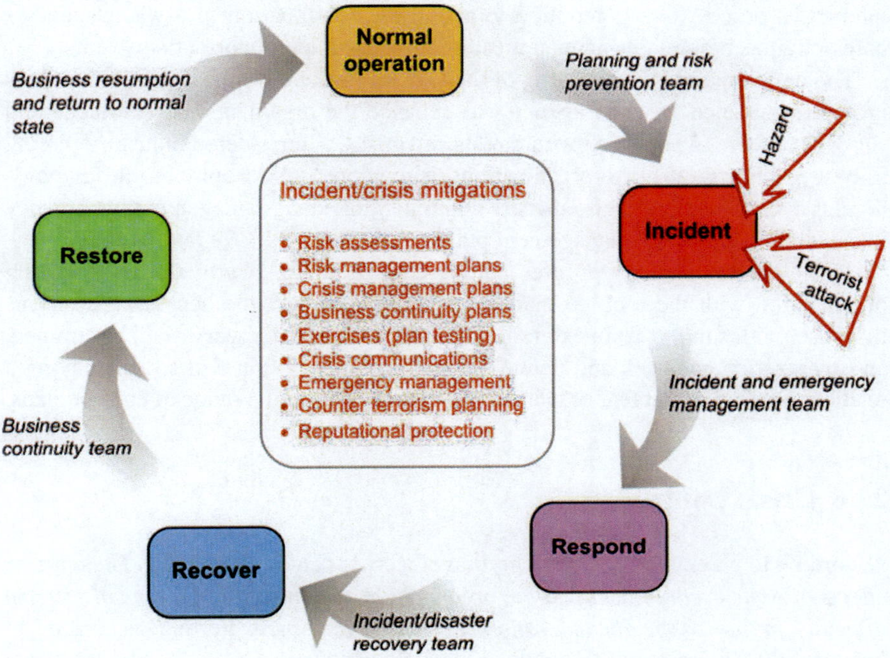

Fig. 23.2 Business resilience cycle

business behavior encapsulates business recovery policy, plans, and procedures. It also encapsulates training and exercising of directors, senior managers, and other personnel.

The full spectrum of possible mitigations as identified in Fig. 23.2 indicates the similarities between planning for terrorist attacks and "all hazard" incidents. As such, normal business continuity and risk management approaches can be applied equally for both terrorist and "all hazard" events to mitigate the likelihood and consequences of business interruptions. The incident risk mitigations are recommended as international best practice.

23.5.3 Practical Business Strategies to Reduce the Threat

Recent global terrorist events have challenged and engaged the world to prepare to manage previously unthinkable situations that may threaten an organization's future. This new challenge goes beyond the normal Emergency Response Procedure (ERP) or disaster management activities which were previously used. Organizations now must engage in a comprehensive process best described generically as resilience. It is no longer adequate to draft a response plan that anticipates naturally, accidentally, or intentionally caused disaster or emergency scenarios, we must *plan for the best and prepare for the worst*. Today's threats require the creation of an on-going, interactive process that is required to assure the continuation of a water agency's core activities before, during, and most importantly after a major crisis event.

The creation and annual testing of the various risk and business continuity plans provides guidance to water agencies to enhance the organizational resilience and survival capacity when faced with threats and crises. A considered industry opinion, as best practice in 2011 for organizations, is to adopt a philosophy of introducing a regular cycle of testing their business continuity, incident management, emergency management, and crisis management plans.

This business strategy ensures cutting-edge training to arm the participating organization with the tools to plan, prepare, and conduct the necessary exercises in-house, with support from external consultants where necessary. Suitably trained, an organization can seek and define the opportunities within a crisis, displaying a resilient image without fear or concern, with the addition of a range of current plans.

23.6 Crisis Control

According to Naylor (2002) the definition of a crisis can be regarded as *The adverse effects of some event in the life of an organisation, which result in a time of extreme difficulty for an unspecified duration for those concerned.* Further, according to Naylor (2002), a crisis has three identifiable elements:

1. An event or occurrence which poses an immediate or definable threat to the organization.
2. An event that occurs with an element of surprise or lack of prediction.
3. An event that presents an organization with a short time frame for decisions and/or action.

The resilient organization incorporates effective linkages and systems which reflect effective proactive strategies to reduce and mitigate future risks as indicated in Fig. 23.3. Progressive thinking may be expressed in this case as: *The future is for sale and it is all about surfing on the front of the wave instead of being in front of the iceberg.*

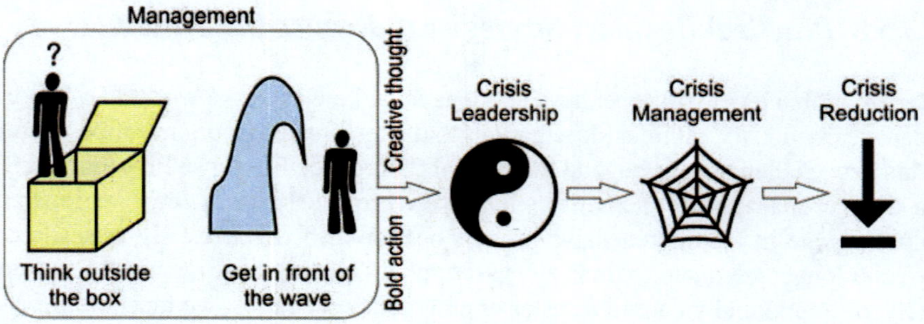

Fig. 23.3 Resilient model reflecting effective crisis control

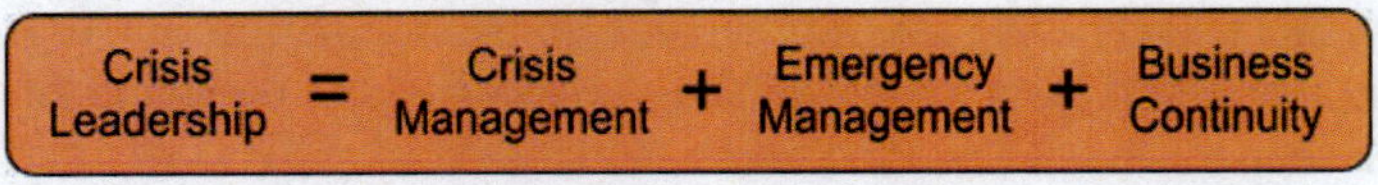

Fig. 23.4 Crisis leadership

23.6.1 Crisis Leadership

The art of business crisis leadership is to be able to not only contain and eliminate crises as they occur but also convey the water agency through to any possible commercial and reputation advantage line exposed by the situation. Displaying the competence to return situations to normal is considered insufficient in multi-competitor times. This visual competency is business continuity or emergency management thinking in isolation. Crisis leadership should therefore be defined as outlined in Fig. 23.4.

What emerged as emergency management and evolved through disaster recovery into business continuity and crisis management has matured to now become crisis leadership and disaster avoidance. *Indeed, it is the position of this chapter that* "Corporate Governance is the strategic response to risk, and Crisis Leadership is your strategic reserve."

Business thrives on instability and there is plenty of room at the top. If a water agency seeks a healthy appetite for risk then there is a need to have a finely tuned crisis leadership capability to stay at the top (Fig. 23.5). Some progressive water agencies invoke their crisis leadership and incident teams, two and three times in one month. It is not recognition of management failure, but rather a mechanism for seizing opportunity and generating business tempo.

The business environment has changed forever. Those that choose to paddle their canoe in the new corporate governance sea are developing their capability to capsize and recover in order to regain their commercial position and claw their way toward

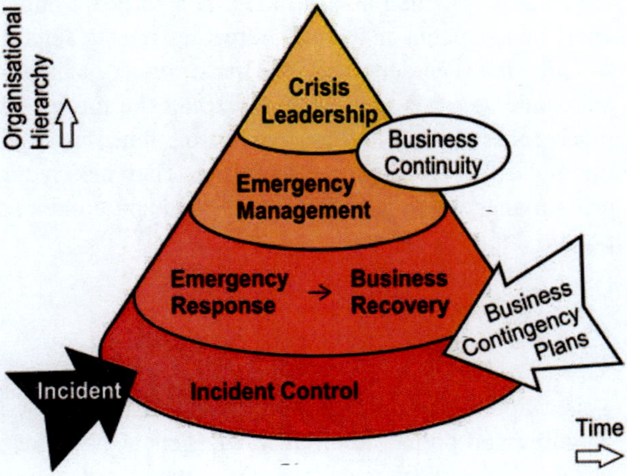

Fig. 23.5 Strategic crisis leadership related to the incident

superiority and even dominance. It is considered that top performing organizations understand that planning to transform from crises is key to corporate survival and commercial superiority.

23.6.2 Crisis Management

Crisis management may be interpreted as the process by which any organization deals with any significant and unpredictable event that threatens to harm or damage an organization. Adequate and proactive risk management involves assessing potential and emerging threats and evaluating the appropriate mitigation approach to reduce the impact and consequences of those identified threats. The art of crisis management is effecting strategies prior to when major incidents occur or are about to occur. Crisis management operates at a higher context of management and is required to display a range of skills and techniques, operating under pressure to assess, understand, and cope with any serious situation, particularly from the time of initial identification to the point of recovery and reversion to normal business state.

Crisis management is interpreted by appropriate administrative structures to effectively deal with external and internal crises. This ensures that all concerned understand who makes the decisions, how the decisions are implemented, and what are the roles and responsibilities of all participants. Personnel used for crisis management are assigned to perform these roles as part of their normal duties. Organizational leadership has a duty to stakeholders to plan for its survival and enhance its resilience.

Crisis capabilities have certainly matured across sectors, driven in part by attitudes to Corporate Governance. To some organizations, crisis management is still just "hot Issues Management in a hurry." To others crisis management has become corporate emergency response in the aftermath of safety and security incidents. To others still, crisis management is a subset of business interruption and disaster recovery. Most organizations perceive it as a tool to treat danger and less so opportunity. The reality is that the majority of these philosophies are just planning for what is regarded as inevitable. A significant number of organizations perceive crisis management as the tool to regain the status quo or the immediate past.

But what is evident in the top performing organizations is to plan for the unthinkable and use crisis leadership to exploit the future. These organizations are not so much focused on fixing the hole in the fence as they are of exploiting the opportunities of the open paddock beyond. They also recognize the difference between crisis management and crisis leadership and build a resilient culture related to any potential terrorist attack across the future.

23.6.3 Plan, Prepare, and Respond

This chapter proposes that in adapting and adhering to a standard series of extensive planning and preparation, all water agencies will have the ability and resources to develop, prepare, plan, and test plans, whether they be incident, emergency, business

Table 23.2 CARVER matrix

Score	Criticality	Accessibility	Recoverability	Vulnerability	Effect on Populace	Recognisability
5	Needed for survival	Public	Difficult	Unable to harden	Mass casualties: high symbolism	Unique
4	Needed for economic	Government or political process recognisable	Admission criteria	One year or more	Can be hardened	Deaths occur; symbolism to some
3	Disruption severe	Screening	One month to one year	Hardened for natural disasters	Injuries; symbolism important	Moderate difficulty to identify
2	Disruption moderate	Inspection of packages	One week	Hardened against snipers and attacks	Major injuries; undetermined symbolism	Very difficult to identify
1	Disruption light	Escort needed	Less than seven days	Hardened against bombs	Minor injuries; not symbolic	Indistinguishable from surroundings

continuity, or crisis-based plans. The preparation of a set of corporate risk, emergency, business continuity, and crisis plans, regularly rehearsed, will contribute to a resilient corporate body, potentially inhibiting terrorists and their possible impacts and ensuring a timely business recovery.

As terrorists tend to conduct extensive Close Target Reconnaissance (CTR) prior to any proposed target identification and in-depth planning, a progressive initiative for water agencies is to conduct similar Red Cell planning. Terrorists may have adopted and utilized the CARVER matrix (Table 23.2) of target identification and analysis. This system was developed by the US Special Forces in the 1950s and used ever since by Special Forces.

This tool illustrates a numerical calculation to identify a suitable cost and politically effective target. The methodology has an identified numerical value below which the target under analysis maybe aborted for an easier hit. Above the nominated value, further intelligence is gathered to confirm the target. If this is in a terrorist tool kit, it equates to proactive business sense to also use it for a Red Cell analysis (outside looking in) versus the traditional Blue Cell analysis (inside looking out).

23.6.4 Business Benefits

In consideration of aspects of crisis leadership, crisis management, and the practical nature of planning, preparing, and responding to incident and crisis, there are tangible business benefits such as

- Crisis leadership as the cheapest insurance policy for counterterrorism and social, financial, and environmental responsibilities.
- Crises as opportunities rather than threats.
- Regular practice, learning after other's mistakes, and benchmarking against competition at a frequency of every 1–2 years.

23.7 Crisis Exercise

To ensure that crisis and business continuity plans are effective and that management at all levels are aware and understand their respective roles and responsibilities, plans need to be enacted as exercises at least once per year to effectively engage, analyze, and review corporate capacity for dealing with all manner of risk and crisis. This concept can be adopted to assist and support water agencies to undertake these annual exercises as a form of assurance that the organization *will survive with reduction and minimization of reputational damage* from any future crisis.

The formulation of relevant plans and regular practicing with annual exercises enable water agencies to lead, coordinate, or support the response to a business continuity, reputation, or liability crisis in any sector. This extends conventional crisis management into crisis leadership.

These actions provide directors and managers with a concise understanding and assurance that the strategies and actions needed to prevent and control corporate crises exist to support the water agency through any terrorist attack or natural crisis. An exercise is a win/win situation for the organization, the team, and participants individually, which builds team confidence, trust, and resilience in the organization's ability to control crisis situation. To achieve these outcomes, exercise ought to be preceded by refresher training.

23.7.1 Types of Exercises

Variable types of exercises have been developed to meet the requirements of specific plans and procedures. As such, all plans and procedures to be tested are able to be identified within the context and framework of the established exercises listed in Table 23.3.

23.7.1.1 Tabletop Exercises

Tabletop exercises involve interactive discussions, based on hypothetical scenario, among members of an incident or emergency response team or representatives of various and multiple teams. These exercises are usually held to refresh or contribute to operational and incident/emergency procedures, which are under review or being formulated. Tabletop exercises additionally examine and review individual and team roles, to identify planning requirements, to set program objectives, and to clearly define the criteria and trigger points for the definition of what is a crisis or emergency for an organization. The process can also redefine individual and team roles rather than validation of skill levels.

The discussion is usually led by an experienced facilitator, utilizing a series of headings to engage the various participants to draw and provide expert advice from the various areas of expertise. These discussions are normally held in an operations room, often with a white board or computer-led scenario on an overhead projector.

Table 23.3 Types of crisis exercises

	Tabletop exercise	Incident/emergency exercise	Live exercise	Business continuity exercise
Objectives	Test incident/emergency responses and review individual and/or team roles	Test plans and incident responses. Usually based on previous incidents or identified risk gap	Usually focused on operational and field response to test and train appropriate personnel to handle crisis and test plans	Test internally business continuity plans and business recovery based on systematic failure
Scope	Identifies what is in scope and out of scope	Establishes what is an acted action and determines what is simulated	Identifies the boundaries for what is and is not a live action	Identifies the boundaries for the tested systemic failure
Timing	Usually 3 h, but may extend to 6 or 8 h	Usually 3 h, but may extend to 1 day	Usually 4 h, but may extend over 1 or 2 days	3 to 4 hours
Stakeholders	Internal with some observers	Internal with some observers	Usually internal only	Usually internal only, but may require advice or discussion with external regulators
External agencies	Routinely invited to produce the networks and liaison necessary in times of future crisis	Routinely invited to produce the networks and liaison necessary in times of future crisis	May be involved to produce an element of reality	Usually excluded
Script	A tight script to control the exercise pace and direction	May or may not have a script. There may be a focus board and discussion	A tight script to control the movement of people and logistics	May or may not have a script. There may be a business continuity plan which is tested
Exercise inputs	E-mail; paper; mobile or land line	E-mail; paper; mobile or land line	Verbal inputs which usually follow existing plans and procedures	Verbal inputs which usually follow existing plans and procedures
Exercise operation	An internal activity, may involve multiple sites	An internal activity, may involve multiple sites	An external (field) activity, may involve multiple sites	An internal activity, may involve multiple sites
Logistics	Venue, catering, administration support, recording software, communications	Venue, catering, administration support, recording software, communications	Venue, catering, administration support, recording software, vehicles, field equipment, (field) communications	Venue, catering, administration support, recording software, communications

Tabletop exercises are clearly distinguishable from other exercise activities by invariably displaying the following elements and actions:

- Allocation and positioning of personnel and resources in a pre-arranged location.
- Establishing a scope and objective for the adequate assessment of the outcome of the exercise.
- Initial presentation of a hypothetical and realistic crisis scenario in a verbal and/or video/power point format with facilitator lead discussion over several hours.
- An absence of a simulation cell or counterplayers, to lead the format of the exercise.
- Recording and documentation of the team and individual discussions to provide an assessment of team performance against the scope and exercise objectives.
- Feedback to teams and individual members, with a two-way process, on the exercise performance as related to teams and roles during the exercise.

23.7.1.2 Incident and Emergency Exercises

The functional aim of incident and emergency (protection of life, property, and the environment) exercises is to enact and test the plan to ascertain and establish its effectiveness and compliance with current organizational requirements and needs. The primary objective is to provide a comprehensive and measurable assessment of the relevant plans and procedures, which would be enacted in an emergency or a major incident. That is a realistic rehearsal of the plans or procedures with personnel taking active roles.

These exercises can include personnel from throughout the water agency at an operational level and representatives from management, in response to a hypothetical scenario in a written and pre-prepared script. They can range from several hours to a day, depending on the complexity of the exercise, and can be "ring fenced" within an office or boardroom scenario or inclusive of diverse operational areas of the organization.

Incident and emergency exercises are distinguishable from other exercise activities by invariably displaying the following elements and actions:

- Assembling and gathering of emergency resources and logistics in one or more geographic locations.
- Formulation of exercise scenario through a credible written textual script and role playing from that script utilizing a team of counterplayers in a separate location from the live players in the exercise.
- The use of exercise controllers to control and alter, if necessary, the pace and cadence of the exercise and liaise with the counterplayer controller. Should multiple sites be used, additional exercise controllers may be deployed to liaise with the main exercise controller in feeding back information as to how the exercise is progressing at those locations.

- The main exercise controller holds a "hot debrief" to capture the immediate feedback from the various elements and locations of the exercise while it is still fresh in all participants minds.
- A report and detailed assessment is compiled and disseminated by the exercise controller as soon as possible after the "endex" or cessation of the exercise.

23.7.1.3 Live Exercises

Live exercises test plans relative to emergency, incident management, and response skills of relevant and appropriate representative elements across the organization. They also encourage the visible interactions and relationships that would occur during an actual incident or emergency. This may involve external agencies, if the organization wishes to simulate an actual emergency or major incident.

These exercises enact, by a script, the interactions of all levels of an organization, ensuring that all are involved in the realistic demonstration and validation of emergency and crisis capabilities as in a real incident. This may incorporate sub-plots of the script where injuries and casualties are identified with the use, for example, of a flour outline on the ground. The indicated example illustrates an activity which simulates injuries/deaths and provides a credible representation of actual events increasing reality within the script.

Live exercises are distinguishable from other exercise activities by invariably displaying the following elements and actions:

- Actual mobilization of most if not all operational response personnel and resources at various geographic locations and sites, with some activated mobility between sites and communication back through the counterplayer's room.
- Scenario information and activities are controlled through the exercise control through the counterplayers' controller and guided by the exercise controller.
- There may be some simulation of non-participants through the counterplayers from the script or actual simulation with the use of controlled and scripted external party involvement.
- Extensive use of exercise controllers to ensure that the exercise remains within the script guidelines, format, and intended parameters, coordinating the script across possibly multiple sites and geographic locations.
- Evaluation and assessment postexercise, with some coaching and mentoring occurring during the exercise by exercise controllers.
- Some selected representational feedback evaluation from designated response personnel to assist the exercise reporting framework.
- An exercise response report from the exercise controller, based on observations and feedback from designated response personnel and guided by "improve, sustain, and fix" parameters.

23.7.1.4 Business Continuity Exercises

Business continuity exercises concentrate on invoking and rehearsing the Business Continuity Plan (BCP) to establish an aim to determine the effectiveness of the BCP in the event of an external or internal incident interrupting the normal flow of business. They may also assess how the organization would continue to operate while recovery operations are occurring subsequent to a significant business interruption or crisis.

Enacting the BCP encourages the visible interactions and relationships that would occur during an actual incident or emergency while the business continues under the BCP, and recovers. This may involve external agencies, if the organization wishes to simulate an actual crisis which affects the business, and in some way interrupts the normal business processes, especially in consideration of interdependencies such as other utilities as power, which may have a business impact in a time of crisis. Business continuity exercises have a script and display the identical elements and actions to incident and emergency exercises as illustrated previously.

23.7.2 Exercise Roles and Responsibilities

Within the context and framework of the types of exercises established, there is a distinct requirement to identify exercise roles and responsibilities to ensure that the various exercises are managed and operated effectively.

23.7.2.1 Exercise Manager

The exercise manager assumes full control and coordination of the exercise function. The exercise manager also delegates some tasks and functions to exercise controllers for the successful operation of the exercise.

23.7.2.2 Exercise Controller

The main exercise controller is responsible for providing support to the exercise manager and typically for observing the live participants as the exercise script unfolds. This position also liaises with the counterplayer controller and any other supplementary exercise controllers who may be placed at other participating locations. The number of exercise controllers varies according to the nature and complexity of the exercise.

23.7.2.3 Crisis Management Team/Incident Management Team

The typical crisis/incident management team consists of variable teams and structures to reflect the nature of the exercise type. These structures furthermore vary according to the composition and identity of the sponsoring organization of the exercise. An example of a typical crisis/incident management team is illustrated in Fig. 23.6.

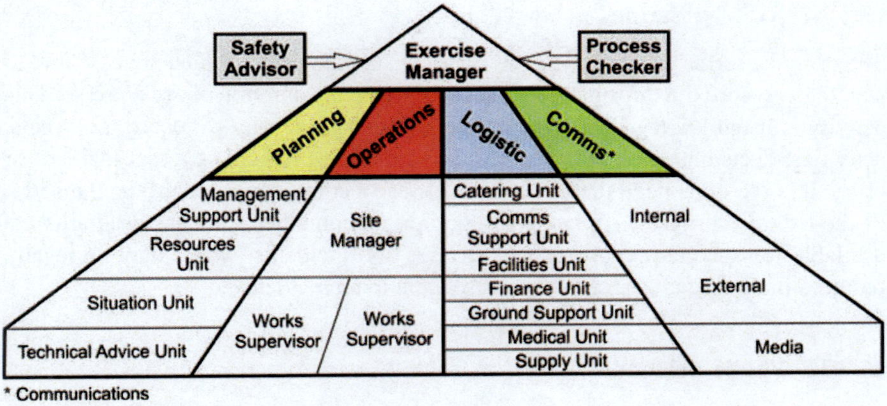

Fig. 23.6 Typical incident or emergency management team structure

23.7.2.4 Counterplayer Team

The counterplayers, depending on the nature of the exercise, assume a variety of roles from the script at times designated, which may range from playing roles of media, bystanders, casualties, relatives, regulatory agencies, joint venture partners, etc. The counterplayer team usually consists of 5–10 individuals, who may adopt multiple roles each during the duration of the exercise.

This team is managed and coordinated by the counterplayer controller, who reports to the exercise controller. The role of the counterplayer controller is to brief and coordinate the functions and activities of the counterplayers. They can also decide, in coordination with the exercise controller, to increase or slow down the pace of the exercise from the observations of the team management of the emerging issues from the script. The pace of the exercise is also dependent on how live participants deal with the incident.

23.7.2.5 Observers

Some organizations, particularly those with regulatory requirements, compliance, or joint venture partners, may wish to have observers from these categories attending the exercise for a variety of reasons. It is important that these people are escorted with boundaries and parameters stated to ensure that the conduct of the exercise is not interfered with, and that the observers view exercise proceedings which are necessary for their visit.

23.7.2.6 Runners

Experience has indicated the need for one or two people in the category of a runner to move messages between groups, to access resources and logistics, and to organize requirements for the incident manager and/or the exercise controller.

23.7.2.7 Exercise Scribe

The exercise scribe ensures adequate lines of communication which are essential to accurately record the information and events across the duration of the exercise. This activity is usually a legislative requirement in most countries to record the events, activities, decisions, and communications across the timeline of a real incident or crisis. It is recommended that the log keeper use a computer to record the transcript of the exercise progress. A range of appropriate computer software programs are available for accurate recording of exercises and incidents, which may be legally required in the event of an enquiry subsequent to an incident.

23.7.2.8 Public Relations

It may be necessary to utilize the media to promote the exercise as a promotional tool and a form of assurance for board, shareholders, and the public.

23.7.3 Exercise Documentation

Exercise documentation is critical for effective planning and execution of a successful exercise. The primary document is the exercise master document which plans the concept of the exercise and incorporates reference to the four other documents. The other documents in relation to an exercise are led by the exercise script, which is illustrated effectively within the counterplayer brief. The exercise hinges on this document as this brief dictates what occurs, when it occurs, and in what order. Additionally, there is also a live participant's brief to set the rules of engagement for exercise participants. The exercise report, which is a documented summary of how the exercise progressed and what risk gaps were identified with areas for improvement, is compiled at the end of the exercise and is detailed further in Section 23.7.5.3.

23.7.3.1 Exercise Master Document

The exercise master document is an integral primary planning guide, which is formulated at the commencement of the exercise process. It identifies all necessary stakeholders, required documentation, administration and planning functions, and communication strategy (Table 23.4) to assure that the exercise is conducted effectively. This document is created with significant input from exercise sponsors and relevant stakeholders.

23.7.3.2 Exercise Script

The purpose of an exercise is to simulate, as closely as is feasible, an emergency or crisis event to accurately gauge how effective existing plans and procedures are and

Table 23.4 Tick box list for exercise master document

Stakeholders	☑ Exercise manager ☑ Exercise controller ☑ Crisis/incident management team	☑ Counterplayer team ☑ Observers ☑ Runners ☑ Exercise scribe	☑ Public relations ☑ Participating organizations ☑ OH&S officers
Documentation	☑ Exercise master document ☑ Exercise script	☑ Counterplayers' brief	☑ Live participants brief
Administration	☑ Preparation authority ☑ Briefing arrangements for counterplayers and incident team	☑ Budget ☑ Exercise control ☑ Exercise debriefing arrangements	☑ Logistics ☑ Equipment ☑ Transport ☑ Accommodation ☑ Toilet facilities ☑ Catering
Planning	☑ Exercise identity/name ☑ Date and time ☑ Location and venue	☑ Reason for exercise ☑ Aim ☑ Scope ☑ Exercise areas and restrictions	☑ Security ☑ Organizational structure ☑ Maps and charts
Communication	☑ Public relations ☑ Media releases	☑ Notification to appropriate agencies	☑ Advice to other personnel who are not involved

to evaluate identified gaps or weaknesses prior to an actual event occurring with a corresponding reduction or mitigation of organizational risk. The major thrust and nature of the exercise script is directly related to the scope and aim of the exercise and the stated requirements of the exercise sponsor or any regulatory or legislative requirement that may have initiated the concept of the exercise initially.

The script is designed in tempo or cadence to produce pressure points for decisions to be made under other pressure from the script, for example, media-persistent and intrusive media enquiries or defined industry issue-motivated groups demanding urgent answers to a range of enquiries. This media and issue group pressure varies, with media and pressure groups interspersed throughout the script, in addition to relatives of injured personnel enquiries, designed to provide the nature and pressure of a real-time incident over a longer time frame.

The script planner is generally an individual with innovative thought capacity with some broad knowledge of an organization and the capacity for research and to gain significant detail of the organization's operations in a short period of time. To maximize the effectiveness, outcomes, and benefits of the exercise, the script is only developed by and known, prior to the exercise date, to a small planning team of individuals, with perhaps a trusted insider to gain technical detail necessary for an exercise. In most cases, if there is a wide awareness of the script details across the organization it tends to produce an artificial understanding of the effectiveness of incident/emergency and crisis plans, defeating the purpose of enacting an exercise. For these reasons, it is imperative to keep most scripts confidential.

The exercise script is provided in a hard copy bound format to exercise controller, counterplayer controller, counterplayers, and exercise controllers. Copies are to be collected and destroyed after exercise to reduce corporate risk from unauthorized access.

23.7.3.3 Counterplayers Brief

The counterplayers brief provides a conduit for the exercise inputs from the counterplayers. The document states the introduction with some background to the incident or main event of the exercise. There is then a précis scenario illustrating the most significant issues building up to and leading to the incident. The scenario may describe in broad terms in a dot point manner the response expectations of the exercise. It is usual for a few trigger events to occur within the initial 5–10 min, which are designed to initiate the incident or emergency planning process of the organization and form the required management team of the organization. Media organizations are also detailed, with the nature of the media and names of reporters and commentators.

23.7.3.4 Live Participants Brief

The live participants brief is the primary document to be collated and populated to set the scene and criteria for the exercise with limitations and protocols required. This crucial document is a short document which identifies and illustrates the aim, which may well be to "Test the Budapest Water Agency's Emergency and Crisis Plans under a simulated emergency and/or crisis in a Water Treatment Plant location." A start state is provided as a brief introductory paragraph, describing the current environment, weather, location, whether it is actual time or another time, exercise boundaries, limitations, exclusions, and management actions.

The functions of the exercise live participants are also described, as to which person or organization is acting as a live operation and as a counterplayer. The exercise controllers are identified with a comprehensive list of contact details and phone numbers, e-mail addresses and fax numbers, radio frequencies and call signs which may be in use over the duration of the exercise.

Additionally, the counterplayer actions are described with the resource limitations and rules outlined, for example, by a scripted or a communications exercise or it may represent some live elements as live field operations. However, all exercise participants are identified as groups and how they engage and interact within the participant's framework.

The passage of information is controlled and outlined to ensure that all personnel involved enforce all communications to be identified as being within the confines of an exercise. There is a requirement for a counterplayer's brief and a live participant's brief to adequately inform all participants to ensure exercise rules and communication boundaries are accurately identified.

23.7.4 *Exercise Planning*

Exercise planning is a crucial process to achieve a successful outcome. The planning commences with the gathering of data usually from the organizational risk register, organizational charts, annual report, the use of an internet search engine for any significant media or government interest in the organizational profile, Wikipedia, media reports, stock exchange reports, relevant industry associations and institutions, and internal organizational cultural intelligence. The objective of this primary research is to identify any potential risk gap or weaknesses that may indicate a focus for the script, providing the foundation for either an incident or crisis and reputational threat.

Some innovation is required by the exercise planner to

- Progressively develop the exercise scenario, over the exercise duration.
- Provide additional information to the live participants via the script.
- Pose problems for live participants to resolve in the script.
- Define any limits on live participant's actions.
- Force action by live participants, by injecting, incidents that require action, questions that require responses, and decisions that require to be made by the management team.

23.7.4.1 Single, Double, or Triple Jeopardy

Within the time frame of an actual incident or crisis, another issue or multiple incidents may well occur. It is considered by experienced crisis practitioners that in order to maximize the benefits of an incident or crisis exercise, it is more productive to add in one or more additional incidents or issues to produce and simulate the level of emotive pressure that is experienced in an actual major incident or crisis. That is, to produce a "double or triple jeopardy" as distinct from a single incident or single jeopardy, by adding in another issue that may or may not be linked to the original exercise incident, as can actually occur in real incidents.

23.7.4.2 Exercise Aim

The aim of an exercise must be very concisely and clearly stated prior to preparing the script or any form of exercise planning. The aim provides the strategic direction of the exercise and should always state its purpose with the use of a verb. A typical example is as follows: *This exercise is to practice the business continuity plan of*

Table 23.5 Exercise scope

In scope	Out of scope
• All participants to be Prague Power Agency personnel • Exercise is confined to the head office only and no communication to go outside the facility • The exercise assesses the relationships between Prague Power Agency divisions in head office • The passage of information communicates and consists of paper and mobile phones only • The exercise tests the validity and currency of existing business continuity plans	• Any live event or operation within The Prague Power Agency • Communication with any of the five call centers used by the Prague Power Agency • Real communication with Prague Emergency Services and the Prague Police Services

the Prague Power Agency and to test for incident response and crisis preparedness, and compliance with business recovery plan policies and procedures.

There should only be one single aim for an exercise and this should be stated at the front of the exercise script. This single aim acts as a reference foundation for the exercise preparation and planning.

23.7.4.3 Exercise Scope

The exercise scope is generally compiled to reflect the needs and requirements of the exercise sponsor, the outcomes expected from the exercise, and the exercise exclusions (if any) from the exercise. The scope should be broad enough to stretch the live participant's intellect, but not too broad to exceed the capacity and capabilities of all participants. A typical scope with reference to the above example is included in Table 23.5.

The scope should be determined between the exercise manager and the exercise owner or the exercise sponsor. This is especially significant in the event of an external regulatory agency which may enforce the exercise as a mandatory requirement under legislation or a government act.

23.7.4.4 Exercise Identity

The exercise identification provides the exercise with an identity to be referred to and remembered more easily. Some exercises have names or numbers, whereas some merely have a date and an organizational name. The identification is generally a reflection of the organizational culture.

Regardless of the exercise identity, however, the exercise must have allocated some identifying code to distinguish it from other sequential exercises in succeeding years or time. Typical names may be associated with the random name or location with a year, for example, "Exercise Viking 2005" or "Exercise Shalom 2006." Each exercise also has a final report stating recommendations and actions. As such, the exercise can be referred to across the future or compared with other exercises of a differing nature or type.

23.7.4.5 Exercise Timing

A target date for the exercise is required to be established at an early stage in planning, with broad consensus and support of the exercise stakeholders and the exercise sponsor. Once the date is agreed and communicated to all, the exercise becomes a real entity with the team working toward exercise completion. This ensures that functions occur in a planned and logical manner, and all involved are aware and informed.

23.7.4.6 Catering and Logistical Support

Both catering and logistical support require to be provided to the exercise participants as it would be the case during a real event, which may well progress over a few days or even weeks. Regarding catering, it is beneficial to appoint a single person to supply meals and beverages to assume responsibility for this task.

Additionally, in accordance with the requirements of the exercise planner and the script there may be other requirements for equipment, vehicles, rooms, wall charts, white boards, computers, fax machines, laser pointers, log sheets, etc. These logistics are also suggested to be coordinated and provided in the correct location in a timely manner by a nominated officer.

23.7.5 Exercise Process

23.7.5.1 Typical Process Model

The typical industry exercise model can incorporate variable time frames and is usually divided into designated sections to trigger and activate the incident teams. Normally, the crisis exercises are held over several hours as per the model in Fig. 23.7. This model is focused on a dynamic duration, with a subsequent "hot wash-up" to capture significant and crucial organizational memories and observations, later used for the exercise controller's exercise report.

Within the initial 60 min period, communications are intense to raise and simulate the pressure which would normally occur in a real crisis. In the real-time event or crisis, the buildup would actually occur over a time period of days, weeks, or materialize suddenly. In the exercise, events and time are compressed to recreate and simulate the psychological pressure of the reality of a real crisis.

After this period, the exercise manager conducts the following actions and activities:

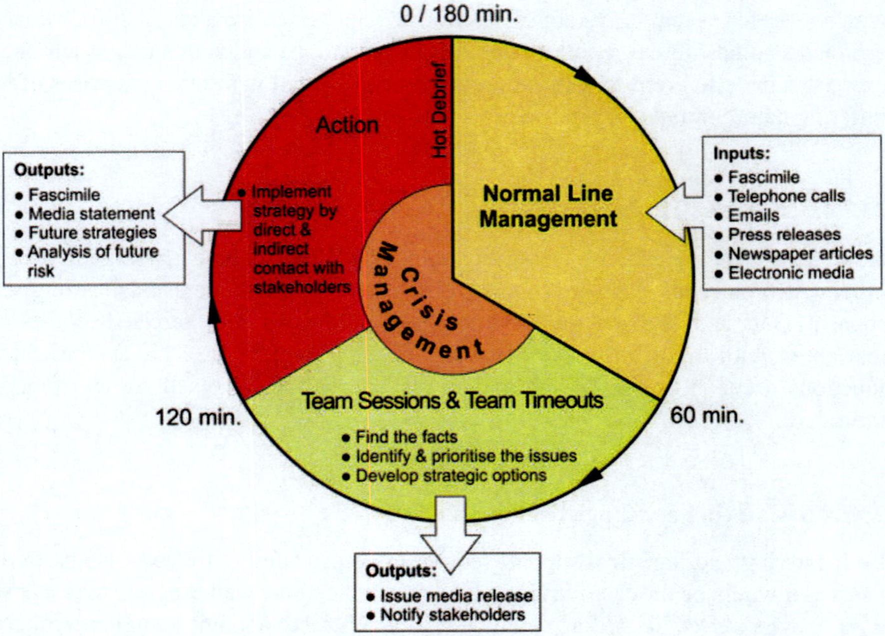

Fig. 23.7 Typical 3-h crisis exercise simulation cycle

- Finds and accurately identifies the facts as distinct from rumor and innuendo.
- Identifies and prioritizes the issues to be addressed and dealt with.
- Develops strategic priorities.
- Issues a press release which transmits "what we know, and what we do not know."
- Ensures that the activities, actions, and communications are captured in a reliable log.
- Notifies key stakeholders.

Later the exercise manager ensures that appropriate strategy is implemented by direct and indirect contact with stakeholders. During this period, reputational damage may be minimized and media can be used as an opportunity to portray a positive image of the organizational achievements. This is achieved by the communication mediums of facsimile, phone, and e-mail, with press briefings.

The exercise process culminates in action demonstrating future strategies related to the crisis, incorporating an analysis of future perceptions and considerations of organizational risk. A "hot debrief" with exercise participants and observers indicates the end of the exercise process. Within the exercise, normal operational line management is reverted to at the end of scripted exercise activity.

23.7.5.2 Exercise Communications and Control

Communications for exercises normally follow a standard format which illustrates and controls the flow of data and information. The exercise commencement or "startex" must start at the indicated time for all participants and is controlled in timing by the counterplayer controller in conjunction with guidance from the exercise controller observing the exercise. Startex is communicated via the initial exercise injects with the primary trigger which leads to the incident, emergency, or crisis management team being called together.

Communications from the counterplayers to the management team are in various forms in conforming to the script and exercise requirements. These may be communicated in the guise of phone calls, e-mails, paper hard copy injects, or radiofrequency message. These messages are transmitted into the organization in various modes, which may be via a call center, to specific public officers, or direct to the management team. Outgoing communications from the counterplayers are always preceded with *Exercise, Exercise, this is . . . and I would like to speak to . . ., and my message is . . ..* This is to ensure that any misdialling or incorrect person receiving the call is aware that it is an exercise, and that it also substantiates to the management team that it is an exercise message only and not in fact reality. Incoming communications are always preceded with *Good morning/good afternoon, this is exercise control. Who do you wish to speak to?* The counterplayer respondent can adopt the player's role or pass the phone to the appropriate counterplayer who is adopting that specific role for the duration of the exercise.

Communications from the management team to the counterplayers are invariably either responses to messages inputting from the script or notifications/instructions to others who are being role played by the counterplayers. These communications can take the form of e-mail, telephone, radio, fax, or hard copy. Other exercise instructions, protocols, rules, and constraints are set down and agreed prior to the commencement of the exercise and are communicated within the live participants brief.

Concerning Occupational Health and Safety (OH&S), each country, state, and province has specific legislation to protect workers in all industries. Exercises should reflect an awareness of the local OH&S acts, legislation, codes of practice, and organizational procedures designed to minimize the risk of injury to all exercise participants and bystanders. The level of realism in the exercise cannot affect or assume priority over safety to all. The exercise timing, pace, and cadence may be altered from instructions by the exercise manager to the exercise counterplayer controller, advising to speed up the injects, slow down, or have a "time-out" for a nominated period of time.

At the end of the exercise or "endex" with instructions given by the exercise manager, all participants either gather or are connected via conference telephone or video conferencing to initiate a "hot debrief." This action is to capture the information while still fresh in people's memory, under the headings of "Sustain, Improve, Fix." Should a "real-time" incident/emergency or crisis occur during the operational time of an exercise, the exercise may be immediately ceased and aborted with the use of the internationally accepted words *Stop, stop.*

23.7.5.3 Exercise Review

A significant component of the end of an exercise involves a "hot brief," an exercise report, and a review and postmortem. The exercise "hot debrief" or exercise "hot wash-up" is necessary to capture the immediate thoughts and suggestions for improvement of the plans being tested, in addition to the conduct of the immediate exercise. It is suggested that all comments be restricted to "Sustain, Improve, and Fix," producing an accent on continual improvement, which is the main focus of conducting exercises.

An exercise report, which captures the comments of all participants and counterplayers, is compiled as soon as possible after the conclusion of the exercise. It summarizes the progress of the exercise, identifies potential risk gaps, and any actions and recommendations which are required to be completed to address adequate risk mitigation. This report is provided to the exercise sponsor and key participants who are identified by the organization as benefitting from receiving this information. It is beneficial after an exercise to conduct an internal review and postmortem of how the exercise progressed and also to identify the benefits and improvements from the exercise as a continual improvement process.

23.8 Conclusion

The chapter has reviewed threats to water and wastewater systems originating from both all hazard and external human interventions. Although there is documented evidence that the all hazard origin incidents have resulted in worse consequences, there is the perceived global threat from the impact of terrorism.

It was highlighted from the business perspective that there is minimal variation in preventative measures to reduce the risk exposure, prevent and control crises, between an all hazard incident and a terrorist attack event. These mitigation strategies can be summarized as crisis leadership, crisis management, and business continuity toward recovery. Prominent among this range of preventative risk mitigations is the practical concept of conducting credible crisis exercises on a regular basis to test the risk plans of the water agencies. The variable types of exercises have been listed to meet the requirements of specific plans and procedures. Operation of exercises has also been detailed to enable insight into these progressive and innovative business crisis mitigation methods.

These strategies enable water agencies to concisely understand the vulnerabilities of their systems and support them through crises. Assurance is then provided that the organization will survive with reduction and minimization of risk from any future crises.

References

Australia Federal Government, Dept of Prime Minister & Cabinet (2004) Protecting Australia Against Terrorism. ACT Canberra Barton 2600.

CIA (2009) The World Factbook 2009. Washington, DC: Central Intelligence Agency. https://www.cia.gov/library/publications/the-world-factbook/index.html

Denileon PG (2001) The Who, What, Why, and How of Counterterrorism Issues. Journal of the American Water Works Association. 93(5):78–85.

Fisher P (2010) Electricity-Hungry Water Providers Need to Get with the Power. The Age Newspaper, Melbourne, Australia, 3 January 2010. Available online at http://www.theage.com.au/opinion/politics/electricityhungry-water-providers-need-to-get-with-the-power-20110102-19czn.html

Gleick PH (1993) Water and Conflict: Fresh Water Resources and International Security. Journal of International Security. 18(1):79–112.

Gleick PH (2006) Water and Terrorism. Water Policy. 8:481–503.

Hayward K (2002) Water Supply's Security Challenge. Water 21 June 2002:15–16. IWA Publishing, Elsevier Science

Isenberg D (2002) Securing U.S. Water Supplies. CDI Terrorism Project. http://www.cdi.org/terrorism/water-pr.cfm

Mays LW (2007) Water Resources Sustainability. McGraw-Hill Companies, New York, NY

Naylor L (2002) Crisis Management. ANZSASI Regional Seminar, Auckland, NZ, April 2002. http://asasi.org/papers/2002/Crisis%20Management.pdf

About the Editors

Robert M. Clark is currently an independent consultant in environmental engineering and public health. He is an adjunct professor in civil and environmental engineering at the University of Cincinnati and recently completed service as a member of the National Research Council's Committee on "Public Water Distribution Systems: Assessing and Reducing Risks." As a consultant Dr. Clark has worked on homeland security issues with Sandia National Laboratories, the US Environmental Protection Agency (USEPA), and Rutgers University (Newark Campus), among others. He served as an environmental engineer in the U.S. Public Health Service and the US EPA from 1961 to August 2002 and was director of the USEPA's Water Supply and Water Resources Division (WSWRD) for 14 years (1985–1999). In 1999 he was appointed to a senior expert position in the USEPA with the title senior research engineering advisor and retired from the USEPA in August 2002. Dr. Clark was a member of USEPA's Water Protection Task Force and was USEPA's liaison for homeland security research. Dr. Clark has published over 380 papers and 5 books and has been professionally active in several organizations where he served in numerous leadership positions. He is a life-time member of both the American Water Works Association (AWWA) and the American Society of Civil Engineers (ASCE). Dr. Clark is recognized both nationally and internationally and has received numerous awards for his work. Dr. Clark holds B.S. degrees in civil engineering from Oregon State University (1960) and in mathematics from Portland State University (1961), M.S. degrees in mathematics from Xavier University (1964), and civil engineering from Cornell University (1968) and a Ph.D. in environmental engineering from the University of Cincinnati (1976). He is a registered engineer in the State of Ohio and can be reached at rmclark@fuse.net.

Simon Hakim is a professor of economics and director of the Center for Competitive Government at Temple University. He is currently editing a book series on *Protecting Critical Infrastructures* with Springer Publisher. He earned M.A. and Ph.D. degrees in regional science from the University of Pennsylvania. He also holds a M.Sc. degree in city and regional planning from the Technion, Israel Institute of Technology, and a B.A. in economics at Hebrew University in Jerusalem. His special areas of research and teaching are privatization, public policy, private/public

R.M. Clark et al. (eds.), *Handbook of Water and Wastewater Systems Protection*, Protecting Critical Infrastructure, DOI 10.1007/978-1-4614-0189-6,
© Springer Science+Business Media, LLC 2011

police, and homeland security. Dr. Hakim has published 58 scientific articles in leading economic, criminal justice, security, and public policy journals. He has written over 40 professional articles and edited 14 books. He collaborated with Professor Blackstone on a major textbook dealing with the security industry. He is constantly invited to teach classes on privatization and international economics in MBA programs worldwide.

Dr. Hakim has conducted several funded research and consulting projects for the U.S. Departments of Justice and Labor, the Commonwealth Foundation, the Independent Institute, the Alarm Industry Research and Education Foundation, the City of Philadelphia, the Philadelphia International Airport, ADT, Vector Security, and other leading security companies. For the complete CV, see http://astro.temple.edu/~shakim; for the Center for Competitive Government see http://www.fox.temple.edu/ccg.

Avi Ostfeld, D.Sc., P.E., D.WRE (www.technion.ac.il/~avi/avi.htm) is an associate professor at the Faculty of Civil and Environmental Engineering at the Technion – Israel Institute of Technology, and the editor-in-chief of the *Journal of Water Resources Planning and Management Division, ASCE*. Dr. Ostfeld was a senior engineer and project manager at TAHAL – Consulting Engineers Ltd. in Tel Aviv from 1997 to 2000; a research associate at the Department of Civil Engineering, the University of Arizona, Tucson, AZ, from 1996 to 1997; and a research associate at the Technion Water Research Institute from 1994 to 1996. During 2008/2009 he spent sabbaticals as visiting professor at the University of Illinois at Urbana Champaign and at the University of Kyoto. Dr. Ostfeld research activities are in the fields of water resources systems, hydrology, and in particular in the area of water distribution systems optimization using evolutionary computation: water distribution systems security, optimal design and operation of water distribution systems, and integrating water quality and reliability into water distribution systems management and control.

About the Principle Contributors

Rakesh Bahadur is senior hydrologist and project manager at Science Applications International Corporation (SAIC). He earned his M.S. (1988) and Ph.D. (1990) in civil engineering from the Colorado State University. He also holds a M.Sc. (1980) in geology from the Punjab University. His experience is in developing and applying models to solve water sector problems related to water security and water quality. He has applied Geographic Information Systems to address various water issues related with risk assessment, environmental assessment, remediation, pathway analysis, and exposure assessment. Dr. Bahadur is working with SAIC since 1992. Prior to joining SAIC, Dr. Bahadur worked as a research scientist at the Atomic Energy Canada Ltd., Canada, from 1990 to 1992; a research assistant at the Department of Civil Engineering, Colorado State University, Colorado, USA, from 1985 to 1990; a visiting faculty at the University of Leads, UK, in 1984; and with Haryana State Minor Irrigation Tubewells Corporation (HSMITC) in India from 1980 to 1985. He can be reached at bahadurr@saic.com.

Dave Birkett BA, Dip Bus (Audit) is currently a manager of regulation and assurance with Grampians Wimmera Mallee Water Corporation in Victoria, Australia, where he is responsible for regulatory compliance, assurance, crisis, and risk advice. He also works as a principal crisis practitioner for Truscott Crisis Leaders, Perth, Australia. As a crisis practitioner, he has consulted for Halliburton KBR, Overseas Development Australia, Santos, Freeport Indonesia, Energex, Pacific Hydro, 15 Victorian Water Corporations, and others.

Dave worked in the public and private sector with the City of Horsham, the City of Adelaide, the City of Unley, the National Archives of Australia, TransAdelaide, the Northern Territory Electricity Commission, the Electricity Trust of South Australia, and Asset Services where he held senior management positions.

Dave now uses his expertise with counterintelligence in Defense, Australia, to advise companies in the prevention and control of emergencies and crises. He is experienced in incident management and disaster planning in the energy, local government, and transport industries. He has a comprehensive understanding of government policies, procedures, and practices, complemented by experience and knowledge of private enterprise methods and practices. His specialty is in leading project teams, implementing processes and physical change often involving multiple stakeholders, and in the emergency response aspects of managing public and private sector assets.

Dave has a B.A. in politics and psychology from the University of Adelaide, Australia, with postgraduate qualifications as a lead auditor and governance studies. He is a member of the National Water Services Infrastructure Assurance Advisory Group in Australia. Dave can be contacted on dbirkett@crisisleaders.com.

Zia Bukhari obtained his Ph.D. in parasitology/microbiology from the University of Strathclyde, Glasgow, UK (1995) and is a senior environmental scientist at American Water. His research interests include feasibility studies for full-scale application of UV technology in treatment of drinking water, determining the impact of environmental contamination with existing and emerging pathogens (i.e., protozoan parasites and *E. coli* O157:H7), epidemiological investigations, and microbial methods development/evaluation for water, wastewater, and water reuse applications. His research also focuses on water security issues, particularly early warning systems, use of hydraulic modeling for emergency response planning, incident responsiveness and characterization. Over the past decade, he has received numerous research awards from various funding agencies, participated in national and international workshops, presented at local, national, and international conferences, and published over 40 articles in peer-reviewed journals. Dr. Zia Bukhari can be contacted at zia.bukhari@amwater.com. Web: www.amwater.com.

Mark D. Burr is currently a research scientist at the Center for Biofilm Engineering at Montana State University, where he has worked since 1996, except for a research appointment from 1998 to 2002 in the Thermal Biology Institute, also at Montana State University. Dr. Burr's research at the Thermal Biology Institute involved the microbial nitrogen cycle in a geothermally heated soil in Yellowstone National Park. Current research interests are in the microbial ecology of drinking water and wastewater treatment. Dr. Burr's expertise is in molecular methods for analysis of microbial communities. His most recent work involves the microbial ecology of denitrification in constructed wetlands used for wastewater treatment. The emphasis of this project is on the gaseous nitrogen emissions from denitrification, especially the ratio of N_2O to N_2 emitted and analysis of the microbial genes involved (abundance and diversity). Dr. Burr received B.S. (1979), M.S. (1984), and Ph.D. (1996) degrees from the Department of Soil, Water, and Environmental Science at the University of Arizona. His Ph.D. research was in DNA fingerprinting of *Salmonella* isolates for microbial source tracking. He can be reached at mark_b@biofilm.montana.edu

Rolf A. Deininger is professor emeritus of environmental health sciences, School of Public Health, The University of Michigan, Ann Arbor, Michigan. At the beginning of his research his interest was focused on the interoceanic canal studies (1970) to search for optimal routes across the Panama Canal. His scientific interests included environmental systems analysis, environmental quality data banks, water quality problems in drinking water distribution systems, early warning systems, security and vulnerability studies, and the rapid detection of biological threats in air, water, and powder samples. His research has been supported in the past by NATO, the US EPA, AWWARF, Michigan Dept. of Environmental Quality, and Michigan Dept. of Community Health, and the State of California. Dr. Deininger has been a consultant to many international agencies such as the Pan American Health Organization, World Health Organization, UNESCO, NATO, and the World Bank. He has served on task forces of the National Academies of Sciences. Department of Defense, and Center for Disease Control on aspects of bio-terrorism. He served as consultant to the US General Accounting Office on Drinking Water Security, to the Office of the Inspector General of EPA on vulnerability studies, and to NSF International on POU and POE treatment devices for contaminant removal. Some of his writings are on the secure Internet site of AWWA. Dr. Deininger also serves on the Water Systems Security Committee of the Michigan Section of AWWA. He has been a consultant to several large water utilities in the USA on vulnerability assessments. Dr. Deininger has published 5 textbooks and over 100 papers. He received his education at the University of Stuttgart, Germany, Dipl.-Ing. in Civil Engineering (1958), holds a M.S. degree in sanitary engineering from Northwestern University (1961), and a Ph.D. from Northwestern University in environmental engineering (1965). He can be contacted by e-mail: rad@umich.edu

Cristiana Di Cristo (dicristo@unicas.it) is currently assistant professor at the Faculty of Engineering at the University of Cassino (Italy) and lecturer for the courses of hydraulics and environmental hydraulics for the students in civil and environmental engineering. She is a member of the Italian Group of Hydraulics (GII) and of the International Association for Hydro-environmental Research & Engineering (IAHR). She is actually secretary and meeting group leader of the Experimental Methods and Instrumentation (EMI) Committee of the IAHR. She is a registered engineer in the Province of Naples.

Dr. Di Cristo holds M.S. degrees in environmental engineering (1997) and a Ph.D. in hydraulics (2002), both from the University of Naples *Federico II*. During 2000 she was visiting researcher at the Iowa Institute of Hydraulic Research of University of Iowa (USA). From 2002 to 2005 she was post doc. researcher at the University of Naples *Federico II*. She has worked on several research projects, financed by the European or Italian governments or by private companies.

Dr. Di Cristo research activities are in the fields of water resources, fluvial hydraulics, and water distribution systems, with particular attention to water distribution systems security and system design and management.

Dr. Hiba S. Ernst is currently the director of the Threat and Consequence Assessment Division in the National Homeland Security Research Center, Office of Research and Development at the US Environmental Protection Agency (USEPA). She is leading research on characterization and sampling for building, outdoors, and water infrastructure following a chemical and biological or radiological contamination event. The research also focuses on exposure assessment for risk-based clearance goals to inform reuse and reoccupation decisions. She has a Ph.D. in environmental science from the Civil and Environmental Engineering Department at the University of Cincinnati and a M.S. in organic chemistry from the American University of Beirut. She joined the National Homeland Security Research Center in 2007 as the associate director for the Water Infrastructure Protection Division that is responsible for research in protection, detection, mitigation, and recovery from homeland security contamination events in water and its infrastructure. Prior to joining the Office of Research and Development, she worked in the Technical Support Center, Standards and Risk Management Division of the Office of Ground Water and Drinking Water. She provided technical support for the development of the microbial/disinfection byproduct regulations and has published numerous peer-reviewed articles in the field. She was the EPA Office of Water representative on the Water Research Foundation Research Advisory Council from 2002 to 2007. She is a member of the American Water Works Association (AWWA) and the Association of Environmental Engineering and Professors. Hiba Ernst can be reached at ernst.hiba@epa.gov

Randy G. Fischer is the health program manager – state liaison to local health departments for the Nebraska Department of Health and Human Services Division of Public Health. Randy Fischer started with the Nebraska Department of Health and Human Services in April 2004. Mr. Fischer was with Nebraska's Drinking Water Program as a Water Security Specialist where his primary responsibilities were to assist public water systems in developing their emergency plans of operation, as well as for securing grant money at the state level for these systems to enhance their emergency response capabilities. Prior to joining the State Randy was in law enforcement for 20 years. Currently Mr. Fischery holds the position as the program manager for the Nebraska Department of Health and Human Services – Division of Public Health, a position that he has held since January 2007. In Mr. Fischer's current position he is the State Liaison to the 21 local health departments across the state. He can be reached at randy.fischer@nebraska.gov.

William E. Hart is a distinguished member of the technical staff in the Discrete Algorithms and Mathematics Department, which is in the Computation, Computers, Information and Mathematics Center at Sandia National Laboratories. He has been involved in a wide range of projects at Sandia. His current research interests are focused on optimization techniques, such as parallel branch-and-bound, heuristic global optimization, and derivative-free local search and optimization modeling tools. He is working on optimization methods that can be flexibly applied on both workstations and on the Department of Energy's large-scale parallel computing resources. Dr. Hart has been a key contributor to a variety of software development efforts including being a member of the DAKOTA project for many years. More recently, he has focused on developing Acro, a framework for developing complex optimization libraries. Acro integrates a variety of software packages to support optimization library development. Dr. Hart is the lead developer of several of these packages: UTILIB, COLIN, Coliny, and GNLP. Dr. Hart has worked on a variety of applications for which this technology has been deployed, including

- computational biology problems such as protein folding and protein–protein alignment,
- flexible ligand docking and protein–protein docking,
- engineering design problems such as levitron stabilization and robust canister design, and
- logistics planning and resource allocation.

He is leading a team of Sandia staff members that is developing computational technologies that protect municipal water supplies. Specifically, the team is developing methods to designing and deploying early warning systems that can rapidly detect contaminants in municipal water supply networks. This project is funded by the US Environmental Protection Agencies National Homeland Security Research Center in collaboration with Argonne National Laboratory and the University of Cincinnati. He is currently manager of the Data Analysis and Informatics Department at Sandia. Dr. Hart received his B.A. from the University of Michigan in 1989 and his Ph.D. from the University of California San Diego in 1994. Dr. Hart can be reached at wehart@sandia.gov

Dan Kroll is currently chief scientist at Hach Homeland Security Technologies and principal investigator for the Hach Advanced Technology Group. Hach company is the world's largest supplier of analytical systems for water and wastewater. Dan has worked at Hach for 22 years in a variety of roles. Dan has been the lead researcher on method development projects for the physical, chemical, and microbiological quality of water and soils for which he holds several patents. Dan has developed both advanced and simplified methods for a variety of crucial water quality parameters. His simplified arsenic testing method is used throughout the world as the standard field method to screen for this toxic metal. Over 10 million of these tests have been performed on wells in South Asia and around the world protecting untold numbers of people from exposure to arsenic. Dan has a bachelors degrees in microbiology(1987) and genetics (1988) and a masters degree in water resource management and environmental engineering (2000) from Iowa State University. Dan has been awarded the R&D 100 Award for the event monitor trigger system. The award was given for developing one of the world's most innovative products in 2005. Dan has given over 100 invited presentations at various industry conferences on the analysis of water. He has written numerous research articles and chapters in books. He is also the author of the book *Securing Our Water Supplies; Protecting a Vulnerable Resource*. Dan can be reached at 970-663-1377 ext. 2637 or e-mail dkroll@hach.com.

Sean A. McKenna, Ph.D. is a distinguished member of the technical staff at Sandia National Laboratories in Albuquerque, New Mexico. Dr. McKenna's broad research interests are in the development and application of numerical models and statistical techniques to solve problems in environmental and earth sciences. Specific research interests are currently focused on time-series analysis, inverse parameter estimation, and multiphysics simulation with applications in water quality event detection, ground water flow modeling, and CO_2 sequestration. Dr. McKenna currently leads a team of researchers from Sandia and the US EPA in development of the CANARY water quality event detection software. CANARY recently received an R&D 100 Award from R&D Magazine as one of the 100 most significant technological developments of 2010. Concurrently with his position at Sandia, Dr. McKenna has held adjunct professor positions at New Mexico Tech, The University of Texas at Austin, and the National University of Singapore. He has a B.A. in geology from Carleton College (1986), an M.S. in hydrology from the University of Nevada, Reno (1990), and a Ph.D. in geological engineering from the Colorado School of Mines (1994). http://www.swcp.com/~mckenna/sean_mckenna.htm

Michael Möderl, D.Sc is a research fellow at the faculty of civil engineering at the University of Innsbruck – Austria Institute of Infrastructure and consultant for water system modeling employed by hydro-IT GesmbH. Dr. Möderl is reviewer for the journals *Water Science and Technology* (IWA Publishing) and *Water Research* (Elsevier). He obtained his master's degree in civil engineering from the University of Innsbruck (Austria) in 2006. Between 2006 and 2009, he worked as a researcher at the University of Innsbruck to commence his doctoral study under the supervision of Professor Wolfgang Rauch. The main goal of his Ph.D. was the development and application of methodical tools for water system analysis. In 2007 Dr. Möderl had an honorary appointment with the University of Exeter – UK Centre, for water systems as researcher. Dr. Möderl's postdoctoral research activities are in the fields of water system security and automatic generation of virtual water systems. He can be reached at michael.moederl@gmail.com.

Srinivas Panguluri is a senior engineer who works for the Shaw Group in the Cincinnati Office. He has over 20 years of hands-on technical and management experience in conducting a variety of applied research. His research experience covers a wide variety of environmental areas including air quality, water quality, online physical/chemical and toxicity monitoring, computer modeling/algorithms, leak detection, water security, drinking water treatment, wastewater treatment, storm water management, and water distribution systems. Many of his applied research projects, especially those conducted on behalf of the U.S. Environmental Protection Agency's (EPA's) Office of Research and Development (ORD) involve the design, implementation, and maintenance of Supervisory Control and Data Acquisition (SCADA) systems and associated Information Technology (IT) components including computer networks, databases, and web sites for online monitoring and process control.

Mr. Panguluri has been involved with the EPA National Homeland Security Research Center's (NHSRC's) water security-related initiatives in Cincinnati since the inception of the EPA NHSRC program. He also serves as a member in the American Society of Civil Engineers (ASCEs) Emerging and Innovative Technology Council. Mr. Panguluri has served as a co-author and lead author for multiple EPA-published reference guides in the water arena. He has also co-authored book chapters and a number of journal/conference publications. Mr. Panguluri holds a B.S. degree in civil engineering from Sri Venkateswara University, India (1990), and an M.S. degree in environmental engineering from the University of Toledo (1993) where he was the recipient of the Outstanding Master's Research Award from the Sigma Xi Scientific Research Society. In addition, Mr. Panguluri possesses certifications in several IT areas including Microsoft Certified Systems Engineer (MCSE), A+, SQL database administration, SAIR Linux. He is also a registered professional engineer in the State of Ohio and can be reached at srinivas.panguluri@shawgrp.com.

Craig L. Patterson is currently an environmental engineer in USEPA's Water Supply and Water Resources Division (WSWRD). Since joining U.S. EPA in 2003, he has published several papers and book chapters and has received numerous awards for his research on drinking water treatment technologies in small communities. Craig is currently editing a book entitled *Assuring Purity of Drinking Water* with Elsevier Publishing. He also served as a project engineer for environmental consulting firms (1984–2003) focusing on water and wastewater treatment processes, landfill technology, pollution prevention, remediation, and environmental software development. He has been active in the American Society of Civil Engineers (ASCE) since 1982. Craig holds a B.S. degree in civil and environmental engineering from the University of Cincinnati (1982) and an M.S. degree in environmental engineering from the University of Cincinnati (1985). He is a registered engineer in the State of Ohio and can be reached at patterson.craig@epa.gov.

Eugene W. Rice is a research microbiologist with the U.S. Environmental Protection Agency's National Homeland Security Research Center. His area of research centers on the detection and inactivation of waterborne bio-threat agents. He holds degrees from Georgetown College (B.S.), Ohio University (M.S.), and the University of Cincinnati (Ph.D.) and has received further training from the U.S. Army Medical Research Institute for Infectious Diseases. He has served as project officer on interagency agreements dealing with inactivation of bio-threat agents with the U.S. Army and the U.S. Centers for Disease Control and Prevention. He is a member of several bioterrorism workgroups with the U.S. Department of Defense and the U.S. Department of Homeland Security and is a subject matter expert for the Chemical, Biological, Radiological and Nuclear Defense Information Center. He is on the joint editorial board for *Standard Methods for the Examination of Water and Wastewater* and is a member of the American Public Health Association, the American Society for Microbiology, and the American Water Works Association. He can be reached at rice.gene@epamail.epa.gov.

Hailiang Shen (shenh@uoguelph.ca) is currently a Ph.D. candidate in water resources engineering at the University of Guelph, specializing in methodologies to select water quality sensor locations and their number, with multi-objective optimization algorithms, and contaminant source identification under nodal demand uncertainties, data mining, and parallel computing. The point of diminishing marginal return is found to be the optimal sensor number, and the toolkit GIS-CSI 1.0 is developed for real-time contaminant source identification.. Hailiang is commencing a post-doc position in geography at the University of Guelph where he will be looking at use of supercomputers to identify optimized agricultural best management practice locations on incorporating hydrologic parameter uncertainties and develop research findings into software. Hailiang earned his M.Sc. in Tianjin University, China, in June 2007. His special research interests include parallel computing (MPI and multithreads) in environmental system modeling and uncertainty analyses, multiple objective optimization, ArcGIS development, water distribution system modeling, environmental risk assessment, and BMPs design for watershed management. He has five published articles and eight conference presentations. For a complete CV, see http://www.uoguelph.ca/~shenh/.

Jeanne M. VanBriesen is a professor of civil and environmental engineering and the director of the Center for Water Quality in Urban Environmental Systems at Carnegie Mellon University. Dr. VanBriesen received her B.S. in education with an emphasis in chemistry from Northwestern University. After teaching high school for several years, Dr. VanBriesen returned to Northwestern for her M.S. and Ph.D. in civil engineering in 1993 and 1998, respectively. Dr. VanBriesen's research is in biological processes in environmental systems, including biotransformation of recalcitrant organics, modeling environmental systems involving complex biogeochemistry, and detection of biological agents in drinking water and natural water systems. Dr. VanBriesen has published 38 scientific articles and given more than 100 professional presentations. She has served on the boards of the Association for Environmental Engineering and Science Professors and the Ohio River Basin Consortia for Research and Education. Her research has been funded by the National Science Foundation, the Department of Defense, the Colcom Foundation, the Heinz Endowments, the Packard Foundation, and the Pennsylvania Infrastructure Technology Alliance. She has supervised 11 Ph.D. dissertations and 8 M.S. theses. Dr. VanBriesen has received numerous awards, including the Pennsylvania Water Environment Association Professional Research Award in 2007, the Best Research Paper in the *Journal of Water Resources Planning and Management* in 2008, the Professor of the Year for the American Society of Civil Engineers (ASCE) Pittsburgh Chapter in 2009, and the McGraw-Hill/Association for Environmental Engineering and Science Professors Award for Outstanding Teaching in Environmental Engineering in 2009. Dr. VanBriesen served on the National Research Council's Committee on Water Quality in Southwestern Pennsylvania in 2002–2004. She was a selected presenter at the National Academy of Engineering Frontiers of Engineering Education in 2010 and a selected presenter at the National Academy of Engineering Indo-US Frontiers of Engineering Symposium on Infrastructure in 2008. Dr. VanBriesen's full curriculum vitae can be accessed from her web site at http://www.ce.cmu.edu/~jeanne/

Laurie Van Leuven is currently serving as a 2010–2011 distinguished fellow in Washington DC at the U.S. Department of Homeland Security (DHS)/FEMA. She is working with FEMA on projects related to her published research on social media technologies to improve community engagement and resiliency during emergencies. She is also working with DHS's National Protection and Programs Directorate on its Enterprise Risk Management Initiative for public/private critical infrastructure owners/operators. Her home agency is Seattle Public Utilities (SPU) where she is a strategic advisor and manager specializing in emergency management and homeland security. As SPU's security manager, she is responsible for physically securing utility assets and ensuring critical infrastructure protection for the regional drinking water system for 1.4 million people in Seattle and neighboring communities as well as protecting the wastewater, drainage and flood control, and solid waste/debris management systems.

Ms. Van Leuven regularly works with multiple first responder disciplines and serves as a planning section chief on a regional, level three Incident Management Team. Her specialties include Critical Infrastructure Protection; Citizen Engagement via Social Networking; Essential Utility Service Delivery (COOP Programs); Enterprise Risk Management and Vulnerability Assessments; Strategic and Emergency Planning; Incident Management within an EOC Environment; and Sector Interdependencies (Local and State Collaboration).

Ms. Van Leuven holds a B.A. in communications from the University of Washington (1998) and a M.A. in homeland security studies from the Naval Postgraduate School, Center for Homeland Defense and Security (2009). She is a member of the Interagency Board, the International Association of Emergency Managers, and works closely with the American Water Works Association, Association of Metropolitan Water Agencies, ASIS International, and WaterISAC. She can be reached at laurie.vanleuven@gmail.com, @laurievanleuven, or http://www.linkedin.com/pub/laurie-van-leuven/11/214/261.

Name Index

A

Abbaszadegan, M., 165
Abdel-Hamid, I., 214, 233
Abdi, M., 206
Abrams, M., 286, 314
Adams, J.Q., 4, 20, 265–283
Alai, M., 89
Allen, H.J., 52
Alocilja, E.C., 217, 225
Al-Zahrani, M., 249
Angulo, F.J., 170
Antonio Canas, J., 119

B

Babayan, A., 423
Babin, S.M., 55
Baeumner, A.J., 209–210, 215, 221
Bahadur, R., 2, 16, 65–83, 87–100, 323, 355
Baranowski, T., 56
Barkdoll, B.D., 323
Bartelt, P., 119, 126
Barton, A., 457
Baxter, C.W., 355
Beecher, J.A., 142
Beniston, M., 121
Benson, P., 287
Berger, B.B., 151
Berry, J.W., 186, 249, 320, 323, 325, 337, 344–345, 355
Birkett, D., 23, 457, 482
Blöschl, G., 125
Boccelli, D.L., 186, 198
Bohman, L., 273
Borst, M., 140
Box, G.E.P., 371
Brashear, J., 120
Bras, R.L., 371, 373
Brewster, J.D., 208, 212, 214, 223
Brown, D., 273
Brown, S.B., 273
Bruins, H.J., 3
Bukhari, Z., 21, 349–366
Burkom, H.S., 55
Burr, M.D., 19, 70, 154, 165–166, 205–226
Burrows, W.D., 70, 154, 165–166
Bushon, R.N., 182
Byer, D., 196, 370–371

C

Call, D.R., 206, 212, 214
Campbell, G.A., 208, 213, 218, 220
Camper, A.K., 205–226
Carlson, K.H., 196, 370–371
Chen, C.-S., 209–211
Chen, J.C., 95
Chen, Y., 119
Christen, M., 126
Clark, R.M., 1–24, 50, 70, 135–149, 159, 163–182, 406
Connell, G.F., 158
Cook, J.B., 370
Copeland, C., 138
Cristopher, G.W., 70

D

Davies, T.R.H., 127
Deb, K., 250
Deininger, R. A., 1, 3–4, 6–8, 18, 69–70, 159, 163–182, 249, 414, 416
Deisingh, A.K., 206
Delahaye, E., 182
De Marinis, G., 397–418
Denileon, P.G., 461
Derzon, M., 208, 222
De Sanctis, A.E., 56
De Toffol, S., 121, 124
Di Cristo, C., 22, 397–418

Dorini, G., 250
Dunbar, S., 211

E
Edwards, K.A., 210
Eitzen, E.M. Jr., 70
Eliades, D., 250
Ellis, P., 285–317
Engelhardt, M.O., 123
Ernst, H.S., 1, 16, 47–62
Eum, N.-S., 220
Ezell, B.C., 120

F
Farabullini, F., 219
Feige, W.A., 137
Fennel, H., 172
Field, M.S., 165
Finke, E.-J., 159
Fisher, P., 461
Fonseca, L.P., 206–207
Fox, K.R., 47–62, 164, 175
Francy, D.S., 55, 156
Fricker, C., 53
Frundzhyan, V., 182
Fujiwara, M., 139

G
Gallardo, V., 56
Gehring, A.G., 214
Geldreich, E.E., 165, 169
Ghimire, S.R., 323
Gillette, J., 286
Gleick, P.H., 2–4, 119, 151, 166, 168, 457, 459–460
Goldberg, D.E., 249
Goldman, E.R., 211
Grayman, W.M., 173
Grigg, N.S., 140
Grimes, D.J., 205, 222, 224
Grindler, B.J., 136
Gruber, U., 119, 126
Grunow, R., 159
Guan, J., 420
Gueli, R., 250
Guestrin, C., 347

H
Haas, C.N., 186
Hahn, M.A., 210–211
Hakim, S., 1–24
Hallam, N.B., 187–188
Hall, J.S., 53, 370
Hamed, M.M., 95
Hanke, S.H., 136
Hart, D.B., 329
Hart, W.E., 77, 94, 319–346
Hasan, J., 351
Hawley, R.J., 70
Haxton, T., 56, 319–346, 369–394
Hayward, K., 460
Hedding, K., 278
Helbling, D.E., 185–201
Henriques, I.D.S., 98
Herold, K.E., 206, 225
Hickman, D., 70
Hoffman, B., 30–31
Ho, J.A., 210, 213, 215
Holland, J.H., 249
Höller, P., 121, 126
Hoover, J.E., 67
Hosni, A.A., 157
Howell, J., 278
Hrudey, E.J., 169
Hrudey. S.E., 169, 394
Huang, J.J., 250
Huang, X.J., 218
Huber, P.J., 342
Hughes, D.M., 142

I
Isenberg, D., 458
Isovitsch Park, S.L., 185–201
Ivnitski, D., 206

J
Jarrett, R., 371
Jayaram, N., 119
Jenkins, G.M., 371
Johnson, L., 266
Johnson-White, B., 221

K
Kessler, A., 249
Khanal, N., 120–121
Khan, A.S., 159
Kim, M., 420
King, K., 371
Kirmeyer, G.J., 186
Klise, K.A., 196, 374–375
Koch, M.W., 394
Kroll, D., 19, 35, 229–244, 371, 491
Kumar, A., 249
Kumar, J., 94
Kumar, M.S.M., 198

Kunze, D.R., 319
Kwon, H.T., 219

L
LaGier, M., 221
Laird, C.D., 56
Lazacka, O., 206–207, 225
LeChevallier, M., 21, 349–366
Lee, B.H., 249, 414, 416
Lence, B.J., 355
Leonard, P., 206
Leopardi, A., 397–418
Liao, W.-C., 210
Ligler, F.S., 224
Lim, C.T., 208
Lim, D.V., 207, 212, 214, 221, 225
Lindquist, H.D.A., 55, 155
Li, Y., 207, 209–210, 216–217
Lu, Q., 206, 219
Lytle, D.A., 164, 175

M
Magnuson, M.L., 59
Mala-Jetmarova, H., 457–482
Mao, X., 213, 218
Mark, O., 120
Marth, E.H., 154
Mays, L.W., 16, 462
Mazenko, R.S., 208, 212, 214, 223
McBean, E.A., 419–433
McKenna, S.A., 54–55, 196, 329, 369–394
Meeusen, C.A., 206
Mehrvar, M., 206
Meier, T., 182
Meinhardt, P.L., 159
Mitscherlich, E., 154
Möderl, M., 17, 119–132
Moied, K., 249
Moore, G.E., 291
Morales-Morales, H.A., 208–209
Morley, K.M., 325
Moser, R.H., 351
Moteff, J., 69
Munavalli, G.R., 198
Munro, N.B., 70
Murphy, B.M., 186, 268
Murray, R., 77, 319–346, 356, 369–394
Mutharasan, R., 208, 213, 218, 220
Myers, R.H., 376

N
Nakashima, E., 312
Naylor, L., 464
Ngundi, M.M., 206
Nilsson, K.A., 120, 406
Noble, R.T., 206
Nocker, A., 205–226
Nuzzo, J.B., 151, 159

O
O'Connell, H.A., 158
Ostfeld, A., 1–24, 186, 198, 247–262, 319, 322, 325, 356, 402, 410, 419–420

P
Panguluri, S., 20, 24, 285–317
Parfomak, P., 69
Parks, S.L.I., 185–201
Patterson, C.L., 4, 20, 265–283
Peckenham, J.M., 140
Perotto-Baldiviezo, H.L., 125
Phillips, C.A., 319–346
Phillips, W.R., Jr., 285–317
Pickus, J., 333
Poitras, C., 213, 219
Polaczyk, A.L., 155
Polycarpou, M.M., 250
Powell, J.C., 187–188
Prasad, T.D., 198
Preis, A., 186, 250
Propato, M., 186, 189, 198, 249

R
Radice, S., 40
Radke, S.M., 217
Raftery, A.E., 371
Rasooly, A., 206, 225
Rauch, W., 119–132
Reid, R.L., 119, 208–209, 220
Reidt, U., 208–209, 220
Renner, S.E., 70, 154, 165–166
Rescher, N., 399
Rice, E.W., 151–159, 165
Rijal, K., 223
Rizak, S.N., 394
Rodriguez-Iturbe, I., 371, 373
Rose, J.B., 205, 222, 224
Rose, L.J., 57, 157–158
Rossman, L.A., 53, 122–123, 172, 325, 405
Rubinstein, R.Y., 250

S
Salomons, E., 249–250, 252, 356, 419
Samuels, W.B., 65–83, 87–100
Sapsford, K.E., 210
Schneider, O.D., 353, 358–359

Scott, J., 47–62, 289
Seger, K.A., 29
Seidenstat, P., 143
Shang, F., 197, 420
Shannon, J., 285
Shen, H., 419–433
Shimoda, T.A., 268, 271
Sinclair, R., 154
Sivaganesan, M., 158
Skadsen, J., 325, 333
Skala, M.F., 171
Soldati, M., 121
Song, J.M., 219
Speth, T., 59
Sreepathi, S., 420
Srinivasan, K., 119
Stamp, J., 81
States, S., 16
Staudinger, T.J., 319
Stenzler, J., 120
Stevenson, A.H., 151
Stoffel, M., 121
Straub, T.M., 208, 213, 216, 220, 222
Sturdivan, G., 268
Subramanian, A., 217
Su, X.-L., 207, 209, 216–217
Szabo, J.G., 52, 58, 159

T
Taylor, A.D., 217, 222–223
Teles, F.R.R., 206–207
Thomas, N., 285
Thompson, S.L., 188, 191
Tims, T.B., 212, 214, 221
Torres-Vera, M.A., 119
Trachtman, G.B., 323
Truscott, J., 457–482
Tryby, M.E., 198
Tucker, J.B., 35, 37
Tufenkji, N., 213, 219
Tu, S.-I., 209, 213, 216, 221

U
Uber, J.G., 56, 189, 198, 356
Ugarova, N., 182

V
VanBriesen, J.M., 185–201
Van der Kooij, D., 182
Vanham, D., 127
Van Leuven, L., 27–45, 435–482
Vreeburg, J.H.G., 120
Vugrin, E., 394

W
Wallace, D., 166
Walpole, R.E., 376
Walski, T.M., 56, 198, 250, 345, 406
Wang, P., 219
Waswa, J., 218
Watson, J.P., 249, 325, 343
Webster, A.H., 182
Weilen, P.W.J.J., 182
Weisberg, S.B., 206
Weiss, J., 286, 314
Westerhoff, G., 142
Woo, H.M., 249
Wu, Z.Y., 250

Y
Yacoub-George, E., 218
Yang, L., 210
Yang, Y.J., 371
Yu, F.-C., 127

Z
Zhang, Y.Y., 208, 218
Zhao, W., 221
Zhu, P., 216, 221, 223
Zordan, M., 209, 220
Zucchino, D., 273, 278

Subject Index

2450 to 2400 BC, 166

A
Abdominal pain, 166
Abnormal traffic load, 124
Above ground structures, 42–43
Acanthamoeba, 165
Access control, 36, 42, 45, 78, 249, 301, 303, 307–309, 313, 401, 436, 445–446, 448–450, 452
Access Control List (ACL), 249
Accidental contamination, 22, 51, 164, 229, 350, 357
Accounting personnel, 269
Acidification, 58
Acidified potassium permanganate, 57
Acridine orange, 177
Activated sludge, 92, 98, 136
Activated sludge treatment, 136
Actual mobilization, 471
Actual or threatened contamination, 66
Acute febrile illness, 154
Adenoviridae, 152
Adenovirus, 165
ADSL-based networks, 415
Advanced Metering Infrastructure (AMI), 359, 365
Aeromonas spp., 165
Aerosol application, 166
Aerosols, 4, 401
Afghanistan, 166
After action reports (AARs), 453
Agency for Toxic Substances and Disease Registry (ATSDR), 70, 91
Aggregated Computational Toxicology Resource (ACToR), 31
Aging infrastructure, 7, 29–30, 39–41, 60, 142, 144, 148, 439
Aging workforce, 39–40
Agricultural, 7–8, 65, 166, 169, 247, 397, 400, 457, 459
Agricultural usage, 247
Ainsworth, NE, 112
Alarm
  event monitors, 446
  points, 436, 445–446, 448, 450
Aldicarb, 239, 353
Alpha–beta radiation detectors, 53
Alpine natural hazards, 121
Al-Qaeda, 35
Amalfi Coast, 403
American National Standards Institute (ANSI), 300
American samoa, 88
American Society of Civil Engineers (ASCE), 249, 278
American society of civil engineers interim volunteer guidelines, 93
American Water (AW), 330–331, 356
American Water Works Association (AWWA), 14–16, 51, 61, 83, 152, 248–249, 267, 278, 310, 313
Ammonia (NH_3), 92, 232, 334, 351, 353, 411
Amperometric free chlorine sensors, 192
Amperometric sensors, 192–194
Amplification, 155, 206, 208–209, 216–217, 219, 221, 460–461
Anacapri, 404
Analog outputs, 359
Analyte, 55, 206–208, 210–212, 214, 217, 219, 222, 230, 235, 364
Analytical guide, 74, 96
Anarchists, 35
Ann Arbor, MI, 333–335
Annular reactors, 58, 159
Anomalous water quality variability, 369
Anomaly detection, 94–95, 196, 200–201, 321
Ant colony optimization, 95

Antennas, 28, 41, 415, 451
Anthrax, 4, 35, 153, 159, 166, 239
Anthrax culture, 239
Anthrax spores, 58–59, 166
Anthrax surrogate spores, 58
Anthropogenic hazards, 124
Antibiotic sensitivities, 155
Antibodies, 181, 206, 208, 210–211, 214–216
Antigen-antibody reactions, 155
Anti-virus software, 249, 287
AODC, 177, 180
Appropriate cleanup, 56
Aquatic environment, 152
Aqueducts, 6, 65
Arava valley, 247
Architecture, 81, 83, 195, 197, 236, 291, 295, 301–302, 358, 408
Area of protection, 242–243
Arenaviridae, 153
Argonne national laboratory, 21, 53, 320
Arizona, 16, 353
Arizona state university, 16
Army navy air force and commandant marine corps field manual, 91
Arsenic mercury, 57
Arthropod-borne encephalitides, 153
Arthropod vectors, 154
Artificial Intelligence (AI), 95
Artificial Neural Network (ANN), 95
Assess, 10, 18, 23, 32, 44, 49–50, 78–79, 82, 224, 266, 272, 311, 329, 334, 346, 435–436, 439, 444–446, 451, 453–454, 472
Assessment methods, 39, 49–50, 71, 78–79, 120, 313, 451
Assessment of radioactivity in sewage sludge, 97
Assessment of vulnerabilities, 104
Asset classification, 451
Association of Metropolitan Sewerage Agencies (AMSA), 79, 88
Association of Metropolitan Water Agencies, 312
Association of State Drinking Water Administrators (ASDWA), 79
ASTM, 175
Astroviridae, 152
Astrovirus, 164–165
ATP bioluminescence, 18, 175–177, 180
Attack scenarios, 82, 295, 300, 357, 398–399, 401, 404–405
Australian commonwealth criminal code act, 460
Austrian, 124, 126–127
AutoCAD®, 329
Autodesk, Inc., 329
Automated Critical Assets Management System (ACAMS), 440
Automated Meter Readers (AMR), 357–358
Automated monitoring, 21, 286
Automated monitoring and control systems, 21, 286
Automated or cyber systems, 41
Automated security, 445
Automatic Meter Readers, 21, 358
Autoregressive (AR), 195, 373
Avalanche damages, 119
Avalanche hazard, 119, 126

B

Bacillus anthracis, 35, 58, 135, 152–153, 157–158, 166
Bacillus anthracis spores, 47, 135
Bacillus atrophaeus subspecies *globigii*, 55, 157–158
Bacillus globigi, 52
Bacillus species, 158
Bacillus subtilis, 57, 280
Bacillus *Yersinia pestis*, 154, 166
Backflow contamination event, 43
Backflow incidents, 350
Backflow preventers, 9, 171, 249
Background water quality signal, 52, 372, 384
Back-siphoned, 419
Back-tracking tool, 56
Bacterial enumeration, 177
Bacterial spores, 55, 156
Bacteriophage, 52
Bacterium *Vibrio cholera*, 166
Baltimore train tunnel fire, 12–13
Banking, 14, 69, 135, 349
Banking and finance, energy, 135
B. anthracis, 55, 57, 153–154, 156–159
B. anthracis Ames strain, 158
B. anthracis Sterne strain, 158
Bare rock, 126
Baseline historical data, 360
Batch mode analysis, 370
B. atrophaeus subsp. *Globigii*, 55, 157–159
Battle of the Water Sensor Networks, 322
Battle of the Water Sensors, 249, 322
Bayesian network, 420
B. canis, 154
B. diminuta, 280
Beatrice, NE, 114

Below ground structures, 42–43
Benchmarking, 462, 467
Bernoulli process, 376–377
Best objective, 335–338
Binomial Event Discriminator (BED), 372, 375–377
Biochemical characterization, 372, 375–376
Biodetection Enabling Analyte Delivery System (BEADS), 208
Biofilm, 57–58, 61, 159, 165, 224, 351
Biofilm growth, 58
Biofouling, 223, 361
Biological agent, 43, 192, 326, 332
Biological agents in warfare, 70
Biological attack, 35, 69, 159, 356–357
Biological contaminant, 9, 34, 52, 70, 146–147, 155, 192, 248–249, 279, 334, 356, 411
Biological threats, 2, 18, 159, 163
Biomass growth, 98
Biomolecules, 206
Bio-safety, 153, 155–156, 158
Bio-security capabilities, 153
Biosensors, 19, 52, 205–226, 410, 413–414, 416
Bioterrorism
  and chemical contamination, 8
  contamination, 8
  events, 151–152, 159, 363
Bioterrorism Act 2002, 287, 349
Biothreat agents, 18, 151–157, 159
Biothreat organisms, 152, 154
Biotoxin, 70, 152, 235
Bio-warfare agents, 18, 151
Bird feces, 171
Black hat, white hat exercise, 443
Blast Vulnerability Assessment (BVA) tool, 50, 61
Bleach, 58, 274
Bloody diarrhea, 166, 169, 274
Blue cell, 467
B. melitensis, 57, 153–154, 157
Bodily secretions, 154
Boil-water advisory, 170
Bonea, 406–407
Booster chlorination, 185
Booster response, 197–200
Boston, MA, 136
Bot-network operators, 293
Botulinum toxin (Botox), 35
B. pseudomallei, 57, 153–154, 157–158
Briefing form, 266
British Columbia Institute of Technology (BCIT), 297
Brucella abortus, 154
Brucella melitensis, 57, 153–154, 157
Brucellosis (*Brucella* spp.), 154
B. suis, 153–154, 157
B. thuringiensis subspecies *israelensis*, 158
Bubonic plague, 154
Building structures, 50
Bulk Electric System, 305
Bulk flow reactions, 187, 189
Bulk Parameter Monitoring, 229–230, 242
Bunyaviridae, 153
Burkholderia mallei, 57, 153, 154, 157
Burkholderia pseudomallei, 57, 153–154, 157
Burlington-Camden-Haddon, 330
Business
  benefits, 467
  continuity exercises, 469, 471–472
  Continuity Plan (BCP), 307, 464, 468–469, 471, 477–478
  network, 44, 81, 288–292, 294, 298, 308, 313, 315
  network electronic connectivity, 313
  strategies, 463–464
BVA tool user manuals, 50

C

Cabool Missouri, 168
Calcium, 58, 413
Calcium chloride, 92, 329
Calibrated water quality models, 371
Calibration, 20, 127, 172, 187, 193, 248, 330, 334, 354–355, 372, 406, 412–413
Caliciviridae, 152
Calicivirus, 164
California, 3, 6–9, 33, 166, 178, 268, 313, 353–354
California aqueduct, 7
California state government, 6
California Utilities Emergency Response Association, 7
Call centre, 33
911 Calls, 55, 354
Cambodia, 166
Campania, 403
Campylobacter jejuni, 164
Campylobacter spp., 173
CANARY, 54–55, 77, 94–95, 320, 329–330, 371–373, 376–377, 379–382, 385–389, 391–392, 394
Canine species, 154

Capital improvement projects, 302, 307–309, 316
Capri island, 397
Carbon adsorption, 281
Carrollton Treatment Plant, 273
CARVER matrix, 467
Cast iron pipes, 5, 140, 170
Catering and logistical support, 479
Cattle, 154, 173, 459
Cattle pastures, 205
C/B/R (Chemical/Biological/Radiological) contamination events, 99
C. burnetii, 154
CDC List of Select Biological Agents, 91
Cell-antibody-enzyme complex, 214
The cellular network, 415
Cellular/satellite modems, 359
Cement-lined, 57–58
Center for Disease Control (CDC), 70
Central IT, 290, 292
Centrifugation, 208–209, 213, 217, 225
Centroid, 374–375
Cesium, 58
Chain of command, 155, 266
Chain of custody, 155
Change points, 371
Cheaper components, 291
Chemical
- abstracts service (CAS), 74
- attack scenarios, 357
- biological or radiological (CBR), 16, 48, 69, 73, 90, 98, 319, 399
- contaminants, 59, 147, 185–186, 192, 279, 281, 356–357, 411
- contamination, 8, 33, 59, 147, 185–186, 192, 279, 281, 327, 331, 334, 356–357, 411, 461
- cyanide, 192
- facilities, 31, 89, 442
- Facility Anti-Terrorism Standards (CFATS), 442
- and microbiological threats, 2, 18
- or biological agents, 43, 164, 185, 192, 397, 404, 419
- and physical stability, 70
- properties, 410
- sector, 441
- spills, 104
- threats, 163
- treatment facilities, 41
- warfare agents, 235
- and Water Security Act of 2009, 442
- weapons convention, 71

Chemiluminescent, 207, 210, 213, 218
Chicago, IL, 136, 151
Chills, 165
Chinese cities, 166
Chloramines, 185, 352
Chlordane, 57–58
Chloride (Cl^-), 52, 92, 140, 329, 334, 351, 353, 413
Chlorination, 57–58, 136, 153, 158–159, 169–170, 173, 185, 282, 442
Chlorine concentration, 18, 137, 186–197, 200–201, 330, 384
Chlorine decay, 187, 190
Chlorine dioxide, 157, 185, 282, 352
Chlorine residual, 9, 18, 52, 58, 168, 185–201, 232, 272, 320, 330, 411, 414
Chlorine Residual Modeling, 186–192
Chlorine sensors, 18, 52, 186, 192–197, 200–201
Cholera (*Vibrio cholerae*), 152
Cincinnati, OH, 56, 334
Cincinnati Pilot Post-Implementation, 328
CIP Reference Documents and Archives, 313
City of Cincinnati Water Works, 354
Cleaning agent, 419
Clean Water Act of 1948, 136
Clean Water and Drinking Water Infrastructure Gap Analysis, 144
Clean Watershed Needs Survey (CWNS), 147, 149
Climate change, 3, 119, 136
Closed circuit television (CCTV), 42, 78, 408, 444, 446, 448
Close Target Reconnaissance (CTR), 467
Coast Guard, 13–14, 91
Cobalt, 58
COD removal, 98
Collection system, 2, 30, 78, 87, 97–99, 143, 148, 457–458
Colorado River, 143
Colorimetric, 192–195
Colorimetric sensors, 192–194
Columbia River, 208, 216
Combined and Separate Sewer Overflows, 138
Commercial airlines, 349
Commercially available herbicides, 235
Commercially Available Sensors/Detectors, 52–53
Commercial software packages, 192
Communications, 7, 11–14, 22, 56, 61, 79, 81, 91, 99–100, 105, 107, 110, 114, 135, 243–244, 266–271, 288–289, 293, 301, 303, 310, 319, 329, 333,

335, 359, 415–416, 453, 462, 469, 471, 474–478, 479–481
Community water supply systems, 4, 135, 139
Community water system (CWS), 8, 39, 104, 106–108, 139–140, 144–145, 185, 287
Comparative efficacies, 56
Comparing methods effectively, 346
Compliance, 60, 81, 137, 146, 231, 301, 304–305, 307, 314–316, 442, 455, 470, 473, 478
Comprehensive National Cybersecurity Initiative, 312
Compromise utility design (CUD), 330–331
Computer, 4, 12–15, 20, 41, 44, 54, 67, 135, 146, 230, 244, 252, 285–288, 291, 294, 296, 303, 308, 312, 319, 323, 356, 399, 423, 449, 468, 474
Computerized maintenance management systems (CMMS), 290–291
Computer systems, 4, 13, 135, 312, 319
Concentration methods, 208–209
Conditional logistic regression, 174
Conductivity, 53, 90, 92–93, 100, 192, 231–232, 320, 328–329, 334–335, 353–355, 369, 378–379, 382, 411–414
Confirmation of positive event, 76
Confirmatory detection, 199
Confirmatory testing, 230
Congressional Budget Office (CBO), 69, 144
Congressional Research Service, 71, 91
Connecticut, 353
Consequence assessment, 44, 325
Consequence Management Plan (CMP), 22, 49, 322, 354, 356, 362–365, 372
Constant potential, 192
Construction incidents, 124
Consumer complaint surveillance, 72, 321
Consumer health, 185, 355, 359, 363, 366
Consumers base demands (BD), 262
Consumption, 92, 108, 137, 189, 253, 268, 336, 349
Contact, 108, 114, 153–154, 194, 208, 214, 222, 269, 271, 277, 400–401, 440, 447, 449, 452, 476, 480
Containment and Mitigation, 56
Contaminant analysis, 356
Contaminant fate and transport, 56
Contaminant intrusion, 10, 18, 197, 200, 250, 365, 405, 419
Contaminants of concern, 57, 59, 69–71, 74, 90, 99, 192, 278, 321, 370, 401
Contaminant source identification (CSI), 23, 419
Contaminant warning system (CWS), 53, 243, 419
Contamination
  detection, 322, 410–411, 416
  Threat Management Guide, 74, 96
  time, 406
  warning system (CWS), 21, 43, 49, 51, 61, 72–73, 250–251, 319–349, 354, 369
Contingency planning, 51, 99, 452
Continuity of Operations Plan (COOP), 31, 33, 453
Continuity threats to workforce and infrastructure, 39
Continuous, 9, 52–53, 89, 94, 100, 106, 194–196, 214, 218, 222, 231, 234, 238, 292, 320, 369, 412
Control Systems Cyber Security Self Assessment Tool (CS2SAT), 99, 311
Control Systems Security Program (CSSP), 294, 311
Copper, 52, 415, 439, 450
Corporate governance, 465–466
Corroded iron pipes, 58
Corrosion, 57–58, 140, 146, 159, 170, 333, 351, 361, 365
Corrosion by products, 361
Cost
  effective, 52–53, 61, 90, 198, 234–235, 279, 333–334, 355, 402, 436–437, 451
  estimates, 12, 278
Counter electrode, 192
The CounterPlayer, 470–474, 476, 481
CounterPlayers Brief, 475–476
CounterPlayer Team, 473, 475
County, 3, 32, 106, 108–109, 111–112, 168, 267–268, 288, 329, 455
Coverage characteristics, 241–244
Coxiella burneti, 154
C. parvum, 55, 156, 165
CP/IP, *see* Transmission control protocol/internet protocol (TCP/IP)
Cramps, 165
Crime Prevention Through Environmental Design (CPTED), 444

Crime scene evidence, 17, 103
Criminal activities, 435, 450
Criminal attack, 405
Criminal reports, 33
Criminals, 39, 398, 439
Crises as opportunities, 467
Crisis control, 23, 462, 464
Crisis exercise, 23, 468–482
Crisis leadership, 23, 465–468, 482
Crisis Management, 464–468
Crisis Management Team, 472, 481
Critical Cyber Asset Identification, 305
Critical Foundations, 285
Critical Infrastructure Protection Advisory Council (CIPAC), 48, 61
Critical Infrastructure Protection (CIP), 1, 7, 11, 22, 48, 50, 61, 82, 135, 285, 300, 397, 418, 435
Cross connections, 43, 172, 185, 342, 350, 361, 419
Cross contamination, 268, 279
Cross Entropy, 250
Cross-sectional study, 174
Cryptosporidium, 55, 156, 164–165, 175, 280, 282, 459–460
Cryptosporidium contamination, 164
Cryptosporidium Oocysts, 280, 282
Cryptosporidium Outbreak, 175
Cryptosporidium parvum, 156, 165
Cryptosporidium spp., 157
CSO, 123–124, 128–129, 143
CT values, 58, 156–158
Cult, 3, 152, 168
Cultural procedures, 155
Cumulative distribution function (CDF), 190, 376, 383
Customer complaint, 319
Customer's service connection, 355, 357
Cuvette, 176, 218
Cyanide, 35, 52, 92, 192, 235, 238–239
Cyber Access Controls, 313
Cyber attack, 310–311, 315–316, 399, 461
Cyber Attack Tools, 299–300
Cyber-dimension, 285
Cyber incident, 10, 311, 314
Cyber infrastructure, 287–290, 293–300, 307
Cyber infrastructure connectivity, 289–291, 293
Cyber security, 10–15, 21, 24, 44, 80–83, 99, 136, 285–317, 454
Cybersecurity Center, 286, 311
Cyber Storm, 311
Cyber Storm I, 311
Cyber Storm III, 311
Cyber threats, 33, 37–38, 293–294, 296, 311–312
Cyber vulnerabilities, 21, 38, 286, 297, 302, 312

D

Daily demand, 127
Dakota City, NE, 114
Dallas, TX, 274
Dalles, Oregon, 3, 152, 168
Damaged reservoirs, 268, 271
Damage to reservoirs, 67
Damage Survey Reports, 272
Daphnia magna, 413
Data analysis, 54, 319, 329, 360
Data analysis methods, 319, 369
Data management, 359–362
Data mining, 23, 95, 250, 360, 370, 420–422, 427, 429, 432
Data mining techniques, 95
Data sets, 22, 95, 125, 186, 371, 378–382, 384–386, 392, 394
Data transmission, 360, 398, 408, 410, 414–416
DBIE, 252, 255, 258
Debris flow, 121, 125–126, 130–131
Decision support, 75, 346, 398
Decision Support System (DSS), 95–96, 398, 409
Decontamination of buildings, 58
Decontamination methods, 57
Decontamination of Wash/Waste Water, 49
Delay, 22–23, 60, 78, 104, 253, 255–257, 259, 334, 356, 387–393, 407–408, 444–445
Delayed action, 70
Demand, 17, 107, 119, 122–123, 127, 132, 146, 169, 172, 174, 187–197, 229, 232, 249, 252–253, 258, 292, 333, 337, 355, 404, 406–407, 411, 419–420, 423, 428, 432, 461
De-Militarized Zone (DMZ), 307
Denial of service scenario, 29
4-Deoxynivalenol, 166
Department of Defense (DoD), 10, 71, 281, 285–286
Department of Energy, 82, 97, 441
Department for Environment, Food and Rural Affairs. UK – Strategic national guidance, 97

Department of Homeland Security (DHS), 10, 15, 27, 47, 61, 89, 286, 294, 310, 361, 439, 441–442, 452–453, 455
Design Base Threats (DBTs), 398, 404–405
Designers, 19, 66, 224–225, 248, 278, 304
Designing Sensor Networks, 322
Design and Renovation of Drinking Water Systems, 51
Design/upgrade of water systems, 76
Destruction of critical assets, 66
Detect
  to protect, 230
  to treat, 230
  to warn, 230
Detectable disinfectant residual, 189
Detected events, 389, 423
Detection
  class requirements, 230
  likelihood (DL), 94, 253, 255–257, 259, 262
  method, 94, 154, 175, 236, 298, 308, 371, 401
  methodology, 154
  response modeling, 251
Deter, 23, 43, 78, 198, 350, 438, 444–445, 450, 452
Deterrence, 78, 407–408, 442, 444
Developed baseline threat information, 72
Development of Strategies Concerning Animal Feeding Operations, 138
Dewaterability, 98
Diacetoxyscirpenol, 166
Diarrhea, 165–166, 169–170, 274
Diazinon, 52
Diesel fuel, 57–58
Difficulties with detection, 70
Difficult recognition of poisoning, 70
Digital data outputs, 359
Digital video recorder (DVR), 446
Diminishing marginal return, 339, 423, 427–428, 432, 494
Direct attack, 248
Direct contact, 153–154, 480
Direct sensors, 369
Discrete (Grab) Sampling, 222–223
Disinfectant Research, 282
Disinfection, 20, 57–58, 107, 114, 136, 146, 152–153, 156, 158–159, 185, 197–200, 205, 265, 267, 277, 279, 282, 365, 411, 459
Disrupted railways, 268
Disruption of Service, 38, 51, 67, 81, 300
Dissolved oxygen, 334–335, 351, 353–355, 411, 413
Distributed Control Systems, 288, 301
Distribution System Modeling, 2, 20–21, 61, 76, 186–187, 199–200, 494
Distribution systems, 2, 5–6, 9, 19–20, 23, 35, 37, 43, 49, 51–55, 57, 61, 72, 76–77, 96, 120, 125, 138, 140, 159, 174, 185–201, 239, 242, 247–262, 276, 279–281, 294, 322–323, 325, 342, 349, 362, 365, 369, 371, 377, 402, 419, 457–458
District of Columbia, 88
DNA, 206, 212–213, 216, 220–221
Dose-response relationships, 356
Double Jeopardey, 477
Downslope hillsides, 126
Drinking Water Infrastructure, 20, 138–142, 144, 265, 268, 278
Drinking Water Security Program, 103
Drinking Water State Revolving Fund, 105
Drinking Water System Assessment Survey, 277
Drinking Water Treatability Database (TDB), 70, 91
Drinking water and wastewater treatment, 68, 75
Drought, 3, 105–106
The Dual Threat, 39–40
Dual Use, 231–234, 239, 241, 322
Ductile iron, 57–58, 140
Dulcitol negative *Salmonella* serovar *Typhimurium*, 170
DVC method, 177
Dynamic duration, 479
Dynamic Frequency, 415
Dynamic response concept, 408–409

E

Early Notification of Community-based Epidemics (ESSENCE), 55, 61
Early warning indicators, 75
Early Warning System (EWS), 19, 89–96, 99, 229–244, 297, 311, 350, 360, 408–409
Earthen dams, 5
Earthquakes, 1, 3–9, 24, 29, 33, 43, 66, 81, 104–106, 266, 268, 271
Ease of use, 233, 439
East Bay Municipal Utilities District (EBMUD), 6
East valley water district, 268
E. coli O157:H7, 165, 168, 173–174, 205–226

Economic impacts, 60, 66, 72, 243, 319–320, 326–327
Economic risks, 400, 460
Ecoterrorists, 35
Edgewood chemical and biological center, 97
EDS Analysis, 377–391
Efficient disaster response, 199
EINSTEIN Program, 311
Electrical conductivity, 53, 320, 369
Electrical grid, 136
Electrical power, 13–14, 274, 276, 279
Electricity, 7, 14, 41, 68, 80, 313, 349, 413, 460
Electricity Sector (ES), 313
Electrochemical signal, 207, 219
Electrode, 192–193, 208, 214, 217, 353, 412–413
Electronic, 12, 37–38, 41–42, 45, 55, 61, 66–67, 78, 207, 225, 278, 287, 301, 305, 312–313, 358, 439, 444–445, 447–448, 460
Electronic control systems, 67
Electronic signal, 207
The Electronic Surveillance System, 55
Elevated tanks, 170, 333
Elevated water towers, 50
Eligible Receiver, 285
ELISA, 214, 255
Elizabeth, New Jersey, 136
Emergency Management Agencies, 267, 452, 455
Emergency Management Assistance Compact (EMAC), 267–268
Emergency Management Programs, 438
Emergency Operations Plan, 107, 269–270, 272
Emergency Response
  efforts, 20, 265, 273
  procedures, 265, 270–272, 419, 463
Emergency Response Plans (ERP), 79, 106–107, 109, 362, 364
Emergency Response Protocols (ERP), 37, 79–80, 99, 108, 287, 315, 349–350, 362, 463
Emergency room visits, 55
Emission, 97, 123–124, 177, 210–212, 410
EMPACT, *see* EPA Environmental Monitoring for Public Access and Community Tracking (EMPACT)
Encysted protozoa, 152, 157–158
Endex, 471, 481
Energy, 20, 28–29, 68–69, 82, 97, 135, 143, 157, 213, 219, 274, 281, 286, 292, 313, 441
Energy sector, 68, 313, 441
Enhanced bio-safety, 153
Enhanced security monitoring, 52, 72, 321–322
Entamoeba Histolytica, 165
Enteric viruses, 156, 164
Enterobacteriaceae, 152
Environmental impacts, 35, 75
Environmental Protection Agency (EPA), 1, 8, 16, 21, 24, 27, 47, 56, 61, 66, 70, 72, 88, 103, 126, 187, 192, 437, 439
Environmental Response Laboratory Network (ERLN), 55, 61, 363
Environmental Response Technical Assistance Document for Bacillus anthracis Intentional Releases, 58
Environmental Technology Verification (ETV), 72–73
Environmental Technology Verification (ETV) Program, 281, 350
EPA Drinking Water Treatability Database, 91
EPA Environmental Monitoring for Public Access and Community Tracking (EMPACT), 331–333
EPANET, 53–54, 56, 75–77, 171–172, 187–188, 197, 243, 249, 252–254, 325, 330, 335, 337–338, 343, 405, 416, 420, 423, 428
EPANET2, 123
EPANET-DPX, 77
EPANET-MCX, 77
EPANET-MSX, *see* EPANET Multi-Species Extension (EPANET-MSX)
EPANET Multi-Species Extension (EPANET-MSX), 77, 187, 197
EPANET-PBX, 77
EPANET-RTX, 77
EPA Region VI emergency response center, 274
EPA's disaster recovery plan, 20, 265
Erroneous measurements, 354
Escherichia coli, 52, 164, 192
Escherichia coli O157:H7, 19, 165, 168, 173–174, 205–226
Escort provisions, 447
Estimated events, 386–387, 393
Estimation error, 373, 377
Ethernet, 296, 359
ETV Program, 72, 281, 350–351
European CORINE database, 125

European Programme for Critical Infrastructure Protection (E.P.C.I.P.), 22, 397, 418
Evaluation of alternatives, 401–402
Event
  detection algorithms, 94–95, 192, 196, 369–394
  detection results, 375, 385–389, 392
  detection system (EDS), 22, 54, 94, 320, 362, 369, 372
  intensity, 406
  Monitor™ Trigger (Hach), 361–362
Evolving Threat Environment, 29–30, 39
EWS, *see* Early Warning System (EWS)
Excavators, 119
Exercise
  aim, 477–478
  communications and control, 480–481
  controller, 470–472, 475–476, 479, 481
  documentation, 474–477
  identity, 475, 478–479
  manager, 472, 475, 478–481
  master document, 474–475
  planning, 477
  process, 474, 479–480
  report, 471, 474, 479, 481–482
  responsibilities, 471
  review, 481–482
  roles, 472
  scope, 478
  scribe, 474–475
  script, 472, 474–476, 478
  timing, 479, 481
Existing countermeasures, 436
Experienced facilitator, 468
Expert opinion, 323
Extent of contamination (EC), 49, 272, 322, 336–337
External Threats, 34–35, 451, 462

F
Failure of Equipment, 105
False alarms, 53, 55, 94, 194, 238, 354, 360, 371, 375, 377, 381–382, 394, 445, 447
False negative rate, 23, 237–238, 423, 426–428, 430, 432
False positive, 22–23, 43, 89, 199, 201, 205, 220, 223–224, 237–238, 240, 308, 336, 353, 387–388, 391–394, 416, 421, 432
FAR, 387–388, 393
Fast, flexible solvers, 325
Fast Fourier Transform (FFT), 374
Fate and transport, 56, 59–60, 74, 187, 243, 355
Fecal contamination, 155, 173
Fecal indicator organisms, 205
Federal agencies, 33, 48, 60, 67, 266, 268, 296, 301, 311–312, 350, 438–439
Federal Bureau of Investigation (FBI), 67, 311, 449, 452, 455
Federal Emergency Management Agency (FEMA), 266, 278
Federal Energy Regulatory Commission (FERC), 441
Federal Information Management Security Act (FIMSA), 288, 301
Federal Response Network, 363
Federal Water Pollution Control Act, 88, 137
Femtocurie, 3, 168
Fences, 42, 45, 349, 401, 408, 445
Fever, 35, 151–154, 165–166, 274
Field assessment crews, 269
Field detection and analysis information, 75
Field manual: treatment of chemical agent casualties and conventional military chemical injuries, 71
Filoviridae, 153
Filter, 12, 136, 166, 176–177, 209, 214, 220, 274, 372–375
Filtration, 59, 136, 146, 176–177, 185, 208–209, 212, 220, 225, 277, 279, 281, 458–459
Filtravette, 176
Finance/administration, 266
Financial penalties, 119
Fingerprint, 244, 361
Finished water reservoirs, 65
Finished water storage facilities, 138
Fire, 4, 6–7, 9, 12–13, 37, 43, 51, 81, 105–106, 110, 136, 138, 163, 171, 174, 267–268, 273–274, 281, 329, 333, 363, 400, 440, 452, 454
Fire District, 267
Firefighting, 29, 48, 67, 268, 286, 349
Firefighting purposes, 349
Fires, 66, 104, 136, 269, 400
Firewalls, 249, 306–309
Flexible polyurethane hose, 6
Floating debris, 136
Flocculation, 98
Flooding, 1, 31, 33, 43, 67, 104, 119, 121, 123–125, 127, 129–131, 268, 273, 276, 300, 459
Flooding hazard, 125
Floods, 4, 81, 104, 106, 109, 266, 269, 457

Flood warnings, 271
Florida, 178, 353
Flow, 6–8, 19, 35, 51, 53, 56–58, 66–67, 76–77, 80, 92, 95, 110, 121, 124–127, 130–131, 142–143, 148, 163, 168, 170, 174, 187–191, 194, 198, 200, 208, 213, 215, 217–218, 222, 236, 244, 249, 253, 258, 262, 272, 288, 308, 324, 336–337, 351, 358–359, 372, 404, 420, 423, 428, 472, 480
Fluctuations in demands, 369
Fluence, 157
Fluorescent dyes, 210
Fluorescently labeled, 208
Fluorescent rhodamine label, 210
Fluoroacetate, 239
Flushing, 56–58, 61, 63, 96, 169–173, 218, 232, 417
FN, 387, 393
Food and Drug Administration (FDA), 70, 206
Food industry, 206, 216, 225
Foreign intelligence services, 294
Foreign nationals, 349
Forests, 125
FP, 387, 393
Frame relay mode, 359
Francisella tularensis, 57, 153, 157
Free chlorine, 53, 57, 77, 156–158, 185–186, 189, 192–197, 277, 328, 334, 352, 354–355, 383–384
Free/total chlorine, 351, 353
Free/total chlorine residual, 52
F. tularensis, see Francisella tularensis
Fungi, 166
Fusarium, 166
Fusion center, 32
Future Investment in Drinking Water and Wastewater Infrastructure, 144
Future Mitigations, 462–463
Fuzzy Sensors, 251

G

Game changer, 287
Gaseous chlorine, 287
Gastroenteritis, 165, 170, 173–174
Gene chip biosensors, 205
General equipment, 50
Generator and roadway failures, 268
Generators, 146, 269–270, 272–273, 276–277, 444
Genetic algorithm (GA), 95, 249–250, 253, 258–259, 420
Genetic algorithm scheme, 249
Genosensors, 206–207, 220–221, 225
Geographic information system (GIS), 23, 290, 329, 420, 427
Geospatial mapping tools, 42
Gestione Ottimale Risorse Idriche (GORI), 402
Giardia lamblia, 137, 165, 282
Giardia spp., 152
Gideon Missouri, 164, 169–173
Girsu, 2, 166
GIS, *see* Geographic information system (GIS)
GIS methodology, 120
GIS toolkit, 420–421, 427, 429, 432
Glanders, 153–154
Global climate change, 3, 136
Global optimum, 420
Global positioning system (GPS), 269, 421, 432
Global water conflict tends, 460
Gloversville, NY, 136
Goats, 154
GoogleEarth, 42
Government agencies, 20, 31, 48, 265, 283, 454
Grab sample, 186, 193–195, 222–223, 234, 354, 402
Grab sampling, 223, 231–232, 236
Gragnano tank, 406
Grand island, NE, 112
Graphical user interface (GUI), 427, 429
Gravity dams, 50
Greater cincinnati water works, 328–329
Great Plains, 3
Great southern california shakeout 2008, 268
Greedy heuristic-based algorithm, 249
Grid/looped system, 139
Ground tanks basins, 50
Ground water sources, 5, 42, 65, 282
Guam, 88
Guelph Ontario, 419
GuideLines Enhancement, 397
Gulf of Naples, 403
Gulf of Salerno, 403

H

Hach's GuardianBlue®, 361
Hackers, 285–286, 291, 293, 299
Hanshin-Awaji region, 5
Hartington Creighton, NE, 211
Hastings, NE, 114
Hazard assessment, 125–127

Hazard events, 120, 463
Hazard identification vulnerability analysis (HIVA), 31–32, 394
Hazard maps, 119–121, 126–127, 131
Hazardous substances, 185, 239
Hazard zone maps (HZM), 125
HAZMAT, 363
Health and safety issues, 275
Health and welfare, 265, 283
Heating ventilation and air condition units (HVAC), 333
Heavy metals, 235, 361
Helix delta-Q, 75
Hemorrhagic fevers, 153
Hepatitis A, 165, 275
Hepatitis A/B, 274, 278
Hepatitis E, 165
Herbicides, 52, 235, 353
Heroin, 235
Heuristic, 240, 249, 325–326, 344
High-impact contamination, 89, 342
High-impact Incidents, 335, 342–343
High-Level Threat, 36–37, 39
High performance radio LAN (HIPERLAN), 415–416
High probability threats, 153
High toxicity, 70
High variability, 330–331, 356
Hiperlan devices, 415
Historical data associated, 33
History window, 374
Hit Rate, 238
Holarctica, 153
Holdrege, NE, 112
Hollow-fiber filters, 55
H2OMAP, 187
H2OMAP/H2ONET, 75
Homeland Security
  act of 2002, 442
  appropriations Act of 2007, 442
  Presidential Directive-7 (HSPD-7), 69, 88–89, 301, 349
  Presidential Directive-9 (HSPD-9), 51, 350
  Presidential Directives (HSPDs), 47, 135, 455
  Research Program Multi-Year Plan (MYP), 48–49
  strategy, 1, 24, 47, 135
H2O Sentinel™ (Frontier Technologies), 361
Hospital water systems, 271
Host-based intrusion protection systems (HIPSs), 308
Hot wash-up, 479, 482
Hourly demand, 174
Household, 5, 20, 65, 265, 272, 279–281
Household devices, 20, 265
HPC, 18, 175, 177–178, 180–181, 351
HPC method, 18, 175
H. pseudoflava, 280
Human-caused Incidents, 32–34
Human health risk, 208
Human interference, 457–458
Humidifiers, 400–401
Hunter Watertech, 286
Hurricane Andrew, 268
Hurricane Hugo, 268
Hurricane Katrina, 4, 13–14, 20, 30–31, 33, 265, 273–279
Hurricane Rita, 277
HVAC, 333
Hybrid, 206
Hybridized, 208, 213, 216, 218
Hydraulic characteristics, 172, 190
Hydraulic control elements, 19, 247
Hydraulic models, 22, 53, 75, 169, 172, 187, 232, 355–356, 365, 406, 409
Hydraulic radius, 187
Hydraulic/water quality network models, 76
Hygiene, 275, 349
Hypochlorite solution, 58

I
Ice storms, 33, 105, 109, 266, 271
ICS, *see* Incident Command System (ICS)
Idaho, 88, 310, 312, 314, 316
Idaho National Laboratory (INL), 312, 314, 316
Identification and verification, 356
Identifying location of contamination, 76
Identify, Screen and Treat Contaminants to Ensure Wastewater Security, 91, 99
IEEE 802.11 (WiFi), 415
Illater, 2, 166
Immunological procedures, 155
Immunomagnetic beads, 214, 216
Immunomagnetic separation (IMS), 208–209, 213, 216, 225
Immunosensors, 206–207, 211–213, 216, 218–221, 225
Implementation Project, 288
Improving the quality of input data, 345
IMS, *see* Immunomagnetic separation (IMS)
Inaccurate meters, 168–169
Inactivation of Anthrax Spores, 58–59

Inactivation of Biothreat Agents, 57
Inadequate engineering design, 457
Incident Command System (ICS), 83, 266–267, 288, 302, 308, 363, 453
Incident and Emergency Exercises, 470–472
Incident management, 106, 266–267, 301, 363, 453–455, 464, 471–473, 475
Incident management team, 267, 455, 472–473, 475, 495
Incident reports, 33, 305, 314, 446, 450, 452
Incidents of national significance, 30
Inclinations, 126
Incubation period, 166, 174, 177
Individual homes, 279
Individual isolates, 155
Industrial, 7, 65, 69, 92, 98, 107, 143, 148, 191, 233, 235, 247, 288, 297–299, 302, 311, 313, 397, 400, 457, 460–461
Industrial Control Systems (ICS), 83, 288
Infected population, 169
Infection risk, 165
Information and Communication, 135, 415
Information technology (IT), 21, 38, 79, 81, 286–288, 290–293, 300–301, 310
InfoWater™ Protector, 75
Infrastructure decontamination, 48, 56–60, 75
Infrastructure failure, 39–40, 457
Infrastructure Vulnerability Assessment Model (I-VAM), 120
Ingestion, 153–154, 164, 243, 400–401
Inhalation, 153–154, 280, 400–401
In-house Monitoring, 445–446
Injection, 9–10, 193–194, 198–199, 213–215, 218, 248–254, 256–259, 262, 324, 329, 336, 339, 345, 356, 383, 406, 419–422, 424
Inoculum preparation, 154
Inorganic contaminants, 52, 58
Inside job, 300
Insiders, 34, 37, 286, 294, 409, 450, 475
Institute and the Security Incidents Organization, 296
Institutional and Management Issues, 2, 21–22
Instrumentation and Control Systems, 288
Instrumentation packages, 229
Intakes, 9, 42, 78, 105, 124, 146, 419
Integer programming scheme, 249
Integrated System, 89, 397, 446
Intelligence Activities, 449
Intelligence reports, 33, 449
Intelligent algorithms, 230
Intentional attacks, 16, 22, 28–29, 397–418, 438, 443
Intentional contamination, 22, 51, 72, 75, 96, 159, 175, 185, 192, 200, 229, 319, 355, 369, 406
Intentional intrusions, 186
Intentional release of toxic chemicals, 66
Interagency Steering Committee on Radiation Standards (ISCORS), 97
Interdependent Infrastructure Failures, 40
Internal Threats, 36–37
International Electrotechnical Commission (IEC), 300
International Instrument User's Association (EWE), 306
International Organization for Standardization (ISO), 300
International Society of Automation (ISA), 288
Internet, 10–14, 24, 34–35, 42, 91, 100, 244, 291, 293–294, 297, 299–300, 308, 311–312, 316, 477
Internet disruptions, 10–11
Internet Recovery, 10–11
Internet Storm Center, 297, 316
Interpretive Algorithms, 242
Interruption of services, 51, 66
Interstate carriers, 136
Interstate participation, 267
Interstate Quarantine Act of 1893, 136
Intrinsic Vulnerability maps, 120–121
Intrusion Detection Systems (IDS), 82, 295, 308
Intrusion Protection Systems (IPS), 308
Intuition, 242, 435, 443
Investigations, 52, 54, 67, 120, 127, 155, 168, 170–171, 173, 236, 239–240, 270, 275, 286, 311, 314, 329, 359, 446–447, 450
In vitro cultivation, 155
Iran, 14–15, 286
Iraq, 166
Iron, 5, 57–58, 140, 170, 209, 333, 413
Iron oxide, 209
ISA SP-99, 303–304
Island of Capri, 403–404
Islands of automation, 289, 292
ISO/IEC 27002, 300–301
Israel, 3, 178, 247, 286
Italy, 22, 397, 403–404, 459
Iterative deepening of pareto solutions, 250
IT Infrastructure, 285, 288, 290

J

Japanese Army, 166
Johnson Atoll, 88
Jordan, 3

K

Kansas, 3
Kearney, NE, 112
Keyless entry, 350
Kismet, 299
Klebsiella pneumonia, 58, 159
Kobe City earthquake, 5
Kobe, Japan, 1, 3

L

Lab analysts, 269
Label-free, 207, 223
Laboratory information management system (LIMS), 290–292
Laboratory Response Network (LRN), 156
Lagomorphs, 154
Lagrangian, 344
Lag spacing, 374
Lake Pontchartrain, 274, 276
Landslide hazard, 119, 125–126
Land use, 119, 124–126, 148
Land use change, 119, 124
Lansing, MA, 175
Laos, 166
Large Networks, 201, 325, 344–345, 420
Large-scale problems, 345
Large transmission water lines, 40
Law Enforcement, 17, 21, 32–33, 78, 103, 106, 109–112, 117, 155, 312, 321–322, 437–438, 440, 445–447, 449–450, 452–454, 490
League of Nebraska Municipalities, 105
Leaks, 21, 106, 168–169, 271–274, 307, 314, 350, 492
Learning after others mistakes, 467
Leased lines, 359
Lectin, 206, 219
Legionella spp., 165
Lethal doses, 3, 168, 236, 356
Levinson–Durbin (LD) recursion, 374
Lexington, NE, 112
Licensed radio frequencies, 359
Lifecycle, 292
Lincoln, NE, 112
Linear filter, 372–374
Linear least squares, 373
Linear prediction-coefficient filter (LPCF), 372, 375, 377, 379–382, 386, 388–389, 391–392
Liposomes, 207, 209–211, 213, 215–216, 221, 274
Live Exercises, 469, 471
Live Participants Brief, 475–477, 481
Livingston Baptist Church, 275
Livingston, LA, 274–275
Loads of industrial chemicals, 98
Local Area Network (LAN), 288, 359, 415
Local Conditions, 93, 269, 420
Local Exchange Carrier (LEC), 291
Local restaurants, 152
Local Village, 267
Locked doors, 408
Loggable packets, 299
Logistics, 266, 469–470, 473, 475, 479, 491
Log keeper, 474
Loma Prieta earthquake, 5
Long-term support, 265, 272–273
LOS, 252, 255–258
Los Angeles, California, 3, 5, 7, 166, 459
Loss of pressure, 7, 146, 163, 276, 279
Low Chlorine, 18, 197–200
Lower values, 381
Low false negatives, 240
Low false positives, 240
Low Level Threat, 36–37, 408
LSD, 235
Luciferase, 177, 413
Luminometer, 175–178, 180, 215
Lyons, NE, 114

M

Magnesium, 58, 413
Magnetic microbeads, 208, 211
Maintenance & Operation, 90
Maintenance and Calibration, 354–355
Malaria, 164
Malathion, 52
Mandates, 20, 286, 439–442, 455
Manufacturing and Control Systems and Industrial Automation and Control Systems, 288
Maroochy Shire Sewage Treatment Plant, 21, 286
Massachusetts, 88
Massa Lubrense, 404
Mass consumed (MC), 336, 338
MATLAB®, 371
Matrix of potential hazards, 268

Maximum contaminant level goals (MCLGs), 137
Maximum coverage, 414, 416
MCSIM, 77
Mean time between false alarms, 238
Mean-zero Gaussian distribution, 373
Measurable signal, 206
Mechanical component failure, 370
Media amplification, 460–461
Medical and toxicity information, 74
Medium-Level Threat, 36–37
Melioidosis, 153–154
Membrane based dissolved oxygen sensors, 354–355
Meta, 404, 406–407
Method Validation, 177–181
Metropolitan Area Network (MAN), 288
Metropolitan Water District of Southern California, 313
Mexico, 40, 88
MHL, 252, 255, 258, 262
Michigan Department of Community Health, 175
Microarray, 208, 212, 214, 222, 225
Microbial Contaminant Detection, 55–56
Microbial pathogens, 18, 55, 151–152, 164, 185, 281–282
Microcapillary, 213, 215
Microchamber, 208, 216
Microcurie, 3, 168
Microfluidics, 208, 214, 218
Micro-luminometer, 177, 180
Microsporidia, 165
Middle East, 3
Midway/Wake Islands, 88
Midwest Assistance Program, 105
MIKE NET, 75
Mild fever, 165
Milli-Joules (mJ) per area, 157
Milwaukee WA, 175
Minden, NE, 112
Minimize disruptions, 355
Minneapolis, 40
Mississippi River, 139, 273
Missouri, 3
Missouri Department of Natural Resources (MDNR), 170, 172
Mitigating contamination threats, 319, 369
Mixed-integer programming (MIP), 249
Mobile treatment systems, 20, 59, 265, 281–282
Modeling Contaminant Propagation and Contaminant Threats, 2, 19–20
Modeling water quality, 19–20, 247–248
Molecular procedures, 155
Monitoring, 2, 9–10, 12, 18–22, 43, 48, 51–53, 55–56, 59–60, 72, 76, 78, 80, 82, 89, 93, 95, 155, 168, 186, 200, 205, 221–223, 229–232, 235–236, 238, 241–244, 248–250, 252–253, 255–262, 269, 272, 274, 278, 282, 286, 291–292, 299, 307, 312, 315, 319–322, 328–335, 349–366, 369–371, 377–378, 380–387, 389, 391–394, 398, 401–402, 408, 413–414, 416, 438, 445–448, 452
Monitoring algorithm, 393
Monitoring for Natural and Manmade Threats, 2, 19
Monochloramine, 57, 157
Monte Carlo, 77, 187, 190, 250, 356
Monte Carlo procedure, 250
Monte Carlo simulations, 77, 187, 190, 356
Moonlight Maze, 285–286
Moore's Law, 291
Mosquitoes, 164, 274–275
MS2, 52, 280
MSD, 253, 255
MSDS, 253, 255, 258–259, 262
MSRD, 253, 255–258
Multi-Disciplinary Emergency Response, 278
Multilayered Security Approach, 23, 435–455
Multi-objective analysis, 325, 342
Multi-parameter on-line monitors, 350–354, 359
Multi-parameter sensors, 331
Multi-parameter units, 355
Multiple demand loading, 249
Multiple sensor stations, 369
Multivariate clustering, 374
Multivariate nearest neighbor (MVNN) algorithm, 372, 374–377, 379, 381, 387–389, 392
Multivariate statistical approaches, 371
Municipal, 6, 9, 18, 28, 65, 136, 138, 140, 151–152, 170–174, 247, 290–291, 331, 455, 491
Mutual Aid Agreements, 5, 454
Mycobacterium spp., 165
Mycotoxins, 166

N

Naegleria, 165
Nalidixic acid, 177
Nanotechnologies, 100
Nanovesicles, 209, 215

Naples Province, 403
National Academies, 48
National Carrier System, 3
National Cyber Investigative Joint Task Force Expansion, 311
National Cybersecurity Center, 311
National Environmental Methods Index, 97, 363
National Environmental Methods Index for Chemical, Biological and Radiological methods (NEMI-CBR), 97, 363
National Homeland Security Research Center (NHSRC), 16, 47–60, 243, 282, 371
National Incident Management System (NIMS), 106, 266, 363, 453, 455
National Institute of Justice Database, 91
National Institute of Justice (NIJ), 71
National Institute of Standards and Technology (NIST), 288, 293, 300–303, 305, 307, 310–317
National monuments, 319
National Research Council (NRC), 48, 70, 137, 140, 351
National Response Framework, 48
National Response Team (NRT), 58
National Rural Water Association (NRWA), 79–80
National Security Agency (NSA), 285, 406
National Water/Wastewater Agency Response Network (WARN), 267
National Weather Service, 271
Natural Disasters, 1, 20, 29–33, 51, 60, 79, 81, 104, 106, 265–283, 452
Natural hazard, 3, 17, 121, 124–125, 131–132, 397
Natural and man-made disasters, 106, 111, 151
Natural and Manmade Threat Response, 2, 22–23
Nausea, 165, 274
Naval Facilities Engineering Service Center, 97
Near-optimal designs, 325
Nebraska, 3, 16, 103–112
Nebraska American Water Works Association, 105
Nebraska Department of Health and Human Services (NE DHHS), 103–104
Nebraska Rural Water Association, 105
Needs Assessment, 17, 115–117, 143–144, 146, 148
Neighborhood groups, 437
Nephelometric Turbidity Units (NTU), 412
NERC CIP, 304–306, 313, 315
Nessus, 299
Network, 4, 7, 9–14, 19, 21–22, 27, 38, 41, 43–44, 48, 54–56, 67, 76–77, 80–83, 94–95, 100, 121, 127, 130, 135, 137–138, 144, 156, 172, 174, 186, 190–191, 197–199, 201, 229, 234, 241–244, 248–251, 267, 285–286, 288–300, 302–303, 306–311, 313–315, 319–346, 354, 357–358, 362–363, 370–371, 378, 382–383, 393–394, 397, 403–404, 406, 415–417, 420–421, 428, 445, 449, 462
Network-based Intrusion Detection Systems (NIDS), 82, 295, 299, 308
Network grids, 190
Network intrusion detection, 370
Neural network, 95, 371, 420
Nevada, 353, 492
New Detectors, 53
New Hampshire, 88
New Jersey, 136, 330–331, 353
New Jersey American Water (NJAW), 330–331
New Mexico, 88, 492
New Orleans, LA, 13, 273, 276, 278
New Orleans Superdome, 278
New York City, NY, 3, 13, 35, 139, 166, 349
NHSRC, *see* National Homeland Security Research Center (NHSRC)
Nicotine, 239, 353
NIE, 252, 255, 258
Nitrate (NO_3^-), 351, 413
Nitrification, 333–334, 351, 361
Nitrogen (N_2), 351
Nivalenol, 166
Nodal demand, 419–420, 423, 428, 432
Noise properties, 237
Noisy Cross-Entropy Sensor Locator (nCESL), 250
Non-community water supply systems, 139
Non-dominated set, 250
Non-radioactive isotopes, 58
Nonspecific event detection, 192–197
Non-transient, non-community water supply systems, 139
Non-viable pathogens, 205
Norfolk, NE, 114
Normal threat environment, 29–31, 39, 449, 453
Normal water quality variation, 330, 370, 376

Norovirus, 165
North American Electric Reliability Corporation (NERC), 300, 304–306, 313–315
North American Energy Regulatory Commission (NERC), 441
North Atlantic Treaty Organization (NATO), 71
North Carolina, 35
Northern Australia, 154
Northern Marianas, 88
North Platte, NE, 114
Northridge earthquake, 4–6
NPDES, 88, 93
NSF International (NSF), 89, 281
Nuclear power infrastructure, 286
Nuclear Regulatory Commission, 12
Nucleic acid amplification, 155
Nucleic acids, 155, 206–208, 216, 218, 221, 223
Nucleic acid sensors, 206
Nuisance alarms, 358
Number of failed detections (NFD), 336, 338, 340–341
Number of Sensors, 91, 100, 239, 326, 330–332, 336, 338–342, 357, 365
Number of trials, 389, 391

O

Oakland, California, 6
O antigen, 206, 219
Observers, 175, 461, 469, 473, 480
Occupational Health and Safety (OH&S), 481
OCMS, 10, 249–250, 401–402, 408, 410, 413–414
OCMS design, 408, 413–414
Office of Ground Water and Drinking Water (OGWDW), 47, 50–51, 60
Office of Infrastructure Protection, 442
Office of Public Health Emergency Response, 112
Office of Research and Development (ORD), 47, 52, 279, 283
Office of Water's Water Security (WS), 51–52, 54, 320–321, 328
Offline, analysis, 371
Offset values, 23, 422, 424–426, 429
Ohio River, 139, 494
Oklahoma, 3, 353
Omaha, NE, 114
One-ton chlorine tanks, 50
On-line analyzers, 351
Online contaminant monitoring system (OCMS), 10, 80, 249–250, 401–402, 408, 410–414
Online monitoring, 51, 328, 331, 333, 350, 354–355, 360, 363, 365, 369–370, 398
Online sensors, 23, 53, 93, 195, 243, 319, 322, 350, 357, 369, 429
On-line Toxicity Monitors, 52
Online water quality monitoring, 43, 186, 236, 320, 322, 353
Operational Attributes, 233–235
Operational and Reliability Requirements, 292
Operational responses, 355, 363, 462, 471
Operations, 7, 13–14, 16, 28–33, 37, 39–40, 51, 53, 60, 68, 71, 79, 83, 88–89, 96, 98, 102, 104–105, 107–108, 112–115, 138, 148, 178, 231, 233–234, 241–242, 244, 265–266, 268–272, 278, 289–290, 301–302, 310, 328, 334–335, 369–370, 394, 399, 410, 444, 453, 455, 468, 472, 475–476
Optical, 177, 207, 209, 216–218, 220, 249, 410, 412, 415
Optical density, 209
Optical fibers, 415
Optical network, 415
Optical properties, 220, 410, 412
OptiDesigner, 75
Optimization, 19–21, 53, 56, 76–77, 95, 215–216, 231, 248, 250–251, 320, 322–326, 335–336, 342–345, 357, 365, 420
Optimization of daily operations, 231
Optimization models, 56, 323, 490
Optimize distribution system operations, 53
Optimize treatment, 53
OptiMQ-S, 252, 257, 262
Order of the Rising Sun, 151
ORP, *see* Oxidation reduction potential (ORP)
Orthomyxoviridae, 153
Outlier detection algorithms, 375–376
Outliers, 340, 342, 360, 377, 379, 381, 388–389, 391–392
Overall Contamination Warning Systems (CWS), 251
Over-the-counter drug sales, 55, 354
Overflows, 121, 138, 143, 168
Owens Valley, California, 3, 166, 459

Owner/operator, 43, 435, 442
Oxidation reduction potential (ORP), 52, 90, 192, 232, 320, 328, 351, 353–355, 411–412

P

Parallel computing, 23, 420–421, 423, 428, 432
Parameter estimation, 373
Parametric analysis, 187, 189
Parasites, 156, 165, 460
Parasitic pathogens, 165
Particle Swarm Optimization, 95
Parts of cells, 206
Pathogenic bacteria, 152, 208
Pathogenic *Escherichia coli*, 164
Pathogenic microorganisms, 185
Patrols and Monitoring, 452
Pattern of illness, 169
PCR, 155–156, 206, 208, 212–214, 216, 218–221
PCR master, 208
PCR products, 208, 214, 218
PD, 387–388, 393
Pentagon systems, 285
Percent of LD50, 238
Performance Attributes, 235
Performance evaluation for urban drainage, 123–124
Performance evaluation for water supply, 122–123
Performance indicators (PI), 122, 124
Persistence, 56–58, 151, 154, 279, 306
Persistence of Contaminants, 57–58
Persistence of target contaminants, 56
Peru, 8, 164, 178
Pesticides, 52, 235, 278–279, 361, 412
Petroleum chemical, 419
PFD, 253, 255, 258, 262
pH, 52–53, 57–58, 77, 90, 92–94, 100, 137, 154, 156–158, 177, 192–194, 219, 231–232, 272, 320, 328, 351–355, 369, 375, 378–380, 384, 386, 411–417
Pharmaceutical sales, 236
pH dependencies, 193
Philadelphia, PA, 52
Phishers, 294
Phi X 174, 280
Physical barriers, 41, 78
Physical countermeasures, 9, 401, 407–408
Physical damage, 13, 36, 66, 277
Physical destruction, 67, 319, 461
Physical disruption, 164
Physical incident, 10
Physical infrastructure, 52, 139
Physical or pathogen properties, 74
Physical parameters, 90–91, 100
Physical Security, 23, 28, 45, 78–80, 82, 93, 248, 292, 305, 310, 329, 401, 436, 438, 444–445, 448, 450–452, 458, 460–461
Physical security components, 45
Physical Security Tools, 78–80
Physical surveillance system, 401, 408–409
Physical threats, 33
Physician Preparedness for Acts of Water Terrorism, 91
Piano di Sorrento, 404
Picornaviridae, 152
Piezoelectric (mass sensitive), 207, 217–219, 225
Pilot-scale, 52, 57, 223, 282
The pilot sites, 402–404, 408, 414
Pilot systems, 415
Pipe bursts, 119–120
PIPE2000/KYPIPE, 75
PipelineNet, 75, 187, 333–334
Pipelines, 6–7, 27–29, 37, 40, 43, 68, 80, 163, 248, 268, 272, 314, 447
Pipes, 3, 5–6, 17, 19, 21, 28, 41–42, 56–58, 65, 87, 119, 121, 124–125, 128–129, 138–140, 142–143, 164, 170, 174, 185, 188, 247, 253, 274, 277, 325, 327, 329, 335–337, 345, 350–351, 357, 406, 416
Pipe surfaces, 57–58
Pipe wall reaction, 187–189, 195
Pittsburgh, 36, 185, 494
Plague, 153–154, 166
Plague bacillus (*Yersinia pestis*), 57, 153–154, 157, 166
Plan, 5, 10–11, 16, 20, 22, 47–48, 50–51, 79, 88, 96, 99, 103, 107–115, 265–267, 269–272, 283, 287, 302, 305, 311, 322, 335, 354, 356, 362, 364–365, 372, 404, 435, 437, 439, 453–454, 462–464, 466–467, 470–471, 477–478
Planning, 6–7, 9, 17, 19–20, 23, 33, 40, 48–49, 51, 57, 59, 74–76, 93, 96, 99, 104–107, 110, 112, 114, 120, 130, 132, 142, 144, 186, 231, 248, 266, 269, 283, 286, 300, 305–307, 437, 444, 450, 452–453, 463, 466–468, 474–479

Plant library, 361
Plaquemines, 273
Platforms, 18, 21, 50, 82, 175, 208, 217, 219, 234, 243, 359, 372, 394, 414, 446
Plutonium, 3, 35, 166
Point-of-Entry (POE) Devices, 279–281
Point-of-Use (POU) Devices, 279–280
Poison control center calls, 354
Poison control centers, 55
Poison control hotlines, 321
Police, 110–111, 268, 273, 281, 329, 363, 416, 445, 447, 450, 478
Pollutant emissions, 123
Pollution event (PE), 253–259
Polymerase chain reaction (PCR) assays, 155–156, 206, 208, 214, 216, 218–221
Polyvinyl chloride (PVC) pipes, 57–58, 140, 193
Population exposed (PE), 336
Population served, 65, 87–88, 140–141, 143, 272, 335
Portable and Mobile Water Treatment, 279–282
Port scanning tool, 300
Possible intrusion, 23, 230, 416, 419, 421, 424, 430
Possible intrusion nodes (PINs), 23, 419
Potable Water CBR Contamination and Countermeasures, 69
Potential Asset/Threat Combinations, 45
Potential chemical agents, 192
Pour-Through Devices, 281
Power or telecommunications systems, 67
Power outages, 14, 31, 358, 448
Power plants, 12, 397
Power surges, 448
Poxviridae, 153
Preconcentration, 208, 219
Predator-prey, 250
Prediction of Chlorine Reactivity, 59
Prediction Errors, 195, 379–381
Prediction tools, 371
Pre-event response scenarios, 76
Prē-Gathering Information, 452
Prepare, 10, 28, 53, 60, 73–74, 82, 93, 103, 106, 176, 268–270, 272, 274, 283, 287, 315, 463–464, 466–467
Preparedness, 1, 23–24, 28, 31, 47, 66, 72, 91, 96–97, 103, 106, 111, 135, 199, 271, 287, 311, 349, 442, 448, 452–455, 462, 478
Presidential Decision Directive No. 63, 89
President's Commission on Critical Infrastructure, 1, 4, 7–8, 135, 285
The President's Commission on Critical Infrastructure Protection, 1, 7, 285
President's National Strategy for Homeland Security, 88
Pressure transients, 419
Pretreatment, 41, 282
Prevention, 5, 50–51, 56, 60, 74, 136, 152, 159, 171, 303, 307, 398, 401, 442, 444, 448–450
Private industry, 230, 266
Proactive tools and methodologies, 60
Probabilistic method, 189, 191
Probabilities, 190–191, 252, 262, 342, 427
Process Control Systems, 99, 288, 313
Process Control Systems Forum (PCSF), 313
Protected wellheads, 168
Protection
  and prevention, 50
  Protection System Design (PSD), 397–398, 404–417
  public health protection, 52, 54, 59, 77, 244, 330–332, 356–357
  source water, 139, 437
  water infrastructure, 66, 243, 397–418
Protective Programs, 436–438
Protocol for the Disposal of Contaminated Water, 97
Protozoan oocysts, 55
Pseudomallei, 57, 153–154, 157–158
Pseudomonas aeruginosa, 165
Publically Owned Treatment and Collection Systems, 143
Public Drinking Water Systems (PWS), 65, 115–116
Public drinking water utilities, 28
Public health, 2, 10, 18, 28, 30, 35, 39–40, 48, 50, 52–55, 59–60, 66–67, 69, 71–74, 77, 87, 89, 96, 98, 104, 108, 112, 136–137, 139, 143, 146–147, 151, 155–156, 159, 164, 170, 192, 220, 225, 243–244, 250, 268–269, 276, 287, 312, 319–322, 329–334, 338–340, 349–350, 354, 356–357, 363, 365, 400, 402, 442, 449, 457, 460
Public health benefits, 356
Public Health and Environmental Impact Response Guide, 96
Public Health Response Guide, 74
Public health risks, 400
Public health and safety, 268

Public Health Security and Bioterrorism Preparedness and Response Act of 2002, 1, 24, 47, 66, 103, 135, 287, 349
Public Health Surveillance, 55, 72, 319, 321–322, 354
Public health surveillance systems, 319
Public Law 107–188, 103
Publicly-owned treatment works (POTWs), 1–2, 87–88, 97, 138
Public officers, 481
Public Relations, 474
Public safety, 4, 8, 38, 81, 265, 283
Public telephone network, 359
Public Water System Inventory Data, 141
Public water utilities, 191, 354
Puerto Rico, 88
Pumping stations, 2, 7–9, 42, 67, 87, 128, 441
Pumps, 7, 14, 19, 38, 43–44, 68, 121, 124, 128–129, 138, 163, 193, 247, 270, 273, 277, 281–282, 335, 369
PVC, *see* Polyvinyl chloride (PVC) pipes

Q
Q fever, 154
Qualitative, 95, 126, 240
Quantitative, 59, 120, 156, 174, 182, 210, 212, 219–220, 241, 399
Quantitative polymerase chain reaction (qPCR), 156
Quantitative Structure Property Relationship (QSPR), 59
Quantum dots, 207, 210–211, 216
Quartz crystal microbalance (QCM), 207, 217
Queensland, Australia, 21, 286
Query Database for individual agents, 91

R
R2A agar, 18, 175, 180
Radiological agent, 90, 363, 399, 438
Radionuclides, 235
Radio propagation, 415
Radio transmitters, 359
Rainfall intensities, 119
Rajneeshee, 3, 152, 168
RAMCAP (Risk Analysis and Management for Critical Asset Protection), 50, 120
RAM-W (Risk Assessment Methodology for Water), 78–79, 120
Random contamination events, 10, 198, 249
Ranking Methods, 323
Rank-ordered list, 443
Ransom, 152
Rapid communication technologies, 319, 369
Rapid Detection of Events, 236
Rapid response and recovery, 60
Rapid techniques, 175
Raw water, 65, 163–164, 170, 208, 222–223, 273, 354
Reaction rate coefficients, 187–189, 195, 197–198
Reagentless, 193, 206, 217
Realistic rehearsal, 470
Real time, 18–19, 21, 52, 54, 56, 61, 73, 75, 77, 89–90, 155, 186, 201, 205, 218–219, 222–224, 244, 248, 292, 328–329, 331, 351, 355, 359–360, 362, 365, 370–371, 394, 402, 409, 421, 424, 426, 428, 431, 445, 452, 475, 479, 481
Real time data, 54, 360, 370
Real-time decision support tool, 75
Receiver Operating Characteristic (ROC) curve, 22, 237, 387
Recirculating pilot-scale simulators, 57
Recovery Cost Estimates, 278
Recovery protocols, 265
Red Cell, 130, 467
Red team, 82, 285
Reduced gradient method, 420
Redundancy (R), 4, 130, 231, 253, 256–259, 262, 415, 451
Reference business process models, 96
Reference electrode, 192
Registry of Toxic Effects of Chemical Substances, 91
Regular practice, 467
Regular water quality sampling and analysis, 51
Regulating agencies, 19, 248
Regulators, 19, 220, 224, 247, 441–442, 469
Regulatory Agencies, 156, 441, 447, 452, 473
Regulatory challenges, 53
Reliable, 7, 41, 89, 93, 100, 142, 198, 205, 229, 234, 287, 304–305, 321–322, 329, 336, 354–355, 359, 413–414, 439, 445, 450, 480
Remediation and Recovery Guide, 74, 96
Remediation response, 155
Remote Data Communications, 359
Remote Telemetry Units (RTU), 359
Reoviridae, 152
Repair and rehabilitation, 140
Reporter probe, 207, 215–216, 221
Reporting relationships, 266

Repository of Industrial Security Incidents (RISI), 297–299
Reproducibility, 239
Reservoirs, 3, 9, 42–43, 65, 67, 121–128, 138, 143, 154, 163–164, 166, 218, 248, 268, 333, 335, 397, 404, 413–414, 457–458
Residual, 9, 18, 49, 52, 58, 155, 168, 185–201, 232, 272, 279, 282, 320, 330, 351, 362, 371–373, 376, 379, 381, 392, 411–414
Residual Chlorine, 351, 411–412
Residual classification, 371–372
Residuals, 18, 49, 168, 185, 190, 198, 279, 282, 330, 362, 379, 381
Respiration rates, 98
Respiratory conditions, 153
Respond, 2, 7, 12, 23, 28, 30–31, 41, 48, 51–53, 66–67, 73, 75, 107–108, 112, 142, 193–194, 199–200, 217, 230, 236–237, 265, 267, 278, 296, 311–312, 321–322, 332, 360, 363, 365, 372, 413, 435, 444, 446–448, 452, 454, 466–467
The response, 416–417
Response Actions, 51, 269, 271–272, 454
Response To Emergencies, 107
Response Protocol Toolbox (RPTB), 73–74, 96
Response Support Corps (RSC), 266
Response time, 230, 236, 244, 329, 332–333, 410, 412
Response to upset event – decontamination, 96
Response to Upset Event in Wastewater – Normal Conditions, 96, 98
Response to Wastewater Treatment from CBR, 97–98
Return on investment (ROI), 231–233
Reverse flow alarms, 358
Richmond, Virginia, 136
Richter scale, 4–5
Ricin, 235, 239
Rickettsial or rickettsia-like organism, 154
R.I.S.E., 35
Risk analysis, 45, 50, 120–121, 131, 398
Risk Analysis and Management for Critical Asset Protection (RAMCAP), 50, 120
Risk assessment, 39, 44, 49–50, 78–79, 119–132, 205, 220, 278, 301–302, 305, 313, 399–401, 405–407, 414, 439–440, 451
Risk Assessment Methodology for small and medium utilities, 79
Risk Assessment Methodology for Water Utilities (RAM-W), 78–79, 120
Risk Assessment Tools, 78, 439–440
Risk-Based Justifications, 439–440
Risk-based performance, 435
Risk Communications, 99
Risk maps, 121, 127–128, 130
Risk Mitigation, 450, 463, 482
Risk Reduction, 45, 248, 327, 339, 342, 398–402, 407–414, 435, 450
Risk reduction options and recommendations, 45
RLU, 177–178, 180
ROC curve, 237–238, 240, 387–394
Rodenticides, 235
Rodents, 154, 166
RO membranes, 281
Root Server Attack, 12
Rotaviruses, 164–165
Roughness coefficient, 187–188
Routers, 249, 288, 290–291, 308
Routine calibration, 354
Routine sampling and analysis, 321
RPM, 253
Rubinacci, 407
Runners, 473, 475
Russian Academy of Sciences, 286

S

Sabotage Reporting, 305
Sabotaging pumps, 7, 163
Saboteur attack, 405
Saboteurs, 35, 398–399, 409
Safe Drinking Water Act (SDWA) of 1974, 17, 105, 136–137, 139, 144, 146–148, 186, 287
SAFETY Act, 361
Salad bars, 3, 152, 168
Salerno Province, 403
Salmonella, 3, 164–165, 168, 170, 172, 219
Salmonella contamination, 164
Salmonella enterica serovar Typhimurium, 152
Salmonella typhi (the causative agent of typhoid fever), 35
Salmonella typhimurium, 3, 166
Salmonellosis, 3, 152, 168–170
SAM document, 55
Sample Filtration, 176
Sample transport conditions, 155
Sampling and Analytical Methods, 55
Sandia National Laboratories, 21, 53, 78, 320, 331, 356, 371
Sandwich assays, 207, 221

San Francisco, CA, 3–4, 6, 52
San Francisco earthquake of, 1906, 4
Sanitary sewer, 17, 28, 142, 333, 355
Santa Clara Valley Water District, 313
Sant'Agnello, 404
Sarin, 235, 239
Savanna River National Laboratory, 53
SCADA Network, *see* Supervisory control and data acquisition (SCADA) networks
SCADA, *see* Supervisory Control and Data Acquisition (SCADA)
SCADA Systems, *see* Supervisory control and data acquisition (SCADA) system
Scattering of light, 410
Scenarios, 3, 6, 18, 23, 76, 82, 106, 110, 113, 119, 169, 171, 174, 192, 194, 199–200, 241, 243, 268–269, 283, 299–300, 310, 356–357, 398–399, 402, 404–406, 409, 415, 422–432, 453, 462–463
Screening People, 450
Script, 469–479, 481
Sean McGurk, 286
Seasonal demand, 345
Section 1433, 287
Sector-Specific Agency, 16, 48, 60
Secured website for members only, 91
Security Administration, 81
Security concerns, 15, 78, 82, 155, 158, 198, 334, 453
Security Countermeasures, 28, 41, 436, 442–448, 450–452
Security Drivers, 438–443, 440
Security and Emergency Management Systems, 80
Security Guidelines, 313
Security incidents consequences, 96–98
Security incidents and response, 66, 96–98
Security Management Programs, 301, 305
Security Organizations, 311, 416
Security Planning, 186, 450
Security regulations, 88–89, 315, 442
Security and Surveillance Systems, 20–21
Security Vulnerability Self-Assessment Guide for small drinking water systems, 79
Select Agents, 152–153, 156
Selection Criteria for an Early Warning System, 89–90
Self-replicating, 153
Semiconductor nanocrystals, 210
Senate Homeland Security Committee, 286
Sensitivity, 52, 81, 90, 95, 122, 124, 153, 175, 177, 187–188, 189, 206–207, 215, 217, 219–221, 224–225, 231, 236–238, 253, 257–262, 334, 370, 372, 377, 387–394, 412
Sensitivity analysis (SA), 122, 187–189, 257–262, 372, 377, 388–391, 393
Sensor
- drift, 330, 354
- fouling, 218, 224, 330, 354
- installation, 357
- location design model, 249
- locations, 18, 93–94, 249, 323–325, 329, 331, 334–337, 344–345, 356, 402, 410, 416
- monitoring/instrument placement, 76

Sensor alarms, 420–422, 427, 429, 431
Sensor Network Design, 21, 243, 319–346
- networks, 9–10, 21, 186, 198–199, 243, 319–346, 428
- packages, 229, 231, 351

Sensor Placement Optimization Tool (SPOT), 21, 53, 77, 320
- placement, 21, 53–54, 62, 76–77, 93, 195, 243, 249–251, 320, 322–327, 329, 335, 341–346, 365
- placement tools, 320

S. enterica serovar Typhi, 152, 156
September 11, 2001, 1, 4, 13, 16, 24, 30, 47, 56, 66, 73, 103, 135, 349
Service connection, 108, 169, 329, 332, 355, 357, 359
Set covering graph theory algorithm, 249
Settleability, 98
Sewage back-ups, 168
Sewer lines, 65, 185
ShakeOut, 268
Shaking Intensity Maps, 268
Shared Expertise, 291
Shared Hierarchical Academic Research Computing Network (SHARCNET), 421
Shared Infrastructure, 291
Sheep, 154
Shigella dysenteriae, 166
Shock, 57–58, 96, 98, 166, 221
Shock chlorination, 57–58
Simulation, 17, 19, 23, 53, 75–77, 99, 120–121, 123–124, 132, 172–173, 187–190, 197–199, 247–248, 253, 323, 326, 330, 337, 345, 356–357, 377, 382–386, 398, 401, 406, 409, 420–425, 427–428, 432, 462, 470–471, 480, 492
Single fluorescent dye molecule, 207

Single Jeopardy, 477
Single pass simulators, 52
Single point failure, 415
Site Characterization and Sampling Guide, 74, 96
Sites for Sensor Deployment, 355–357
Skeletonization, 341, 344–345
The Slammer Worm, 12, 14
Slope, 125–126, 388, 392–393
Small crustaceans, 413
Smallpox, 4
Smart water meters, 358
Software malfunction, 10
Solar powered equipment, 359
Solar Sunrise, 285–286
Solids, 92, 136, 281, 412–413
Solid surface, 206
Solver scalability, 325–326
Soman, 235
Somatic cell releasing agent, 176
Sorrento Peninsula Water Supply System (SPWSS), 403–405
Sorrento (Sorrentine) Peninsula, 397, 403
Source Water Protection, 139, 437
South Dakota, 3
Southeast Asia, 154
Southern Hyogo Prefectural Earthquake (the Great Hanshin-Awaji Earthquake), 5
Southern Italy, 403
Spammers, 294
Spatially diverse, 163
Spatial profiles, 187
Special Publication (SP), 288, 293, 300–303
Specification Project 99 (SP99), 288
Specific chemical agent, 8, 29, 71, 152, 235, 326
Specific conductance, 52, 351, 353, 355, 384
Specificity, 206–207, 214, 219, 221, 237–239
Spore-forming bacterium, 166
Spores, 47, 55, 58–59, 135, 153–154, 156–159, 166
Spyware/Malware Authors, 294
Stable signal, 391
Standard 17799, 300
Standard deviation, 22, 195–196, 330, 379–381, 383–384, 388, 391, 393, 423, 433
Standard IP, 289
The Standardized Analytical Methods for Environmental Restoration following Homeland Security Events (SAM), 55
Standardized Analytical Methods (SAMs), 55, 73, 364
Standardized assessment forms, 269
Standard Methods, 175, 177, 219
STANET®, 75
State agencies, 47, 440–441
State estimation, 371–372, 374
State Revolving Fund (SRF), 105, 143
State's Intelligence/Fusion Center, 32
State of Victoria, Australia, 462
Statistical analysis for Guelph WDS, 430
St. Bernard, 273, 275
Steady-state hydraulic models, 75, 187
St. Louis, Missouri, 152, 177
Storage, 5, 17, 28, 38, 41–43, 50, 78, 105, 116, 128, 136, 138–139, 143, 145–146, 170–171, 174, 188, 195, 198, 248, 253, 270, 273, 277, 279, 282, 307, 333, 349, 361, 427, 441–443, 446–447, 451, 457–458, 461
Storage and distribution facilities, 17, 41, 136
Storage and handling, 41
Storage reservoirs, 43, 248, 333
Storm runoff, 205
Storm water discharges, 138
Strain variations, 156
Stratford Ontario, 419
Streams, 17, 19, 21, 54, 72, 126, 137, 196, 205, 230, 286, 321–322, 394, 437
Street drugs, 235
Structural query language (SQL), 421
St. Tammany, 273, 275
Stuxnet, 14–15, 286
Stuxnet Virus, 15
S. typhimurium, 3, 168, 170–171
Success, 6, 45, 225, 375–376
Suite of sensors, 231, 355
Supervisory Control and Data Acquisition (SCADA), 2, 14, 20–21, 37–38, 43–45, 80–83, 90, 98, 174, 248, 286, 288–296, 300–301, 303, 306, 308–310, 313–314, 319, 359–360, 370, 375, 409, 414, 461
Supervisory control and data acquisition (SCADA) networks, 38, 44, 80–82, 288–293, 295, 308–310, 314, 359
Supervisory control and data acquisition (SCADA) system, 2, 14, 20–21, 37–38, 43–45, 80–81, 83, 90, 98, 174, 248, 286, 289–296, 300–301, 303, 306, 308–310, 313–314, 319, 360, 370, 375, 409, 414, 461
Supply deficits, 119

Support vector machines, 371
Surcharging, 123
Surface plasmon resonance (SPR), 207, 217–218, 220, 222–223
Surface water sources, 139, 335, 442
Surfactants, 57–58
Surrogate parameters, 91, 93, 100, 370–371, 410–412
Surrogate sensors, 369
Surveillance of consumer complaints, 52
Surveillance and monitoring systems, 51, 72, 350
Surveillance of public health, 52
Surveillance systems, 2, 20–21, 319, 369, 401, 448
Suspicious Behavior Reporting, 437
Suspicious circumstance reports, 33
Suspicious circumstances, 447
Swine, 154
Swine flu (H1N1), 40
Swinnex filter holder, 176
SWMM, 123–124
Syndromic surveillance, 52, 55, 236, 321, 357
SysAdmin, Audit, Network, Security (SANS), 296–297, 307, 310, 313
System Data, 81, 253, 294
System fault detection, 370
System maintenance, 120, 447–448
System Status Codes, 277
System Wide and Topological factors, 93–94

T

Tabletop Exercises, 110–111, 468–470
Tanks, 5, 19–20, 28, 39, 42, 50, 75, 127–128, 138, 146, 163, 170–174, 198, 247, 252–253, 255, 259, 265, 269–270, 272–274, 277, 282, 288, 333, 335, 351, 355, 369, 397, 404–406, 416, 419, 451, 457
Target cells, 206, 209, 217, 222, 224
Target Compounds (USGS), 91
Target genes, 208
Taste and odor, 139, 170, 171–172
Technical Assistance Team, 105
Technology Testing and Evaluation Program (TTEP), 59, 72–73, 350
Telecommunications, 7, 11, 13, 28, 41, 67–69
Telecommunications infrastructure, 11
Teledyne Isco, 55, 186
Temperature, 3, 7, 33, 80, 90, 92–93, 100, 137, 154, 156–158, 170, 172, 195, 275, 320, 351, 353–355, 401, 413–414
Temporal profiles, 187
Terrorism, 1, 3–4, 8–9, 18, 24, 30, 32–33, 47–48, 57, 67, 72, 80, 103, 105–106, 135, 151–152, 155, 157–159, 165, 287, 312, 349, 362–363, 438, 442, 449, 452–453, 457–482
Terrorist activity, 159, 229, 241, 312, 358
Terrorist attacks, 1, 4, 7–9, 13, 23–24, 28, 30–31, 39, 47–49, 60, 65, 71–73, 98, 103, 119, 135, 151, 153, 287, 319, 350, 369, 404, 460–463, 466, 468, 482
Terrorists, 3–4, 8, 13, 29, 31, 34–35, 66, 81, 135, 294, 398, 442, 462, 467
Test and Evaluation Facility, 334
Testing data sets, 22, 379, 382, 384–386, 392
Testing Exercises, 453
β-Testing phase, 361
Tetanus shots, 274
TEVA-SPOT, *see* Threat Ensemble Vulnerability Assessment Sensor Placement Optimization Tool (TEVA-SPOT)
Thermal expansion, 358
Thermal inversion, 170, 172
Threat assessment, 31–32, 44
Threat Ensemble Vulnerability Assessment Sensor Placement Optimization Tool (TEVA-SPOT), 21, 54, 320, 322, 324–326, 328–330, 332–334, 336–337, 341, 344–345
Threat Ensemble Vulnerability Assessment (TEVA), 21, 53, 77, 243, 320, 324, 356–357
Threat Level Planning, 453–454
Threshold, 189, 191, 198, 237–238, 240, 358, 372, 375–377, 381–382, 388–389, 392, 441
Threshold Value, 376, 381–382, 392
Tick Box List, 475
Tier 1 sensors, 90
Tier 2 sensors, 90
Tier 3 sensors, 90
Tier 4 sensors, 90
Time of detection (TD), 336, 387
Time for effective treatment, 356
Time to onset of symptoms, 356
Time series, 195, 248, 370–372
Time-series data, 370
TN, 387, 393
TOC on-line detector, 53
TOC, *see* Total organic carbon (TOC)
Toeplitz matrix, 374
Togaviridae, 153

Toluene, 52
Tornadoes, 33, 104, 106, 109, 266, 271, 273
Total chlorine, 52, 186–187, 276, 335, 353
Total maximum daily loads (TMDLs), 138
Total number of monitoring stations (TNOMS), 255–259
Total organic carbon (TOC), 52–53, 59, 77, 90, 93, 100, 192, 231–232, 320, 328, 334, 351, 353–355, 359, 375, 378–381, 384–386, 391, 411–412
Total rRNA, 208
Township Level, 267
Toxic bacteria, 3, 168
Toxic chemicals, 66–67, 70, 279
Toxicology Data Network Toxnet, 91
Toxic shock, 96, 98
TP, 387, 393
Tracer study, 329
Tradeoff, 250, 332
Tradeoff curves, 332
Traffic accidents analysis, 370
Training Activities, 109–111
Training data sets, 379–381, 384, 386
Training exercises, 269, 453
Transducer, 206, 208, 217
Transient water supply systems, 139
Transmissible, 166
Transmission control protocol/internet protocol (TCP/IP), 291
Transmit Power Control (TPC), 415
Transportation, 29, 33, 40, 68, 80, 105, 107, 269, 297, 314, 349, 356, 442
Transportation infrastructure, 40
Transpose of Z, 373
Treatment of contaminated wash water, 56–57
Treatment facilities, 5, 31, 41, 44, 65, 87, 89, 97, 105, 143, 163–164, 276, 281, 442, 451
Treatment plants, 5, 9, 16–17, 28, 42, 68, 78, 96, 138, 142, 146, 163–182, 185, 223, 247–248, 269, 276, 329, 351, 457
Tribal agencies, 266
Trichothecene mycotoxins, 166
Trigger Threshold, 238
Triple Jeopardy, 477
True events, 375, 386–389, 391, 393, 422, 424, 426–427
True intrusion node, 420, 422, 425, 428, 431
Trusted Internet Connections Initiative, 311
Tsunami detection, 370
T-2 toxin, 166
Tuberculation and corrosion, 170
Tucson Water, 331–333
Tularemia, 153
Turbidity, 53, 90, 92–93, 100, 155, 209, 232, 272, 281, 320, 328, 351–355, 411–412, 421
Twisted copper pairs, 415
Two-stage event detection, 371
Typhoid fever (*Salmonella enterica* serovar Typhi), 35, 151–152
Typical Process Model, 479–480
Tyrol, 121
Tyrolean Alps, Austria, 127

U

Ultraviolet 254 spectrum, 240
Ultraviolet (UV)-254, 334
Ultraviolet (UV) irradiation, 157
Umma, 2, 166, 459
Uncertainty analysis, 421–423
Uncontrolled water pollution, 136
Uncorrelated Gaussian noise, 377
Underlying management processes, 82–83
Undulant fever, 154
Unique genetic signatures, 155
University of Cincinnati, 21, 53, 320
University of Innsbruck, Technikerstr in Innsbruck, Austria, 17
Unlicensed radio frequencies, 359
Unmanaged deterioration, 457
Unmetered connections, 168
Unsecured facilities, 446–447
UN's Office of Foreign Disaster Assistance, 3
Unspecified system contamination events, 104
Unsteady water quality propagations, 249
Unusual effects of poison, 70
Upper bound, 251, 336
Urban flooding, 119
Urlama, King of Lagash, 2, 166
U.S. Army Combined Arms Support Command, 8
U.S. Army Corps of Engineers, 271
U.S. Army Engineer Research and Development Center, 50
U.S. Army ERDC, 97
U.S. Army Soldier and Biological Chemical Command (SBCCOM), 97
U.S. Capital, 153
U.S. Centers for Disease Control and Prevention (CDC), 91, 152
U.S. Coast Guard. 2001. "Chemical Hazards Response Information System", 91
U.S. Computer Emergency Readiness Team (US-CERT), 296

U.S. Congress, 1, 24, 47, 89, 135, 159, 349
U.S. critical infrastructure, 319
U.S. Department of the Army, 97
U.S Department of Defense, 10, 71, 281, 285
U.S. Department of Energy, 97
U.S. Department of Homeland Security (DHS), 10, 15, 27, 47, 89, 286, 294, 310–312, 361, 439, 441–442, 452–453, 455
U.S. Dept. of Defense, 10, 71, 281, 285, 288
U.S. Environmental Protection Agency (USEPA), 1, 8, 17–18, 24, 28, 43, 47, 51, 54–55, 58, 70, 104–106, 109, 135, 143, 148, 168, 189, 192, 229, 247–249, 253–254, 350–351, 354, 356, 361, 363–365
U.S. EPA Environmental Technology Verification (ETV) Program, 281
USEPA, *see* U.S. Environmental Protection Agency (USEPA)
U.S. EPA's Maximum Contaminant Levels, 91
U.S. Food and Drug Administration, 206
U.S. Forest Service, 271
U.S. Geological Survey (USGS), 91, 268, 271, 330–331, 356
The US Government Accountability Office (GAO), 10, 24
U.S. Postal Service facilities, 153
Utah, 353
Utilities, 6–7, 12, 14–15, 17, 19–22, 27–31, 33, 35–40, 44, 49–56, 59–61, 65–68, 72, 74–75, 77–80, 87, 95–96, 99–100, 107, 136–137, 139–140, 142–143, 146, 159, 164, 178, 191, 194, 197–200, 205, 224, 242, 244, 248, 265, 267–273, 276, 285–287, 289, 291–292, 294, 299, 302, 309–310, 312–315, 319–322, 325–328, 341–342, 349–351, 353–354, 359–360, 362–365, 372, 381, 394, 435–440, 442–444, 446, 448–454, 472
Utility vulnerabilities, 29
UV-254, 231, 335
UV, 92, 157–158, 231, 281–282, 334–335, 351, 355
UV dose, 157
UV-ozone simultaneous oxidation process, 281
UV spectrometry, 351, 355

V

Vaccinations, 274–275
Valencia hotel, 3
Validation framework, 350
Value of Statistical Life (VSL), 327
Valves, 5–7, 19, 38, 44–45, 51, 138, 163, 232, 247, 269–270, 357, 359, 369, 447
Vandalism, 9, 66, 107, 435, 438, 447
Vandals, 34–35, 39, 398–399, 408
Variable time frames, 479
V. cholerae, 156
Vector, 154, 164, 371, 374–375
Verbal threat, 447
Verification methodologies, 56
Very High-Level Threat, 36
Vesuvian Water Supply System (VWSS), 403–404, 416–417
Vesuvian water system, 397
Viable cells, 177, 215–216, 220–222
Vibrio, 165
Vibrio fischeri, 413
Vico Equense, 404
Video for law enforcement personnel, 17, 103
Video surveillance, 350
Vinegar, 58
Viral agents, 152
Viral pathogens, 165
Viral taxonomic, 152–153
Virtual private networks (VPNs), 307
Visual assessment, 445
Visualization, 128
Vital Human Services, 4, 135
Vitek Boden, 286
VNC (Virtual Network Computing), 244
Volume consumed (VC), 336, 338
Voluntary Water Infrastructure Security Enhancement (WISE) Initiative, 80
Vomiting, 165–166
VSAT (Vulnerability Self Assessment Tool), 79, 88, 120, 439
Vulnerabilities, 4, 16–18, 21, 24, 27–45, 50, 56, 75, 79–83, 120–121, 186, 191, 229, 285–286, 291, 293–300, 302, 307, 312, 314–315, 350, 435, 438, 451, 458, 461
Vulnerability assessment (VA), 16, 18, 38, 41, 44–45, 50, 53–54, 66, 71–72, 79–80, 104–107, 120, 122, 124, 186–192, 287, 307, 312, 349, 362, 364, 397, 399, 435, 439, 442–443, 451
Vulnerability maps, 119–121, 128–129
Vulnerability maps for UDSs, 129
Vulnerability maps for WSSs, 128–129
Vulnerability Self Assessment Tool (VSAT^{TM}), 79, 88, 120, 439

VulNet, 17, 121, 124, 132
VulNetWS (Vulnerability of Networks, Water Supply), 121
VX, 235, 239

W
WADISO SA, 75
Walkerton, Ontario, 173–174
Wall 1: Proactive Software Assurance, 307
Wall 2: Blocking Attacks (Network based), 307
Wall 3: Blocking Attacks (Host Based), 307
Wall 4: Eliminating Security Vulnerabilities, 307
Wall 5: Safely Supporting Authorized Users, 307
Wall 6: Tools to Manage Security and Maximize Effectiveness, 307
War dialer, 300
Warfare agents, 18, 52, 151, 165, 235, 361
Washington, DC, 12–13, 166–167, 178, 241, 275, 314, 353–354
Wastewater
 collection systems, 98, 457–458
 collection and treatment facilities, 65, 88, 94, 99–100, 349
 contaminants of concern database, 90
 infrastructure, 4, 17, 20, 23, 27–45, 58
 treatment
  facilities, 5, 31, 87, 97, 149
  plant, (WWTP), 42, 68, 96, 123, 136, 163–182, 457
Wastewater Response Protocol Toolbox (WWRPTB), 96
Wastewater Security Regulations, 88–89
Wastewater Security Research, 98–99
 sensors, 90–93, 99
Waste Water Sensors and Early Warning Systems, 89–96, 230, 234, 236, 243–244, 297, 311, 350, 360, 408–409
Wastewater Treatment Works Security Act of 2009, 89
 utilities, 14, 21, 28–29, 31, 68, 80, 87, 93, 96, 99–100, 136, 285–286, 288–290, 302, 310, 312–315, 438, 440, 444, 449, 454–455
Wastewater Utility Planning Guide, 96
Water
 reservoir, 9, 35, 43, 65, 120, 163
 shortages, 3, 268
 source, 8, 51, 65, 105, 107, 113, 125, 129, 139, 163, 168, 174, 247, 281, 335, 437, 442, 457
Water age, 127, 130, 187, 190, 322, 334, 351
Water Agency Response Network, 7, 267
Water aging, 361
Water-based pathogens, 164
Waterborne disease, 18, 43, 136, 152, 156, 164, 168, 175, 270–271, 280
Waterborne disease outbreaks, 136, 152, 271, 280
Waterborne pathogens, 18, 151–154, 156, 164–165, 270, 272, 279
Waterborne route, 350
WaterCAD, 75, 174, 187–188
Water collection, 41
Water Contamination Information Tool (WCIT), 70, 74–75, 91, 97–98, 363–364, 439
Water distribution
 pipes, 6, 140
 systems, 2, 5–6, 9, 19–20, 23, 35, 37, 43, 49, 51–55, 57, 61, 72, 76–77, 96, 120, 125, 138, 140, 159, 174, 185–201, 239, 242, 247–262, 276, 279–281, 294, 322–323, 325, 342, 349, 362, 365, 369, 371, 377, 402, 419, 457–458
Water emergencies, 103–104, 107–108, 110–111, 233
Water Environment Federation (WEF), 47, 398, 401
Water Environment Research Foundation (WERF), 88–91, 94–96, 98–99, 310–311
Water flow, 7, 51, 174, 187
Water hammer effect, 7
Water Health and Economic Analysis Tool (WHEAT), 439
Water infrastructure, 4, 16–17, 20, 23, 27–45, 48–49, 65–66, 80, 88, 98–100, 119–132, 136, 138–147, 243, 265, 267–268, 270–271, 279, 285–317, 397–418, 435–482
Water Infrastructure Network (WIN), 144
Water Infrastructure Protection, 66, 397–418
Water Infrastructure Protection Division (WIDP), 243
Water infrastructure systems, 65–66, 80, 279
Water ISAC (Information Sharing, Analytic Center), 34, 50, 59–60, 67, 71, 312–313, 449

WaterISAC, *see* Water ISAC (Information Sharing, Analytic Center)
Water Laboratory Alliance (WLA), 56, 363
Water meters, 168–169, 272, 358
Water Protection Task Force (WPTF), 248
Water quality
- aberrations, 22, 351, 355
- events, 22, 192, 244, 360, 365, 369–394
- excursions, 354–355, 357, 362–363
- goals, 53
- legislation, 136
- modeling research, 61
- monitoring, 43, 59, 72, 186, 205, 221, 236, 244, 320, 322, 328, 334, 350, 353–354, 357, 386, 438
- security, 71–75
  - models, 75–77
- sensors, 52–53, 198, 319, 330, 332, 355–357, 359, 370, 383

Water Sector
- coordinating council, 48
- dependencies, 67–68
- Government Coordinating Council, 48
- interdependencies, 68
- roadmap, 313

Water Sector Coordinating Council's Cyber Security Working Group (WSCC-CSWG), 313
Water Security Initiative (WSI), 21, 72, 328, 354, 361
Water Security Research, 1, 16, 47–62, 88, 98–99, 319, 369
Water Security Research and Technical Support Action Plan, 47, 88
WaterSentinel, 320, 354
Watershed land, 139
Watersheds, 19, 139, 205, 437
Water System Emergency Response, 107–111, 268–279
Water System Infrastructure, 57, 269
Water System Security Program, 109
Water System Vulnerabilities, 41–45
Water Treatment Decontamination, 56–61
Water treatment plants (WTPs), 16, 42, 68, 78, 96, 123, 136–137, 163–182, 223, 273, 351, 457, 476
Water type, 154
Water UK, 97
Water Utility Planning Guide, 74, 96
Weak encryption technology, 300
Weaponization, 8, 166
Weeping Water, NE, 114
Well head, 50
WERF 03-CTS-2S, 91
WERF 92-OPW-1, 91
Wetlands, 2, 87, 138
WHO Database, 71, 91
Whole cells, 206
Wide Area Network (WAN), 288–289, 291, 359
Wildfires, 271
Wild rodents, 154
WiMax, 415
Window size, 377, 379–381
Wired networks, 415
Wireless access points, 288, 294, 299–300
Wireless modems, 359
Workforce Illness, 40–41
Workforce/Infrastructure Threats, 242
Working electrode, 192, 214
World Health Organization (WHO) Database, 71
World Trade Center, 13, 30, 349
World War II, 140

X

X-ray adsorption spectroscopy, 58

Y

Yersinia enterocolitica, 164
Yersinia pestis, 57, 153–154, 157, 166
YSI®, Incorporated, 93, 331, 352
Yule–Walker equations, 374

Z

Zoonotic pathogens, 153–154